Alberta's Lower Athabasca Basin

Alberta's Lower Athabasca Basin

Archaeology and Palaeo environments

EDITED BY BRIAN M. RONAGHAN

AU PRESS

Copyright © 2017 Brian Ronaghan

Published by AU Press, Athabasca University
1200, 10011 – 109 Street, Edmonton, AB T5J 3S8
ISBN 978-1-926836-90-4 (print) 978-1-926836-91-1 (PDF) 978-1-926836-92-8 (epub)
doi: 10.15215/aupress/9781926836904.01

Cover and interior design by Marvin Harder, marvinharder.com.
Printed and bound in Canada by Friesens.

Library and Archives Canada Cataloguing in Publication
    Alberta's lower Athabasca basin : archaeology and palaeoenvironments / edited by Brian M.
Ronaghan.

(Recovering the past : studies in archaeology, ISSN 2291-6784)
Includes bibliographical references and index.
Issued in print and electronic formats.

    1. Athabasca Basin (Sask. and Alta.)—Antiquities. 2. Athabasca Tar Sands (Alta.)—Antiquities.
3. Paleontology—Athabasca Basin (Sask. and Alta.). 4. Paleontology—Alberta—Athabasca Tar Sands.
I. Ronaghan, Brian M., editor of compilation II. Series: Recovering the past (Edmonton, Alta.)

QE748.A4A43 2016   560.9712'32   C2012-901896-1   C2012-901897-X

We acknowledge the financial support of the Government of Canada through the Canada Book Fund
(CBF) for our publishing activities.

Canadä

Assistance provided by the Government of Alberta, Alberta Media Fund.

Government

Contents

Part Three   Lithic Resource Use

Part Four   Archaeological Methods

Alberta's Lower Athabasca Basin

Introduction | The Archaeological Heritage of Alberta's Lower Athabasca Basin

BRIAN M. RONAGHAN

From its headwaters in Jasper National Park, Alberta's Athabasca River travels some 1,500 miles northeastward across the province, its waters ultimately flowing into Lake Athabasca. The river drains a vast region—roughly 159,000 square kilometres—along the southern margin of Canada's Boreal Plains ecozone, which encompasses much of central and northern Alberta. In contrast to the grasslands further south, the Boreal Plains ecozone is heavily forested, forming part of Canada's more broadly defined boreal forest.[1] The Lower Athabasca basin lies wholly within the Boreal Plains ecozone, and, although some notable uplands occur, the area consists for the most part of glacial lake bed or till plain on which large tracts of muskeg and fen have developed between areas of modest elevation. Major fish-bearing lakes—Lake Athabasca and Lac La Biche— are situated along its northern and southwestern margins, respectively, and smaller lakes and ponds are scattered throughout the area. As with similar regions across the northern hemisphere, food resources are relatively meagre and widely distributed, traditionally supporting only comparatively small populations of hunter-gatherers. With the exception of the access afforded by the river, travel throughout the region is difficult.

Although the Lower Athabasca basin features a broad spectrum of natural resources, two classes of resource have proved to be of major commercial interest to outsiders. In 1778, explorer and trader Peter Pond arrived in the Athabasca region and established a fur trade post, thereby inaugurating an activity that would serve as the basis of the region's commercial economy for a century and a half. Pond noted the occurrence of a fluid tar-like substance (Stringham 2012, 21), which explorer and trader Alexander Mackenzie described in his journal a decade later, in 1788:

> At about 24 miles from the fork [of the Athabasca and Clearwater Rivers] are some bituminous fountains into which a pole of 20 feet long may be inserted without the least resistance. The bitumen is in a fluid state and when mixed with gum, the resinous substance collected from the spruce fir, it serves to gum the Indians' canoes. In its heated state it emits a smell like that of sea coal. (Mackenzie 1970, 129)

As Mackenzie noted, the Athabasca River and some of its tributaries had cut down through the surrounding bedrock, exposing bitumen along river banks in various locations to the north of modern Fort McMurray.

Aboriginal people had known of these bitumen deposits for millennia. Long before Pond arrived in the oil sands area, a Cree chief, known in historical records as "the Swan," had brought them to the attention of the Hudson's Bay Company traders. In 1715, the Swan visited York Factory, where he described the Athabasca region, telling of a river on whose banks could be found "Gum or pitch." Four years later, the Swan returned to York Factory, bearing a sample of "that Gum or pitch that flows out of the Banks of that River," which he presented to HBC governor Henry Kelsey.[2] Other resources in the region were, however, of greater value to the indigenous inhabitants. As the chapters in this volume demonstrate, a conjunction of natural and human factors has resulted in an unusually rich record of landscape development and human use that makes the Lower Athabasca basin exceptionally valuable for understanding the early history of Canada's north. The archaeological record currently being revealed represents perhaps the most intense pattern of prehistoric human landscape use yet identified in Canada's boreal forest region.[3]

Investigation into the character, extent, and value of Alberta's bitumen resources began in the 1890s. Early in the twentieth century, in an account of an expedition to the Mackenzie basin in connection with Treaty 8 (signed in 1899), Charles Mair wrote: "The tar, whatever it may be otherwise, is a fuel, and burned in our camp-fires like coal. That this region is stored with a substance of great economic value is beyond all doubt, and, when the hour of development comes, it will, I believe, prove to be one of the wonders of Northern Canada" (1908, 121). Mair's instinct would prove to be correct. Research was accelerated by the demand for fuel during both world wars, and in the early 1960s a commercially viable process for recovering bitumen was industrially implemented.[4] Only in the late 1990s, however, did technical advances and high oil prices combine to make large-scale investment in recovery attractive. Interest in understanding and developing bitumen resources originated among academic researchers and in government circles, although the role of these sectors has since evolved into providing research support for, as well as oversight of, the industrial activities now underway. Information relating to the effects of oil sands development on both the natural and the social environment continues to be collected by industry for compliance and planning purposes, but this mass of information has not been synthesized or widely disseminated. It is the intent of this volume to provide an overview of the information available on a subject that bears significantly on some of the current issues in the region: the historic value of the landscapes that have already been, or may still be, irrevocably altered by development activities.

Oil sands are deposits of bitumen, a molasses-like, viscous oil that will not flow unless heated or diluted with lighter hydrocarbons. For reasons discussed in this volume, bitumen deposits occur close to the surface only in a limited area within the oil sands region.[5] Owing to its viscosity and the depth at which it is usually buried, extracting the bitumen from the sands in which is embedded requires large-scale industrial operations. In addition to on-site separation facilities, the process demands the creation of an infrastructure to supply power, water, and materiel, and some of the by-products, including overburden and tailings, require large storage areas prior to reclamation. In addition, the bitumen must undergo further processing and then be transported before it becomes a marketable product. Not only do these activities require major capital investment, massive amounts of equipment, and substantial workforces, but they transform large areas of formerly forested land.

Like other natural resources in Canada, petroleum reserves are owned by the people of the province in which they occur, and virtually all of the land that contains oil sands resources is provincial Crown land. Consequently, the public has a vested interest in how development proceeds. Since the onset of modern oil sands development in the 1970s, environmental and historical resources legislation has established comprehensive procedures to determine whether specific development projects are in the public interest and to licence and monitor such projects as they progress through various stages. These procedures typically include efforts to address the impact of development on the natural and cultural heritage.

As a result of the review and approval processes now in place and the unprecedented levels of development in the oil sands region over the past two decades, many studies have been undertaken to meet the conservation requirements laid out in the Alberta Historical Resources Act (2000). These studies have produced a remarkably abundant and detailed record of prehistoric land use. It is ironic to realize that much of the information presented in this volume might very well never have been collected were it not for the processes involved in planning for regional oil sands development and assessing its environmental effects.

To date, close to 3,400 archaeological sites have been recorded in the oil sands area, and numerous major excavations have recovered evidence of intensive prehistoric human use of the region's resources. In addition, key geological and palaeoenvironmental studies have provided important contextual information that sheds crucial light on the reasons underlying changes in the prehistoric use of this landscape, as well as the nature of those changes. In the decades to come, if the large-scale surface mines and related developments currently planned are completed, they will erase a critical portion of this irreplaceable record. Although these archaeological resources are non-renewable, and the impact of industrial development on them is permanent, mitigation measures implemented in advance may help to offset these losses. While challenges exist, the expanding information base produced by ongoing studies enhances our ability to limit the effects of industrial development on the region's natural and cultural heritage.

THE ARCHAEOLOGICAL STUDY OF THE LOWER ATHABASCA BASIN: KEY ISSUES

The record of prehistoric human use of the forested landscapes of northern Alberta consists of the remains of materials lost, discarded, or abandoned by the small groups of hunter-gatherers that lived in these environments for more than ten thousand years. Archaeological resources can range from a single artifact lost along a trail to dense concentrations of materials that represent a complex series of tasks undertaken by large groups of people, perhaps during repeated use of a specific resource-rich location over long periods of time. As important as the physical remains of these activities are, the relationships among them can provide equally valuable information on such matters as group structure and linkages between the activities required to complete certain tasks. The range in scientific and historic values reflected in this archaeological evidence is considerable, and because it represents past activity in response to past conditions, it cannot be replicated.

Determining the scientific significance of the archaeological resources recovered in northern Alberta poses certain challenges, however. Archaeological resources identified in boreal forest environments typically occur in shallow soil horizons that are acidic as a result of the decomposition of the coniferous vegetation that covers most upland sites. Given such conditions, many of the organic components of the archaeological record originally present—including any wood, hide, feather, grass, or reed artifacts, all bones, and most residues from animals or plants consumed or processed—have long since decayed. With the exception of microscopic residues that may be present on stone tools or fragments of ceramic vessels, and possibly mineralized remnants of organic materials that were subjected to fire (calcined bone or carbonized seeds, for example), there is little direct evidence of the resources on which prehistoric groups subsisted, and many classes of artifacts that were employed in economic, social, and cultural activities are missing from the record. What most frequently remains is only the evidence of stone tool manufacture and use, along with occasional remnants of hearths and cooking fires.

Although many of these conditions also apply in regions that possess more neutral soil conditions, and although stone artifacts can reveal much about ancient cultural practices, the absence of bone and other organic materials places significant limitations on the interpretive potential of the archaeological record of the boreal forest. For the most part, subsistence strategies can only be inferred, and ascribing absolute dates to archaeological assemblages is virtually impossible given the analytical techniques currently available to us. Dating boreal forest site occupations is almost entirely a matter of inference, one that depends on the presence of diagnostic artifacts—projectile points, ceramics, and occasionally other tool types—the style of which of which corresponds to the style of artifacts recovered elsewhere, the age of which has been established by radiocarbon methods. Diagnostic artifacts typically make up only a small proportion of the materials present at any site, however, and may not be represented at all in the remains of small, task-specific activities.

As will become clear in the course of this volume, despite these frustrations, we have learned an enormous amount about the prehistory of the oil sands region since archaeological investigations began there in the early 1970s. In the course of these investigations, attention has focused on several topics that come up repeatedly throughout the book. One concerns the origins and nature of the landscape that became available for human occupation at the end of the Pleistocene Epoch, following the retreat of glacial ice.[6] As the ice receded, large lakes (termed proglacial lakes) formed along its edges. We know that somewhere between 9,800 and 9,600 years ago, a catastrophic discharge of water from one such proglacial lake—Glacial Lake Agassiz—scoured the deglaciated Lower Athabasca basin. Possibly in combination with an earlier flood event, this massive deluge of water created a landscape that fundamentally influenced prehistoric human use of the region.

Those familiar with glacial floods will realize that the Glacial Lake Agassiz event reflects processes that have occurred in many other situations associated with retreat of glacial ice. These events varied in the extent and degree of their effects, and each was conditioned by the specific environmental and geological circumstances in which it occurred. Given the intricacies of these location-specific variations, a comparative analysis of the Agassiz flood is beyond the scope of this book. The Glacial Lake Agassiz flood is, however, crucial to understanding many of the analyses presented in the volume, particularly those that highlight geological and palaeoenvironmental information. A number of the chapters thus include discussions of the Lake Agassiz flood, offering various perspectives on both the timing and significance of this event.

For reasons mentioned above, the majority of the evidence of prehistoric occupation of the Lower Athabasca basin exists in the form of remnants of the manufacture and use of stone tools. These artifacts were produced in vast quantities almost exclusively from a single type of locally obtained stone, originally named Beaver River Quartzite but now most commonly called Beaver River Sandstone (BRS). Issues surrounding the origin and use of this ubiquitous stone material constitute another major theme in oil sands archaeology. BRS, which occurs within the Cretaceous-age McMurray Formation, largely consists of sand grains fused in a matrix. It ranges in granularity from exceedingly coarse material that looks like a variety of quartzite or sandstone to very fine-grained material in which the original grains have been subsumed in a silica matrix—variations that reflect the degree of post-depositional heat and pressure applied to the formation. For the most part, the artifacts that have been recovered are made of relatively fine-grained material, which is better suited to the manufacture of stone tools.

The question thus arose as to the source of this material, given that the raw BRS found in the area, whether in situ (notably at a site known as the Beaver River Quarry) or in the form of boulders or cobbles, was typically quite coarse in grain. In 2003, the discovery of a complex of archaeological sites now called the Quarry of the Ancestors shed new light on this issue, as the stone available there is generally of finer grain.[7] However, numerous questions surround the role of the Quarry of the Ancestors in prehistoric patterns of land use in the area and in the distribution of BRS throughout the region. In addition, the question remains whether the fine-grained BRS found in artifacts reflects human intervention in the form of heat treatment, a process that can transform even relatively coarse-grained stone into a material more suitable for tool manufacture. Fortunately, in 2012, the Quarry of the Ancestors was designated a Provincial Historic Resource and will thus be permanently preserved for future study.

TIME AND PLACE: SOME CONVENTIONS

For the benefit of readers for whom archaeology is relatively new terrain, some basic background information may be in order. It is sometimes possible to assign an age to prehistoric materials on the basis of radiocarbon dating of directly associated organic residues. This technique measures the proportional decay of a radioactive isotope of carbon (carbon 14, or ^{14}C) subsequent to the death of living entity, be it plant or animal. It is most often applied to bone or wood but is suitable for a wide range of organic substances, all of which contain carbon. Originally developed in 1949 at the University of Chicago's Institute for Nuclear Studies, radiocarbon analysis has since been refined through the use of accelerator mass spectrometry (AMS), which has increased its accuracy and decreased the amount of material required. Laboratories that conduct radiocarbon analysis report ages in radiocarbon years "before present" (BP), with the "present" defined as 1950 (reflecting, of course, the time at which the technique originated). From an archaeological perspective, the relatively short span of time between 1950 and the present day is for all practical purposes insignificant. Because levels of atmospheric carbon have not remained constant over time, however, ages reported in radiocarbon years BP begin to deviate from ages expressed in calendar years as one moves back in time from 1950: 9,850 radiocarbons years BP is, for example, roughly 11,250 calendar years BP. Calibration curves have thus been developed that, by taking into account past variations in the levels of atmospheric carbon, allow radiocarbon ages to be converted into calendar dates.

Archaeologists generally express dates in radiocarbon years rather than in calendar years. When laboratories report radiocarbon dates, they do so giving the age yielded by the analysis, the uncertainty range, and the sample number, which identifies the specimen on which the analysis was conducted: $9,410 \pm 280$ ^{14}C yr BP (UCR-2430B), for example. In general discussion, such dates are typically abbreviated by omitting both the uncertainty range and the sample number: 9,410 ^{14}C yr BP. As the original uncertainty range indicates, radiocarbon dates are not absolutely precise—nor is radiocarbon analysis perfectly reliable, as samples may occasionally be contaminated with hydrocarbons from an external source. All the same, dates founded on the scientific analysis of organic remains are regarded as relatively firm.

Given that boreal forest settings are characterized a paucity of organic remains, radiocarbon analysis is rarely possible. Instead, archaeologists must rely on stylistic comparisons of the sort described above. Dates derived through comparative methods are necessarily in the nature of estimates, and they depend for their reliability on the observational skills of the person examining the artifacts, as well as on the depth and breadth of the knowledge base that this person brings to the analysis. Such dates are not, however, inherently subjective. Assuming an experienced analyst, dates generated by comparative methods can in fact be credited with a fair degree of reliability. In this volume, dates that are in the nature of estimates founded on some form of comparative analysis are labelled with a simple "BP," while those that are grounded in radiocarbon analysis of a specific specimen are labelled "^{14}C yr BP." When dates in calendar years are included (usually in addition to dates in radiocarbon years), these are labelled "cal yr BP."

An archaeological site may represent only a single occupation, whether relatively brief or of long duration, or it may reflect successive occupations by different cultural groups, stretching over many millennia. In other words, archaeological sites cannot meaningfully be assigned a single, discrete date but rather occupy a span of time. Sites do, however, exist in a specific place. In Canada, archaeological sites are identified using the Borden system—an alpha-numeric system developed in 1952 by Charles Borden at the University of British Columbia. The system establishes a grid based on longitude and latitude that extends across the entire country and divides it into major and minor blocks. Major blocks, which are designated by capital letters, correspond to areas of 2 degrees of latitude by 4 degrees of longitude. Minor blocks, designated by lower-case letters, represent areas of 10 minutes of latitude by 10 minutes of longitude. Each minor Borden block thus represents an area of approximately 16 square kilometres. Within each minor block, archaeological sites are then numbered consecutively as they are recorded. Thus, for example, a specific site might be designated HhOv-73, with "Hh" representing the major and minor blocks of latitude and "Ov" the major and minor blocks of longitude. This system makes it possible to know the approximate location of a site purely on the basis of its designation.

AN OVERVIEW OF THIS VOLUME

The contributions to this volume reflect two fundamental approaches to the study of the past, the first relating to palaeoenvironmental conditions and the second to prehistoric human adaptations to these conditions. The chapters in part 1 discuss changes in the postglacial landscapes and environments of the Lower Athabasca basin, factors that are essential to our understanding of past human use of the area. This discussion is followed, in part 2, by chapters that focus on the patterns of prehistoric human occupation that emerge from the archaeological evidence so far recovered. Given that most of this evidence consists of stone tools and remnants of their manufacture and use, the chapters in part 3 discuss aspects of the origin, processing, and distribution of the region's lithic resources. The volume concludes, in part 4, with two chapters that examine the effectiveness of the field and analytical methods presently in use, including current approaches to cumulative effects assessment, which seeks to forecast the impact of regional development on archaeological resources.

Postglacial Environments

The volume opens with a consideration of the postglacial landscape in the Athabasca region. In the first chapter, "A Tale of Two Floods," James Burns and Robert Young propose an alternative explanation for the extreme dearth of Pleistocene mammalian fossils in northern Alberta. In the oil sands area, only two vertebrate fossils have so far been recovered: pelvic bones from a mammoth, which appear to date to roughly 32,150 ^{14}C yr BP, and the skull of a wapiti, or elk, which has been dated to 5,550 ^{14}C yr BP and thus to the early Holocene Epoch. Burns and Young describe these specimens and consider the depositional contexts in which they were found. With regard to the question of why so few such remains have survived, they introduce the idea that, as during the most recent period of deglaciation, the Lower Athabasca basin may have been extensively modified not once but twice by large-scale floods. In addition to the well-known Glacial Lake Agassiz event, which occurred around 9,800 to 9,600 ^{14}C yr BP and flooded recently deglaciated terrain, a less well-defined event is believed to have taken place some millennia earlier. This flood would have occurred beneath the retreating ice cap, when the pressure of the meltwater that had built up under the ice reached the point that the water escaped and a massive subglacial flood ensued. By washing away massive amounts of Pleistocene sediments across a broad swath of the Lower Athabasca valley, these floods left bitumen deposits relatively close to the surface, thereby bequeathing to us ready access to the oil

sands. But these floods would also have washed away virtually all existing megafaunal remains.

The second chapter, by Timothy Fisher and Thomas Lowell—"Glacial
Geology and Land-Forming Events in the Fort McMurray Region"—discusses
the timing and geomorphic effects of the retreat of the Laurentide Ice Sheet at
the end of the Wisconsinan glaciation, more than 10,000 years ago. Drawing on
digital elevation data from the Shuttle Radar Topography Mission and on radiocarbon dates obtained from sediment at the bottom of lakes in the vicinity of
newly defined moraines, Fisher and Lowell offer new insights into the palaeogeography of the proglacial lakes that formed in the Churchill and Lower
Athabasca valleys along the edges of the retreating ice sheet. As they suggest, the
immediate source of the flood waters that scoured the Lower Athabasca valley
was Glacial Lake Churchill, a proglacial lake that formed temporarily in the
Churchill River valley as deglaciation proceeded. Building on earlier research,
they review the geomorphic and sedimentological evidence for the massive discharge of water along the Clearwater–Lower Athabasca spillway, concluding that
the spillway was first occupied sometime between 9,800 and 9,600 ^{14}C yr BP,
with water continuing to flow along the channel for several hundred years afterward as Glacial Lake Agassiz drained through its northwestern outlet. In combination, they argue, evidence suggests that the processes of deglaciation that
shaped the Lower Athabasca landscape took place somewhat later than had previously been thought.

In "Raised Landforms in the East-Central Oil Sands Region," Robin
Woywitka, Duane Froese, and Stephen Wolfe shed further light on the landscape that supported intense prehistoric human activity by examining the formation and character of the elevated landforms within the flood-modified Cree
Burn Lake–Kearl Lake lowland. An analysis of landform shape and orientation,
as revealed by LiDAR imaging, coupled with sedimentary observations, confirms that a majority of these features were formed as gravel bedforms related
to the catastrophic flooding during deglaciation. These features are frequently
mantled with windblown sand, indicating that windy, dry conditions prevailed
following the deposition of sediments by the flood, and the occurrence of
archaeological materials in these aeolian sands points to a human presence
during this period and/or shortly thereafter. Subsequent to aeolian deposition,
peat began to accumulate in the intervening lowlands, suggesting that, by this
time, the surfaces of raised landforms had been stabilized by vegetation. The
combination of a burgeoning wetland community and stable uplands would
have provided an attractive habitat for human occupation. In addition, these

well-drained landforms would have differed sharply from the surrounding till-based plain, with its dense, silty soils.

The final chapter in part 1, Luc Bouchet and Alwynne Beaudoin's "Kearl Lake: A Palynological Study and Postglacial Palaeoenvironmental Reconstruction of Alberta's Oil Sands Region," presents the results of an analysis of pollen recovered from a sample core taken from Kearl Lake, located within the Athabasca-Clearwater plain along the margins of the Glacial Lake Agassiz outwash zone. As this pollen record indicates, spruce-dominated woodland had become established in the area prior to the flood, perhaps as early as 10,250 ^{14}C yr BP, during the terminal Pleistocene. Between about 9,820 and 7,580 ^{14}C yr BP, this landscape gave way to relatively more open deciduous woodland, characterized by birch, a shift that reflects the advent of the warmer, dryer conditions that prevailed during the Hypsithermal interval.[8] The proliferation of birch might additionally have been encouraged by the destructive impact of the flood, given that, as the authors note, birch often dominates the new tree cover in areas that have been disturbed. The shift in climate is also visible in the increased presence of non-arboreal pollen types, variously indicative of reduced lake levels and of greater openness in surrounding upland vegetation. As the Hypsithermal waned, jack pine became a more prominent member of the vegetation community, as did spruce, while the growing abundance of peat moss spores probably signals the development of muskeg in lowland areas—alterations that essentially represent the establishment of modern boreal forest in the region. The pollen record thus provides crucial evidence of the shifting environmental conditions to which the region's human inhabitants reacted.

Human History

In part 2, the focus shifts to prehistoric human occupation of the Lower Athabasca basin. The opening chapter, "The Early Prehistoric Use of a Flood-Scoured Landscape in Northeastern Alberta," written by Grant Clarke, Luc Bouchet, and myself, offers an interpretive model of the Early Prehistoric human occupation of the Lower Athabasca valley in the wake of the Glacial Lake Agassiz flood. Drawing on palaeoenvironmental data, we argue that the scouring effects of the flood, in combination with the Hypsithermal climatic conditions that prevailed during the early postglacial period, created a highly productive regional microenvironment that stood in striking contrast to the surrounding higher-elevation forest. This parkland-like landscape, dominated by grasses and herbs, with open deciduous forest along ridge tops and meadows in intervening channels, would have been attractive to caribou and bison—grazing species that, because of their herding behaviour, are well suited to communal hunting. A review of the chronological sequence of human occupations developed in archaeological studies to date appears to support the contention that the most intense prehistoric use of the Athabasca oil sands region coincided with the period during which this microenvironment existed. As temperatures cooled and forests began to close in, animal populations would have shifted, with browsers such as moose increasing in number, and a pattern more typical of the boreal forest would have emerged, one characterized by smaller, more dispersed human settlements.

In the following chapter, "A Chronological Outline for the Athabasca Lowlands and Adjacent Areas," Brian Reeves, Janet Blakey, and Murray Lobb lay out a sequence of occupations in the Lower Athabasca lowlands region. Drawing on Alberta's database of archaeological sites, in tandem with a comprehensive review of the existing literature and collections, the authors offer detailed descriptions of the series of cultural complexes that, in their analysis, characterize the history of human occupation in the region, from its beginnings some 10,000 years ago through to the arrival of Euro-Canadian fur traders toward the end of the eighteenth century. Given that the paucity of organic remains typical of boreal forest settings largely rules out radiocarbon dating, the authors rely instead on established archaeological methods of comparative stylistic analysis, focusing on the identification of chronologically diagnostic artifacts, notably projectile points. As is standard practice, the individual archaeological complexes within the proposed sequence are given local names, but relationships with more widely distributed cultural traditions are both identified and explored. The resulting framework, the product of an extraordinary exercise in synthesis, will serve as a basis for establishing the relative age and/or cultural affiliation of assemblages recovered in the future, at least until techniques for the absolute dating of archaeological materials recovered from boreal forest settings can be developed.

Robin Woywitka's "Lower Athabasca Archaeology: A View from the Fort Hills" provides an introduction to the prehistory of the area surrounding the Fort Hills, a moderately elevated series of uplands situated to the northeast of Fort McKay, on the northern periphery of the Lower Athabasca archaeological "heartland." During the Lake Agassiz flood, outwash waters surrounded this uplands area but left its flanks untouched, producing a landscape that offers something of a contrast to the flood zone immediately to the south. The area now encompasses five major geographic features, which include, in addition to the uplands, the Late Pleistocene Athabasca braid delta and two prominent wetlands. Archaeological sites in the Fort Hills region, while less densely concentrated than those to the south, have yielded assemblages that are again dominated by Beaver River Sandstone. As is the case in the Athabasca lowlands, assemblages consist primarily of debitage but do include both formed and expedient tools. Woywitka offers a detailed consideration of the diagnostic projectile points recovered from sites in the Fort Hills area and of the evidence for microblade technology found at the Little Pond site, together with a review of the very few radiocarbon dates currently available. Drawing on these discussions, he provides a chronological and palaeoenvironmental account of the Fort Hills region, as well as an analysis of subsistence and land use patterns, and situates the Fort Hills in a broader regional context. As he points out, although the Fort Hills archaeological record is closely tied to that of the heartland area, it also exhibits characteristics more akin to boreal forest assemblages recovered throughout the southern Canadian subarctic.

The final chapter in part 2, "The Early Human History of the Birch Mountains Uplands," by Jack Ives, discusses the prehistoric human history of a major contrasting ecosystem adjacent to the Lower Athabasca valley. Rising some 525 to 850 metres above the plain below, the Birch Mountains were among the first landforms to be exposed in the Athabasca region as the Laurentide Ice Sheet retreated. The ecology differs from that of the lowlands, featuring vegetation communities that developed under colder, drier conditions, as well as significant fish-bearing lakes. Drainage forms a radial pattern, linking the Birch Mountains not only to the Lower Athabasca region but also to the Peace River area and the Wabasca River drainage, to the northwest and southwest, respectively. Evidence of continued human occupation is present from the earliest postglacial times down to the historic period, with groups resident elsewhere travelling to the Birch Mountains as part of their seasonal round. As Ives notes, Beaver River Sandstone—so ubiquitous in the Lower Athabasca valley—is relatively rare in Birch Mountains assemblages, with the notable exception of a cache of tools discovered at the Eaglenest Portage site that appear to be the contents of a container used for transport. Ives reviews the key archaeological findings for the Early, Middle, and Late Prehistoric periods, including several radiocarbon dates, while cautioning against the temptation to define specific phases or complexes on the basis of scant and/or potentially ambiguous evidence. Our understanding of prehistoric land use in the Athabasca lowlands, he argues, would benefit from further research conducted in the surrounding regions, as this would allow us to investigate crucial questions pertaining to patterns of human movement and variations in the use of lithic resources.

Lithic Resource Use

The vast bulk of the archaeological record throughout Canada's boreal forest region consists of the remnants of stone tool manufacture and use. Understanding how this evidence is distributed across the landscape and what kinds of human activity it represents is thus a critical component of regional archaeological study. The third part of this volume is accordingly devoted to the lithic record as it has emerged in the Lower Athabasca basin.

Part 3 opens with a chapter by Eugene Gryba, "Beaver River Sandstone: Characteristics and Use, with Results of Heat Treatment Experiments." In it, Gryba reviews the physical and chemical characteristics of Beaver River Sandstone, as well as its origins and stratigraphic position within the regional geological sequence. In addition, he considers both known and potential sources of the stone and the frequency of its occurrence in archaeological assemblages that lie at some distance from source locations. As is well known, the material properties of a particular stone have a significant influence on the uses to which the stone can be put. These properties can to some extent be altered, however, by the application of heat. The heat treatment of raw stone material to improve its workability has been attested in a wide range of archaeological contexts, in North America and elsewhere. Gryba reports the results of experiments in which BRS was heated to temperatures in the range of 400°C to 450°C, which pre-historic peoples would also have been able to achieve. At such temperatures, BRS recrystallizes, often developing a thin, rust-coloured rind on its cortex and a smoother, more lustrous fracture surface. It also becomes considerably easier to work by both percussive and pressure methods. The possible application of this technique to BRS clearly has important implications for our understanding of lithic technology in the Lower Athabasca region, while it also provides a new angle on the question of why the BRS in artifacts sometimes appears to be of higher quality than the naturally occurring stone.

In the following chapter, "The Organization of Lithic Technology at the Quarry of the Ancestors," Nancy Saxberg and Elizabeth Robertson present the first detailed discussion of an archaeological site of such significance that it has been set aside for permanent preservation. Discovered in 2003, the Quarry of the Ancestors is a complex of archaeological sites centred around two exposures of Beaver River Sandstone in the Muskeg River basin, to the east of the Athabasca River. The quarry was used intensively for several millennia roughly 9,000 to 6,000 years ago, when the climate was comparatively warm and dry, and data collected during initial archaeological investigations reveal that the inhabitants practiced a flexible, opportunistic approach to lithic reduction. Saxberg and Robertson surmise that the physical characteristics of BRS in its natural state facilitated extraction and reduction and that easily transportable packages of stone were then removed to sandy uplands, where the stone may have been heat-treated to improve its workability. The discovery of the Quarry of the Ancestors raises a series of questions—about the relationships among specific components of the site complex, about the lithic reduction methods employed at the quarry, about the variety of tools that occur and their distribution across the site, about the relative dearth of finished tools thus far recovered, and about the role of the quarry in regional patterns of occupation and resource use. Analyzing the evidence, Saxberg and Robertson propose that, rather than representing a place where groups out on their seasonal round stopped to procure lithic materials, the Quarry of the Ancestors served as a home base for peoples of relatively low residential mobility. A more complete understanding of the significance of the quarry will have to await future excavations, which may also shed light on the intriguing question of why the quarry apparently fell into disuse.

Microblade technology—that is, the manufacture of tiny stone blades, typically designed to be inset into projectile points made of bone, antler, or ivory but sometimes hafted onto handles—is highly characteristic of adaptations to arctic and subarctic environments. Microblades are common in Alaska and the Yukon, and they also occur in assemblages from sites along the northern Pacific coast and in the interior of British Columbia. They are, however, seldom seen in Alberta. In "Microblade Technology in the Oil Sands Region: Distinctive Features and Possible Cultural Assoications," Angela Younie, Raymond Le Blanc, and Robin Woywitka examine the evidence for microblade technology at sites in the Lower Athabasca region. The technology was first identified in the early 1980s at the Bezya site, located in the Lower Athabasca valley not far northeast of Cree Burn Lake and, more than two decades later, at the Little Pond site to the north of the Fort Hills. However, with the onset of intensive oil sands development and the consequent upsurge in archaeological impact assessments, reports of the discovery of microblades and other evidence of microblade technology proliferated. On the basis of a re-examination of many of the specimens recently identified, including several from the Quarry of the Ancestors, the authors develop a critical approach that turns on the need to distinguish genuine microblades from blade-like flakes. As the authors point out, because microblades can be produced in a number of different ways, the analysis of microblades in Asian, Alaskan, and Beringian studies has focused on identifying the underlying technology—that is, the sequence of reduction of a microblade core—rather than on the simple presence or absence of microblades themselves. The authors argue that microblade technology is defined by an entire suite of distinctive features, not by only a few of these features in isolation. Evidence of true microblade technology in northern Alberta remains very scant, but careful analysis of this evidence in the light of traditions of microblade production recognized elsewhere allows us to trace potential cultural relationships between peoples in the Lower Athabasca and those in the far northwest of North America.

Archaeological Methods

Historic resource management in Alberta is based on principles enshrined in legislation, which requires that commercial developers undertake archaeological assessment studies in advance of a proposed project (unless the impact of the project will clearly be negligible). If the initial assessment indicates that significant archaeological resources exist in the area slated for development, developers are further required to arrange for mitigative excavations, which are intended to recover a representative sample of the site in question. The goal of the assessment process is thus to determine the relative importance of the archaeological materials that exist within an area proposed for development, prior to making decisions about whether a proposed project will be allowed to proceed and, if so, on what conditions.

Because assessment studies are typically prompted by specific plans for development, they are fundamentally reactive, focusing on the immediate and relatively localized effects of a proposed project. Moreover, archeological assessment of these plans must often proceed without the benefit of a broader regional perspective founded on extensive prior archaeological study, of the sort that would provide contextual knowledge. As a result, the methods employed to assess the actual and potential effects of development are designed to be applied in a broad set of circumstances. As the knowledge base within a given region builds through continued study, these methods often prove to fall short of their intended goal, namely, to strike an optimal balance between the economic and social needs that drive development and the historical and cultural interest in preserving the archaeological record. These failures can, however, be instructive, as they often point the way to new, more effective assessment strategies. The final section of this volume offers two discussions of some of the issues surrounding current approaches to assessment.

"Quarries: Investigative Approaches in the Athabasca Oil Sands," by Gloria Fedirchuk, Jennifer Tischer, and Laura Roskowski, examines the effectiveness of some of the archaeological techniques currently employed in impact assessment studies. One critical question concerns the fragmentation of data—that is, the difficulty inherent in efforts to construct an accurate picture of prehistoric land use patterns on the basis of widely dispersed studies undertaken in connection with specific project proposals, which typically stop at the borders of the area slated for development. Another concerns the philosophy underlying mitigative excavations, which privileges strategies designed to maximize the number of artifacts recovered, to the neglect of areas of lesser density that could well contain artifacts of greater interpretive significance. In addition, the authors critically evaluate the effectiveness of predictive models, such as those generated for the Quarry of the Ancestors, that attempt to determine the most promising locations for excavations by forecasting archaeological potential on the basis of landforms. In response to the perceived shortcomings of existing approaches, the authors propose a shift toward a more interpretive approach that would place greater emphasis on context, on identifying relationships among sites, and on developing specific research questions to guide archaeological study. By way of illustration, they explore three main areas of theoretical inquiry regarding ancient quarry sites: technology and mining, economic interactions, and social organization. In the light of this analysis, the authors suggest a number of ways in which these new approaches could be integrated to existing impact assessment procedures.

Development projects are additive, and so, of course, is their impact. In the closing chapter, "Cumulative Effects Assessment," I examine the issues surrounding our efforts to assess the combined effects of regional oil sands development on archaeological resources. In Alberta, cumulative effects assessments (CEAs)—which, as in most jurisdictions, are mandated by regulatory processes—are incorporated within the framework of Environmental Impact Assessments, embedded in which are requirements that the effects of a proposed project on historical resources be described. CEA procedures are rooted in ecological studies, and, in addition to a number of more general shortcomings, the effectiveness of these procedures is significantly reduced when they are applied to archaeological resources, the character of which clearly differs from that of natural resources. In boreal forest settings, the situation is complicated by the nature of surviving archaeological materials and by the methods employed to find them and to evaluate their significance. Turning to issues of impact, I describe the long-term effects, both direct and indirect, that mining operations and planned in situ projects are likely to have on archaeological resources and also review our current understanding of the distribution and significance of known archaeological resources within the three officially demarcated oil sands regions. Using the categories developed for the CEA process, I then evaluate the significance of the combined effects of development activities on archaeological resources, focusing on the degree of confidence with which effective predictions can be made. As I point out, in its present form, the CEA process is initiated by developers, on a project-specific basis, for purposes of gaining approval from regulatory agencies. While this does not necessarily imply deficiencies in the quality of information provided, the restricted scope and objectives of CEAs have arguably limited their utility on a regional scale, as well as their capacity to guide regulators who consider the public interest in matters of development. Effective assessment is also limited, however, by our baseline knowledge, which accumulates slowly and is constantly evolving. As I suggest in closing, ongoing review and synthesis of archaeological information may therefore prove to be one of the most powerful methods available to offset the cumulative effects of development, whether in the oil sands region or elsewhere.

CLOSING THOUGHTS

As I write, development of Alberta's bitumen reserves has slowed considerably, largely in response to the worldwide oil economy, and, especially given global imperatives to reduce our dependence on petroleum products, the extent to which the pace will pick up again remains to be seen. All the same, the Lower Athabasca basin continues to be the locus of widespread industrial activity, and, in one form or another, further commercial development in the region seems likely. In addition to its other effects, this activity will disturb intact boreal forest, including the as yet undiscovered archaeological resources that lie not far below the surface. These resources are fragile, and they cannot be reclaimed, much less replaced. Archaeological research in the Lower Athabasca valley to date—the overwhelming majority of it occasioned by oil sands development—has revealed an exceptional record of intense prehistoric human use, one that appears unparalleled in the Canadian boreal forest. The information generated by these studies and the physical materials recovered have already immeasurably enriched our understanding of the early human presence in the region and hold significant value for future scientific research.

What is less often recognized is the value of this information for public education. Thus far, the findings of archaeological studies are for the most part scattered across unpublished compliance reports on specific research permits, while the results of geophysical and palaeoenvironmental studies typically appear as articles in scientific journals. This volume represents an effort to draw some of this information together, to take stock of the current state of our knowledge, to offer some provisional interpretations of the evidence, and to discuss some of the issues that complicate our efforts to develop a more comprehensive picture of prehistoric lifeways in the Lower Athabasca region. It is my hope that the chapters in this book will contribute to a deeper and more nuanced public understanding of the history embedded in this landscape and, in so doing, will help to build interest in the rich prehistoric heritage of northern Canada, while also serving to illustrate how archaeological knowledge evolves. I also hope that, by encouraging a more complex appreciation of the origins and archaeological significance of the Lower Athabasca basin, this volume will provide the impetus for improved conservation of the resources, both archaeological and natural, of the oil sands region.

NOTES

1   Canada's boreal forest covers more than a third of the country, spanning upwards of 1.5 million square miles, or some 3.9 million square kilometres (Henry 2002, xiii), roughly 38.9% of Canada's total area. On the Boreal Plains ecozone, see "Boreal Plains," Canadian Biodiversity Web Site, n.d., http://canadianbiodiversity.mcgill.ca/english/ecozones/borealplains/borealplains.htm; and, on the Athabasca River drainage, "About the Athabasca River Basin," Athabasca Basin River Research Institute, Athabasca University, n.d., http://arbri.athabascau.ca/About-the-Athabasca-River-basin/Index.php.

2   L. H. Neatby, "Swan," *Dictionary of Canadian Biography Online,* 1982 [1969], http://www.biographi.ca/en/bio/swan_2E.html.

3   The term *prehistoric* has come under criticism for tending to imply that oral cultures have no history (or no sense of history)—which is, of course, manifestly untrue. Some authors thus prefer the term *precontact,* and both terms occur in this volume. Although the latter term avoids reinforcing stale images of oral cultures as timeless and static, it also tends to suggest that the arrival of Europeans was an event of pivotal importance, one that fundamentally transformed these cultures. While no one would debate the destructive effects of colonization, these cultures in fact possess a resilience that is visible in the continuity of language and traditions despite the depredations wrought by the Euro-Canadian presence. In short, neither term is ideal. We understand *prehistory* simply as history preceding written history, with no value judgment implied.

4   The history of oil sands development has been covered in numerous publications. See, for example, Chastko (2005); Ferguson (1986); and Hein (2000).

5   Strictly speaking, the term *oil sands region* refers to three administrative areas created by the Energy Resources Conservation Board (now the Alberta Energy Regulator), an independent quasi-judicial agency of the Government of Alberta, in order to manage applications for the development of heavy oil deposits under the provisions of the Oil Sands Conservation Act and its attendant regulations. Under this legislation, three administrative areas—Athabasca, Cold Lake, and Peace River—were created in 1984, by independent orders, each specifying both the geological provenance of the relevant deposits and the area of land within which applications will be considered. Since then, the boundaries of the areas have been extended somewhat, such that the three areas now span a total of 142,200 square kilometres. See "Facts and Statistics," Alberta Energy, 2016, http://www.energy.alberta.ca/oilsands/791.asp, and, for the three orders, "Rules and Directives: Oil Sands," Alberta Energy Regulator, 2016, https://www.aer.ca/rules-and-regulations/by-topic/oil-sands.

6   The Pleistocene Epoch extended from about 2.58 million to 11,700 years ago. During this time, several episodes of glacial advance and retreat occurred, separated by warmer,

dryer interglacial periods. The last of these glacial episodes, the Wisconsinan, which began roughly 110,000 years ago, is of the greatest relevance for our understanding current landscapes and subsequent human use. The Pleistocene Epoch was succeeded by the Holocene Epoch, which extends down to the present day; together, the two constitute what geologists define as the Quaternary Period. For an outline of geological time, see *International Chronostratigraphic Chart,* International Commission on Stratigraphy, 2015, http://www.stratigraphy.org/ICSchart/ChronostratChart2015-01.pdf.

7 Following the discovery of the Quarry of the Ancestors, the term Muskeg Valley Microquartzite (MVMq) was coined to describe the relatively fine-grained material that occurs at the quarry (which is situated in the Muskeg River valley). BRS is also sometimes called Beaver River Silicified Sandstone, with reference to the process whereby it was formed. These variations in nomenclature reflect geological analyses of the structure of the stone, which give rise to differing opinions as to how it should be classified. Several contributors to this volume thus speak of Muskeg Valley Microquartzite rather than Beaver River Sandstone, although the two are essentially the same material.

8 The Hypsithermal interval, also known as the Holocene Climatic Optimum or the Holocene Thermal Maximum (as well as by several other names), began early in the Holocene and persisted for several millennia, into the mid-Holocene, ending by roughly 5,000 BP. In the high Arctic, temperatures rose by several degrees centigrade, but the increase declined rapidly with latitude. As Bouchet and Beaudoin indicate in chapter 4, evidence for Hypsithermal conditions also varies with altitude, such that both the timing and the effects of this period of global warming differ depending on the location under consideration.

REFERENCES

Chastko, Paul
> 2004   *Developing Alberta's Oil Sands: From Karl Clark to Kyoto.* University of Calgary Press, Calgary.

Ferguson, Barry G.
> 1986   *Athabasca Oil Sands: Northern Resource Exploration, 1875 to 1951.* Canadian Plains Research Center, Regina.

Hein, Francis J.
> 2000   *Historical Overview of the Fort McMurray Area and Oil Sands Industry in Northeast Alberta.* Earth Sciences Report 2000-05. Alberta Energy and Utilities Board and Alberta Geological Survey, Edmonton.

Henry, J. David
> 2002   *Canada's Boreal Forest.* Smithsonian Natural History Series. Washington DC: Smithsonian Institution Scholarly Press.

Mackenzie, Sir Alexander
> 1970   *The Journals and Letters of Sir Alexander Mackenzie.* Edited by W. Kaye Lamb. Cambridge University Press for the Hakluyt Society, Cambridge.

Mair, Charles
> 1908   *Through the Mackenzie Basin: A Narrative of the Athabasca and Peace River Treaty Expedition of 1899.* William Briggs, Toronto.

Stringham, Greg
> 2012   "Energy Developments in Canada's Oil Sands." In *Alberta Oil Sands: Energy, Industry and the Environment,* edited by Kevin E. Percy, pp. 19–34. Elsevier, Oxford and Amsterdam.

1 Postglacial Environments

1 A Tale of Two Floods | How the End of the Ice Age Enhanced Oil Sands Recovery—and Decimated the Fossil Record

JAMES A. BURNS AND ROBERT R. YOUNG

Although Ice Age vertebrate fossils occur in abundance in some regions of Alberta, the oil sands area, to the north of Fort McMurray, has yielded only a few remains of extinct mammals.[1] Historically, the dearth of such fossils in northern Alberta has been attributed to the slower rate of settlement and development of natural resources—in-ground and above-ground—in comparison with the more southerly parts of the province. While Ice Age fossils from the northern portion of the province are indeed demonstrably rare in museum collections, we propose an alternative explanation for their scarcity.

The Athabasca oil sands have been known and used by local inhabitants for centuries, and their potential for commercial development has been an object of interest for decades. Until recently, factors limiting access and extraction kept the oil locked up in the sands, but over the past several decades extraction techniques have made mining of the sands an economically supportable enterprise. However, the oil sands would not have been accessible at all without the extraordinary events that took place during the latter part of the Wisconsinan glaciation, the most recent of the glacial episodes that occurred over the course of the Pleistocene Epoch. During that period, we argue, two major floods, one subglacial and one proglacial, discharged in different directions, scouring the landscape and eating through Pleistocene deposits and into the Mesozoic bedrock. Massive removal of these sediments by flood waters rendered underlying oil sands deposits in the Athabasca River valley readily accessible, but the action of these floods also offers a cogent explanation for the dearth of late Pleistocene vertebrate fossils.[2]

THE GEOMORPHOLOGICAL CONTEXT

The past extent of glaciers and their paths of flow can be reconstructed by mapping the occurrence of rocks in areas where they are not normally found, as well as by the presence of characteristically glacial landforms, such as drumlins, flutings, hummocky topography, and tunnel channels, that is, channels thought to have been cut beneath a glacier (Rains et al. 2002). During the most recent glaciation, which reached its maximum extent approximately 25,000 years ago, the massive Laurentide Ice Sheet became so thick in northeastern Alberta that it was able to push its way *up* the slope of land to the west and southwest of present-day Fort McMurray, ultimately terminating at positions over 1,000 metres higher in the Porcupine Hills, southwest of Calgary, and all the way down into Montana (Young et al. 1994, 1999; Rains et al. 2002). In the Porcupine Hills, specimens of rock from the Northwest Territories and an array of subglacially eroded valleys occur at approximately 1,500 metres above sea level, along the margin of the Laurentide Ice Sheet (Rains et al. 1993). From this we can infer that the ice reached a depth of at least 1,500 metres in the southern part of the province. Given that the Lower Athabasca valley, in which Fort McMurray now lies, is situated at approximately 245 metres above sea level and the surrounding plains, above the valley, at about 350 metres, the city and region would have been submerged under a *minimum* of 1,255 and 1,150 metres of ice, respectively. It is very unlikely the ice sheet was flat, however: it would have been thicker toward the dispersal centre. Theoretically, then, the ice over the Fort McMurray region could have been as much as twice as thick—greater than 2,000 metres.

During the early stages of deglaciation, as the ice sheet began to melt, reservoirs formed on the surface of the ice, as a result of solar heating, as well as at its base, owing to internal friction and geothermal heat. Both reservoirs would have possessed a large amount of potential energy. A basal reservoir, or a series of such reservoirs, would have been confined and pressurized by the overlying kilometre or two of ice, while the water bodies on top of the ice would have contained tremendous amounts of potential energy by virtue of their degree of elevation above the land surface. The solar-heated water in surface reservoirs would gradually have eroded tunnels and crevasses, allowing it to flow into pre-existing subglacial reservoirs. Eventually, the pressure of water beneath the ice would have exceeded the pressure of the overlying ice, causing the water to escape and drain catastrophically, eroding enormous channels across Alberta, from the Fort McMurray region as far south as Montana (fig 1.1; and see Rains et al. 1993).

As these regions drained, the landforms and sediments characteristic of large turbulent water flows were left behind across the landscapes of Alberta (Rains et

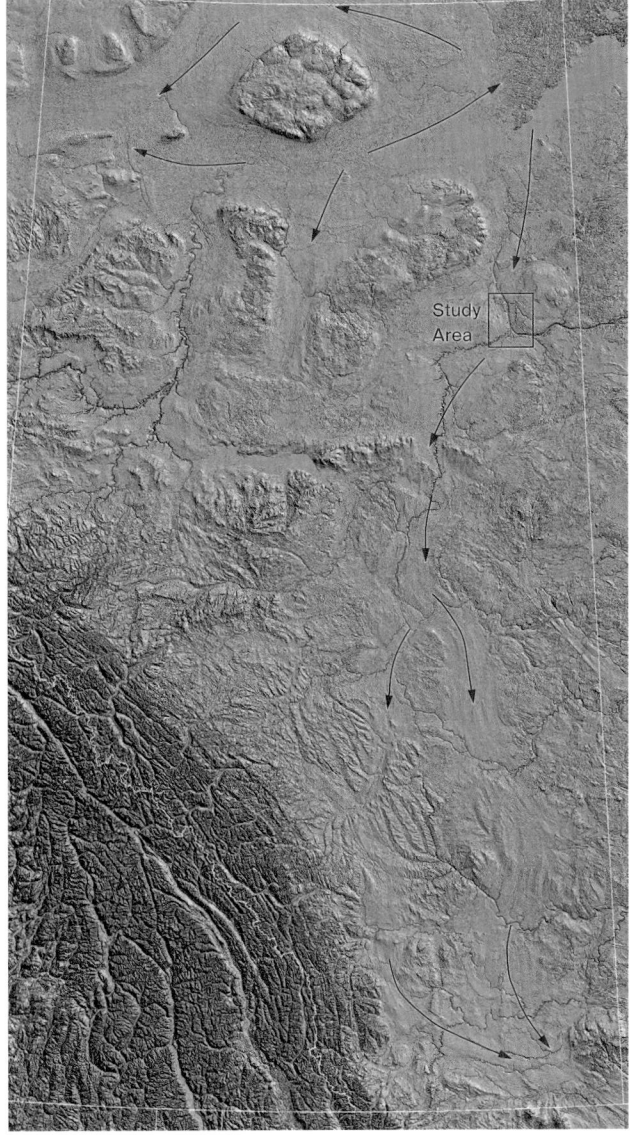

Figure 1.1. Digital elevation model (DEM) of Alberta, showing the channels cut by Late Wisconsinan subglacial megafloods

al. 1993; Sjogren and Rains 1995), while the outflow caused global sea levels to rise by 5 to 8 metres. Currently, similar but much smaller floods occur in Alaska, as well as in Iceland, where they are called *jökulhlaup*s. The result in the Fort McMurray region was the formation of a wide erosional swath that stripped off some of the Ice Age deposits, leaving behind a level plain in the areas surrounding the Athabasca valley (fig 1.2, the black arrows). That swath can be traced to the

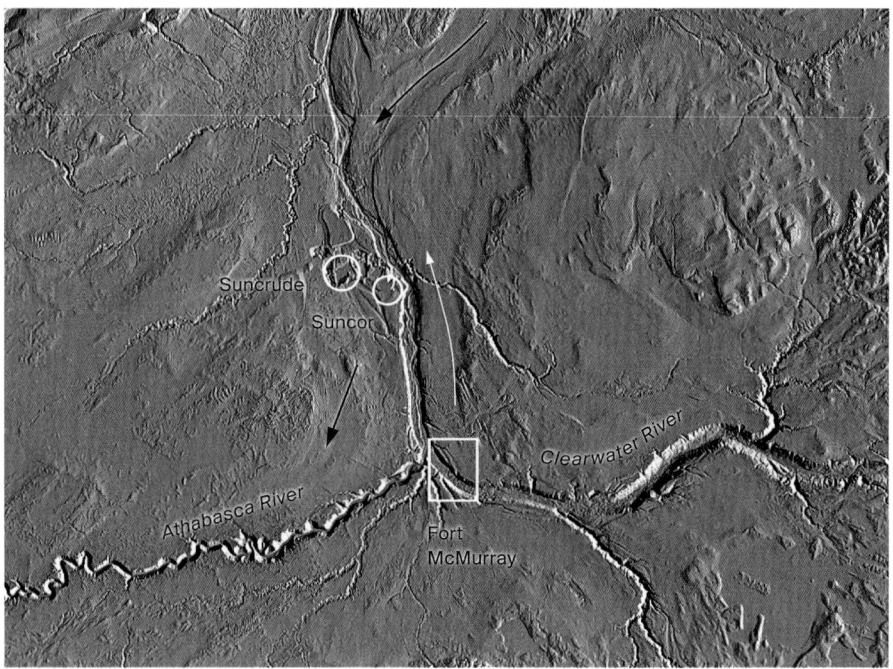

Figure 1.2. The flow direction of the two flood events in the Fort McMurray area. The black arrows pointing south indicate the local subglacial flood paths, while the arrow pointing north shows the subsequent path of Lake Agassiz flood waters entering the region through the Clearwater River.

south (see fig 1.1), where its elevation rises nearly a kilometre before it exits from Alberta into Montana. The flowing water was able to follow a path uphill because it was pressurized and flowed under confinement beneath the ice.

Several thousand years later, during deglaciation, large lakes were impounded against the margins of the retreating Laurentide Ice Sheet. One of the largest, and later, of these lakes—Glacial Lake Agassiz—covered most of the Hudson Bay drainage at one point or another in its existence. Rather than spanning a relatively fixed area, however, the lake occupied a series of areas, its shorelines shifting, and drained in a number of directions over the course of its lifetime, causing its water levels to rise and fall during specific phases in its history. At one early period in the evolution of Glacial Lake Agassiz, its water was diverted into the Missouri-Mississippi drainage, where it flowed into the Gulf of Mexico (Kennett and Shackleton 1975). Later, the failure of the ice dam along the eastern edge of the lake allowed a large discharge of meltwater through what is now the St. Lawrence Seaway, where the frigid water cooled the North Atlantic and may have caused a temporary relapse into ice age conditions (see, for example, Lowell et al. 2005; but see also Fisher, Lowell, and Loope 2006).

At a subsequent high stand, known as the Emerson Phase, the lake drained to the northwest through the Fort McMurray area, into Glacial Lake McConnell

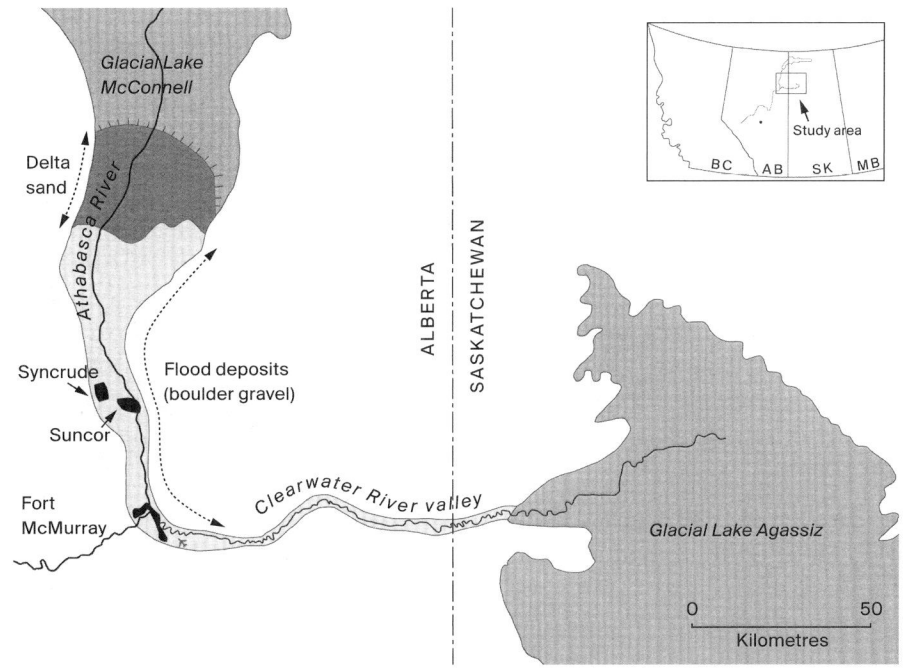

Figure 1.3. The northwestern drainage path of Glacial Lake Agassiz, about 9,800 ^{14}C yr BP. Flood waters carved a broad, deep valley along the Clearwater River, joined the Athabasca River, and then headed north to Glacial Lake McConnell and on to the Arctic Ocean. Two vertebrate fossils, one dating to the late Pleistocene Epoch and the other to the early Holocene, were recovered in 1976 and 1997, respectively, from the Suncor mine north of Fort McMurray.

(see fig 1.2, the white arrow), and then through the Mackenzie River drainage into the Arctic Ocean (Smith and Fisher 1993). This flood, originally thought to have occurred around 9,900 ^{14}C yr BP (about 11,335 cal yr BP), lowered the level of Lake Agassiz some 52 metres (Fisher and Smith 1994). Although it raised global sea levels by only about 6 centimetres, this flow would have caused the Arctic Ocean to rise 6 metres and would also have effected a freshening of the water. The lighter fresh water would have flowed over the denser ocean water, and since the fresh water was more likely to freeze, it would have served to increase the thickness of pack ice in the Arctic. The thickened pack ice and somewhat altered ocean circulation in the North Atlantic might have been responsible for a brief, 100- to 150-year cooling trend at that time, known as the Preboreal oscillation (Fisher, Smith, and Andrews 2002).

This last flood also formed the valley of the Clearwater River, which rose in Saskatchewan, flowed westward into Alberta, and joined the Athabasca River on its way north (fig 1.3). The valley is much larger than it should be, given the size of the Clearwater River today. It is much wider than the Athabasca River valley upstream of Fort McMurray, even though the Athabasca channel carries a great deal more water. Downstream of Fort McMurray, where many of the current oil sands projects are located (including the Syncrude and Suncor operations), the valley retains the broad dimensions inherited from the Clearwater flood. The widest parts of the valley experienced the initial decreases in velocity during the waning stages of the flood and became sediment traps for large amounts of flood gravel and sand. Deposition of thick gravel-and-sand sequences and the establishment of roughly modern flow conditions set the final stage for the current geological conditions in the valley. Several thousand years later, a wetter and cooler climate saw the growth and expansion of the boreal forest (MacDonald and McLeod 1996), as well as the formation of the bogs that preserved the mid-Holocene wapiti antlers, described below, at the Suncor site.

THE GREAT CANADIAN OIL SANDS MAMMOTH

To date, the oil sands area has yielded only two significant Quaternary vertebrate fossils. The first was discovered in July 1976, when Steve Vayda and Dennis Olson, two employees of the Great Canadian Oil Sands Company (now part of Suncor), recovered three large pieces of bone from overburden removed during mining operations. Although the bones were not found in situ, a site supervisor estimated that they had come from a gravel pit about 50 feet (15.2 m) below the surface, where they had apparently been deposited by fluvial action.[3] The bones are from the pelvis of a mammoth, most probably a woolly mammoth (*Mammuthus primigenius*), and consist of the nearly complete right innominate, as well as the pubic and iliac portions of the left innominate (fig 1.4). Dark mottling, ranging from mid-brown to black, suggests impregnation by hydrocarbons from the tar. The fossils are retained by the Royal Alberta Museum (accession no. P97.6.1).

The first radiocarbon date obtained for the specimen positioned it squarely in the mid-Wisconsinan interstadial: $32,150 \pm 1,950$ [14]C yr BP (S-3005). From the outset, though, this date was questioned by University of Calgary geomorphologist Derald Smith. In the early 1990s, Smith and Timothy Fisher (then Smith's graduate student) had been studying postglacial landscape modification in the Fort McMurray region. They posited a flood event, brought about by the catastrophic draining of Glacial Lake Agassiz westward through the Clearwater River from Saskatchewan (see fig 1.3), which had inundated the area at the confluence of the Clearwater and Athabasca rivers, not far from present-day Fort McMurray. In the space of only a few months to a year, the flood dumped an estimated 21,000 cubic kilometres of water into the Athabasca River, which flowed into Glacial Lake McConnell (which survives today as Lake Athabasca) to the north (Fisher and Smith 1994; Smith and Fisher 1993).

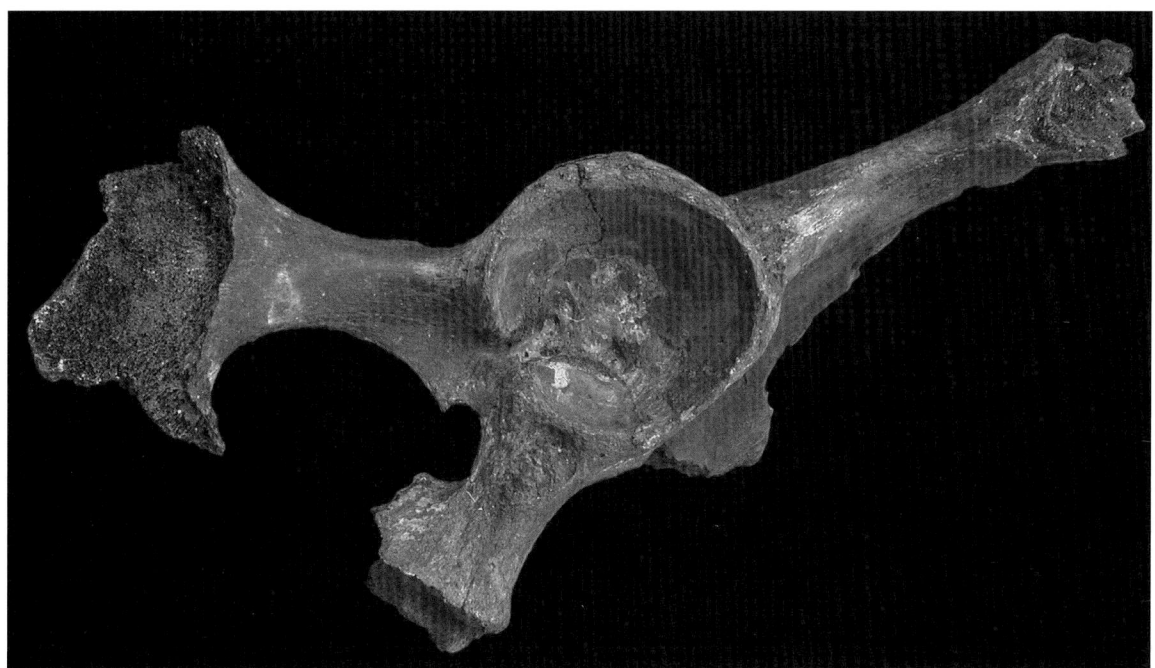

At the time, Smith and Fisher argued that the Lake Agassiz flood occurred at approximately 9,900 ^{14}C yr BP, during the lake's Emerson Phase, when water levels were relatively high. Their dating of the event was based on radiocarbon dates obtained from eleven samples of wood and peat deposited in areas scoured by the flood, which yielded an average age of 9,869 ^{14}C yr BP. The dates for these samples coincide with dates for wood samples from elsewhere on the Emerson Phase shorelines of Glacial Lake Agassiz (Bajc et al. 2000; Smith and Fisher 1993). But in order to have been deposited by the Agassiz flood in the fluvial sediments of the Athabasca River, Smith and Fisher argued, the mammoth bones must necessarily date to this postglacial period. Hence, Smith ordered a second radiocarbon assay.

Smith may also have rejected the initial date because he suspected contamination, as the sample used to obtain that date had not been pre-treated for what appeared to be hydrocarbons introduced by bitumen from the surrounding oil sands. The second assay employed a method of hydrocarbon extraction developed at the University of California at Riverside, in which benzene was used to remove hydrocarbon contamination (in this case caused by tar from the La Brea tar pits in downtown Los Angeles). Using only the organic fraction of the Fort McMurray mammoth bone, the lab performed hydrocarbon extraction, with negligible

Figure 1.4. The Great Canadian Oil Sands mammoth. The nearly complete right innominate (pelvic) bone, from what was probably a woolly mammoth, found in 1976 on the site of the Great Canadian Oil Sands project (now part of Suncor). The bone was initially dated to roughly 32,150 ^{14}C yr BP.

residue, and reported a date of 9,410 ± 280 ^{14}C yr BP (UCR-2430B; *in litt.*, R. E. Taylor to D. G. Smith, August 1990). For Smith, this younger date for the specimen was acceptable, as it indicated a postglacial deposition that coincided with the other available dates, from the wood and peat samples. All the same, the date should give one pause, as it suggests the occurrence of a very late-surviving mammoth in northern Alberta. This is not impossible, but it is unlikely, and so far unmatched.

In contrast, the date of 32,150 ^{14}C yr BP is not unreasonable given the suite of dates that were subsequently derived from mammalian megafaunal fossils recovered in various parts of the province (see, for example, Burns 1991, 1996a, 1996b, 2004; Burns and Young 1994; Matheus et al. 2004; Wilson and Burns 1999; Young et al. 1994, 1999). If the earlier date can be accepted (and we have no problem with that estimate), then, at the very least, it serves as a datum point, in both time and space, in the late Pleistocene fossil record of Alberta. Such a date is in fact nothing remarkable in Alberta, as mammoths have been dated from sites spanning an interval from roughly 43,000 ^{14}C yr BP in the Edmonton area (Burns, unpublished data) to 10,240 ± 325 ^{14}C yr BP from a site near Sundre, 95 kilometres north-northwest of Calgary (Burns 1996a).

In 2006, a third sample was removed from the mammoth pelvis and submitted for AMS radiocarbon dating to the Oxford University Accelerator Unit. Pretreatment included hydrocarbon decontamination that, again, yielded negligible residue. The date returned—32,140 ± 230 ^{14}C yr BP (OxA-16322)—handily validates the initial date of 32,150 ± 1,950 ^{14}C yr BP and suggests that the process of tar extraction carried out at Riverside may have introduced younger carbon into the sample.

THE SUNCOR WAPITI SKULL

In March 1997, an impressive skull specimen, complete with a full rack of antlers (fig 1.5), was recovered from overburden at a site on Suncor's property to the south of Mildred Lake.[4] Loader operator Joe Revol spotted the large, creamy-white specimen lying next to the place from which he had taken his previous load of dirt, some 29 feet (8.8 m) below the current surface. We subsequently visited the site to collect the skull, which was identified as that of a wapiti, or elk (*Cervus elaphus*), and now resides at the Royal Alberta Museum (accession no. P97.3.1). Some minor desiccation cracks were evident on the bone, and the nasal cavity was blocked with a marly deposit containing tiny mollusc shells.

Stratigraphy of the site. Although we initially assumed that the marl on (and in) the skull meant that it had come from marly deposits, we reconsidered the

Figure 1.5. The Suncor wapiti skull, with antlers. Found in 1997 in the vicinity of Mildred Lake, the skull has been dated to 5,550 ^{14}C yr BP.

assumption upon seeing the stratigraphy of the find's locality. The actual locality had been obliterated by subsequent overburden removal, and we had to settle for examining a section 50 metres east of the find site. Even then, the depth of the face of this "proxy" site had been reduced by overburden removal. Working from the current top of the proxy section, which consisted of a layer of relatively dry golden peat, we determined that the skull had probably come from about 1.8 metres below the surface, toward the bottom of a progression of decomposed peat (muskeg), with some deep roots extending into the underlying sediment, that gave way to a dark grey marl from 1.84 to 1.96 metres below the surface (fig 1.6). From 1.96 to roughly 3.00 metres lay a stratum of medium planar-bedded sand with some cross-bedding in the lower portion; this layer also contained low-grade, friable, sandstone rip-up clasts and bitumen. Below that, there was a layer of extremely poorly sorted clastic fragments of local bedrock.

The 1.8-metre-thick peaty layer, in which the skull is thought to have lain, graded from fine, dry, golden peat at the top through to rooty peat, and then into a 0.6-metre-thick layer of marl, shelly and beige above, less shelly toward the bottom. Fairly rapid stream flow, with a high sedimentation rate, is indicated by the inclusion of the rip-up clasts below 3.20 metres. The flow direction was downstream relative to the Athabasca River. The A-axis, or longest dimension, of the larger, angular, highly weathered clasts was in excess of 35 centimetres, and their orientation was parallel to the flow—evidence of the relatively high-velocity

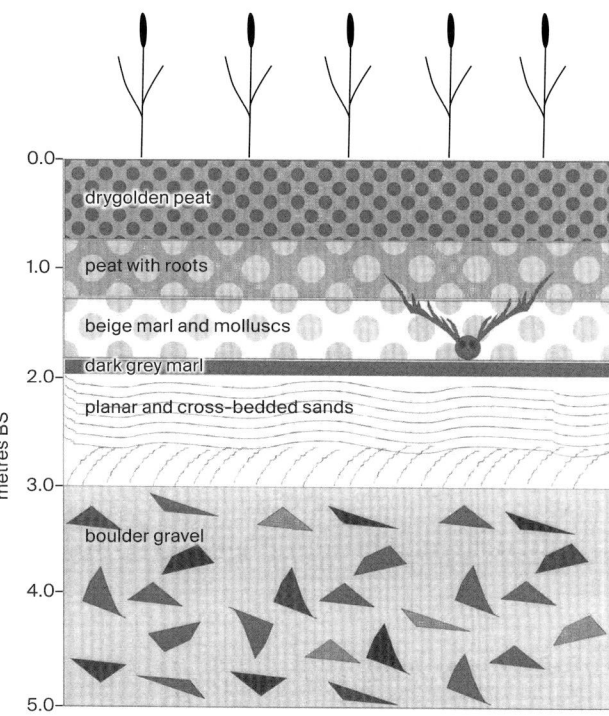

Figure 1.6. Stratigraphy of the sediments at the Suncor wapiti site. The skull was probably located toward the bottom of the layer of beige marls, which, in this proxy section, ends at about 1.8 metres below the surface.

discharge from Lake Agassiz as it poured into Glacial Lake McConnell to the north of Fort McMurray. The deposition of sands above the higher energy stratum indicates a return to less energetic stream flow, and the marls and peats above signal subsequent paludification and muskeg formation. The skull had not suffered any significant abrasion or stream rolling and was probably deposited in the marl- and peat-bottomed water body in late fall or early winter.

Radiocarbon dating of the wapiti skull. Three plugs of compact bone from the basiocciput (5.9 g total; Royal Alberta Museum sample QP-119) were sent to Thomas Stafford, Jr., director of the Laboratory for Accelerator Radiocarbon Research in Boulder, Colorado. Stafford extracted collagen by XAD-resin methodology and sent the sample to the Lawrence Livermore National Laboratory at the University of California at Berkeley for accelerator dating. The result—5,550 $\pm$ 50 ^{14}C yr BP (6,350 $\pm$ 50 cal yr BP; Calib 7.10) (CAMS-38689)—corresponds to the mid-Holocene.

Description of the skull. The wapiti skull is indistinguishable from a modern wapiti skull, as it is essentially modern itself (see the radiocarbon results above). With six points on both sides, it spans a maximum distance of 45$^6/_8$ inches using the Boone and Crockett Club method of measurement.[5] The unofficial "score" for this rack, based on the full set of prescribed measurements, is 341$^6/_8$. While this is a very respectable score, the Boone and Crockett record for a "typical" North American elk rack (from Arizona) is 442$^5/_8$, and the record for an Alberta rack stands at 419$^5/_8$. (In fact, a rack must score at least 360 to be included in the Boone and Crockett Club Awards book.)

Much of the rostral portion (that is, the snout) of the skull is absent, and the rear of the skull appears somewhat battered and eroded. Although some cracking of the surface of the antlers has taken place, the specimen is remarkably well preserved. It did not require any special conservation measures and was not coated with any surface consolidants.

Mollusc and arthropod identification. Table 1.1 summarizes the findings of James Van Es, who identified the shells and tests (internal shells) found in the marly deposit as closely as possible. Most were common aquatic snail species accompanied by small pelecypod valves and ostracode tests. A single terrestrial snail shell was identified as *Catinella* cf. *C. avara.*

Seed and charophyte identification. The plant seeds recovered from the wapiti skull represent emergent, aquatic types of plants that are rooted to the pond bottom and send up leaves and flowers on or above the surface (Alwynne Beaudoin, pers. comm., 1998). Seeds of sedges (*Carex* spp.), unusually numerous waternymph (*Najas* spp.), and one pondweed species (*Potamogeton* sp.) were present, along with several others as yet unidentified. In addition, hundreds of oögonia of undetermined alga-related charophytes were recovered. The seeds, charophytes, clams, and snails from the skull were compared to a sample from the seed- and shell-rich marl from the assumed level of the skull. The samples were identical, containing the same organisms in numbers of a generally similar magnitude.

Interpretation of the invertebrate fauna and aquatic flora. Pelecypod taxa include diminutive members of the Sphaeriidae family. The globular peaclam (*Pisidium ventricosum*) is found throughout central Canada south of the tree line and has the same general habitat preferences as the aquatic snails. The ostracodes, attributed to the order Podocopa (seed shrimp), are another type of freshwater aquatic that lives among vegetation.

Among the gastropod Mollusca, Van Es identified one terrestrial and three aquatic taxa. The aquatics (*Physa jennessi skinneri* and *Gyraulus deflectus*, as well as species of *Helisoma*) are widespread throughout much of Canada. They prefer

TABLE 1.1 Invertebrate taxa associated with the Suncor wapiti skull

Phylum MOLLUSCA

Order Basomatophora (aquatic snails)	
Family Physidae	
Genus *Physa*	
SPECIES *P. jennessi skinneri*	1
Family Planorbidae	
Genus *Gyraulus*	
SPECIES *G. deflectus*	108
Genus *Helisoma*	
SPECIES unidentified	11
Order Stylomatophora (land snails)	
Family Succineidae	
Genus *Catinella*	
SPECIES cf. *C. avara*	1
Order Eulamellibranchia (clams, mussels)	
Family Sphaeriidae	
SPECIES unidentified	many
Genus *Pisidium*	
SPECIES *P. ventricosum*	many

Phylum ARTHROPODA

Subphylum Crustacea	
Order Podocopa (seed shrimp)	many

SOURCE: Analysis by James Van Es, April 1997.

permanent-water habitats in lakes, ponds, rivers, and streams with moderate to abundant vegetation and—most commonly—muddy bottom substrates, as deep as 5 metres in the case of *Physa* (Clarke 1981).

The dominant gastropod taxon is the land snail *Catinella* cf. *C. avara*. In many of its habits, it is like all northeastern Albertan mollusc fauna. It is an inhabitant of shallow-marsh environments characterized by plentiful vegetation and muddy substrates. However, it is a "riparian terrestrial gastropod" (Bajc et al. 2000) that would have occupied wetlands bordering on the pond or lake in which the Suncor wapiti was interred.

All of the floral species identified have fairly broad tolerances within their habitats. So they tell us almost nothing significant about parameters such as air temperature, precipitation, depth or temperature of the water, or rate of flow. We can suggest that a small, shallow lake or pond existed at the site at the time the wapiti was deposited, but we cannot read in these data anything about the surrounding lands.

In sum, the wapiti site, which lies near present-day Mildred Lake, was clearly a permanent-water habitat supporting abundant vegetation and possessing a mud bottom. Although this possibility admittedly rests on a single shell, the occurrence of *Catinella* cf. *C. avara* suggests a near-shore aquatic site to which the terrestrial wetland snail was carried. As the suite of molluscs is not large, sampling errors may have skewed the inferred habitat in the direction of a more aquatic environment. The regional environment may in fact have been a true muskeg with interspersed wetland and aquatic habitats.

DISCUSSION

Evidence Concerning the Regional Environment

The wapiti is considered a "flexible" animal. It prefers "open areas such as alpine pastures, marshy meadows, river flats, open prairies, and aspen parkland, but occasionally it is found in coniferous forests" (Banfield 1974, 400). We can ask whether the Suncor wapiti was one of the animals that occasionally frequented boreal forest habitats.

Other evidence from the region is found at Eaglenest Lake. There, the pollen record suggests that, by 7,500 ^{14}C yr BP, the area surrounding what is now Fort McMurray was essentially modern boreal forest. Eaglenest Lake is located at the northeast end of a chain of lakes in the Birch Mountains, about 85 kilometres northwest of the Suncor project. The lake was cored through the ice in April 1982, and a spectrum of the pollen from the core was constructed on the basis of the identified taxa (Vance 1986). The dates returned on bulk organics from the core were 10,740 ± 150 ^{14}C yr BP, at 4.9 to 5.0 metres, and 7,240 ± 80 ^{14}C yr BP, at 2.6 to 2.7 metres below the surface of the lake muds. At the time the Suncor elk was alive, the Eaglenest Lake pollen influx figures show that spruce and pine pollen each constituted about 35% of the sample, while tree birch pollen accounted for about 10%, and alder was on the rise at around 15%. These are definite indicators of boreal forest cover; this is not conspicuously open country.

Vance (1986) concluded from his pollen study that, by 9,000 ^{14}C yr BP, the warm, dry conditions of the Hypsithermal had given way to a more moderate climate in the region of Eaglenest Lake, although conditions remained warmer and drier than at present. Pine reached the Birch Mountains by 7,500 ^{14}C yr BP, and no major changes in vegetation are registered in the pollen spectrum after that time. As Vance noted, this suggests that modern climatic conditions were established by 7,500 ^{14}C yr BP (although certain characteristics of pollen rain in the boreal forest may mask minor climatic fluctuations). The Suncor elk, at 5,500 ^{14}C yr BP, may thus very well have found its niche in the forest.

Moreover, let us suppose that the 9,400 ^{14}C yr BP for the Great Canadian Oil Sands mammoth is correct. If so, then consider two facts: Fort McMurray was hundreds of kilometres south and southwest of the glacial margin, and mammoths are grazers, not browsers. These two facts lead us to argue that, at the latitude of Fort McMurray, the boreal forest had not yet closed in on the area at that time, and the landscape still offered enough open grassland habitat to support megafaunal species like the Great Canadian Oil Sands mammoth. Four thousand years later, however, when the Suncor elk lived and died, the Eaglenest Lake pollen record shows that the area supported boreal forest, most likely dotted with sloughs and small lakes. The big difference between the two landscapes is that, at 9,400 ^{14}C yr BP, the succession of vegetation following Laurentide glaciation was still subject to variable climate. At 5,500 ^{14}C yr BP, the vegetation was probably not much different from what it is today, with a stable though perhaps slightly warmer and drier climate.

How Did the Elk Die?

We are partial to a drowning scenario. The articulated skeleton of an elk—the presumed victim of drowning—was excavated in 1985 from a commercial gravel pit in a terrace 50 metres above the modern Smoky River, near Watino, Alberta, and subsequently radiocarbon dated to about 9,900 ^{14}C yr BP (Burns 1986).[6] The complete skeleton of the mature, six-point stag was recovered from low-energy fluvial sediments, and the presence of full-grown antlers indicated that the animal had died late in the year. Together, these facts suggested that the animal had drowned in the late fall or early winter as it tried to cross the frozen surface of the slow-flowing river. Sediments in the subsequent spring flood buried the skeleton rapidly and preserved it articulated for more than 10,000 years.

Similarly, the Suncor wapiti may have fallen through the ice on a pond where it drowned and sank to the bottom. There, peat and marl accumulations eventually covered it over. It is interesting that only the antlered partial skull was found; none of the postcranial skeleton was recovered. In explanation, we offer two possibilities, one modern and one ancient. On the one hand, the bucket of the loader could have damaged the skull as it scooped up the rest of the skeleton, which was later dumped elsewhere. On the other hand, if the carcass had originally lain in a relatively shallow slough or pond, it would have been accessible to predators and scavengers at the time of death or the following spring. In addition, it would have taken a few years for the elk to become covered with peat and marl, and in the meanwhile predators could have scavenged and scattered the bones. It is also possible that predators *killed* the animal and scattered the skeleton far and wide, or even consumed it. However, no carnivore (or small rodent) bite marks were detected on the skull. So, while the evidence is not conclusive, it is not unreasonable to propose that the remaining parts of the skeleton were accidentally hauled away in a truck.

Regional Affirmation for the Mammoth

The mid-Wisconsinan age of the Great Canadian Oil Sands mammoth is supported by two AMS dates, each around 32,000 ^{14}C yr BP. The specimen is the sole large-mammal representative of mid-Wisconsinan fauna in the region, and its presence suggests that ice-free conditions prevailed there at that time. This conclusion accords well with data from a recently discovered site in the Birch Mountains, to the south of Jean Lake. There, Paulen, Beaudoin, and Pawlowicz (2005) examined a peaty deposit 9 metres below the surface that was sandwiched between two layers of till. The deposit (roughly 80 centimetres thick) contained macrofossil fragments of wood, bark, and moss, as well as conifer needles, charcoal, and a few seeds, the last identified as Cyperaceae. Fragments of wood (*Pinus* sp.) yielded an AMS date of 32,690 ± 340 ^{14}C yr BP.

SUMMARY

Two different flood mechanisms, a few thousand years apart, probably account for the form and size of the valleys of both the Clearwater River and that portion of the Athabasca River downstream of Fort McMurray. Clearly, however, neither of the two fossils we have considered in this chapter has any direct connection to either of the late-glacial flood events in the region. Assuming the older radiocarbon dates are correct, the Great Canadian Oil Sands mammoth lived long before these two floods occurred. Likewise, the Suncor wapiti, firmly dated to a mid-Holocene time (5,550 ^{14}C yr BP), missed them by several millennia. The question remains, however, why the oil sands area has yielded only two significant Quaternary vertebrate fossils to date.

The dearth of vertebrate fossils that coincide with subglacial megaflooding and with the earliest Holocene flooding of Lake Agassiz is most plausibly explained by the scouring and cleansing effects of the floods themselves within the two river valleys. Most megafaunal remains deposited prior to the earlier megaflood would have been flushed out of the Fort McMurray area by that event.

As we have seen, the bones of the mammoth were found in gravels, which in all likelihood indicates postglacial redeposition. By the time of the subsequent Lake Agassiz flood event, postglacial mammalian megafaunal species had been locally extirpated, and the force of the water removed virtually all traces of their presence. But if these floods left us with little by way of palaeontological materials, they also exposed the bitumen-laden bedrock, allowing us ready access to the oil sands.

ACKNOWLEDGEMENTS

We are very grateful to Suncor Energy Inc. for their freely offered co-operation, their marvelous hospitality while we were on site, and their generous reimbursement of all costs that we incurred in the course of the study, including the AMS radiocarbon dating of the wapiti skull. Reclamation Officer Steve Tuttle and his assistant, Leo Paquin, are to be particularly acknowledged in this regard. Suncor's John Davis and Dave Ryan also helped expedite the project with arrangements and publicity. Duane Froese, of Earth and Atmospheric Sciences at the University of Alberta, funded the Oxford AMS date on the mammoth pelvis, and we are extremely grateful to him. The late Jim Van Es, of the Department of Zoology at the University of Alberta, examined the invertebrate fauna associated with the Suncor wapiti and prepared a report of his findings: a posthumous thank-you, Jim. Alwynne Beaudoin, of the Royal Alberta Museum, kindly summarized her findings with respect to the plant and charophyte identifications. We are also grateful to Bill Weimann and Gary Erickson, vertebrate technicians at the Royal Alberta Museum, who scored the wapiti skull to Boone and Crockett Club standards. Photographs of the fossil specimens were provided by Peter Milot, of the Royal Alberta Museum, and Gregory Baker, of Phantom Ink, Edmonton. The greatest thanks, however, are reserved for Joe Revol, without whose quick thinking there would have been neither trophy elk nor palaeo-interpretation, nor any of this new knowledge of the Ice Age history of the oil sands area.

NOTES

1   We use the term "Ice Age" to refer to the Pleistocene Epoch, which began roughly 2.58 million years ago and was characterized by a series of glacial episodes, during which ice sheets advanced and then receded. The end of the Pleistocene, about 11,700 years ago,

marked the beginning of the Holocene Epoch, which continues down to the present day. Together, these two epochs constitute the Quaternary Period.

2　In the oil sands region, virtually all pre-Pleistocene fossil material consists not of vertebrate remains but of tar-impregnated wood of Cretaceous age, preserved by bitumen from deep in the oil sands seeping into the grain of the wood. Such fossils are occasionally uncovered by earth-moving associated with mining activities at the Syncrude and Suncor plants. When the non-petrified wood is exposed to air, the volatile hydrocarbons in the tar escape from it, causing the wood fibres to shrink and separate. Splinters may then be propelled into the air, like whining bullets.

3　The site from which the bones were removed is located at 56º 58.5' N, 111º 31.5' W (in legal subdivision 9 of section 15 in township 92, range 10 west of the 4th meridian), at an elevation of about 330 masl. See NTS map 74D, *Surficial Geology, Waterways* (Bayrock and Reimchen 1973). The Great Canadian Oil Sands Company became part of Suncor in 1979, as the result of a merger with the Sun Company of Canada.

4　Relative to the gravel pit in which the mammoth bones were found, the site that yielded the wapiti skull lies a little further north, at 57º 1.5'N, 111º 33.1'W (also in township 92, range 10 west of the 4th meridian, but in legal subdivision 14 of section 33), at an elevation of roughly 320 masl. See NTS map 74E, *Surficial Geology, Bitumount* (Bayrock 1971).

5　For details of the methodology, see "Scoring Your Trophy: Typical American Elk," *Boone and Crockett Club,* 2014, http://www.boone-crockett.org/bgRecords/bc_scoring_typical-elk.asp?area=bgRecords&type=Typical+American+Elk, as well as the associated score chart at http://www.boone-crockett.org/pdf/SC_elk_typical.pdf.

6　The published date was 9,075 ± 305 ^{14}C yr BP (S-2614, uncorrected), but a subsequent sample was dated to 9,920 ± 220 ^{14}C yr BP (AECV: 272c, uncorrected) (Burns, unpublished data).

REFERENCES

Bajc, A. F., D. P. Schwert, B. G. Warner, and N. E. Williams
2000　A Reconstruction of Moorhead and Emerson Phase Environments Along the Eastern Margin of Glacial Lake Agassiz, Rainy River Basin, Northwestern Ontario. *Canadian Journal of Earth Sciences* 37: 1335–1353.

Banfield, A. W. F.
1974　*The Mammals of Canada.* University of Toronto Press, Toronto, for the National Museums of Canada.

Bayrock, L. A.
1971　*Surficial Geology, Bitumount, NTS 74E.* Map 34, Alberta Research Council, Edmonton. Available as Map 140, Alberta Geological Survey, Edmonton. http://ags.aer.ca/publications/MAP_140.html.

Bayrock, L. A., and T. H. F. Reimchen
1973　*Surficial Geology, Waterways, NTS 74D.* Alberta Research Council, Edmonton. Available as Map 148, Alberta Geological Survey, Edmonton. http://ags.aer.ca/publications/DIG_2005_0525.html.

Burns, James A.
1986　A 9000-Year Old Wapiti (*Cervus elaphus*) Skeleton from Northern Alberta, and Its Implications for the Early Holocene Environment. *Géographie physique et Quaternaire* 40(1): 105–108.

1991　Mid-Wisconsinan Vertebrates and Their Environment from January Cave, Alberta, Canada. *Quaternary Research* 35: 130–143.

Burns, James A.

1996a　Vertebrate Paleontology and the Alleged Ice-Free Corridor: The Meat of the Matter. *Quaternary International* 32: 107–112.

1996b　Review of Pleistocene Zoogeography of Prairie Dogs (Genus *Cynomys*) in Western Canada with Notes on Their Burrow Architecture. In *Palaeoecology and Palaeoenvironments of Late Cenozoic Mammals*, edited by Kathlyn M. Stewart and Kevin L. Seymour, pp. 34–53. University of Toronto Press, Toronto.

2004　Pleistocene Lemmings (*Lemmus trimucronatus* and *Dicrostonyx groenlandicus*); Muridae; Rodentia) from Alberta, Canada. *Journal of Mammalogy* 85(3): 379–383.

Burns, James A., and Robert R. Young

1994　Pleistocene Mammals of the Edmonton Area, Alberta. Part I: The Carnivores. *Canadian Journal of Earth Sciences* 31: 393–400.

Clarke, Arthur H., Jr.

1981　*The Freshwater Molluscs of Canada*. National Museum of Natural Sciences, National Museums of Canada, Ottawa.

Fisher, Timothy G., and Derald G. Smith

1994　Glacial Lake Agassiz: Its Northwest Maximum Extent and Outlet in Saskatchewan (Emerson Phase). *Quaternary Science Reviews* 13: 845–858.

Fisher, Timothy G., Thomas V. Lowell, and Henry M. Loope

2006　Comment on "Alternative Routing of Lake Agassiz Overflow During the Younger Dryas: New Dates, Paleotopography, and a Re-evaluation" by Teller et al., 2005. *Quaternary Science Reviews* 25: 1137–1141.

Fisher, Timothy G., Derald G. Smith, and John T. Andrews

2002　Preboreal Oscillation Caused by a Glacial Lake Agassiz Flood. *Quaternary Science Reviews* 21: 873–878.

Kennett, James P., and Nicholas J. Shackleton

1975　Laurentide Ice Sheet Meltwater Recorded in Gulf of Mexico Deep-Sea Cores. *Science* 188 (11 April): 147–150.

Lowell, Thomas V., Timothy G. Fisher, Gary C. Comer, Irka Hajdas, Nicholas Waterson, Katherine Glover, Henry M. Loope, Joerg M. Schaefer, Vincent Rinterknecht, Wallace Broecker, George Denton, and James T. Teller

2005　Testing the Lake Agassiz Meltwater Trigger for the Younger Dryas. *Eos, Transactions, American Geophysical Union* 86(40): 365–373.

MacDonald, Glen M., and T. Katherine McLeod

1996　The Holocene Closing of the "Ice-Free" Corridor: A Biogeographical Perspective. *Quaternary International* 32: 87–95.

Matheus, Paul E., James A. Burns, Jacobo Weinstock, and Michael Hofreiter

2004　Pleistocene Brown Bears in the Mid-continent of North America. *Science* 306 (12 November): 1150.

Paulen, Roger C., Alwynne B. Beaudoin, and John G. Pawlowicz

2005　An Interstadial Site in the Birch Mountains, North-Central Alberta. Paper presented at the biennial meeting of the Canadian Quaternary Association (Water, Ice, Land, and Life: The Quaternary Interface), Winnipeg, Manitoba, 5–8 June. Abstract available in *Program and Abstracts*, p. A66.

Rains, R. Bruce, John Shaw, Darren B. Sjogren, Mandy J. Munro-Stasiuk, K. Robert Skoye, Robert R. Young, and Robin T. Thompson

2002　Subglacial Tunnel Channels, Porcupine Hills, Southwest Alberta, Canada. *Quaternary International* 90: 57–65.

Rains, R. Bruce, John Shaw, K. Robert Skoye, Darren B. Sjogren, and Donald R. Kvill

1993   Late Wisconsin Subglacial Megaflood Paths in Alberta. *Geology* 21: 323–326.

Sjogren, Darren B., and R. Bruce Rains

1995   Glaciofluvial Erosional Morphology and Sediments of the Coronation-Spondin Scabland, East-Central Alberta. *Canadian Journal of Earth Sciences* 32: 565–578.

Smith, Derald G., and Timothy G. Fisher

1993   Glacial Lake Agassiz: The Northwestern Outlet and Paleoflood. *Geology* 21(1): 9–12.

Teller, James T., Matthew Boyd, Zhirong Yang, Phillip S. G. Kor, and Amir Mokhtari Fard

2005   Alternative Routing of Lake Agassiz Overflow During the Younger Dryas: New Dates, Paleotopography, and a Re-evaluation. *Quaternary Science Reviews* 24: 1890–1905.

Vance, Robert E.

1986   Pollen Stratigraphy of Eaglenest Lake, Northeastern Alberta. *Canadian Journal of Earth Sciences* 23: 11–20.

Wilson, Michael C., and James A. Burns

1999   Searching for the Earliest Canadians: Wide Corridors, Narrow Doorways, Small Windows. In *Ice Age Peoples of North America: Environments, Origins, and Adaptations of the First Americans,* edited by Robson Bonnichsen and Karen L. Turnmire, pp. 213–248. Oregon State University Press, Corvallis.

Young, Robert R., James A. Burns, R. Bruce Rains, and D. B. Schowalter

1999   Late Pleistocene Geomorphology and Environment of the Hand Hills Region and Southern Alberta, Related to Middle Wisconsin Fossil Prairie Dog Sites. *Canadian Journal of Earth Sciences* 36: 1567–1581.

Young, Robert R., James A. Burns, Derald G. Smith, L. David Arnold, and R. Bruce Rains

1994   A Single, Late Wisconsin Laurentide Glaciation, Edmonton Area and Southwestern Alberta. *Geology* 22: 683–686.

2 Glacial Geology and Land-Forming Events in the Fort McMurray Region

TIMOTHY G. FISHER AND THOMAS V. LOWELL

Deglaciation events in the area around Fort McMurray are central to our understanding of the archaeological sites in the Lower Athabasca valley. Viewed from the perspective of physical geography, however, our area of interest is somewhat broader, extending from northwestern Saskatchewan along the Clearwater River system into northern Alberta. The Clearwater River, which joins the Athabasca River at Fort McMurray, is unique in Alberta, as it is the only major westward-flowing river in the province. Relief in the hills surrounding Fort McMurray reaches 430 metres, and locally 235 metres within the Clearwater valley. Climatically, the region is immediately north of the discontinuous permafrost line (Brown 1967) and lies within the Central Mixedwood ecoregion of the boreal forest. Short winter and long summer days characterize the area, with July being the wettest month and most of the precipitation occurring during summer (Natural Regions Committee 2006, 31–32; see also 136–40). The landscape is mantled with deciduous forests, white spruce, balsam fir, jack pine, and peatlands. The geology generally consists of a stacked set of Quaternary-age sediments unconformably overlying Lower Cretaceous sandstones (some oil bearing) and marine shales, all of which overlie Devonian-age rocks (Langenberg et al. 2003). Beneath the Palaeozoic strata are Archean Precambrian Shield rocks consisting of granites and gneisses with varying degrees of metamorphism (Mossop and Shetsen 1994; Tremblay 1960).

The exploitation of resources in the Fort McMurray area has been significantly influenced by glacial and deglacial processes that culminated at the end of the last ice age. These resources include Beaver River Sandstone, which early First Nations hunter-gatherers used to manufacture tools, and the Athabasca tar sands, which supply oil to satisfy the insatiable thirst of modern societies for energy. Preceding the onset of numerous episodes of glaciation during the Pleistocene Epoch, subaerial weathering and fluvial incision over millions of years shaped the landscape in northern Alberta, developing the first-order hills and mountain physiography. During the Pleistocene, which began about 2.58 million years ago, glaciers advanced from a northerly direction and, over multiple separate glacial cycles, sculpted and modified the landscape (Andriashek and Atkinson 2007). The complexity of these multiple distinct glaciations is evident from a close examination of figure 2.1, where, in the broad valleys and on the highest hills, numerous sets of streamlined ridges, indicative of glacial activity, are oriented in different directions, sometimes cross-cutting one another. Such a variety of flow directions within the same region but at different points on the landscape provides evidence for distinct events during one or more past glaciations, as well as for the complexity of the subglacial environment and the likelihood of the differential erosion and preservation of palaeoglacial landscapes.

Here we examine events associated with deglaciation from the Late Wisconsinan maximum advance position of the Laurentide Ice Sheet in the foothills of Alberta, with a focus on the oil sands region of northeastern Alberta. According to dates from the bones of now mostly extinct ungulates that have been recovered from sand and gravel deposits beneath till from the Laurentide Ice Sheet, the margin of the glacier reached the Edmonton area sometime after 21,300 ^{14}C yr BP (Young et al. 1994). Older wood samples from the Fort McMurray area, dating to 35,000 ^{14}C yr BP (Fisher et al. 2009; Paulen et al. 2005), may be used to establish an uppermost age for the advance of the ice sheet over the Fort McMurray area. The meeting of the continental Laurentide Ice Sheet with outlet glaciers extending eastward from the Cordilleran Ice Sheet along the Rocky Mountain foothills is recorded by the foothills' erratic train, a series of orthoquartzite blocks, extending from Jasper into Montana, that have been dated, using cosmogenic isotopes, to 16,000 to 11,000 calendar years ago (Jackson et al. 1997). Recession of the ice sheet across Alberta from its maximum extent in southwestern Alberta or northern Montana is weakly constrained by radiometric dates. In addition, the absence of significant recessional moraines has been taken as evidence of widespread glacial stagnation, perhaps following large-scale subglacial meltwater flooding (Rains et al. 1993; Shaw et al. 2000; Sjogren et al. 2002; see also Burns and Young, this volume), or of a slow, continuous retreat of the ice across Alberta, without the pauses or readvances that would

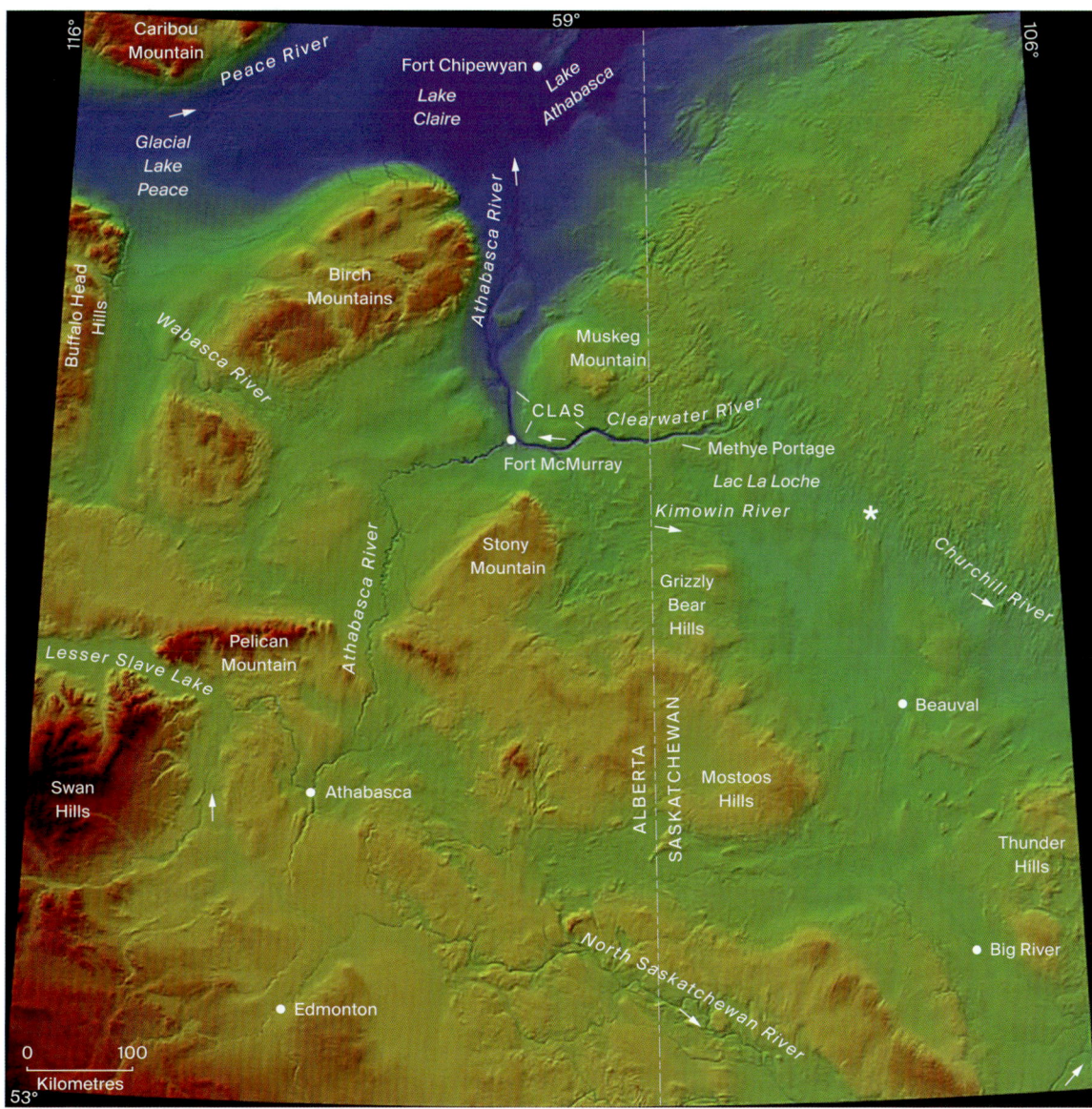

Figure 2.1. Digital elevation model (DEM) map of northeastern Alberta and northwestern Saskatchewan. Highest elevations are indicated by the dark red colours, and lowest by the deep blue colours. Note the steeper slopes on the hills and mountains that face northeast, the direction from which the Pleistocene glaciers came. A careful examination of the tops of many of these hills will reveal linear to curvilinear ridges that developed beneath the glacier and are used as indicators of past ice flow direction. The star marks the location of the channels between the Wycherley and Leboldus lakes in northwestern Saskatchewan. CLAS = Clearwater–Lower Athabasca spillway.

generate large moraines. The release of the digital elevation data gathered by the Shuttle Radar Topography Mission (SRTM) sparked a resurgence of interest in glacial geology and has resulted, for example, in the discovery of new moraines and landforms in the Fort McMurray area (Waterson et al. 2005) and in Saskatchewan (Campbell 2005).

In Alberta, Quaternary and glacial geology mapping and research provide a basis for interpreting the regional glacial history, as well as for identifying resources to support tar sands mining and related infrastructure. The first surficial geology maps, of Bitumount, NTS 74E, and Waterways, NTS 74D, were published by Bayrock (1971) and Bayrock and Reimchen (1973), respectively. Shortly afterwards, a more detailed surficial geology map, focusing on locating aggregate for mining areas, was produced by McPherson and Kathol (1977). More recently, Andriashek and Atkinson (2007) compiled borehole logs for the Fort McMurray region, identifying aquifers in buried channels and valleys, which laid the groundwork for a hydrogeological characterization of the subsurface and tunnel valleys system (Atkinson, Andriashek, and Slattery 2013).

When glaciers retreat down elevation gradients, proglacial lakes develop from glacial meltwater and from rivers dammed against the glacier. As ice margins continue to recede, and lower outlets are uncovered, lakes evolve in size, extent, and elevation. While simple in concept, the uplift of the Earth's crust from glacioisostatic adjustment complicates matters because the land tilts differentially, causing lake levels to fluctuate, which can in turn cause outlets to shift. Radiocarbon dating of basal organics in scour lakes within spillways yields minimum ages, from which we can estimate when a lake outlet was abandoned (see, for example, Fisher 2007).

Taylor (1960) recognized that a large lake, which he referred to as Lake Tyrell, had once occupied the Athabasca basin, forming when northward-flowing rivers were dammed by the retreating ice sheet. However, his palaeogeographic reconstructions did not take into account glacioisostatic rebound, and subsequent researchers divided Lake Tyrell into Glacial Lakes McMurray, Peace, and McConnell (see Dyke and Prest 1987; Raup and Argus 1982; Rhine 1984; Rhine and Smith 1988). Others have suggested an eastern diversion of meltwater up the Clearwater River in response to an ice dam north of Fort McMurray (see Christiansen 1979; Lemmen et al. 1994; and Schreiner 1983, 1984). In this model, the water would have drained eastward into Saskatchewan, past the present-day town of Big River (see fig 2.1), and then into the North Saskatchewan River. Although the history of glacial lakes in this region is complex, the existing palaeocurrent data presented by Smith and Fisher (1993) demonstrated that flow in the Clearwater system was, however, from east to west. According to Rhine and Smith (1988), much of the sediment that was eroded to form the Clearwater and Lower Athabasca river valleys was deposited as a large delta (named the Late Pleistocene Athabasca braid delta) in Glacial Lake McConnell (see fig 2.3). Glacial Lake McConnell was a large proglacial lake that encompasses the current basins of Lake Athabasca, Great Bear Lake, and Great Slave Lake.

In addition, as the Laurentide Ice Sheet retreated from the centre of the continent, a massive lake, Glacial Lake Agassiz, began to form. Over a period of 7,000 years, it grew in size, beginning in Minnesota and North Dakota and eventually extending across most of present-day Ontario and Manitoba and west into northern Saskatchewan (Fisher et al. 2011). That meltwater from Glacial Lake Agassiz may have drained from Saskatchewan down the Clearwater River was suggested first by Upham (1895), who was seeking a possible drainage route for the Lake Agassiz in the vicinity of the Methye Portage, located on a fur trade route not far north of Lac La Loche (see fig 2.1). Following the publication of new 1:50,000 topographic maps, Elson (1967) suggested that Lake Agassiz may have drained through a series of channels between Wycherley and Leboldus lakes in northwestern Saskatchewan (marked by a star in fig 2.1). In subsequent work, however, the northwestern extent of Lake Agassiz was not mapped as far northwest as these channels (see Schreiner 1983; Teller et al. 1983). Interest in a possible northwestern outlet returned with research by Smith (1989), who suggested that the Clearwater–Lower Athabasca spillway, now occupied by the underfit Clearwater River (Sproule 1939), was cut by meltwater from Lake Agassiz. Subsequent work by Fisher and Smith documented strandlines and glaciolacustrine sediments at elevations higher than those previously mapped in northwestern Saskatchewan, which were interpreted to represent a further northwestern extension of Lake Agassiz to the head of the spillway, well above the elevation of the channels identified by Elson in 1967 (see Fisher 1993a; Smith and Fisher 1993; Fisher and Smith 1994). Fisher and Smith (1993) inferred that Lake Agassiz had been dammed by the Beaver River Moraine, which extends northward discontinuously from the position shown below on figure 2.2 to the Clearwater River, parallel to the 11.0 (11,000 ^{14}C yr BP) ice margin. The moraine failed when it was overtopped by the transgressing Lake Agassiz, releasing the flood that flowed through the spillway, an event then estimated to have occurred around 9,900 ^{14}C yr BP. The sediments associated with the flood are similar to deposits from other large, catastrophic Pleistocene-age floods (see Fisher 1993a; Fisher and Smith 1993; Fisher et al. 1995; and, for comparisons, Baker 1973; Kehew and Lord 1986; Kehew et al. 2009).

Deriving a better deglacial chronology and spillway age has been the focus of research over the past fifteen years. Dyke and Prest (1987) and Dyke, Moore, and Robertson (2003) mapped isochrones across the area at 11,000 ^{14}C yr BP (see fig 2.2A) on the basis of bulk sediment ages from the lakes shown below in figure 2.3. Fisher and Souch (1998) recored a scour lake situated at the head of the Clearwater–Lower Athabasca spillway that had previously been dated by Anderson and Lewis (1992) and then AMS-dated organics found in the same stratigraphic position. The new ages turned out to be approximately 2,000 years younger, suggesting that the spillway formed later than had previously been thought. Similarly, Fisher et al. (2009) recored Mariana Lake, southwest of Fort McMurray, and dated terrestrial macrofossils from the same stratigraphic interval at which Hutton et al. (2004) had earlier dated bulk lake sediment. The Mariana Lake radiocarbon date proved to be a thousand years younger, again suggesting that the ice sheet did not recede from this area until more recently. A review of the deglacial ages associated with glacial lakes in northwestern Saskatchewan by Fisher (2007) also points to a later date for deglaciation than do the reconstructions by Dyke, Moore, and Robertson (2003). More recently, Murton et al. (2010) suggested that the Clearwater–Lower Athabasca spillway was occupied twice during deglaciation, the first time about 11,000 ^{14}C yr BP and then again at approximately 9,900 ^{14}C yr BP. However, following Fisher, Lowell, and Loope (2006), Fisher et al. (2009), and Fisher and Lowell (2012), we note that no field data exist to support the hypothesis that Lake Agassiz extended this far north at 11,000 ^{14}C yr BP. As we will see below, Fisher et al. (2009) proposed that the large proglacial lake (Glacial Lake Churchill) in the Fort McMurray region that extended southeast into the upper Churchill valley was formed from the coalescing of Glacial Lake McMurray, Glacial Meadow Lake, and the body of water in the upper Churchill valley earlier identified by Fisher and Smith (1994).

A moderate amount of research is thus in place that helps us to understand Quaternary events in the Fort McMurray area. However, the number and chronology of floods in the Clearwater spillway are matters of debate, the deglacial chronology of ice margin recession in the region remains weakly defined, and the palaeogeographic reconstructions of proglacial lakes in the Churchill, Beaver, and Athabasca river valleys are poorly understood. In what follows, we will review numerical ages for a series of newly discovered moraines in the Fort McMurray area and then briefly comment on the palaeogeography of the proglacial lakes in the light of the new age data for the moraines, before summarizing the landscape-forming events associated with deglaciation and the subsequent paraglacial time (Church and Ryder 1972).

DEGLACIAL MORAINE SEQUENCE

During deglaciation, the Laurentide Ice Sheet retreated northeastward and began forming moraines west of Fort McMurray. Until the ice retreated down the Athabasca valley beyond the Birch Mountains and Muskeg Mountain, ice-contact glacial lakes remained in the Fort McMurray area, and, until these lakes drained, the Clearwater–Lower Athabasca spillway was not occupied. Unravelling the deglacial history of the Fort McMurray area is hindered by poorly accessible sediment exposures and relatively few radiocarbon dates from exposed late glacial stratigraphic sections. Work within the past few years has resulted in an increase in radiocarbon dates from small lakes adjacent to newly mapped moraines (Fisher et al. 2009) and lithostratigraphic logging of remote stream banks (Andriashek and Atkinson 2007), but without dating control. While the site-specific details of these lake core sites are presented elsewhere (Fisher et al. 2009), we show here the ages of a series of moraines that constrain the age of any proglacial lake in the Fort McMurray region. These moraine ages were determined from seventy-three separate radiocarbon dates obtained at thirty-four core locations. At each site, multiple cores relating to the contact between organic lacustrine sediments (equivalent to contemporary lake sediments) and inorganic sediment (glacial lake or subaqueous ice-marginal sediment) were recovered with a hydraulically assisted modified Livingstone corer. All reported radiocarbon dates represent AMS ages from terrestrial macrofossils and are uncalibrated (for calibrated ages, see Fisher et al. 2009). The isochrones of the retreating ice sheet are shown in figure 2.2A, which depicts an earlier reconstruction based on data from Dyke and Prest (1987) and Dyke, Moore, and Robertson (2003), and figure 2.2B, which provides the updated ice margins based on our own recent work.

Figure 2.2B shows a stepped recession of a lobate geometry ice margin retreating northward down the Athabasca valley. We suggest that an ice margin at the Fort Hills and Firebag moraines, which block the Lower Athabasca valley, would have dammed in any proglacial lake. The oldest organic material from a core site on the proximal (northern) side of the Firebag Moraine was $9,600 \pm 70$ [14]C yr BP (ETH-30174) (Fisher et al. 2009), which we consider our best minimum age for drainage of the lake and presumably for the incision of the Clearwater–Lower Athabasca spillway. Continued ice retreat was to the Richardson Moraine and then to the Old Fort Moraine; the latter is usually considered a northwestern extension of the well-developed Cree Lake Moraine in Saskatchewan. These new results suggest that deglaciation took place considerably later than Dyke, Moore, and Robertson (2003) suggest. These results also agree well with the ages of two

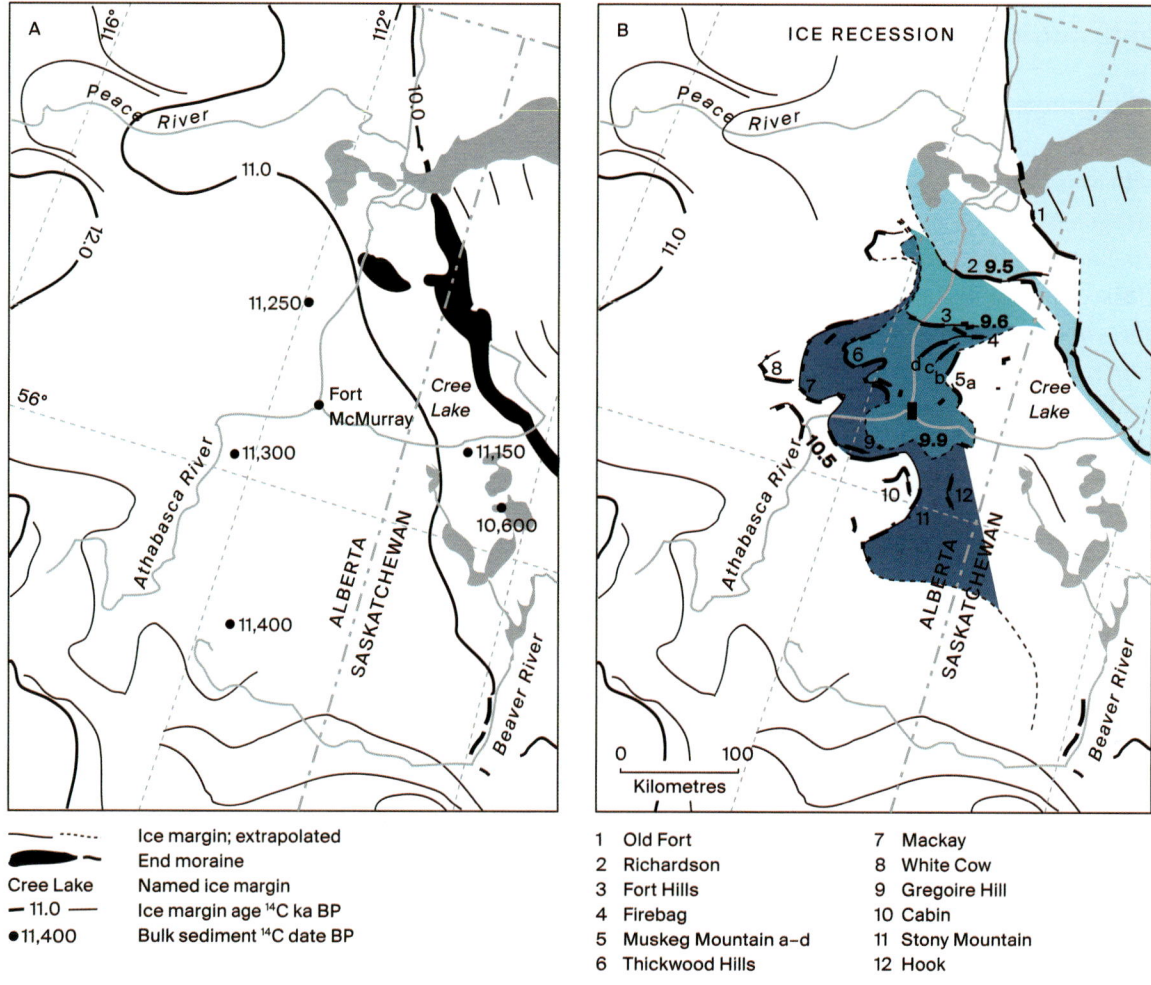

Ice margin; extrapolated
End moraine
Cree Lake — Named ice margin
— 11.0 — Ice margin age ^{14}C ka BP
●11,400 — Bulk sediment ^{14}C date BP

1	Old Fort	7	Mackay
2	Richardson	8	White Cow
3	Fort Hills	9	Gregoire Hill
4	Firebag	10	Cabin
5	Muskeg Mountain a–d	11	Stony Mountain
6	Thickwood Hills	12	Hook

Figure 2.2. Two models of deglaciation in the Fort McMurray region, with, in "A," deglacial isochrones and bulk sediment dates from Dyke and Prest (1987) and Dyke, Moore, and Robertson (2003), and, in "B," from Fisher et al. (2009). Regarding the latter, note that deglaciation in the Fort McMurray region is estimated to have occurred up to 1,500 radiocarbon years later than in initial estimates.

logs found in the Late Pleistocene Athabasca braid delta, which have been dated to 9,900 ± 20 ^{14}C yr BP (GSC-4301) and 9,700 ± 130 ^{14}C yr BP (AECV-1183C) (Smith and Fisher 1993).

In summary, between 10,500 and 9,600 ^{14}C yr BP, the low-lying area around the Fort McMurray region was being deglaciated. At elevations below the strandlines of Glacial Lake McMurray (see fig 2.3), a lake replaced the glacial ice, but above these elevations vegetation had begun stabilizing the landscape.

PROGLACIAL LAKES

As the Laurentide Ice Sheet retreated northeastward down the regional slope, ephemeral ice-contact lakes formed where drainage pathways were blocked either by ice or by higher topography. The location of the most important of these lakes, as well as of the principal moraines in the area, is shown in figure 2.3. In Alberta, Glacial Lake McMurray formed over the Fort McMurray region (see Fisher 1993b), while, in Saskatchewan, Glacial Meadow Lake formed west of the Beaver River Moraine (see fig 2.2). Outlets for Glacial Meadow Lake and Glacial Lake McMurray are not well understood. For Glacial Lake McMurray, earlier pro-glacial lake stages must certainly have existed as ice retreated northward from the Pelican Mountains. Outlets at that time were likely to the east, south of the Mostoos Hills (see fig 2.3). With continued ice recession, water may have drained northwestward in the gap across the Birch Mountains now occupied by the Wabasca River into Glacial Lake Peace (Lemmen et al. 1994), unless Lake Peace was draining into Lake McMurray through this spillway. Alternatively, these pro-glacial lakes may have continued to drain to the southeast, and, as the ice retreated north of Stony Mountain, drainage was redirected through Glacial Meadow Lake southward through the channel at Big River and into the North Saskatchewan River system (see fig 2.3). Drainage of Glacial Meadow Lake west-ward to form the Clearwater–Lower Athabasca spillway was refuted by Smith and Fisher (1993) on the grounds that the head of the spillway is located further east than the Beaver River Moraine.

Reconstructing the palaeogeography of lakes that no longer exist or that once were higher in elevation requires the identification of landforms and sediments associated with a series of lake levels. Such data may include now-abandoned shorelines (beaches), higher-elevation spillways that drained the lake, perhaps in more than one direction, and lake-bottom sediment. Fisher and Smith (1994) identified high-elevation shorelines near the head of the Clearwater–Lower Athabasca spillway and high-elevation lake sediment further to the east between the Cree Lake Moraine and the Churchill River. On the basis of this evidence, they proposed that Glacial Lake Agassiz stood at its relatively high Norcross level when it drained over the Beaver River Moraine to initiate the Clearwater–Lower Athabasca flood. However, this reconstruction can no longer be accepted, given that, as Fisher (2005) pointed out, the Norcross strandline cannot be traced from the main basin of the lake in Ontario across Saskatchewan to the high-elevation strandlines mapped by Fisher and Smith (1994). Moreover, the Norcross beaches have since been dated by OSL (optically simulated luminescence) to about 13,600 calendar years ago, or about 11,750 ^{14}C yr BP (Lepper et al. 2013), making

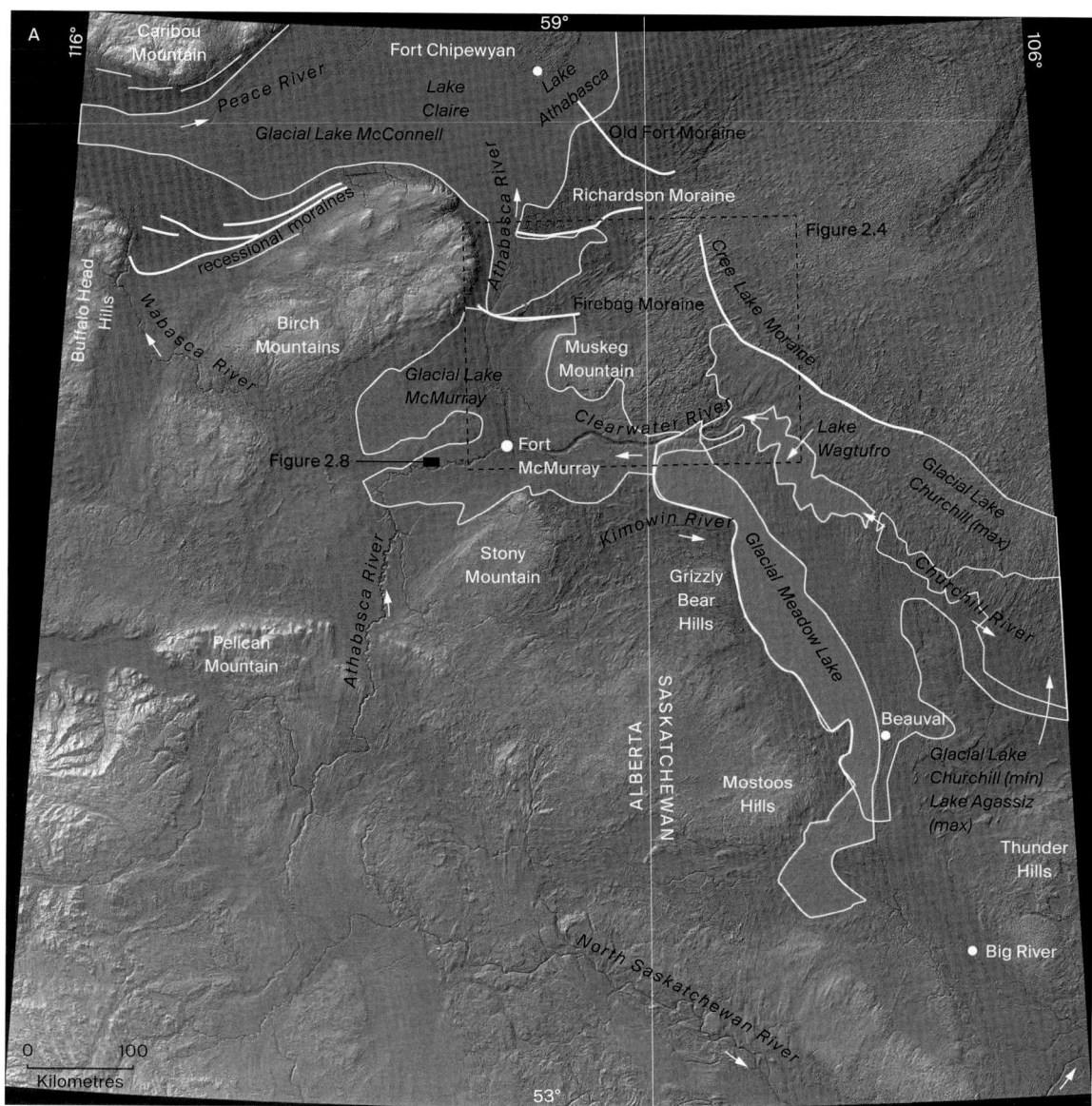

Figure 2.3. Location of major proglacial lakes and moraines in northeastern Alberta and northwestern Saskatchewan

them much older than the flood deposits at the Syncrude mine site, which date to roughly 9,900 ^{14}C yr BP (Fisher and Lowell 2012). To explain the high-elevation strandlines and lake sediment, Fisher et al. (2009) proposed that, as the ice margin retreated east from the Beaver River Moraine into the headwaters of the Churchill River, a glacial lake developed, which they named Glacial Lake

Churchill. Glacial Lake Churchill was constrained by ice to the east and to the north, in the Lower Athabasca valley, and probably overflowed to the south until the ice in the Athabasca valley gave way, resulting in a catastrophic flood.

At elevations below the Norcross strandline, the Campbell beaches of Glacial Lake Agassiz have been mapped throughout much of the lake basin, including all along the western shore of the lake from the southern outlet in Minnesota and North Dakota, northwestward into Saskatchewan, and ending at a series of small meltwater channels between Wycherley and Leboldus lakes (see fig 2.1) (Fisher and Lowell 2012). These channels, which are lined with boulders, record flow to the northwest, into Lake Wagtufro (which spanned the modern lakes Wasekamio, Turnor, and Frobisher) and then into the Clearwater–Lower Athabasca spillway (Fisher and Souch 1998). Many radiocarbon and OSL ages from the Campbell shoreline date it to around 9,400 ^{14}C yr BP (Lepper et al. 2013), which is well after the flood event in the Clearwater–Lower Athabasca valley. Even though Murton et al. (2010) projected high shorelines of Glacial Lake Agassiz to the northwest to force drainage out of a northwest outlet, the available strandline data does not support their palaeogeographic reconstruction (Fisher and Lowell 2012). The available geological record supports northwest drainage from Glacial Lake Agassiz only when it was at the Campbell level, at approximately 9,400 ^{14}C yr BP.

The current understanding of the source of water in the Clearwater–Lower Athabasca spillway flood is that Glacial Lake Churchill drained catastrophically when the ice margin retreated from the Firebag Moraine, at about 9,800 ^{14}C yr BP, in the Lower Athabasca valley, a process that would have been facilitated by the lake overtopping the moraine and/or by subglacial drainage beneath the ice sheet's western edge. Meanwhile, in the Glacial Lake Agassiz basin, water levels were rising to the Campbell level, and approximately five hundred years after the flood, Glacial Lake Agassiz would have transgressed up the Churchill valley, where it could drain to the northwest through the channels described earlier as a tributary to the Clearwater River. Minimal fluvial erosion would be expected from the northwest routing of water, as it inherited a fluvial system adjusted for much higher flow volumes when Glacial Lake Churchill drained. The date at which a northwest outlet for Glacial Lake Agassiz was abandoned can be inferred from basal AMS dates obtained on terrestrial macrofossils found on the bottom of lakes and channels entering or draining Lake Wagtufro, which range between 9,450 and 9,120 ^{14}C yr BP (Fisher 2007).

This section will outline the geomorphic and sedimentological evidence for the large flood in the Clearwater–Lower Athabasca spillway. Beyond scientific reasons for examining the spillway, its formation not only exposed the Beaver River Sandstone used by Palaeoindian groups but also eroded away much of the glacigenic overburden, exposing the oil-bearing bedrock and depositing the aggregate necessary to support mining operations.

The geomorphology of the large spillway extending from northwestern Saskatchewan into northeastern Alberta and its concomitant boulder-gravel deposits provide evidence for a large flood. The spillway extends 148 kilometres from its head westward to Fort McMurray, where it bends northward for another 85 kilometres (see fig 2.4). Along most of its length it ranges in width from 2.5 to 5 kilometres and is wider at each end. Spillway depth varies from over 200 metres, near its head, to 80 metres just before it bifurcates northward around the Fort Hills (figs. 2.4 and 2.5). Data from Alberta Transportation and Utility Bridge Engineering Branch indicate that 49 metres of sediment now lies above Devonian-age bedrock at the Highway 63 bridge north of the Syncrude plant, indicating that the northern end of the spillway has filled in over time. The sediment beneath the Athabasca River may be explained by decreasing gradients in response to glacioisostatic adjustment that is still active in the region today (Fisher and Souch 1998). Mass movement processes are ubiquitous along much of the spillway within the Lower Cretaceous bedrock, especially along the Clearwater River. A longitudinal profile of the spillway shows how the modern stream gradient lessens downstream, in part a function of the underlying bedrock (fig 2.5).

Water entered the spillway at its eastern end as Lake Churchill dropped from its maximum elevation of approximately 500 metres to its minimum level at 438 metres, the elevation of a well-developed strandline of Lake Wagtufro above Lake Wasekamio at the head of the spillway (Fisher and Souch 1998). In figure 2.4, the two arrows southeast of the spillway head represent water flow from Lake Wagtufro (see fig 2.3), recording steady-state flow through the outlet. The spillway head consists of large (1 to 24 km²) erosional residual islands of variable but often streamlined form. Channels between residual bedrock islands are 1.5 to 2.0 kilometres wide and often hanging. The main channel occupied by the Clearwater River is 2.5 to 3.0 kilometres wide, nearly double that of the spillway floor downstream. The channel marked with a "T" on figure 2.4 represents Tocker Lake, in which a geophysical survey revealed laminated glaciolacustrine sediment within the bedrock channel (Gilbert et al. 2000). The presence of the

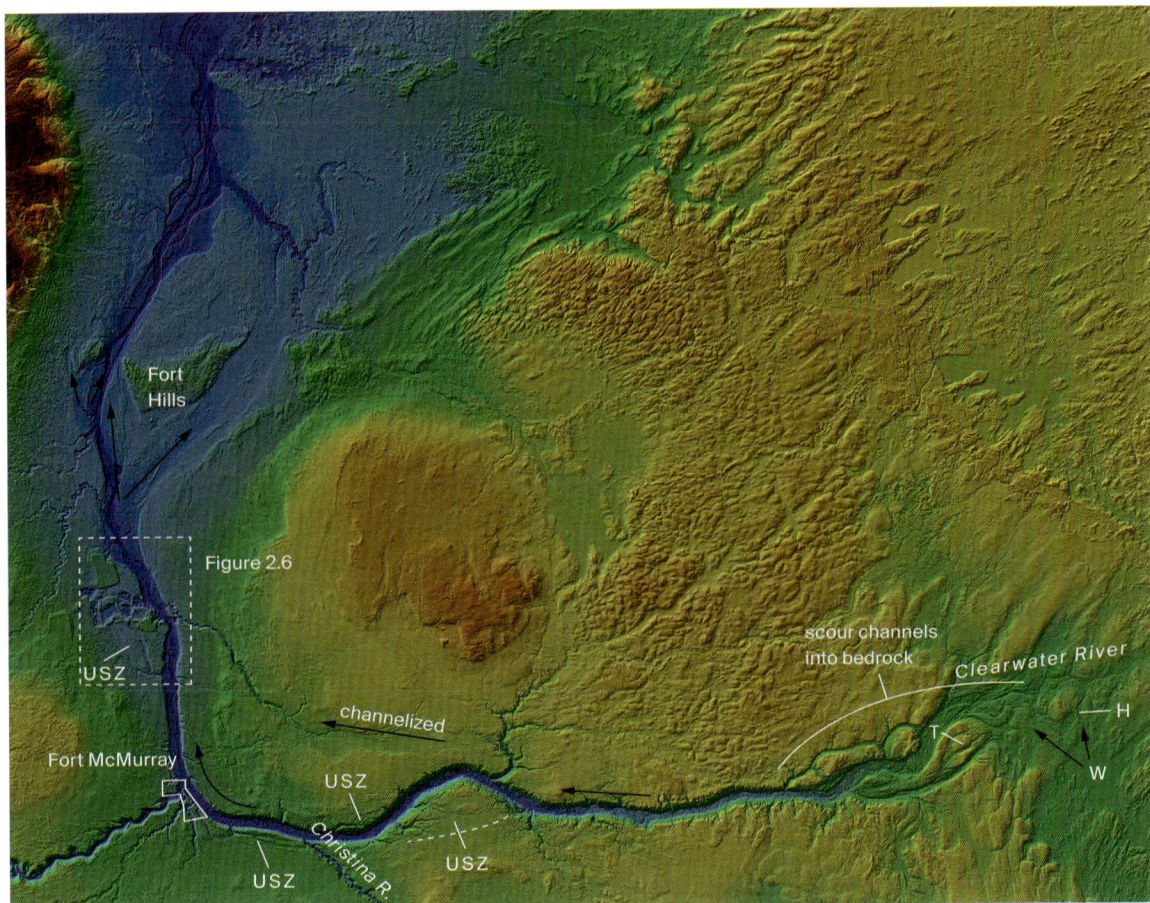

lacustrine sediment within the channel well above the Clearwater River and the presence of till on the spillway floor at the head of the channel (Fisher 1993) are interpreted to represent the pre-existence of the spillway before the last glacial lake formed and incomplete scouring of the spillway at its head during its last occupation.

Given that cubic kilometres of Archean-age shield rocks have been eroded from the head of the spillway, it seems likely that spillway evolution extends back through many glacial cycles, perhaps even to pre-Quaternary time. Wood predating the late Wisconsinan period has been recovered from the Fort McMurray area (Fisher et al. 2009; Paulen et al. 2005), and a sample of wood more than 46,000 years old was collected from flood gravel within the spillway (Teller et al. 2005), documenting the incomplete glacial erosion of the earlier sedimentary record and the reworking of older organic material into more recent sediment.

Figure 2.4. Geomorphic map of the Clearwater–Lower Athabasca spillway, with upper scoured zones (USZ) indicated. (For the location of this area, see figure 2.3.) The two arrows in the southeast corner indicate flow from Lake Wagtufro, which provided water to the spillway during steady-state flow conditions following its incision. T = Tocker Lake, W = Wasekamio Lake, and H = Hass Lake.

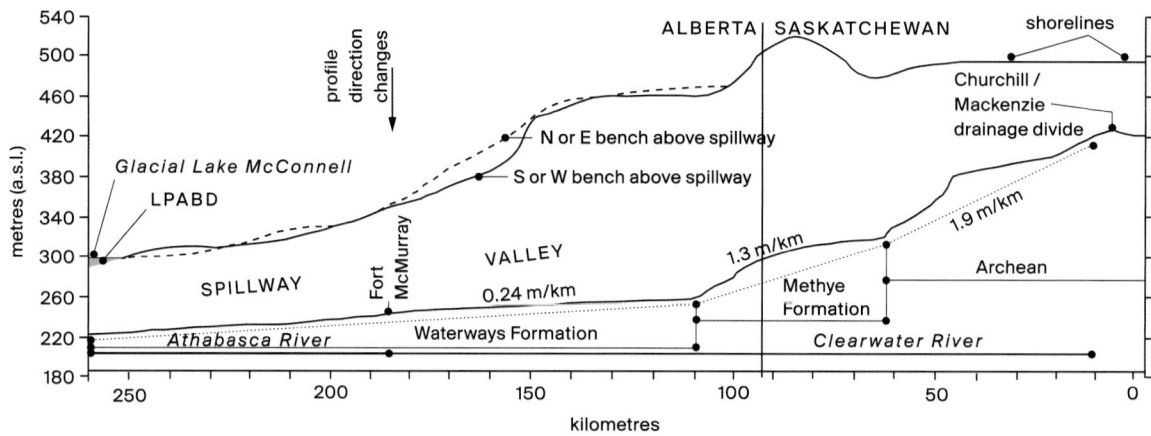

Figure 2.5. Topographic cross-section of the Clearwater–Lower Athabasca spillway from Lake Wasekamio and the Clearwater River westward to Fort McMurray, then northward to the Fort Hills. Modified from Fisher (1993). LPABD = Late Pleistocene Athabasca braid delta.

Downstream of the spillway head, ten areas have been mapped as fluvial upper scour zones outside and above the spillway channel (Fisher 1993). The Steepbank River is located within a smaller spillway channel, parallel to the Clearwater–Lower Athabasca spillway, that probably formed as Lake McMurray drained. The most obvious zones are indicated on figure 2.4, around the city of Fort McMurray (these are associated with the location of the original Syncrude and Suncor mines), as well as further downstream, where a series of bars and channels indicate flow divergence around the Fort Hills. The upper scour zones consist of streamlined hills, often with a gravelly surface with intervening channels incised into bedrock. These scoured areas closely resemble those observed along the Souris River in the central Great Plains (Kehew and Lord 1986) and areas elsewhere in formerly glaciated regions that were subjected to high-energy floods (Kehew et al. 2009, for example). Kehew and Lord (1986) suggested that such scouring occurred where the cross-sectional area of the channel was unable to accommodate the flood volume, resulting in sheet flows parallel to the main channel. It is unclear whether these areas were active at the beginning of the flood flow and are thus representative of a lower flood volume than would be the case if the zones were active after much of the spillway channel had been cut. The economic importance of such scoured areas is exemplified by the zone located 40 kilometres north of Fort McMurray, in the Poplar Lake area (see fig 2.6), which has been an important source of aggregate for industrial growth in the region (Fisher 1993; Fisher and Smith 1993).

Detailed descriptions of the sedimentary flood facies associated with flow through the Clearwater–Lower Athabasca spillway may be found in Fisher (1993) and Fisher and Smith (1993). The following is a brief summary of their

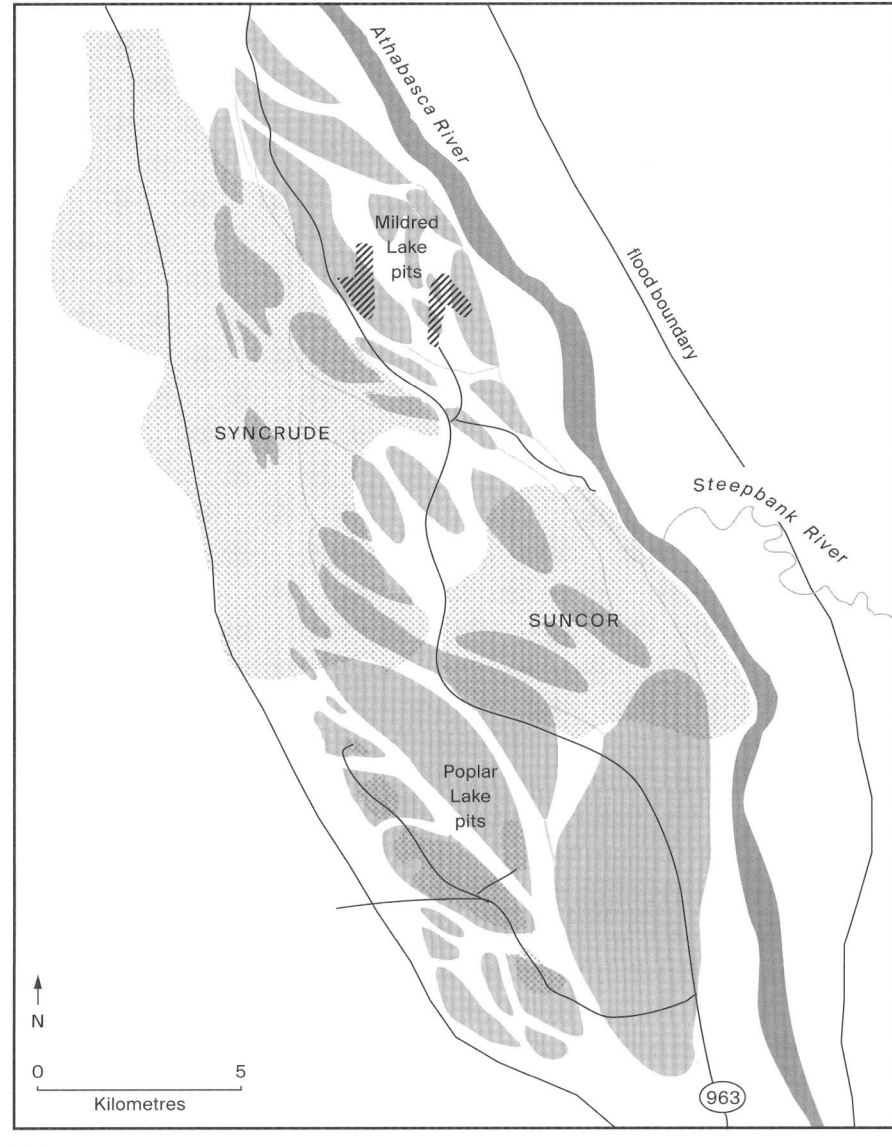

Figure 2.6. Detail of the upper scour zone, 40 kilometres north of Fort McMurray, the location of which is indicated on figure 2.4. Note that where the flood zone widens, gravel was deposited that was subsequently channelized. The flood bars have now mostly been mined for aggregate to support commercial mining operations. Modified from Fisher (1993).

findings. The older Quaternary stratigraphy into which the spillway is cut into may be observed along river and road exposures. It consists of rhythmically laminated silty-clay sediment of glacial lacustrine origin, often pink in colour, conformably overlying massive glacial diamicton (till) (fig 2.7A), with

Figure 2.7. Examples of sediment in the Fort McMurray area. A: laminated glaciolacustrine sediment of Glacial Lake McMurray conformably overlying (see the white arrows) massive glacial diamicton, or till, exposed along an access road to the golf course in the Thickwood subdivision of Fort McMurray. B: Plane-bedded sand and fine pebbles with interbedded oil sand "tar balls" a few centimetres thick (see the horizontal black areas), overlain by cross-bedded fine gravel and sand, from the western Mildred Lake pit (see fig 2.6). Note the rare cobbles and boulders within the sand. C: Massive clast-supported gravel (above the broken white line), overlying well-sorted, cross-bedded gravel above bedrock, from the eastern Mildred Lake pit. Cross-beds record a northerly flow direction and dip to the right in the figure. D: Intraformational conglomerate from the western Mildred Lake pit, here consisting of glacial diamicton "clast" (outlined with a broken white line), which is encased within gravel.

stratigraphically older tills with interbedded sand and gravel also described from outcrop and borehole logs (Andriashek and Atkinson 2007). In the upper scour zone adjacent to Fort McMurray and in the Poplar and Mildred Lake pits (see fig 2.6), plane-bedded sand truncates the glacial lake sediment. Where gravel was encountered, generally at the proximal ends of the bars shown in figure 2.6, it was either massive, clast-supported gravel or a poorly to well-sorted cross-bedded gravel (fig 2.7C and D), with individual cross-beds commonly 3 metres high (Fisher and Smith 1993), recording large-scale bedforms in the flood flow. Clast dimensions for the massive gravel ranged up to 15 metres in exposed dimension for rafts of bedrock or Quaternary sediment (fig 2.7D), while, for the cross-bedded gravel, clasts are rarely larger than 1 metre. Occasional clasts with a B-axis of 2 to 4 metres have been observed in the floor of pits (Fisher 1993; Fisher and Smith 1993) that are often sandstone concretions. In the Mildred Lake pits, the volumetrically most common facies are trough or planar cross-stratified, gravelly sand and plane-bedded sand. These facies have extensive lamina of oil sand "tar balls" and rare boulder-size clasts that indicate highly competent flow (fig 2.7B). Dated wood associated with the gravel found in the upper scour zones (Smith and Fisher 1993; Fisher and Lowell 2012) suggests that the Clearwater–Lower Athabasca spillway was occupied at approximately 9,800 ^{14}C yr BP, much as Fisher et al. (2009) conclude. In summary, the geomorphology and sedimentology of the spillway and its associated deposits in the upper scour zones are similar to features described from other large-scale flows from the interior plains (Kehew et al. 2009), the Bonneville flood in Idaho (Jarrett and Malde 1987), and the Channeled Scablands of eastern Washington (Baker 2009).

PARAGLACIAL ENVIRONMENTS

Following deglaciation, the drainage of Glacial Lake Churchill, and the flow of water through the Clearwater–Lower Athabasca spillway, other important geomorphic processes were still operating on the landscape. These included fluvial incision, aeolian activity, and permafrost and periglacial processes. Elsewhere in the region, vegetation had already been established, as is evident from spruce wood associated with spillway deposits and within the Late Pleistocene Athabasca braid delta. Once the proglacial lakes drained, streams confluent with the Clearwater and Athabasca rivers in the spillway must have begun adjusting to a base level that had been lowered by approximately 100 metres. Streams such as the Christina River (see fig 2.4) continue to erode headward as they try to reach grade. Stream incision thus occurred in response to the higher stream gradient. Today, upstream of Fort McMurray, the Athabasca River lies in a V-shaped valley, with rapids, and without a significant floodplain, all evidence that incision is continuing. Northeast- and north-flowing rivers will have their gradients reduced because of ongoing glacioisostatic adjustment raising land to the northeast.

Sudden drainage of the glacial lakes in the region also would have exposed vast areas of sediment to deflation processes. Sand dunes have been mapped in the region (David 1977; Fisher 1996; Wolfe et al. 2007). Forested parabolic dunes open to the southeast are characteristic of the region and in areas further south (Smith 1987). The dunes on the delta built by the Athabasca River into Glacial Lake McMurray (fig 2.8; see also Fisher 1993b), as well as dunes further to the southeast in the area once covered by Glacial Lake Churchill, record palaeowind directions from the southeast (see David 1981; Fisher 1996; Wolfe et al. 2004). The southeast airflow direction from the dunes has always been interpreted as anticyclonic flow off the Laurentide Ice Sheet, which suggests that the dunes formed in relative close proximity to the ice sheet after drainage of Glacial Lake Churchill. Optical ages from dunes at North Battleford, Saskatchewan (to the south of the area shown in figure 2.3), are mid-Holocene, recording dune activity well after deglaciation (Wolfe et al. 2006). The dunes in the Fort McMurray area are within the "Preserved Late Glacial (relict)" zone described by Wolfe et al. (2007), suggesting stability throughout the Holocene. However, small, active dune fields along the Athabasca River between Fort McMurray and Lake Athabasca indicate that not all dunes have been stabilized since deglaciation. Well-developed ventifacts associated with the sand dunes in northwestern Saskatchewan are another indicator of strong winds following the drainage of Lake Agassiz. Sand wedges, a permafrost feature found today in continuous permafrost with mean annual air temperatures ranging from –5°C (Washburn

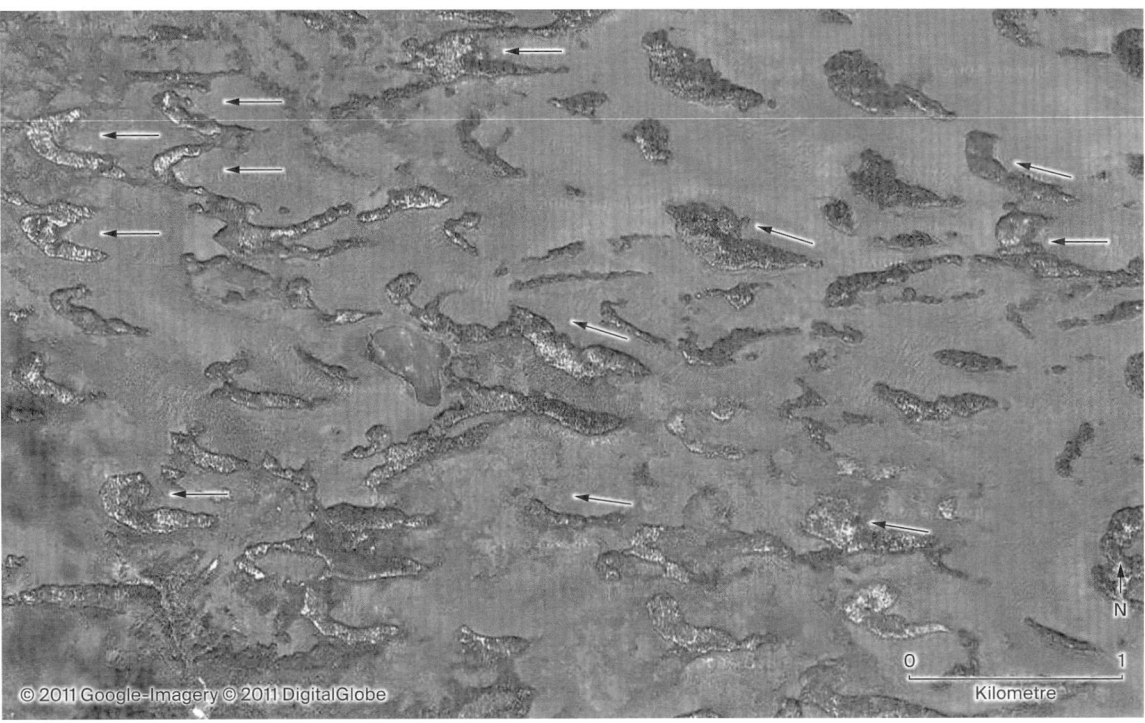

Figure 2.8. Relict parabolic dunes on an abandoned delta west of Fort McMurray that once built into Glacial Lake McMurray (Fisher 1993). White arrows indicate the former wind direction. (For the location of these dunes, see figure 2.3.)

1979) to –20°C (Karte 1983), have also been observed in northwestern Saskatchewan (Fisher 1996). Thus, at the time deglaciation occurred, the region was within the continuous permafrost zone, indicating that not only was it windy but that it was cold and windy.

One final geomorphic process warrants mention, namely, the shifting of the subcontinental drainage divide in northwestern Saskatchewan that separates the Arctic Ocean and Hudson Bay watersheds. Today, the divide passes between the Clearwater River and Lake Wasekamio ("W" on fig 2.4). A core taken in Hass Lake ("H" on fig 2.4), to the south of the modern divide, contains cross-bedded gravel, with the youngest wood dated at 5,000 ± 80 ^{14}C yr BP (Beta-104543). This is indicative of a fluvial channel that once existed between Wasekamio Lake, to the south, and Klap Lake and the Clearwater River, to the north (Fisher and Souch 1998). Today, Wasekamio Lake is part of a complex of lakes that drains to the south into the Churchill River. For the period leading up to about 5,000 years ago, the drainage divide must have been further south, in the Wycherley Lake area, with glacioisostatic adjustment forcing the divide to migrate northward (Fisher and Souch 1998). Other such migrations of the subcontinental drainage divide over significant distances are likely to have occurred elsewhere in the Fort

McMurray region, and, with glacioisostatic adjustment still continuing, future shifts in the divide locations are probable. Thus, archaeological sites located along northward-flowing streams may have been submerged as lakes transgressed southward and/or as a result of increased rates of alluvial aggradation near the mouths of streams.

SUMMARY

An accurate analysis of the nature and timing of deglacial events in the Fort McMurray region is important to our understanding of the history of human occupation in the area. A downslope recession of the Laurentide Ice Sheet resulted in a series of proglacial lakes that formed between the ice and higher land to the southwest and that merged into the large proglacial lake known as Glacial Lake Churchill. The palaeogeography of these lakes is poorly known, in part because of the masking effect of the boreal forest vegetation and partly because, until fairly recently, only small-scale topographic maps were available. With the release of SRTM digital elevation data, landform recognition and thus geomorphological interpretations of past land-forming events have improved substantially.

Many new moraines have been identified along a southwest-northeast transect from Fort McMurray to Lake Athabasca, and our results indicate that deglaciation in this region took place significantly later than previously published results suggest. It is likely that the Clearwater–Lower Athabasca spillway did not form until approximately 9,800 to 9,600 ^{14}C yr BP and was followed by steady-state flow for a few hundred years afterward. Sand dunes near the head of the spillway in Saskatchewan and to the west of Fort McMurray record katabatic palaeowinds from the southeast continuing after the spillway had formed. Ventifacts and sand wedges further indicate paraglacial conditions under a windy and continuous permafrost regime. Throughout the Holocene, rebound of the Earth's crust has reduced stream gradients in northerly flowing rivers, resulting in migration of the subcontinental drainage divide and aggradation in lower stream reaches.

REFERENCES

Anderson, T. W., and C. F. M. Lewis
    1992   Evidence for Ice Margin Retreat and Proglacial Lake (Agassiz?) Drainage by About 11 ka, Clearwater Spillway Area, Saskatchewan. Geological Survey of Canada, Paper 92-1B,

Current Research, Part B, Interior Plains and Arctic Canada, pp. 7–11. Minister of Supply and Services Canada, Ottawa.

Andriashek, Laurence D., and Nigel Atkinson

2007   *Buried Channels and Glacial-Drift Aquifers in the Fort McMurray Region, Northeast Alberta.* Earth Sciences Report 2007-01. Alberta Energy and Utilities Board and Alberta Geological Survey, Edmonton.

Atkinson, Nigel, Laurence D. Andriashek, and Shawn R. Slattery

2013   Morphological Analysis and Evolution of Buried Tunnel Valleys in Northeast Alberta, Canada. *Quaternary Science Reviews* 65: 53–72.

Baker, Victor R.

1973   *Paleohydrology and Sedimentology of Lake Missoula Flooding in Eastern Washington.* Geological Society of America Special Paper no. 144. Geological Society of America, Boulder, Colorado.

2009   The Channeled Scabland: A Retrospective. *Annual Review of Earth and Planetary Science* 37: 393–411.

Bayrock, L. A.

1971   *Surficial Geology, Bitumount, NTS 74E.* Map 34, Alberta Research Council, Edmonton. Available as Map 140, Alberta Geological Survey, Edmonton. http://ags.aer.ca/publications/MAP_140.html.

Bayrock, L. A., and T. H. F. Reimchen

1973   *Surficial Geology, Waterways, NTS 74D.* Alberta Research Council, Edmonton. Available as Map 148, Alberta Geological Survey, Edmonton. http://ags.aer.ca/publications/DIG_2005_0525.html.

Brown, Roger J. E.

1967   *Permafrost in Canada.* Geological Survey of Canada Map 1246A and National Research Council of Canada, Division of Building Research, NRC 9769. Geological Survey of Canada and National Research Council of Canada, Ottawa.

Campbell, Janet E.

2005   SRTM DEM Imagery: Previously Unrecognized Regional-Scale Ice Streams, Ice Flow Indicators and Glacial Landforms in Saskatchewan. Paper presented at the biennial meeting of the Canadian Quaternary Association (Water, Ice, Land, and Life: The Quaternary Interface), Winnipeg, Manitoba, 5–8 June. Abstract available in *Program and Abstracts,* p. A12.

Christiansen, Earl A.

1979   The Wisconsinan Deglaciation of Southern Saskatchewan and Adjacent Areas. *Canadian Journal of Earth Sciences* 16: 913–938.

Church, Michael, and June M. Ryder

1972   Paraglacial Sedimentation: A Consideration of Fluvial Processes Conditioned by Glaciation. *Geological Society of America Bulletin* 83: 3059–3072.

David, Peter P.

1977   *Sand Dune Occurrences of Canada: A Theme and Resource Inventory Study of Eolian Landforms of Canada.* Department of Indian Affairs and Northern Development, Canadian National Parks Branch, contract no. 74-230.

David, Peter P.

1981   Stabilized Dune Ridges in Northern Saskatchewan. *Canadian Journal of Earth Sciences* 18: 286–310.

Dyke, Arthur S., Andrew Moore, and Louis Robertson

2003   *Deglaciation of North America.* Geological Survey of Canada, Open File 1574, doi: 10.4095/214399.

Dyke, Arthur S., and Victor K. Prest

    1987   Late Wisconsinan and Holocene History of the Laurentide Ice Sheet. *Géographie physique et Quaternaire* 41: 237–263.

Elson, John A.

    1967   Geology of Glacial Lake Agassiz. In *Life, Land and Water*, edited by W. J. Mayer-Oakes, pp. 37–96. University of Manitoba Press, Winnipeg.

Fisher, Timothy G.

    1993   Glacial Lake Agassiz: The Northwest Outlet and Paleoflood Spillway, N.W. Saskatchewan and N.E. Alberta. PhD dissertation, Department of Geography, University of Calgary, Calgary.

    1996   Sand-Wedge and Ventifact Palaeoenvironmental Indicators in Northwest Saskatchewan, Canada, 11ka to 9.9 ka BP. *Permafrost and Periglacial Processes* 7: 391–408.

    2005   Strandline Analysis in the Southern Basin of Glacial Lake Agassiz, Minnesota and North and South Dakota, USA. *Geological Society of America Bulletin* 117: 1481–1496.

    2007   Abandonment Chronology of Glacial Lake Agassiz's Northwestern Outlet. *Palaeogeography, Palaeoclimatology, Palaeoecology* 246: 31–44.

Fisher, Timothy G., and Thomas V. Lowell

    2012   Testing Northwest Drainage from Lake Agassiz Using Extant Ice Margin and Strandline Data. *Quaternary International* 260: 106–114.

Fisher, Timothy G., and Derald G. Smith

    1993   Exploration for Pleistocene Aggregate Resources Using Process-Depositional Models in the Fort McMurray Region, NE Alberta, Canada. *Quaternary International* 20: 71–80.

    1994   Glacial Lake Agassiz: Its Northwest Maximum Extent and Outlet in Saskatchewan (Emerson Phase). *Quaternary Science Reviews* 13: 845–858.

Fisher, Timothy G., and Catherine Souch

    1998   Northwest Outlet Channels of Lake Agassiz, Isostatic Tilting and a Migrating Continental Drainage Divide, Saskatchewan, Canada. *Geomorphology* 25: 57–73.

Fisher, Timothy G., Harry M. Jol, and Derald G. Smith

    1995   Ground-Penetrating Radar Used to Assess Aggregate in Catastrophic Flood Deposits, Northeast Alberta, Canada. *Canadian Geotechnical Journal* 32: 871–879.

Fisher, Timothy G., Thomas V. Lowell, and Henry M. Loope

    2006   Comment on "Alternative Routing of Lake Agassiz Overflow During the Younger Dryas: New Dates, Paleotopography, and a Re-evaluation" by Teller et al. (2005). *Quaternary Science Reviews* 25: 1137–1141.

Fisher, Timothy G., Ken Lepper, Allan C. Ashworth, and Howard C. Hobbs

    2011   Southern Outlet and Basin of Glacial Lake Agassiz. In *Archean to Anthropocene: Field Guides to the Geology of the Mid-Continent of North America*, edited by James D. Miller, George J. Hudak, Chad Wittkop, and Patrick I. McLaughlin, pp. 379–400. Geological Society of America, Boulder, Colorado.

Fisher, Timothy G., Nicholas Waterson, Thomas V. Lowell, and Irka Hajdas

    2009   Deglaciation Ages and Meltwater Routing in the Fort McMurray Region, Northeastern Alberta and Northwestern Saskatchewan, Canada. *Quaternary Science Reviews* 28: 1608–1624.

Gilbert, Robert, Timothy G. Fisher, and Ted Lewis

    2000   *A Subbottom Acoustic Survey of Six Lakes near the Former Outlet of Glacial Lake Agassiz in Northwestern Saskatchewan.* Open File Data Report G00-02. Department of Geography, Queen's University, Kingston.

Hutton, Michael J., Glen M. MacDonald, and Robert J. Mott

    1994   Postglacial Vegetation History of the Mariana Lake Region, Alberta. *Canadian Journal of Earth Sciences* 31: 418–425.

Jackson, Lionel E., Fred M. Phillips, Kazuharu Shimamura, and Edward C. Little

    1997   Cosmogenic ^{36}Cl Dating of the Foothills Erratics Train, Alberta, Canada. *Geology* 25: 195–198.

Jarrett, Robert D., and Harold E. Malde

    1987   Paleodischarge of the Late Pleistocene Bonneville Flood, Snake River, Idaho, Computed from New Evidence. *Geological Society of America Bulletin* 99: 127–134.

Karte, J.

    1983   Periglacial Phenomena and Their Significance as Climatic and Edaphic Indicators. *Geo-Journal* 7: 329–340.

Kehew, Alan E., and Mark L. Lord

    1986   Origin and Large-Scale Erosional Features of Glacial-Lake Spillways in the Northern Great Plains. *Geological Society of America Bulletin* 97: 162–177.

Kehew, Alan E., Mark L. Lord, Andrew L. Kozlowski, and Timothy G. Fisher

    2009   Proglacial Megaflooding Along the Margins of the Laurentide Ice Sheet. In *Megaflooding on Earth and Mars,* edited by Devon M. Burr, Paul A. Carling, and Victor R. Baker, pp. 104–127. Cambridge University Press, Cambridge and New York.

Langenberg, C. W., F. J. Hein, and H. Berhane

    2003   *Three-Dimensional Geometry of Fluvial-Estuarine Oil Sand Deposits of the Clarke Creek Area (NTS 74D), Northeastern Alberta.* Earth Sciences Report 2001-06. Alberta Energy and Utilities Board and Alberta Geological Survey, Edmonton.

Lemmen, Donald S., Alejandro Duk-Rodkin, and Jan M. Bednarski

    1994   Late Glacial Drainage Systems Along the Northwestern Margin of the Laurentide Ice Sheet. *Quaternary Science Reviews* 13: 805–828.

Lepper, Kenneth, Alex W. Buell, Timothy G. Fisher, and Thomas V. Lowell

    2013   A Chronology for Glacial Lake Agassiz Shorelines Along Upham's Namesake Transect. *Quaternary Research* 80: 88–98.

McPherson, R. A., and C. P. Kathol

    1977   *Surficial Geology of Potential Mining Areas in the Athabasca Oil Sands Region.* Open File Report 1977-04. Alberta Research Council, Edmonton.

Mossop, Grant D., and Irina Shetsen (compilers)

    1994   *Geological Atlas of the Western Canada Sedimentary Basin.* Canadian Society of Petroleum Geologists, Calgary, and Alberta Research Council, Edmonton.

Murton, Julian B., Mark D. Bateman, Scott R. Dallimore, James T. Teller, and Zhirong Yang

    2010   Identification of Younger Dryas Outburst Flood Path from Lake Agassiz to the Arctic Ocean. *Nature* 464 (1 April): 740–743.

Natural Regions Committee

    2006   *Natural Regions and Subregions of Alberta.* Compiled by David J. Downing and Wayne W. Pettapiece. Government of Alberta Publication no. T/852. https://www.albertaparks.ca/media/2942026/nrsrcomplete_may_06.pdf.

Paulen, Roger C., Alwynne B. Beaudoin, and John G. Pawlowicz

    2005   An Interstadial Site in the Birch Mountains, North-Central Alberta. Paper presented at the biennial meeting of the Canadian Quaternary Association (Water, Ice, Land, and Life: The Quaternary Interface), Winnipeg, Manitoba, 5–8 June. Abstract available in *Program and Abstracts,* p. A66.

Rains, R. Bruce, John Shaw, K. Robert Skoye, Darren B. Sjogren, and Donald R. Kvill

    1993   Late Wisconsin Subglacial Megaflood Paths in Alberta. *Geology* 21: 323–326.

Raup, Hugh M., and George W. Argus

    1982   *The Lake Athabasca Sand Dunes of Northern Saskatchewan and Alberta, Canada.* Publications in Botany no. 12. National Museums of Canada, Ottawa.

Rayburn, John A., and James T. Teller

    2007   Isostatic Rebound in the Northwestern Part of the Lake Agassiz Basin: Isobase Changes and Overflow. *Palaeogeography, Palaeoclimatology, Palaeoecology* 246: 23–30.

Rhine, Janet L.

    1984   Sedimentology and Geomorphological Reconstruction of the Late Pleistocene Athabasaca Fan-Delta, Northeast Alberta. MSc thesis, Department of Geography, University of Calgary, Calgary.

Rhine, Janet L., and Derald G. Smith

    1988   The Late Pleistocene Athabasca Braid Delta of Northeastern Alberta, Canada: A Paraglacial Drainage System Affected by Aeolian Sand Supply. In *Fan Deltas: Sedimentology and Tectonic Settings*, edited by W. Nemec and R. J. Steel, pp. 158–169. Blackie and Son, Glasgow and London.

Schreiner, Bryan T.

    1983   Lake Agassiz in Saskatchewan. In *Glacial Lake Agassiz*, edited by James T. Teller and Lee Clayton, pp. 75–96. Geological Association of Canada Special Paper no. 26. Geological Association of Canada, St. John's.

    1984   *Quaternary Geology of the Precambrian Shield, Saskatchewan*. Geological Report no. 221. Energy and Mines, Government of Saskatchewan, Regina.

Shaw, John, Dennis M. Faragini, Donald R. Kvill, and R. Bruce Rains

    2000   The Athabasca Fluting Field, Alberta, Canada: Implications for the Formation of Large-Scale Fluting (Erosional Lineations). *Quaternary Science Reviews* 19: 959–980.

Sjogren, Darren B., Timothy G. Fisher, Lawrence D. Taylor, Harry M. Jol, and Mandy J. Munro-Stasiuk

    2002   Incipient Tunnel Channels. *Quaternary International* 90: 41–56.

Smith, Derald G.

    1987   *Landforms of Alberta, Interpreted from Airphotos and Satellite Imagery*. Canadian Society of Petroleum Geologists, Calgary.

    1989   Catastrophic Paleoflood from Glacial Lake Agassiz 9,900 Years Ago in the Fort McMurray Region, Northeast Alberta. Paper presented at the biennial meeting of the Canadian Quaternary Association (Late Glacial and Post-glacial Processes and Environments in Montane and Adjacent Areas), Edmonton, Alberta, 25–27 August. Abstract available in *Program and Abstracts,* p. 47.

Smith, Derald G., and Timothy G. Fisher

    1993   Glacial Lake Agassiz: The Northwestern Outlet and Paleoflood. *Geology* 21(1): 9–12.

Sproule, John C.

    1939   The Pleistocene Geology of the Cree Lake Region, Saskatchewan. *Transactions of the Royal Society of Canada,* 3rd ser., vol. 33, sec. 4: 101–109.

Taylor, R. S.

    1960   Some Pleistocene Lakes of Northern Alberta and Adjacent Areas (Revised). *Journal of the Alberta Society of Petroleum Geologists* 8(6): 167–178, 185.

Teller, James T., Matthew Boyd, Zhirong Yang, Phillip S. G. Kor, and Amir Mokhtari Fard

    2005   Alternative Routing of Lake Agassiz Overflow During the Younger Dryas: New dates, Paleotopography, and a Re-evaluation. *Quaternary Science Reviews* 24: 1890–1905.

Teller, James T., L. Harvey Thorleifson, Lynda A. Dredge, Howard C. Hobbs, and Bryan T. Schreiner

    1983   Maximum Extent and Major Features of Lake Agassiz. In *Glacial Lake Agassiz,* edited by James T. Teller and Lee Clayton, pp. 43–45. Geological Association of Canada Special Paper no. 26. Geological Association of Canada, St. John's.

Tremblay, Leo Paul

    1960   *Geology, La Loche, Saskatchewan.* Map 10-1961, Geological Survey of Canada, Ottawa.

Upham, Warren

    1895   *The Glacial Lake Agassiz.* Monographs of the United States Geological Survey, Department of the Interior, vol. 25. Government Printing Office, Washington, D.C.

Washburn, Albert L.

    1979   *Geocrylogy: A Survey of Periglacial Processes and Environments.* Edward Arnold, London.

Waterson, Nicholas, Thomas V. Lowell, Timothy G. Fisher, Katherine Glover, and Irka Hajdas

    2005   The Deglaciation of the Fort McMurray Area, Alberta: Implications for Melt-water Drainage. Paper presented at the biennial meeting of the Canadian Quaternary Association (Water, Ice, Land, and Life: The Quaternary Interface), Winnipeg, Manitoba, 5–8 June. Abstract available in *Program and Abstracts,* p. A102.

Wolfe, Stephen A., David J. Huntley, and Jeffrey Ollerhead

    2004   Relict Late Wisconsinan Dune Fields of the Northern Great Plains, Canada. *Géographie physique et Quaternaire* 58: 323–326.

Wolfe, Stephen A., Jeffrey Ollerhead, David J. Huntley, and Olav B. Lian

    2006   Holocene Dune Activity and Environmental Change in the Prairie Parkland and Boreal Forest, Central Saskatchewan, Canada. *The Holocene* 16: 17–29.

Wolfe, Stephen A., Roger C. Paulen, I. R. Smith, and Michel Lamothe

    2007   *Age and Paleoenvironmental Significance of Late Wisconsinan Dune Fields in the Mount Watt and Fontas River Map Areas, Northern Alberta and British Columbia.* Geological Survey of Canada Current Research 2007-B4. Natural Resources Canada, Ottawa.

Young, Robert R., James A. Burns, Derald G. Smith, L. David Arnold, and R. Bruce Rains

    1994   A Single, Late Wisconsin, Laurentide Glaciation, Edmonton Area and Southwestern Alberta. *Geology* 22: 683–686.

3 Raised Landforms in the East-Central Oil Sands Region | Origin, Age, and Archaeological Implications

ROBIN J. WOYWITKA, DUANE G. FROESE, AND STEPHEN A. WOLFE

A strong correlation exists between raised landforms and archaeological sites in the lowland between Cree Burn Lake and Kearl Lake, to the south of the Fort Hills portion of the Firebag Moraine (see fig 3.1). The origin of such landforms figures prominently in archaeological models proposed for the surface-minable oil sands region north of Fort McMurray. Saxberg and Reeves (2003) have interpreted similar features as remnant shorelines and beaches related to the recession of catastrophic flood waters from the northwest outlet of Glacial Lake Agassiz, while others propose that these features are gravel bedforms formed during peak outflow from the lake through its northwest outlet (Clarke and Ronaghan 2000; Clarke, Ronaghan, and Bouchet, chapter 5 in this volume). According to the first model, human occupation of the region occurred fairly early, as the flood waned, with groups drawing on the rich resources of the littoral zones associated with the floodwaters and with remnant ponds and lakes (Saxberg and Reeves 2003). In the second model, human occupation would have been limited until the floodwaters had fully receded and a productive grassland environment had recolonized the scoured landscape, including the newly exposed gravel bedforms.

The origin of the raised landforms in the Cree Burn Lake–Kearl Lake lowland has implications for how people used the landscape and for the timing of human occupation following the Lake Agassiz flood. However, no formal examination of landform morphology, sediment, or age has been conducted in support of archaeological models of human habitation. In what follows, we address this issue by examining the origin and age of a number of these landforms, using Light Detection and Ranging (LiDAR) images as well as geomorphological and sedimentary observations based on fieldwork. We then discuss the genesis of these deposits in relation to the northwestern outlet of Glacial Lake Agassiz and the aeolian landforms in the area.

GEOLOGICAL SETTING AND METHODS

Smith and Fisher (1993, 10) observe that "sheets and long bars of boulder gravel" mantle the area south of the Fort Hills. As they note, the gravel is poorly sorted (ranging from sand to boulders) and poorly stratified and contains large rip-ups of oil sand and rounded diamicton blocks. This gravel unit has been interpreted as a high-velocity fluvial deposit related to catastrophic flows through the Lower Athabasca valley associated with the northwestern outlet of Lake Agassiz (Smith and Fisher 1993; Fisher 2007; Fisher et al. 2009; Fisher and Lowell, chapter 2 in this volume).

The route of the flood can be traced in the Clearwater–Lower Athabasca spillway as defined by Smith and Fisher (1993). The headwaters of the spillway lie in the upper reaches of the Clearwater River just east of the Alberta-Saskatchewan border (see figure 2.4 in the preceding chapter). The spillway follows the Clearwater River west to its confluence with the Athabasca River at Fort McMurray, where it turns north, widens, and grades into the Late Pleistocene Athabasca braid delta, north of the Fort Hills (Rhine and Smith 1988). South of the Fort Hills, in the area of the Cree Burn Lake–Kearl Lake lowland, the spillway bifurcates, with one path of flow following the Muskeg River valley to the northeast, toward the Firebag River, and another following the present course of the Athabasca River through the Bitumount Gap (see fig 3.1). Flow divergence to either side of the Fort Hills is indicated by the direction of imbricate clasts in the flood gravels and by the orientation of a series of large bars and channels located along the southern margins of the upland (Smith and Fisher 1993). The landforms we examined are superimposed on one of these large gravel bars at the point of flow divergence (fig 3.1, box 2). Radiocarbon dates from lake cores and deltaic sediments initially suggested that the Clearwater–Lower Athabasca spillway became active somewhere between 9,850 and 9,660 ^{14}C yr BP (Fisher et al. 2009); Fisher and Lowell (chapter 2 in this volume) revise these dates slightly, to 9,800 and 9,600 ^{14}C yr BP. However, an alternate chronology suggests that flooding may have begun somewhat earlier, prior to 10,000 ^{14}C yr BP (Teller et al. 2005; Froese, Smith, and Reyes 2010; Murton et al. 2010).

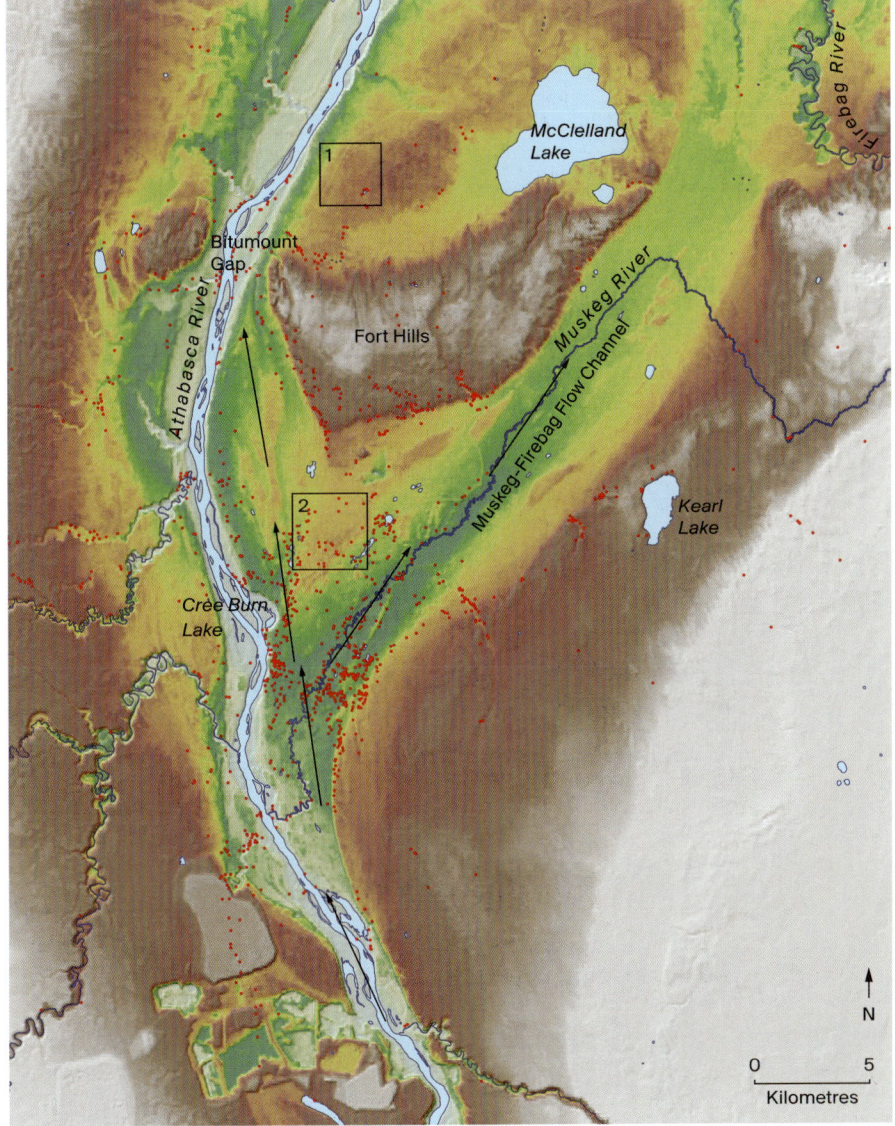

Figure 3.1. The Cree Burn Lake–Kearl Lake lowlands. To date, approximately four hundred archaeological sites (marked by red dots in the figure) have been identified in this area. Box 1 indicates the location of aeolian landforms north of the Fort Hills (see figure 3.2). Box 2 shows the location of flow bifurcation to the north-northwest and to the northeast of the Fort Hills (see figures 3.3 and 3.4).

SRTM ELEVATION

High: 832.000000

Low : 220.000000

Figure 3.2. LiDAR image of aeolian landforms (parabolic sand dunes) in the oil sands region north of the Fort Hills

To the north of the Fort Hills, archaeological sites are again concentrated on raised landforms, similar in morphology to those in the Cree Burn Lake–Kearl Lake lowland (Unfreed, Fedirchuk, and Gryba 2001; Woywitka, chapter 7 in this volume). However, these more northern landforms are aeolian features, consisting largely of stabilized sand dunes, both transverse and parabolic (see fig 3.1, box 1, and fig 3.2; and see David 1981; Rhine and Smith 1988; Wolfe, Huntley, and Ollerhead 2004).

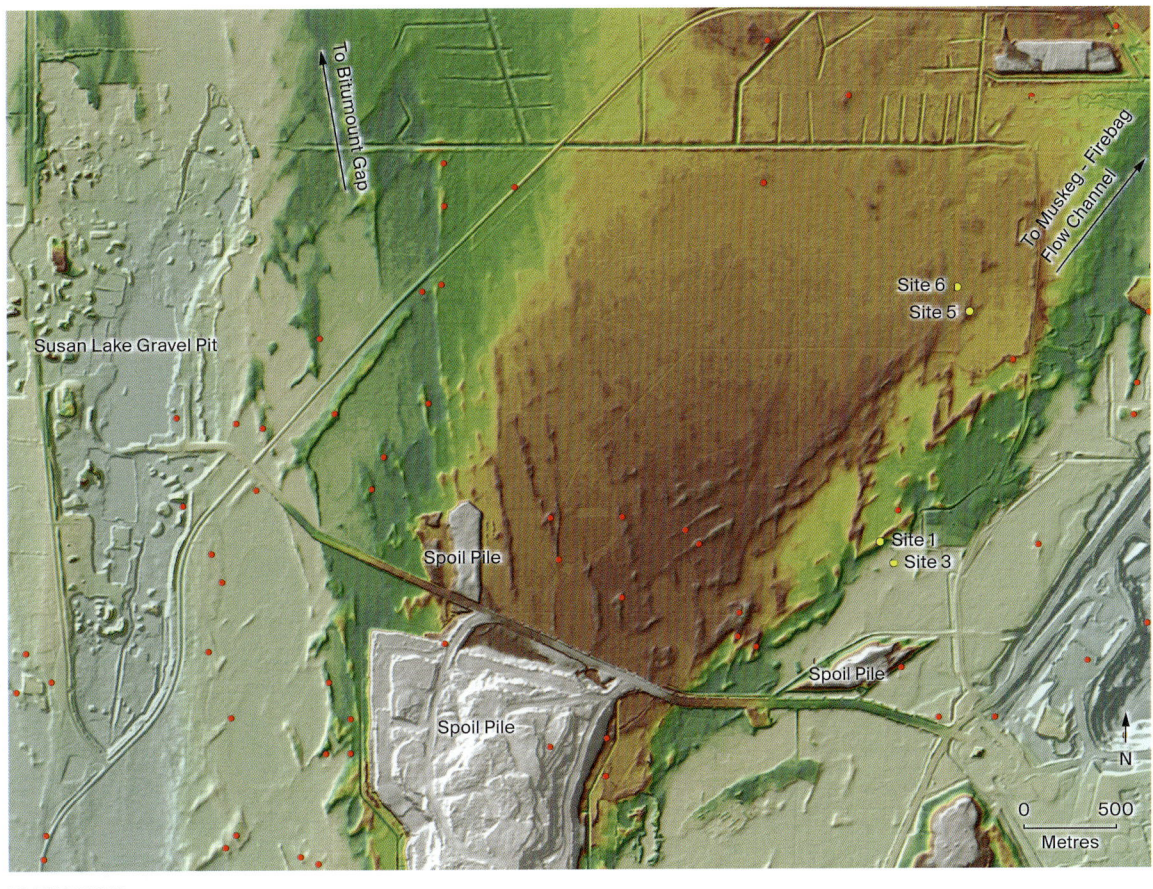

ELEVATION

354 masl

210 masl

○ Sedimentary observation site
● Archaeological site

Figure 3.3. LiDAR image of the Cree Burn Lake–Kearl Lake lowland. Prominent raised linear landforms are common in the study area, with long axes oriented toward the northwest, in the direction of the Bitumount Gap, on the west side of the area, and, on the east side, toward the northeast along the Muskeg-Firebag flow channel.

The landforms we studied were initially identified by a visual examination of shaded relief images derived from a LiDAR bare-earth digital elevation model (DEM) (fig 3.3). The cell size of the DEM grid was 1 metre, and the vertical accuracy 0.6 metres. Relief images were calculated with a 1-metre cell size, an azimuth angle of 315°, and an altitude angle of 45°. Relief is estimated from 1-metre interval contours interpolated from the bare-earth DEM. Sediment exposures in the area had been directly examined several times since 2006, and in August 2010 we conducted field studies at some of the raised landforms identified in the LiDAR images. First-hand observations were made at locations where access road rights-of-way, clearings for well pads, and other mining disturbances had already cut into the landforms. The sediment exposures were described and photographed, and samples of the sediment collected.

LANDFORM SHAPE AND ORIENTATION

Raised landforms within the area of study are predominantly streamlined and range from linear to rhomboidal in shape. We recognize three primary landform types: linear landforms that trend to the northeast; linear landforms that trend to the north-northwest; and composite ridge-to-rhomboidal landforms (see fig 3.4, a, b, and c, respectively).

Northeast-trending linear landforms range from 30 to 350 metres in length and from 20 to 30 metres in width. One prominent northeast-trending landform, known as Ronaghan's Ridge ("RR" on the figure), considerably exceeds these dimensions, however, with a length of roughly 1,200 metres and a width of about 70 metres. Most of these landforms are located on the east side of the study area, with long axes oriented in the same direction as the Muskeg-Firebag flow channel (see fig 3.4, a). Relief on these features is up to 4 metres, and crests have a sharper gradient on their northwest-facing slopes, with crest elevations ranging from 297 to 302 metres above sea level.

The north-northwest-trending linear landforms are largely located on the west side of the study area, with a main long-axis orientation toward the Bitumount Gap, running roughly parallel to the Athabasca River (see fig 3.4, b). These features are generally somewhat longer and slightly wider than the northeast-trending ones, with lengths of up to 600 metres and widths of up to 50 metres, although most range from 30 to 35 metres in width. Relief on these features is up to 3 metres, with many rising only 1 metre above the surrounding peatland. The crests of these landforms have a sharper gradient on their east-north-east-facing slopes, and crest elevations range from 299 to 302 metres above sea level. In view of their lower relief, these north-northwest-trending features are generally more subtle on the LiDAR image than the northeast-trending ones.

Composite ridge-to-rhomboidal landforms (see fig 3.4, c) occur throughout the study area but are more common along the southeastern boundary. The landforms along this boundary are oriented to the northeast, toward the Muskeg-Firebag flow channel, and are variously linear, arcuate, or rhomboidal in shape, with pronounced heads.[1] Their length varies between 150 and 300 metres, with head widths of 50 to 60 metres and horn widths of 30 to 40 metres. Relief is up to 5 metres, and horns are of closely matching length, with sharper gradients on their northwest-facing slopes. On the west side of the study area, composite

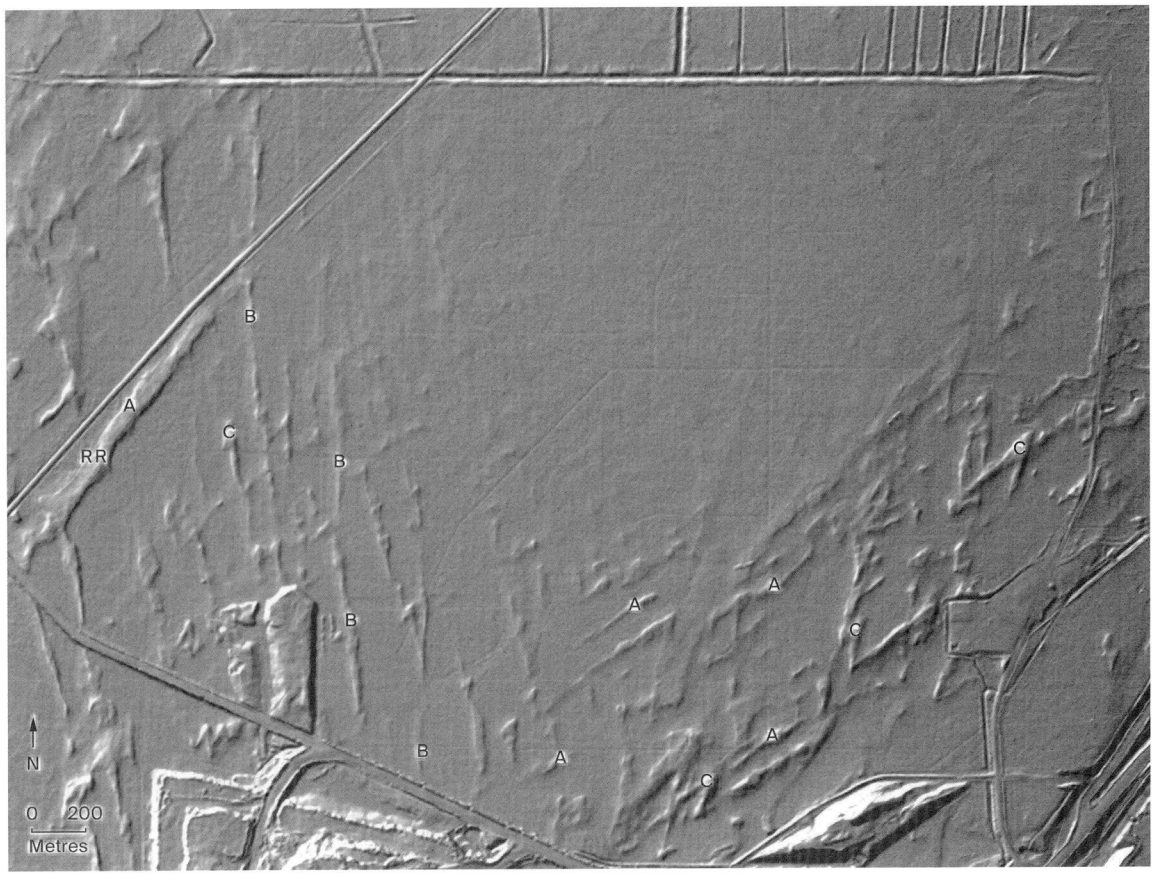

ridge-to-rhomboidal landforms are oriented to the north-northwest, toward the Bitumount Gap. Their length varies between 120 and 150 metres, with head widths of 40 to 50 metres and horn widths of 20 to 30 metres. These features are of a less pronounced arcuate or rhomboidal shape and frequently have one horn that is longer than the other. Relief is up to 3 metres, and the horns have a sharper gradient on their east-northeast-facing slopes. Crest elevations range from 299 to 302 metres above sea level.

SEDIMENTARY OBSERVATIONS

Poorly sorted, coarse-grained imbricate gravels are present at the base of all the sediment exposures we examined in the field. (The location of several of our field sites—1, 3, 5, and 6—is shown in figure 3.3.) Rip-ups of bitumen occur in this

Figure 3.4. Detail of LiDAR image, showing raised landforms. At least three prominent varieties of landform can be recognized: (a) linear landforms that trend to the northeast; (b) linear landforms trending north-northwest; and (c) composite ridge-to-rhomboidal landforms. All the sites examined in this study consist of boulder-gravel cores with an aeolian mantle of sand on their surface. RR = Ronaghan's Ridge.

Figure 3.5. Photographs of sites in the Cree Burn Lake–Kearl Lake lowland: (a and b) exposures in the Susan Lake gravel pit showing catastrophic flood deposits with oil sand rip-up clasts; (c) approximately 2-metre relief on a linear landform at site 3; (d) plane-bedded sands overlying imbricate boulder gravel at site 1; (e) lag boulders in the study area dislodged by mining activities; and (f) a ventifacted cobble on the surface of a linear landform recording aeolian processes following the deposition of the landform. The relatively shallow relief on many of these landforms, such as that illustrated in (c), makes these features rather subtle and thus difficult to observe in the field.

gravel unit (fig 3.5, a, b), and well-rounded boulders as large as 2 metres in diameter can be found in spoil piles nearby (fig 3.5, e). At site 1, the gravel is locally overlain by approximately 1.2 metres of plane-bedded sands (fig 3.5, d). At site 5, the plane-bedded sand is absent, and the gravel is instead covered with massive, well-sorted, fine-grained sand. This massive sand layer is up to 1.1 metres thick and is overlaid by a thin cover of organic litter. A ventifacted cobble was also found in a surface exposure of the gravel-sand contact at this site (fig 3.5, f). In areas of low relief, peat deposits occur, typically overlying sands or a thin veneer of sands on gravel between raised landforms. At site 6, the peat is 1.5 metres thick and lies on top of a thin bed of clay over a massive layer of grey sand.

INTERPRETATION

The coarse-grained gravel unit and plane-bedded sands observed at our field sites indicate deposition by high-velocity flows. The imbrication, grain size, and presence of bitumen rip-ups are all consistent with the Agassiz flood deposits described by Smith and Fisher (1993). The superposition of the landforms on the large gravel bar south of the Fort Hills indicates that they formed following the initial deposition of the bar.

The two differing landform orientations (northeast and north-northwest) are indicative of the divergence of flow around the Fort Hills, with the northeast-trending landforms representing flow toward the Muskeg-Firebag valley and the north-northwest-trending landforms representing flow that followed the Athabasca River through the Bitumount Gap (see fig 3.1). These orientations are consistent with gravel fabric data presented by Smith and Fisher (1993). The shape of these landforms, their orientation, and the gravel fabric data suggest two possible interpretations. These landforms may reflect a sequence of events, in which an initial flow along one of the channels was followed by occupation of the second. Alternatively, they may represent synchronous flow both to the northeast and to the north-northwest, with the construction of ridge-to-rhomboidal landforms occurring during bifurcated flow. The closely similar crest elevations on landforms of both orientations (297 to 302 m) suggest that flow was synchronous. These elevations are slightly above estimates of the divide elevation between the Firebag and Muskeg rivers (289 to 292 m). The prevalence of the higher-relief combined ridge-to- rhomboidal pattern in the northeast-trending landforms suggests that flow was stronger down the Muskeg-Firebag flow path.

The hydraulic processes underlying the formation of meso- to micro-scale depositional landforms during catastrophic floods is poorly understood (for a discussion, see Carling et al. 2009), and the use of LiDAR data to image these and other geomorphic features is still in its early stages. We know of no landform-scale examples that exhibit a form immediately similar to the ridge-to-rhomboidal pattern observed in the study area. However, a variety of bed-scale features display a similar pattern. These include rhomboid ripples found on bar tops and commonly in beach swash zones, which are typically on the order of a few grain-diameters high and are associated with very shallow supercritical flows

(Allen 1982, 404). In these flows, rhomboid ripples are the product of two hydraulic jumps oblique to the main direction of flow, with the crests of the ripples corresponding to the hydraulic jumps (Allen 1982, 404). As recent experimental work has shown, with more viscous flows, the rhomboid pattern may also be produced at subcritical velocities (Devauchelle et al. 2010). When they occur, however, these rhomboid bedforms tend to be quite homogenous, and the irregular ridge and rhomboidal forms in this study are at best partial analogues.

As far as we are aware, the forms most similar to those observed in the present study are the flow-aligned ridges and the ridge and rhomboidal patterns produced experimentally by Karcz and Kersey (1980). In these experiments, flow-aligned ridges were associated with laminar, subcritical flow, while combined ridge and rhomboidal patterns were associated with laminar, supercritical flow. There are several caveats to the form analogy, however, given the differences in the scale of bedforms produced from these experimental sand channels and the landforms observed in this study, not the least of which is the difference in grain size, as well as the presence of presumably sediment-laden waters during catastrophic flooding. However, despite the uncertainties surrounding the details of flow conditions, the bedform examples all consistently indicate high-velocity flows during bedform generation.

The massive layer of well-sorted sand overlying the gravel unit at some of our field sites is best interpreted as aeolian sediment deposited after subaerial exposure of the landforms. The deposition of this sand unit, as well as the ventifacted cobble found at the sand-gravel contact, indicates that windy conditions prevailed after the recession of flood waters from the study area. Following the cessation of aeolian deposition, surfaces were stabilized by vegetation, with boreal communities eventually becoming established on uplands. The sediment exposure at site 6 indicates that peatlands also began to accumulate in lowlands following aeolian activity in the area.

DISCUSSION AND CONCLUSION

The raised landforms in the study area appear to be gravel bedforms created by flow related to the catastrophic flooding of Glacial Lake Agassiz through the Clearwater–Lower Athabasca spillway. The source and timing of flooding through the spillway is an area of active research. Fisher et al. (2009) originally suggested that the spillway was created between 9,850 and 9,660 ^{14}C yr BP and may have drained a glacial lake independent of Glacial Lake Agassiz. The spillway would have accommodated flow from this source until roughly 9,450 ^{14}C yr BP. Murton et al. (2010) postulate an earlier Lake Agassiz–related chronology for the flood, with an initial outburst event occurring before 10,000 ^{14}C yr BP. Our study does not provide information on the precise timing of landform development. However, the position of the study area near the upper margins of the Muskeg-Firebag and Bitumount Gap flow channels and the evidence of high-velocity flows suggest that the landforms were formed during the early stages of the flood, when large discharges occupied both channels.

Although flood-related bedforms are well known from glacial outburst events in a large variety of settings (see Burr, Carling, and Baker 2009), most of these previously studied landforms are of greater size and relief than the features we have described here (for comparisons, see Pardee 1942; Baker 1973; O'Connor 1993). For the most part, the landforms we examined appear as subtle features in the field (see fig 3.5, c). It is likely that forms of the relatively small magnitude that we describe here are in fact more common elsewhere but are difficult to recognize without the exceptional resolution of LiDAR-derived DEM data.

A thin, discontinuous mantle of windblown sediments was deposited across the newly exposed landscape following the recession of flood waters. The lack of well-defined aeolian landforms (such as sand dunes) in the area could be due to the dominance of very coarse-grained material, although it could also reflect the prevalence of low to moderate wind speeds. The former explanation is more likely, given the ubiquity of coarse-grained gravel deposits in the local environment and the well-established presence of strong postglacial wind regimes throughout northern Alberta (Wolfe, Huntley, and Ollerhead 2004). The exact age of landscape stabilization following the cessation of aeolian deposition is unknown, but pollen records in the region indicate that forests were established in upland areas shortly after flooding (Bouchet and Beaudoin, chapter 4 in this volume) and that wetlands began to accumulate approximately 8,000 to 6,000 years ago (Halsey, Vitt, and Bauer 1998).

The oil sands region preserves diverse landforms that are closely associated with the archaeological record. Raised landforms in the lowland areas hold special significance, however, since the majority of archaeological sites in areas such as the Cree Burn Lake–Kearl Lake lowland are found on these features. In contrast to areas to the northwest of the Fort Hills, in which aeolian landforms predominate, these landforms are subaqueous gravel bedforms created by the extraordinary flows associated with the northwest outlet of Glacial Lake Agassiz. The lack of directly dated archaeological deposits in the area prevents a definitive determination of the timing of initial human occupation of the Cree Burn Lake–Kearl Lake lowland. However, these landforms would have been available for human occupation only after recession of the floodwaters. On the basis of other studies of flood chronology (Fisher et al. 2009; Murton et al. 2010), the emergence of these landforms can be dated to the transitional terminal Pleistocene–early Holocene period. The frequent preservation of archaeological materials in the aeolian sands that drape these landforms also suggests that human occupation occurred during and/or following the deposition of the windblown sediment. Future research aimed at determining the timing of landscape stabilization will help us to develop a precise chronology for human occupation of the oil sands region.

ACKNOWLEDGEMENTS

We are grateful for the generous assistance of Ayo Adediran, of Shell Canada Energy, Chris Doornbos, formerly of Petro-Canada, and Dale Nolan and Rolly Boissonnault, of Athabasca Minerals Inc., who facilitated access to field sites. We benefitted from useful discussions with John Shaw, at the University of Alberta, about the form analogues of the features we describe in this chapter. We also thank Hazen Russell, of the Geological Survey of Canada, for his critical but constructive comments, which improved the chapter. Portions of this work were funded by Natural Sciences and Engineering Research Council (NSERC) and Canadian Research Chairs (CRC) grants to Duane G. Froese.

NOTE

1 The "head" is the point of highest elevation on a landform, while "horn" refers to the trailing arms of arcuate or rhomboidal landforms.

REFERENCES

Allen, John R. L.
  1982 *Sedimentary Structures: Their Character and Physical Basis.* Vol. 1. Developments in Sedimentology, vol. 30, part A. Elsevier, Amsterdam.
Baker, Victor R.
  1973 *Paleohydrology and Sedimentology of Lake Missoula Flooding in Eastern Washington.* Geological Society of America Special Paper no. 144. Geological Society of America, Boulder, Colorado.
Burr, Devon M., Paul A. Carling, and Victor R. Baker (editors)
  2009 *Megaflooding on Earth and Mars.* Cambridge University Press, Cambridge.

2009 A Review of Open-Channel Megaflood Depositional Landforms on Earth and Mars. In *Megaflooding on Earth and Mars*, edited by Devon M. Burr, Paul A. Carling, and Victor R. Baker, pp. 33-49. Cambridge University Press, Cambridge.

Clarke, Grant M., and Brian M. Ronaghan

2000 *Historical Resources Impact Mitigation, Muskeg River Mine Project: Final Report (ASA Permit 99-073)*. Copy on file, Archaeological Survey, Heritage Resources Management Branch, Alberta Culture, Edmonton.

David, Peter P.

1981 Stabilized Dune Ridges in Northern Saskatchewan. *Canadian Journal of Earth Sciences* 18: 286-310.

Devauchelle, O., L. Malverti, É. Lajeunesse, C. Josserand, P.-Y. Lagrée, and F. Métivier

2010 Rhomboid Beach Pattern: A Laboratory Investigation. *Journal of Geophysical Research* 115, F02017, doi: 10.1029/2009JF001471.

Fisher, Timothy G.

2007 Abandonment Chronology of Glacial Lake Agassiz's Northwestern Outlet. *Palaeogeography, Palaeoclimatology, Palaeoecology* 246: 31-44.

Fisher, Timothy G., Nicholas Waterson, Thomas V. Lowell, and Irka Hajdas

2009 Deglaciation Ages and Meltwater Routing in the Fort McMurray Region, Northeastern Alberta and Northwestern Saskatchewan, Canada. *Quaternary Science Reviews* 28: 1608-1624.

Froese, Duane G., Derald G. Smith, and Alberto V. Reyes

2010 Revisiting the Northwest Outlet of Glacial Lake Agassiz: Catastrophic Floods, Permafrost and Mega-deltas. Paper presented at the 52nd annual meeting of the Western Division of the Canadian Association of Geographers (WDCAG 2010: A Spatial Odyssey), Edmonton, Alberta, 25-27 March.

Halsey, Linda A., Dale H. Vitt, and Ilka E. Bauer

1998 Peatland Initiation During the Holocene in Continental Western Canada. *Climatic Change* 40: 315-342.

Karcz, Iaakov, and David Kersey

1980 Experimental Study of Free-Surface Flow Instability and Bedforms in Shallow Flows. *Sedimentary Geology* 27: 263-300.

Murton, Julian B., Mark D. Bateman, Scott R. Dallimore, James T. Teller, and Zhirong Yang

2010 Identification of Younger Dryas Outburst Flood Path from Lake Agassiz to the Arctic Ocean. *Nature* 464 (1 April): 740-743.

O'Connor, James E.

1993 *Hydrology, Hydraulics and Geomorphology of the Bonneville Flood*. Geological Society of America Special Paper no. 274. Geological Society of America, Boulder, Colorado.

Pardee, Joseph T.

1942 Unusual Currents in Glacial Lake Missoula, Montana. *Geological Association of America Bulletin* 53: 1569-1600.

Rhine, Janet L., and Derald G. Smith

1988 The Late Pleistocene Athabasca Braid Delta of Northeastern Alberta, Canada: A Paraglacial Drainage System Affected by Aeolian Sand Supply. In *Fan Deltas: Sedimentology and Tectonic Settings*, edited by W. Nemec and R. J. Steel, pp. 158-169. Blackie and Son, Glasgow and London.

Saxberg, Nancy, and Brian O. K. Reeves

2003 The First Two Thousand Years of Oil Sands History: Ancient Hunters at the Northwest Outlet of Glacial Lake Agassiz. In *Archaeology in Alberta: A View from the New Millennium*,

Smith, Derald G., and Timothy G. Fisher

    1993   Glacial Lake Agassiz: The Northwestern Outlet and Paleoflood. *Geology* 21(1): 9–12.

Teller, James T., Matthew Boyd, Zhirong Yang, Phillip S. G. Kor, and Amir Mokhtari Fard

    2005   Alternative Routing of Lake Agassiz Overflow During the Younger Dryas: New Dates, Paleotopography, and a Re-evaluation. *Quaternary Science Reviews* 24: 1890–1905.

Unfreed, Wendy J., Gloria J. Fedirchuk, and Eugene M. Gryba

    2001   *Historical Resources Impact Assessment, True North Energy L.P. Fort Hills Oil Sands Project: Final Report (ASA Permit 00-130).* 2 vols. Copy on file, Archaeological Survey, Historic Resources Management Branch, Alberta Culture, Edmonton.

Wolfe, Stephen A., David J. Huntley, and Jeffrey Ollerhead

    2004   Relict Late Wisconsinan Dune Fields of the Northern Great Plains, Canada. *Géographie physique et Quaternaire* 58: 323–336.

4 **Kearl Lake** | A Palynological Study and Postglacial Palaeoenvironmental Reconstruction of Alberta's Oil Sands Region

LUC BOUCHET AND ALWYNNE B. BEAUDOIN

Although it is the largest ecoregion in Alberta, the Boreal Forest Natural Region has received much less attention from archaeologists than the Parkland and Grassland regions to the south. The archaeological record of the oil sands area north of Fort McMurray was first explored in some detail in the 1970s (Ives 1993). With the increased pace of industrial development in the area, however, a corresponding increase in impact-assessment archaeological work has resulted in the discovery of a large number of intact archaeological sites. Although the precise antiquity of these sites has not yet been confirmed by radiocarbon evidence, Saxberg and Reeves (2003) suggest that many are Early Prehistoric, and they thus consider these sites to be of both provincial and national significance. If this suggestion is eventually supported by analytical evidence, then the human history of this area has more time depth and complexity than hitherto suspected. There is therefore a compelling need for greater information about the postglacial climate and vegetation history of this area, which would help to provide the landscape context for human occupation.

As has also been the case with archaeological work, palaeoecological studies of areas in northeastern Alberta, including the oil sands region, have been sparse until recently in comparison to those conducted in other areas of the province (see Beaudoin 1993). In the Birch Mountains, to the west of the oil sands area, palaeoecological records have been obtained from Eaglenest Lake (Vance 1986) and Otasan Lake (Prather and Hickman 2000), and a record has also been recovered from Mariana Lake (Hutton, MacDonald, and Mott 1994), about 100 kilometres

**Figure 4.1. Location of pollen
sampling sites and localities:
Kearl Lake (KL); Eaglenest Lake
(EL) and Otasan Lake (OL), in the
Birch Mountains; Wild Spear Lake
(WSL), in the Caribou Mountains
to the northwest; and, to the south,
Mariana Lake (MaL), Lofty Lake (LL),
and Moore Lake (MoL). The position
of the Cree Lake Moraine (CLM) is
taken from Dyke and Prest (1987);
the positions of the Firebag Moraine
(FBM) and the Stony Mountain
Moraine (SMM) are based on Lowell
et al. (2005). Also shown are the
positions of the Survive Moraine
(SM) and Don's Moraine (DM).**

southwest of Fort McMurray (see fig 4.1). Further afield, pollen records have been
obtained from Wild Spear Lake (MacDonald 1987), in the Caribou Mountains of
northern Alberta, and Lofty Lake (Lichti-Federovich 1970) and Moore Lake
(Hickman and Schweger 1996), in central Alberta, considerably to the south of Fort

McMurray. Although these studies are useful in outlining broad, regional palaeoecological histories of the postglacial interval, they are not necessarily useful for constructing more detailed, localized sequences directly relevant to the oil sands region. This is because differences in both altitude (such as the rise of roughly 450 metres to the Birch Mountains from the Lower Athabasca valley) and latitude (such as the 100 kilometres that separate Mariana Lake from the Fort McMurray area) may result in differences not only in the composition of the vegetation community but also in the chronology of various biotic events. To provide a more relevant subregional context, we report here on an early-to-mid-Holocene pollen record from Kearl Lake, which is situated within the oil sands area, immediately adjacent to the large concentrations of archaeological sites described by Saxberg and Reeves (2003).

SAMPLE SITE AND STUDY AREA

Kearl Lake (57°17' N, 111°14' W, elevation 330 masl) lies approximately 65 kilometres north of Fort McMurray (see fig 4.2), in the Central Mixedwood Natural Subregion of the Boreal Forest Natural Region (Natural Regions Subcommittee 2006). It is located to the southeast of the Fort Hills, dissected uplands that rise about 60 metres above their surroundings (McPherson and Kathol 1977), and is separated from them by the Muskeg River valley. Measuring roughly 2 kilometres by 4 kilometres, Kearl Lake occupies a shallow depression and does not much exceed 2.5 metres in depth (fig 4.2, inset). Because of its surface area and inlet stream, the major sources of pollen are likely to be extralocal (between 20 and 100 metres from the lake edge) and regional (greater than 100 metres) (Jacobson and Bradshaw 1981).

Several different forest types occur in the area surrounding Kearl Lake. On well-drained high ground, such as the high sandy banks of Muskeg Creek, open jack pine (*Pinus banksiana*) forest mixed with aspen (*Populus tremuloides*) occurs. On well-drained but mesic sites, closed white spruce (*Picea glauca*) forest dominates. Mixed white spruce and balsam fir (*Abies balsamea*) forest is found in a strip along the Athabasca River, where the wetter areas support stands of balsam poplar (*Populus balsamifera*). In the numerous poorly drained lowland muskeg areas, black spruce (*Picea mariana*) thrives with an understorey of peat moss (*Sphagnum* sp.), Labrador tea (*Ledum groenlandicum*), and dwarf birch (*Betula glandulosa*).

The Kearl Lake area is underlain by Lower Cretaceous bedrock of the McMurray Formation to the north and the Clearwater Formation to the south, comprising mainly sandstones, mudstones and siltstones (Prior et al. 2013), mantled by late

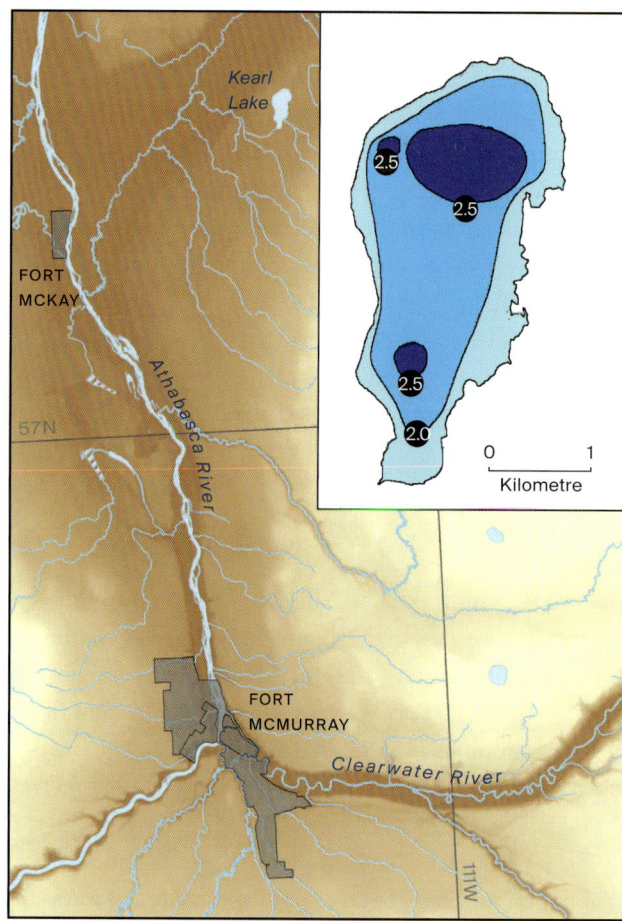

Figure 4.2. The regional setting of Kearl Lake. The inset shows the bathymetry of the lake, with relative depths divided at 2.0 and 2.5 metres.

Quaternary surficial sediments, primarily glacial diamicton (Bayrock 1971; McPherson and Kathol 1977; Fenton et al. 2013). The lake lies within the region affected by glaciation by the Late Wisconsinan Laurentide Ice Sheet (Dyke and Prest 1987; Dyke, Moore, and Robertson 2003). From the perspective of vegetation history, it is the style and timing of deglaciation that is most important. In an effort to clarify the history of deglaciation in the region, considerable research effort has been expended to establish chronological control for the emplacement of a series of moraines that run roughly southeast to northwest across the region, parallelling the retreat of the ice front. The most prominent and extensive of these is the Cree Lake Moraine to the east of the oil sands area, with Kearl Lake lying between the Firebag and Survive moraines (see fig 4.1). Dyke and Prest (1987) had the Cree Lake

Moraine aligned with their 10,000 ^{14}C yr BP Laurentide ice retreat isoline. However, more recent studies from the oil sands region have yielded a sequence of progressively younger radiocarbon dates, which together suggest a somewhat later deglacial chronology. Fisher et al. (2009) constrain the emplacement of the Survive Moraine at around 9,900 ^{14}C yr BP, the Firebag Moraine at about 9,700 ^{14}C yr BP, and, further west, the Stony Mountain Moraine and its northward continuation, Don's Moraine, at about 10,500 ^{14}C yr BP. These are all minimum limiting dates, however, and so do not preclude somewhat earlier deglaciation.

Generally, these data suggest that the Kearl Lake area was deglaciated toward the very end of the late Pleistocene or the early Holocene, sometime between 9,700 and 9,000 ^{14}C yr BP, or about 11,500 to 10,940 cal yr BP. During deglaciation, the Clearwater and Athabasca river valleys, south and west of Kearl Lake, formed part of a northwest outlet for Glacial Lake Agassiz, known as the Clearwater–Lower Athabasca spillway (Smith and Fisher 1993). The flow from the lake appears to have bifurcated around the Fort Hills, with the Muskeg River valley channelling part of the palaeoflood (Smith and Fisher 1993). Given their new chronological data for glacial retreat, Fisher et al. (2009) suggest that the northwest outlet could not have been open until about 9,850 to 9,660 ^{14}C yr BP (about 11,250 to 11,130 cal yr BP), that is, until after the ice had retreated from the position of the Firebag Moraine.

Core Description

The Kearl Lake core (KEARL2, length 483 cm) was one of three obtained in 1995 from the deepest part of the basin (see fig 4.2, inset), using a vibracoring system (Beierle 1996). The core was sealed and kept unfrozen, to prevent deformation of sediments, until it was split in half. Each half was wrapped in plastic cling wrap to prevent drying and contamination. One half of the core was used by Beierle (1996) to study sedimentological, geochemical, and macrofossil facies, while the other half was used for this pollen study.

Beierle divided the core into five zones. Zone 1, the oldest and lowest zone in the record, extends from the base to 220 centimetres and is subdivided into subzones 1A, 1B, and 1C. Subzone 1A (0 cm up to 69 cm; depths for this and other zones in Beierle's record are measured up from the base of the core) is composed of basal Laurentide till containing reworked McMurray Formation tar sands and abundant large (5 cm) angular glacially transported clasts in a matrix of sand, silt, and clay. Subzone 1B (69 cm up to 180 cm) is "massively bedded with some weak stratification, tar sand rip-up clasts and . . . bedrock clasts" and is interpreted by Beierle (1996, 4) as proglacial sediments deposited during deglaciation. Subzone

1C (180 cm up to 220 cm) is characterized by fine-grained sandy silt, where pea clam (*Pisidium* sp.) mollusc macrofossils make a first appearance in sediment that otherwise has a very low organic content. In this subzone, Beierle (1996) also notes an increase in macroscopic charcoal values, which he interprets as the effect of a warmer, more arid climate following the end of the cooler Younger Dryas interval.

Zone 2 (220 cm up to 234 cm) marks an abrupt increase in organic carbon and *Pisidium* sp. mollusc count values. Beierle (1996, 6) describes it as an extremely shelly banded gyttja, containing upwards of 40% organic carbon, measured through loss-on-ignition (LOI). The mollusc fossils indicate the presence of high levels of dissolved oxygen and organic carbon in the waters, probably caused by an increase in aquatic vegetation in response to the warmer temperatures reflected in Subzone 1C. In Zone 2, macroscopic charcoal is absent, perhaps signifying more open forests that are less prone to burning.

Zone 3 (234 cm up to 379 cm) mostly consists of brown-green gyttja with a high organic carbon content (above 60%). Macroscopic charcoal counts remain low. Zone 4 (379 cm up to 414 cm) is characterized by banded gyttja yielding increased mollusc and charcoal macrofossil counts. Several different mollusc taxa are present in this zone, which Beierle (1996) suggests reflect shallower waters able to support photosynthesizing aquatic plants. Beierle (1996, 13) interprets increased charcoal counts as indicating warmer and drier conditions. The sediment record contains no evidence of depositional hiatuses or erosional events, which suggests that, even during the Hypsithermal interval of maximum warmth and dryness (now more often called the Holocene Thermal Maximum), the lake never entirely dried up. Finally, Zone 5 (414 cm up to 482 cm), the topmost zone, represents a modern sedimentary facies. Low macroscopic charcoal counts imply cooler and wetter conditions.

Chronological Control

Chronological control for the Kearl Lake sediments was established by two AMS radiocarbon dates obtained from charred spruce needles: 5,340 ± 70 ^{14}C yr BP (Beta-94235), at roughly 385 centimetres up from the base of the core (in Beierle's sediment Zone 4), and 10,100 ± 60 ^{14}C yr BP (Beta-94234), at roughly 215 centimetres up from the base (Subzone 1C) (Brandon Beierle, pers. comm., 2000). Calendrical estimates, derived using CalPal with the InterCal04 data set (Weninger, Jöris, and Danzeglocke 2005), yield values of 11,590 ± 180 cal yr BP for the earlier date and 6,120 ± 100 cal yr BP for the later date, indicating that the record spans the interval from the late Pleistocene to the mid-Holocene.

Given that Beierle identified no discontinuities in the sediment between the lower and upper radiocarbon-dated samples, we assume that the rate of deposition was relatively uniform. Working on this assumption, we obtain a sedimentation rate of 0.357 millimetres per year, or the accumulation of 1 centimetre every twenty-eight years. Pending additional chronological data, this rate was applied to the Kearl Lake core to generate estimated values in radiocarbon years BP, as well as approximate calendrical ages (cal yr BP), for the upper and lower boundaries of the pollen zones.

Fisher et al. (2009) report an AMS radiocarbon date of $9,395 \pm 75$ [14]C yr BP (ETH-32326) on "seed pods" obtained from a near-shore core from Kearl Lake. Their stratigraphic diagram shows that this date was obtained from organic material found in laminated silt, just above a sand layer, with the sand underlain by diamicton. This date occurs about 12 centimetres above a marked increase in organic content, as estimated by LOI. At the dated level, LOI values are around 30% (Fisher et al. 2009, 1615). Although correlations are tentative, this level appears to be roughly equivalent to the lower part of Beierle's Zone 2. Thus, the $9,395 \pm 75$ [14]C yr BP date is stratigraphically consistent with the older radiocarbon date for our Kearl Lake record ($10,100 \pm 60$ [14]C yr BP, in Subzone 1C).

Laboratory Methods for Pollen Analysis

The top metre of the core consisted of wet and unconsolidated sediments that appeared to have been disturbed by vibracoring. Hence, no pollen samples were taken from this span. Similarly, no samples were taken from the diamicton or the proglacial lake sediments at the base of the core. This study concentrated on the 190-centimetre to 378-centimetre core section, where the sediments were most organic-rich. Comparisons may therefore be made only to Beierle's (1996) Subzone 1C and Zones 2 and 3.

Sample depths were measured from the bottom of the core. Samples were obtained at 10-centimetre intervals up to 220 centimetres, at which point the organic content of the core increases. From this point up, sampling was carried out at 2-centimetre intervals. All samples comprised 1 cubic centimetre of sediment taken with an open-ended plastic hypodermic syringe and a metal spatula. Samples were stored in plastic vials until processed.

Pollen residues were extracted from samples following standard methods as described by Faegri, Kaland, and Krzywinski (1989). Samples were inoculated with known quantities of *Lycopodium* (clubmoss) spores ($27,822 \pm 975$ spores, 2 tablets, batch number 710961) to enable calculation of pollen concentration values (Stockmarr 1971). Residues were stained with safranin and suspended in silicone oil. Pollen grains were identified and counted using a Leitz Laborlux 12 light microscope. To avoid distributional bias (see Brookes and Thomas 1967), whole slides were scanned with regular traverses. Samples were analyzed at 10-centimetre intervals along the entire length of the organic section of the core. In places where pronounced differences in pollen assemblages occurred between adjacent 10-centimetre samples, intervening samples were also counted so as to better characterize the changes. In total, thirty-seven samples were analyzed. A target pollen sum of 500 grains per sample was established; the pollen sum comprised counts for tree, shrub, forb, and graminoid taxa. Counts for pollen of emergent and aquatic taxa and spores were not included in this sum. Except in those cases where pollen concentration was low (for example, at 210 cm), this grain count was achieved (mean of 640, range from 134 to 1140).

Counts and identifications were made at 400x magnification, with 1000x magnification and oil immersion used for more difficult identifications. Pollen grains were identified to the lowest possible taxonomic level, which in the majority of cases was to genus. Pollen taxa were identified using published sources (see, for example, McAndrews, Berti, and Norris 1973) and the pollen reference collection at the Royal Alberta Museum (fig 4.3). Most of the *Pinus* pollen was assigned to diploxylon type, characterized by the absence of verrucae on the distal portion of the corpus (Habgood 1985, 8) and probably derived from the locally growing jack pine (*Pinus banksiana*). The *Picea* pollen could be from either of two species: *Picea mariana* (black spruce) or *Picea glauca* (white spruce). We did not attempt to distinguish between the two species because doing so on the basis of palynological criteria is a labour-intensive operation that is not always reliable (see Birks and Peglar 1980). For the early part of the record, however, it is probable that white spruce is represented (see Ritchie and MacDonald 1986; Vance 1986, 17). During counting, measurements were taken on *Betula* (birch) pollen grains from some levels. Although size measurements are not definitive (Ives 1977), previous work in western Canada shows that shifts in the mean values may indicate the relative proportions of pollen derived from shrub and from tree birch in the assemblage (White and Mathewes 1986). Generally speaking, *Betula* grains larger than 20 mm in diameter tend to be derived from tree taxa, whereas those less than 20 mm in diameter are more likely to be derived from shrub birch (see Ives 1977). The Chenopodiaceae-*Amaranthus* (Cheno-*Am*) category comprises the tally of undifferentiated periporate pollen types. Many plants in the Chenopodiaceae (goosefoot) family and *Amaranthus* (amaranth) genus produce periporate pollen grains that cannot reliably be distinguished from each other.

Pollen data were plotted as percentage values (fig 4.4) and as concentrations (fig 4.5), with the boundaries of three pollen assemblage zones (K-1, K-2, and K-3)

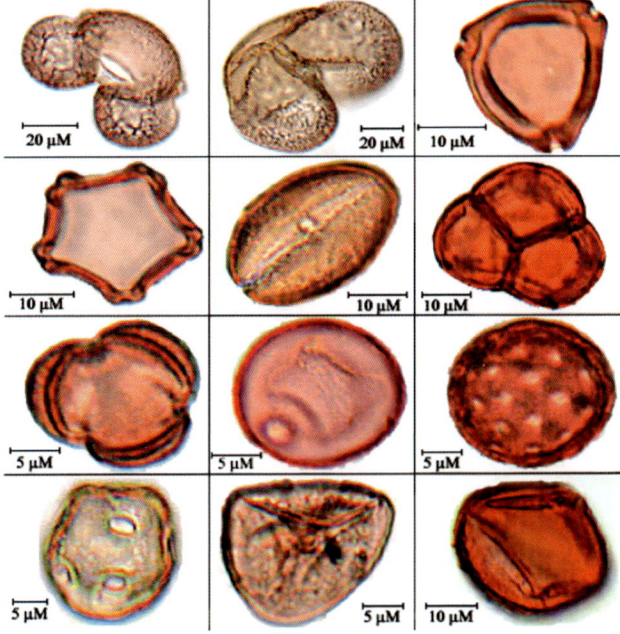

Figure 4.3. Modern pollen and spore types representative of those identified at Kearl Lake, drawn from the Royal Alberta Museum's Pollen Reference Collection. Top row, left to right: jack pine, *Pinus banksiana* (R00238); white spruce, *Picea glauca* (R00787); and paper birch, *Betula papyrifera* (R00242). Second row, left to right: green alder, *Alnus crispa* (R00726); buffaloberry, *Shepherdia canadensis* (R00110); and an example of an Ericaceae pollen type, Labrador tea, *Ledum groenlandicum* (R00491). Third row, left to right: sage, *Artemisia campestris* (R00592); an example of a grass (Gramineae) pollen type, slough grass, *Beckmannia syzigachne* (R00719); and an example of a Cheno-*Am* pollen type, strawberry blite, *Chenopodium capitatum* (R00735). Bottom row, left to right: veiny meadow rue, *Thalictrum venulosum* (R00024); peat moss spore, *Sphagnum* sp. (R00674); and prickly rose, *Rosa acicularis* (R00474).

determined subjectively on the basis of the location of concurrent changes in a number of different pollen types (Moore, Webb, and Collinson 1991, 178–179). For Zone K-3, the uncertainties involved in the concentration estimates (see Maher 1981) preclude them from being used for interpretation.

Unfortunately, no archived material or specimens from this work are now available. Any remaining core material, sediment samples, pollen residues, and slides appear to have been lost or destroyed subsequent to this study. Therefore, additional analyses or counts are not possible.

RESULTS

Zone K-1 (Spruce-Birch-Herb Zone), 210 to 225 cm Above Base of Core

Description and estimated age. Zone K-1 is initially dominated by *Betula* (birch) pollen, with a maximum value of 55% declining to 25% by the end of the zone. *Picea* (spruce) pollen also dominates this zone, steadily increasing from 25% to 70% at the end. *Salix* (willow) pollen is at a maximum for the core with values of 5% to 10%. *Artemisia* (sage) pollen is well represented at 5%, while some Compositae (daisy family) and Cheno-*Am* values reach 3% to 4%. *Pinus* (pine)

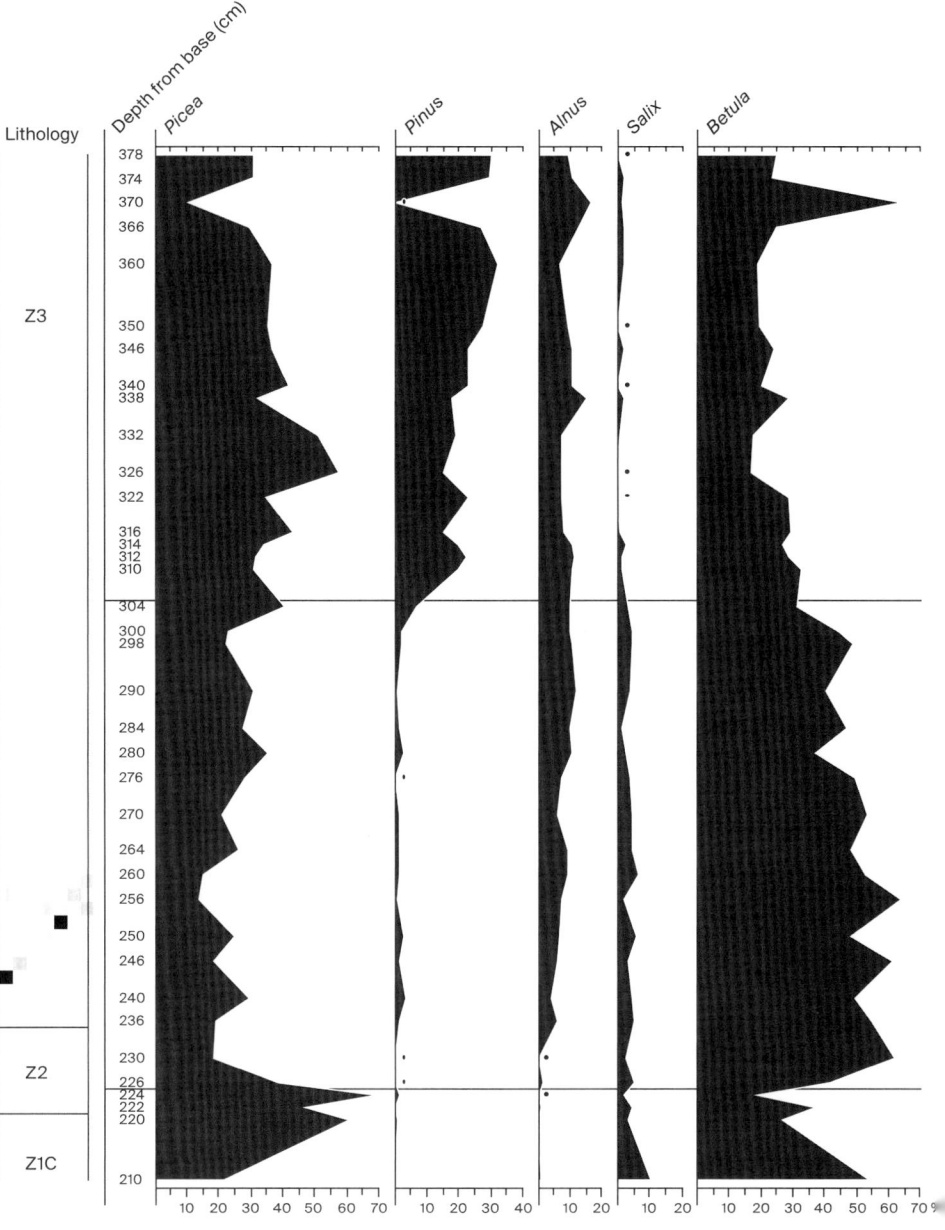

Figure 4.4. Pollen percentage record from Kearl Lake, showing selected taxa. The "Lithology" column (on the left) indicates the sediment types: Z1C = fine-grained sandy silt; Z2 = shelly banded gyttja; Z3 = brown-green gyttja. The horizontal scales are percentage values, derived by comparing pollen counts for individual taxa to the pollen sum; values of less than 1% are indicated in the figure by a small black dot. The pollen sum used for computing percentages does not include the counts for *Typha*, Other wetland, *Myriophyllum*, *Sphagnum*, Ferns, and *Lycopodium*. (Hence, the total of the percentage values for a specific sample may exceed 100%.) For *Shepherdia*, Compositae (HS), Compositae (LS), Cheno-Am, *Thalictrum*, Other NAP, and Cyperaceae, the areas shown in outline represent a 10x exaggeration of the areas in black, so that changes will be more visible. Pollen and spore taxa that occur in small amounts or in only a few samples are grouped into categories as follows: Other tree = *Abies, Larix, Populus, Corylus, Amelanchier, Juglans*; Other shrub = *Shepherdia argentea*, Ericaceae undiff., *Chamaedaphne, Vaccinium*, Rosaceae undiff., and *Potentilla*; Other NAP (non-arboreal pollen) = *Cornus, Fragaria, Agoseris, Galium boreale, Rumex, Stellaria, Impatiens*, Scrophulariaceae, Cruciferae, and Leguminosae; Other wetland = *Caltha*, Hippuridaceae, *Sparganium, Sium suave, Myrica*, and *Equisetum*; Ferns = *Dryopteris, Athyrium*, and *Cystopteris*.

percentage values

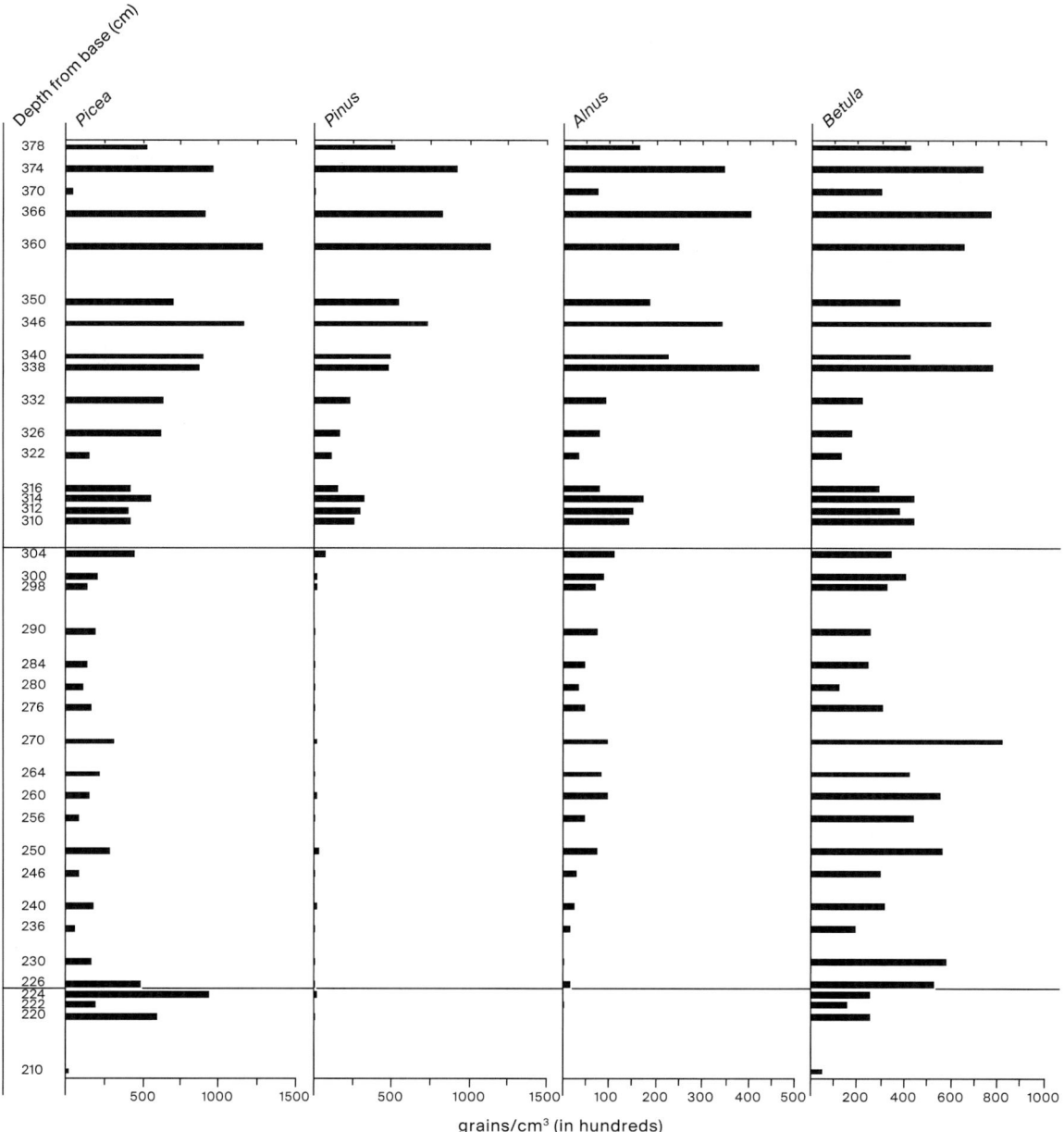

Figure 4.5. Pollen concentration record from Kearl Lake, showing selected taxa. Concentrations are computed as the number of grains per cubic centimetre, where the volume is the volume of sediment processed. The numbers on the horizontal scale are in hundreds (so, for example, "500" represents 50,000 grains/cm³); note that the scaling varies for different taxa. The "Total" column is based on the pollen sum, that is, the subset of taxa used for computing percentages (see figure 4.4).

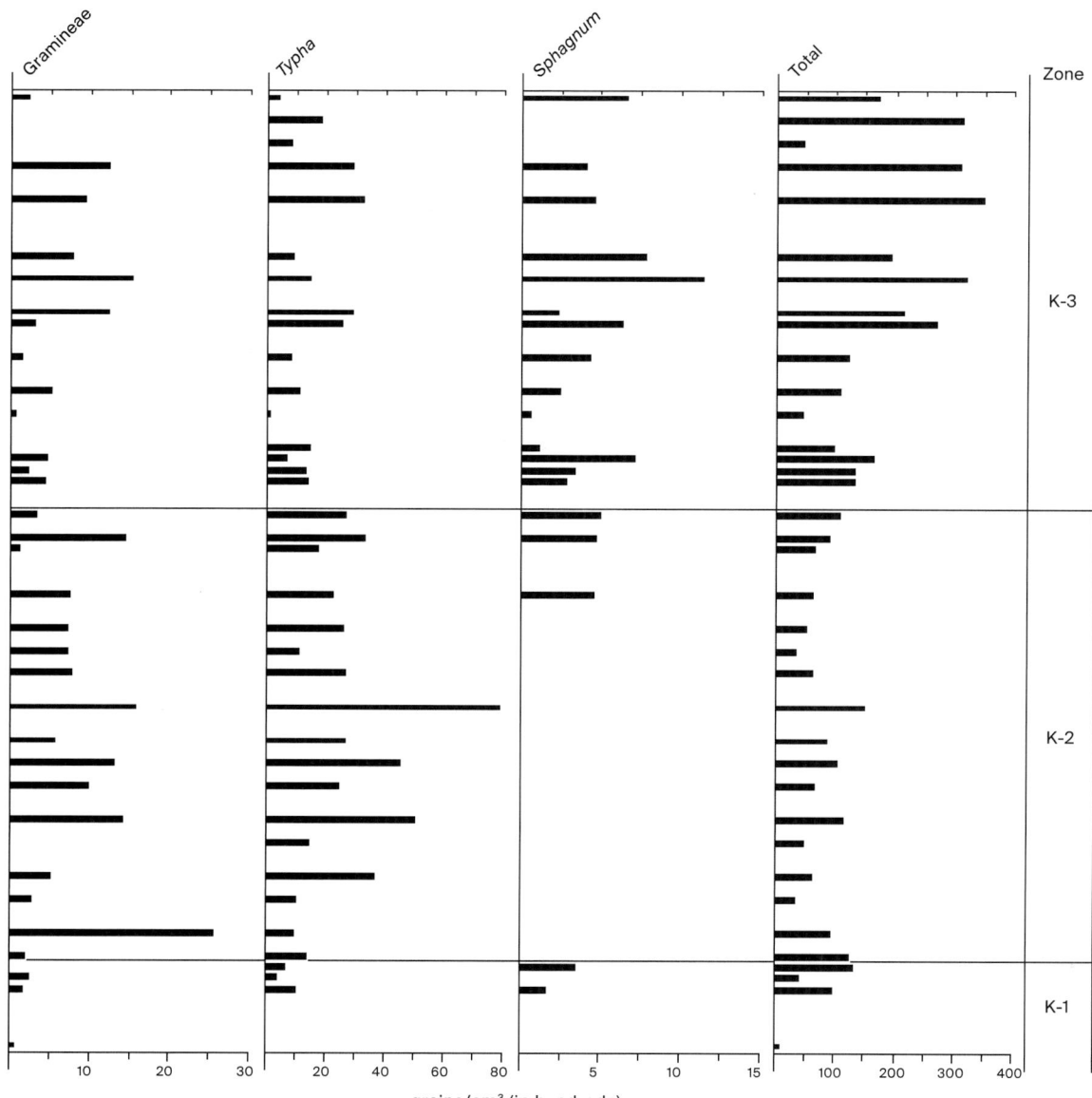

grains/cm³ (in hundreds)

pollen is present in minute quantities only, probably indicating long-distance transport. This zone is estimated to span the interval from about 10,250 to 9,820 ^{14}C yr BP, or 11,890 to 11,210 cal yr BP.

Interpretation. This pollen assemblage indicates the establishment of northern woodland, probably replacing the initial sparse shrub and herb community that would probably have first colonized the landscape in the wake of the retreating ice sheets. The significant presence of *Artemisia*, Compositae, and Cheno-*Am* pollen in this zone indicates that the spruce forest was open.

Discussion. Zone K-1 may be compared to Zone 2 of Mariana Lake (10,500 to 9,000 ^{14}C yr BP), which is located about 100 kilometres southwest of Fort McMurray (Hutton et al. 1994), and Zones 1 and 2 of Wild Spear Lake, situated about 280 kilometres northwest of Kearl Lake (MacDonald 1987). Mariana Lake recorded an abrupt rise in *Picea* and *Betula* pollen values at the beginning of Zone 2, indicating forest establishment, following a sharp decline of the previous zone's high *Artemisia*, Gramineae (grass family), and Cheno-*Am* counts (fig 4.6). *Populus* pollen is also present at Mariana Lake, whereas it does not occur in Zone K-1. At Wild Spear Lake, Zone 2 is likewise marked by an abrupt increase in *Betula* pollen values, although the increase in *Picea* pollen occurs considerably later. The underlying Zone 1 was dominated by *Artemisia* pollen up to 50% and Gramineae pollen up to 25%. Both these sites had earlier pollen zones dominated by non-arboreal pollen (NAP), as does the Eaglenest Lake record (Vance 1986), followed by marked increases in arboreal pollen (AP). The transition between the NAP- to AP-dominated assemblages is time-transgressive, occurring earlier at about 10,500 ^{14}C yr BP at Mariana Lake, in the south, and about 10,000 ^{14}C yr BP at Wild Spear Lake, in the north (fig 4.6). Zone K-1 is thus probably not the first vegetation community established around Kearl Lake following the retreat of the ice sheet. Moreover, these data suggest that, in all likelihood, the transition to arboreal vegetation at Kearl Lake did not occur much before the base of the record.

Following Beierle (1996, 7–8), the high *Picea* pollen values near the end of Zone K-1, which decline markedly shortly after 10,000 ^{14}C yr BP, probably represent spruce stands that were established in the cooler and wetter Younger Dryas interval, estimated to have spanned the interval between 11,000 and 10,000 ^{14}C yr BP (Rutter et al. 2000). That this period was cooler and wetter than the subsequent interval is corroborated by the presence of *Sphagnum* spores. Although it is possible that the spruce grains are the result of long-distance transport, we feel this is unlikely because spruce macrofossils were found in this zone and used to

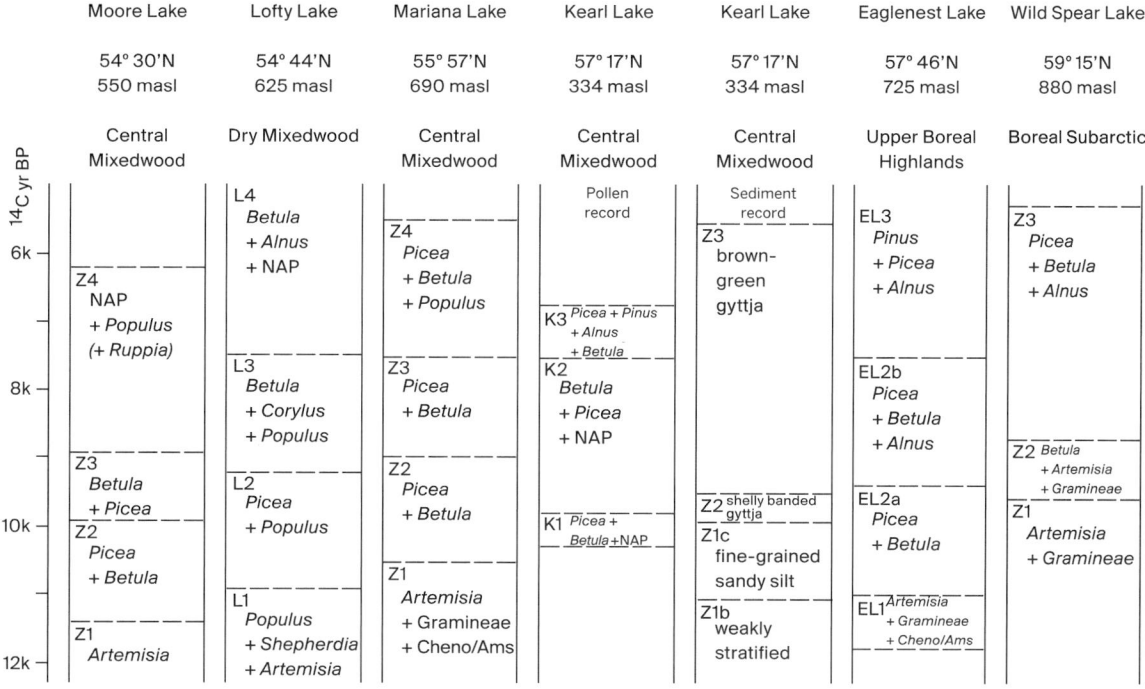

Figure 4.6. Major vegetation changes inferred from late Pleistocene to early Holocene pollen records in northeastern Alberta. Sites are arranged from southernmost (Moore Lake) to northernmost (Wild Spear Lake); natural subregion assignment follows Natural Regions Subcommittee (2006). Sources: Hickman and Schweger 1996 (Moore Lake); Lichti-Federovich 1970 (Lofty Lake); Hutton, MacDonald, and Mott 1994 (Mariana Lake); Beierle 1996 (Kearl Lake sediment record); Vance 1986 (Eaglenest Lake); MacDonald 1987 (Wild Spear Lake).

obtain a basal date, providing unequivocal evidence for its local presence. Previous researchers (Ritchie and MacDonald 1986; McLeod and MacDonald 1997) have noted that in the western interior of Canada, the spread of white spruce was very rapid. Geologically speaking, it was almost instantaneous: white spruce apparently migrated across the 2,000 kilometres from south-central Alberta to the Mackenzie Delta in roughly a single millennium. Ritchie and MacDonald (1986) suggest that strong adiabatic winds blowing clockwise off the retreating ice sheets promoted spruce's rapid migration from populations centred in areas south of the Laurentide Ice Sheet. The Younger Dryas interval may be of importance here, because the associated cooler and wetter conditions may have provided better growing conditions for mesic spruce stands. If the retreat of the ice sheets slowed during the Younger Dryas, this would have allowed a longer interval for adiabatic winds to operate in the same location, which could have sped up the northwesterly spread of spruce. However, it is worth noting that the rapidity of white spruce migration has recently been challenged by new work from the northern Great Plains, which suggests that migration across this region was slower by some thousand years (Yansa 2006). Moreover, estimated migration rates may appear rapid owing to the apparent compression of ages resulting

Figure 4.7. Comparison of *Picea*, *Pinus*, and *Betula* curves and the distribution of birch grains, by size, in selected samples from the Kearl Lake record. On the horizontal scale, *Picea*, *Pinus*, and *Betula* values are percentages (as in figure 4.4), while values for B16.8, B19.6, B22.4, B25.2, and B28.0 are the actual numbers of *Betula* grains measured, binned into 16.8 μm, 19.6 μm, 22.4 μm, 25.2 μm, and 28.0 μm grain diameter categories, respectively. Only well-preserved *Betula* grains were measured; torn, crumpled, obscured, or degraded ones were not. *Betula* counts were not made at every sample; thus, the absence of counts does not reflect preservation.

from the "plateau" in radiocarbon values in the late Pleistocene and early Holocene (see Fiedel 1999). In this regard, the presence of spruce macroremains at Kearl Lake is especially significant because it provides evidence to support the relatively rapid spread of the taxon to the northwest.

The size measurements of *Betula* pollen grains suggest a transition between shrub and tree birch at Kearl Lake (fig 4.7). In the lowest level (210 cm from base of core), the *Betula* pollen sizes suggest an even mixture of tree and shrub species, while by the end of Zone K-1 the shrub species seems to dominate the *Betula* assemblage. These results indicate that at the end of Zone K-1, shrub birch was present within mesic spruce stands. Shrub birch (*Betula glandulosa*) is a mesic species, preferring wetlands and moist depressions (Johnson et al. 1995, 40). One possible explanation for the decrease in tree birch at the end of Zone K-1 is that, given the extremely high spruce pollen values here, spruce competition may have successfully crowded out paper birch (*Betula papyrifera*), which does not tolerate shade conditions (Farrar 1995, 283). Remaining birch populations may have been restricted to nearby more open, gravelly terraces. It is interesting to note that the end of Zone K-1, which falls around 9,820 ^{14}C yr BP, could very well coincide with the 9,900 ^{14}C yr BP Lake Agassiz flood event (Smith and Fisher 1993; Lowell et al. 2005; Fisher 2007), which affected areas only 2 kilometres from Kearl Lake. This destructive flood might well have benefitted birch populations, because birch is known to form a large part of the new tree cover in areas that have been disturbed (Farrar 1995, 283). Alternatively, the higher proportion of tree-birch-type pollen in the early part of the record could reflect long-distance transport and the low local pollen productivity.

Zone K-2 (Birch-Spruce-Herb Zone), 225 to 305 cm Above Base of Core

Description and estimated age. Zone K-2 is characterized by a sudden decrease in *Picea* pollen from 70% to between 20% and 40%. At the same time, *Betula* pollen increases from a low of 25% in Zone K-1 to values ranging from 35% to 63%. Significant amounts (10% to 12%) of *Alnus* (alder) pollen appear in the record for the first time. This zone is characterized by an increased presence of shrubs and herbs. *Salix* pollen values remain high at 5% to 8%, and *Shepherdia* (buffaloberry) pollen reaches 1% to 2%. *Artemisia*, Cheno-*Am,* and Compositae (high-spine) pollen types remain constant at around 5%, 3% to 5%, and 1% to 2%, respectively. *Sphagnum* spores are absent throughout most of this zone (see fig 4.3). Gramineae pollen increases to 2% to 4%, while Compositae (low-spine), *Thalictrum* (meadow rue), and *Sarcobatus* (greasewood) are largely confined to this zone. Zone K-2 is estimated to span the interval from about 9,820 to 7,580 ^{14}C yr BP, or 11,200 to 8,390 cal yr BP.

Interpretation. Zone K-2 is interpreted as reflecting a reduction in mesic growing conditions that affected spruce more than birch. That this period was warmer and drier than the previous one is indicated as well by the presence of *Sarcobatus*, which grows in dry habitats, usually along saline sloughs and flats, and today is confined to the Grassland Natural Region of Alberta (Moss 1983). In view of its present distribution and habitat requirements, the relatively persistent presence of *Sarcobatus* pollen at Kearl Lake in Zone K-2 is highly informative. However, other high salinity indicators, such as pollen from *Ruppia* (widgeongrass), are not present in Kearl Lake, in contrast to sites further south, such as Moore Lake (Hickman and Schweger 1996).

The absence of *Sphagnum* spores attests to drier conditions in the landscape around the lake. Likewise, an increased presence of *Typha* (cattail) pollen in Zone K-2 is probably indicative of a drop in the lake's water level, resulting in lake-edge vegetation that may have been more tightly focused around the sample site. The bathymetry of the lake, with gently sloping margins (see fig 4.2, inset), suggests that falling water levels would also expose more extensive littoral areas with shallow water or wetlands. Therefore, habitat for wetland or emergent aquatic vegetation may have been more prevalent; cattail, being a wetland plant, would have been able to take advantage of these conditions.

Combined with the abundance of pollen from deciduous tree taxa, relatively high Zone K-2 pollen percentage values of *Artemisia*, Cheno-*Am*, and Compositae, as well as an increased presence of Gramineae pollen, suggest open deciduous woodland in the area at this time.

Discussion. High values of birch pollen have been noted in the early Holocene in many records from interior western Canada (MacDonald 1993). In the Mariana Lake record, Hutton, MacDonald, and Mott (1994) found in their Zone 3 (9,000 to 7,500 ^{14}C yr BP) a decrease in *Picea* pollen values, accompanied by an increase in *Betula* pollen counts (see fig 4.6). Zone K-2 mirrors this pattern. However, these changes apparently occur some 800 radiocarbon years earlier in the oil sands region.

Citing increased charcoal counts in sedimentary Subzone 1-C (Zone K-1, above), Beierle (1996, 7–8) suggests that the dense spruce stands established during the Younger Dryas died once warmer and drier conditions became prevalent and that the resulting availability of fuel caused an increase in fire activity. However, dead standing timber would have been rapidly eliminated by initial fire events, reducing fuel sources. Therefore, this suggestion does not seem tenable as an explanation for higher fire frequencies that lasted for centuries. Rather, the decrease in *Picea* pollen values of Zone K-2 may indicate the transition to the warmer and drier climate of the Hypsithermal interval. Increased charcoal counts in Zone K-1 may therefore represent only the presence of available fuel sources in the form of denser, though perhaps localized, spruce stands, whereas the subsequent forests inferred for Zone K-2 would likely not have burned as readily because of their more open nature. Moreover, unlike Beierle (1996, 8), we cannot assume that the vegetation of the area changed to open stands of pine with grassy forest floors because *Pinus* pollen begins to enter the Kearl Lake record only in Zone K-3 (see below), some 2,500 years after the end of the Younger Dryas.

Betula pollen grain size measurements indicate a fairly even mixture of tree and shrub birch existing near the boundary of Zones K-1 and K-2, with the tree form becoming more firmly established as Zone K-2 progresses (see fig 4.7). This increase in tree birch abundance is concomitant with a marked decrease in *Picea* pollen abundance. A similar shift between *Picea*-dominated and *Betula*-dominated pollen assemblages in the early postglacial has been noted in many records from western interior Canada (see MacDonald 1993). Initially, we hypothesized that this pattern resulted from increased fire frequencies. Jack pine (*Pinus banksiana*) is a fire-successional species, but birch is also known as an aggressive colonizer in fire-disturbed areas (Farrar 1995, 283). Because Zone K-2 occurs earlier than the known jack pine migration into the area, we theorized that birch was the sole fire-successional tree species, which allowed it to gain dominance at the expense of spruce in this period of warmer and drier conditions. However, Beierle (1996) indicates maximum macroscopic charcoal counts only in his Subzone 1-C (Zone K-1), which then decrease to zero in Zone K-2. This indicates that fire activity in Zone K-2 was much reduced over that of Zone K-1, making the *Betula* fire-succession hypothesis untenable.

This does not mean, however, that warmer and drier conditions in Zone K-2 did not occur. MacDonald (1989), in his study of Toboggan Lake, located in the southern Alberta foothills, also noted a lack of charcoal macrofossils in a warm and dry interval. He ascribed this phenomenon to the fact that deciduous stands are less prone to burning than coniferous stands, arguing that "the lower availability of resinous conifer wood for fuel and decreasing forest density may have led to a decrease in fires and/or charcoal production" (MacDonald 1989, 164). At Kearl Lake, between roughly 9,820 and 7,580 ^{14}C yr BP, increased *Betula* and *Alnus* pollen values, combined with higher frequencies of herb pollen, suggest a more open and deciduous forest. This might explain the lack of charcoal fragments in Zone K-2, despite warmer and drier conditions.

Although *Alnus* grows on wet sites (Farrar 1995, 297), the presence of *Alnus* pollen for the first time in Zone K-2 is probably not indicative of moister conditions in comparison to the preceding period. Rather, the sharp increase in alder pollen percentage suggests that the taxon is now present in the local area, as a consequence of migration. Interestingly, alder's local presence may actually be an indicator of drier conditions. Ritchie (1984, 146) proposed that alder migrated northward from areas south of the ice sheets at the same time as spruce. Because white spruce tends to establish itself in dense stands, and alder cannot tolerate shady conditions, the latter became established only in small numbers at first. Ritchie therefore interpreted the abrupt increase in *Alnus* pollen seen in records from sites near the western Arctic settlement of Tuktoyaktuk around 6,700 ^{14}C yr BP as a response to the opening up of the spruce forest as its margin shifted with the slow cooling of the climate following the Hypsithermal. An increase in *Alnus* pollen was thus associated with the presence of relatively more open spruce forest. This interpretation is supported by the Kearl Lake data, which shows a large increase in *Alnus* pollen at the beginning of Zone K-2, reflected in both the percentage and concentration values (see figs. 4.4. and 4.5). In contrast to the far north, in this case the opening of the spruce forests was caused by the warmer and drier conditions of the Hypsithermal.

Other sites in the region, however, indicate that the arrival of *Alnus* may not necessarily be related to the opening up of forests. For example, Vance (1986, 18) gives a date of 8,450 ^{14}C yr BP for the first appearance of *Alnus* at Eaglenest Lake, in the Birch Mountains, despite dating the start of the Hypsithermal interval to roughly 11,000 ^{14}C yr BP. Farther south, at Mariana Lake, *Alnus* appears later, at around 7,700 ^{14}C yr BP (Hutton, MacDonald, and Mott 1994, 421), despite signs of the Hypsithermal having begun shortly after 9,000 ^{14}C yr BP. Yet further south, at Lofty Lake, *Alnus* appears later still, starting after 7,480 ^{14}C yr BP, long after the pronounced decline of spruce around 9,200 ^{14}C yr BP, at the end of Zone L2 (Lichti-Federovitch 1970, 941). It thus seems likely that, instead of representing the opportunistic colonization of open spruce forests, the variability in *Alnus* arrival dates merely signifies vagaries in its migration path and its ability to establish itself.

A comparison of the pollen values at these sites with those at Kearl Lake suggests regional differences in the expression of Hypsithermal conditions (see fig 4.6). Between around 9,500 and 7,500 ^{14}C yr BP, both Kearl and Mariana lakes have *Betula* pollen frequencies of around 55% to 60%, with 20% to 35% *Picea* pollen (Hutton, MacDonald, and Mott 1994, 421). This contrasts with the Eaglenest Lake record, which consistently has more *Picea* pollen (around 50%) and less *Betula* pollen (20% to 40%) (Vance 1986, 14), and with the Lofty Lake record, which indicates *Picea* pollen values of under 10%, *Betula* pollen frequencies of about 60%, and much higher herb representation (Lichti-Federovitch 1970, 941). These results demonstrate latitudinal and altitudinal differences in the effects of the Hypsithermal. In the northern woodlands, little impact was felt in the highlands, while deciduous trees assumed greater importance at lower elevations. Further south, woodlands were primarily deciduous and more open, and they contained more non-arboreal vegetation.

Zone K-3 (Arboreal [Spruce-Pine-Alder-Birch] Zone), 305 to 380 cm Above Base of Core

Description and estimated age. Zone K-3 is once again dominated by *Picea* pollen values of up to about 60%. *Pinus* pollen now makes a significant appearance, reaching 30%. *Alnus* values remain constant from Zone K-2, at about 10% to 12%, and *Betula* pollen decreases slightly to between 20% and 30%. *Salix* pollen decreases to 1% to 3%, while *Shepherdia*, Gramineae, Cheno-*Am,* and Compositae (high- and low-spine) pollen is reduced to trace quantities. *Sphagnum* spores re-enter the record. Pollen concentration is greatest in K-3, and values generally increase toward the end of the zone. Zone K-3 is estimated to span the period from around 7,580 to 5,900 ^{14}C yr BP, or about 8,390 to 6,720 cal yr BP.

Interpretation. The arrival of *Pinus* pollen in this zone, assumed to be from jack pine, represents the establishment of a vegetation assemblage that is more recognizable as the modern boreal forest. *Alnus* pollen, which appeared first in Zone K-2, remains fairly constant in its percentage values throughout Zone K-3. This lack of any apparent change could be due to the abundance of *Pinus* pollen, however, which may have masked any increase in *Alnus* values. At the outset, this does seem to be the case. Despite the relatively stable percentage values of *Alnus,* its pollen concentration values do reveal a marked change: starting at around 320 centimetres, *Alnus* concentrations increase from a previous low of around 100 grains per cubic centimetre to four times that amount (see fig 4.5). Although, as noted above, the concentration values for Zone K-3 should be used with extreme caution, this increase in *Alnus* pollen could be interpreted as evidence of a greater availability of moist growing conditions.

In this zone, *Picea* pollen percentage values increase, which again indicates a return to cooler and moister conditions. Although *Betula* pollen abundance shows an apparent decrease, this probably reflects the increased amount of

Pinus pollen, and therefore we cannot infer a reduction of mesic growing conditions.

Artemisia and Gramineae pollen percentage values decrease in Zone K-3, as do those for *Sarcobatus* and Compositae pollen values. Without a clear indication that their concentration values also decline (see fig 4.5), however, we cannot infer an actual reduction of NAP (non-arboreal pollen) input in this zone. All the same, the increased presence of *Picea* pollen and *Sphagnum* spores would indicate a change in local conditions, possibly induced by climatic factors, which resulted in a reduction of the forb and herb components. With this reduction in NAP representation, coupled with a large increase in the various arboreal pollen concentrations, we may envision closed northern woodland, becoming more akin to the modern boreal forest.

The presence of *Sphagnum* in this zone is evidence of renewed paludification, that is, peat growth made possible by water tables that have risen to the surface, or close to it. Such increased water availability may have caused lake levels to rise, resulting in the slight reduction in *Typha* pollen values.

Discussion. The arrival of *Pinus* at Kearl Lake is consistent with the records at Eaglenest Lake (Vance 1986) and Mariana Lake (Hutton, MacDonald, and Mott 1994). A broader analysis of pollen records from regions across the western interior of Canada consistently shows the slower migration and postglacial establishment of jack pine (McLeod and MacDonald 1997). It has been suggested that the northward migration of jack pine was encouraged by frequent fires caused by Hypsithermal conditions (see, for example, Vance, Beaudoin, and Luckman 1995, 84–86). If this were the case, one might initially wonder why *Pinus* seems to enter the record only near the *end* of the Hypsithermal. At Kearl Lake and Mariana Lake, for example, the maximum Hypsithermal effects are seen to have occurred much before the arrival of *Pinus* at 7,500 ^{14}C yr BP. This is probably due to the fact that, as indicated above, fires are less frequent in open, deciduous forests, such as existed at the height of the Hypsithermal at these sites. Therefore, even if warmer and drier conditions did exist at the height of the Hypsithermal, extensive *Pinus* migration might have had to wait until closer to the end of this interval, when coniferous forests started to become established again but conditions still remained warm and dry enough to promote higher fire frequency. Indeed, at Kearl Lake, Beierle (1996, 6) sieved occasional charred spruce needles from between the 295- and 305-centimetre level of the core, which contrasts with the negative evidence for fire he found between the 220- and 295-centimetre level, which covers all but the final 10 centimetres of our Zone K-2. This indicates that it was only near the very end of the Hypsithermal (Zone K-2) that fires became prevalent again around Kearl Lake, probably as the result of increased *Picea* cover. It was probably this complex interplay between moisture conditions and fire frequency that encouraged the spread of jack pine, including its arrival in the Kearl Lake area.

Zone K-3 is consistent with other records from the Boreal Forest Natural Region, which indicate a return to cooler and moister conditions starting between 7,700 and 7,200 ^{14}C yr BP (MacDonald 1987; Vance 1986; White and Mathewes 1982, 1986). At Otasan Lake, Prather and Hickman (2000, 193) point to rising lake levels and increased nutrient concentrations between 7,300 and 5,000 ^{14}C yr BP, based on diatom assemblages, a pattern that is consistent with wetter conditions. One slight difference is that, starting at 7,500 ^{14}C yr BP at Mariana Lake, *Typha* is present in abundant quantities but *Sphagnum* is not, leading Hutton, MacDonald, and Mott (1994, 423) to conclude: "It is likely that moist sites existed, but *Sphagnum* growth and peat accumulation was precluded by fluctuating water levels promoting periodic drying or high decomposition rates due to warmer conditions." It was not until around 6,000 ^{14}C yr BP, when *Sphagnum* spore counts increased, that the development of modern vegetation and the establishment of extensive peatlands occurred in the vicinity of Mariana Lake. At Kearl Lake, we see this process happening earlier. Already at the very end of Zone K-2, trace amounts of *Sphagnum* spores occur, which become more abundant and more consistent throughout Zone K-3. At Kearl Lake, then, *Sphagnum* is estimated to have increased at about 7,100 ^{14}C yr BP, which presumably reflects the site's latitudinal positioning relative to Mariana Lake. Indeed, Zoltai and Vitt (1990) determined from a series of basal peat dates that peatlands started to form earlier in the northern boreal forest than in the more southern parts. In this case, the 150 kilometres that separate Kearl Lake from Mariana Lake probably resulted in earlier peat formation at the former. At Eaglenest Lake, *Sphagnum* is found throughout the record (Vance 1986), indicating that in northern Alberta the highlands remained relatively cooler and moister throughout the Holocene.

Further south, at Lofty Lake, Lichti-Federovitch (1970) found evidence of grassland expansion, indicating warm and dry Hypsithermal conditions, beginning around 7,500 ^{14}C yr BP and continuing to at least 5,000 ^{14}C yr BP. Other studies indicate that, although Lofty Lake was filled at 11,400 ^{14}C yr BP, the sediment record shows that it had shrunk to a small, shallow saline pond by 8,700 ^{14}C yr BP and then started refilling again around 6,300 ^{14}C yr BP (Schweger and Hickman 1989, 1828). At Moore Lake, reduced lake levels, associated with significant values of NAP and occurrence of *Ruppia* pollen, lasted until about 6,200 ^{14}C yr BP (Hickman and Schweger 1996). On the basis of records from several lakes, Schweger and Hickman (1989) conclude that in central Alberta the main impact of the Hypsithermal persisted until about 6,000 ^{14}C yr BP. The effects of the Hypsithermal, then, lasted about 1,000 years longer in central Alberta than farther north in the woodlands, again most likely as a result of latitudinal differences in climate.

Near the end of Zone K-3 is a brief anomaly, where a marked increase in *Betula* pollen (reaching up to 65%) and a slight increase in *Alnus* pollen (15%) exceed the *Picea* and *Pinus* pollen values (see fig 4.2). Non-arboreal pollen values remain low, as in the rest of Zone K-3. At first, we thought this might have been an isolated forest fire incident in which birch—recognized to be a fast grower (Farrar 1995, 283)—was able to recolonize the affected landscape more quickly than competing jack pine. Pollen-size measurements for this episode (see fig 4.7) suggest a preponderance of tree birches, which are known to dominate the landscape between 26 and 50 years following a burn (MacDonald et al. 1991, 68). This interpretation is not supported by macroscopic charcoal counts, however, which indicate low values for this zone (Beierle 1996). This episode might instead represent some other type of disturbance, such as wind throw or an insect outbreak.

SUMMARY AND IMPLICATIONS FOR HUMAN OCCUPATION

Overall, these results generally parallel Beierle's (1996) reconstruction from the Kearl Lake sedimentary record (see the discussion in Bouchet-Bert 2002, 65–67), with some exceptions, notably the interpretation of the charcoal record. Around 10,250 ^{14}C yr BP (about 11,890 cal yr BP), a spruce-birch-herb-dominated vegetation was in place, probably having replaced the sparse shrub-and-herb vegetation that colonized the area immediately following the retreat of the ice. This spruce-birch-herb community was representative of early northern woodland, probably established during the Younger Dryas interval, when a colder and moister climate (also indicated by the presence of peat moss) encouraged the growth and quick spread of spruce. This forest was relatively open, judging by higher percentage values of NAP, but not as open as the succeeding woodland of the Hypsithermal. Charcoal abundance in this zone probably reflects the availability of denser, though localized, fuel stands, rather than an interval of warmer and drier climate.

Around 9,800 ^{14}C yr BP (about 11,210 cal yr BP), spruce stands seem to have dwindled and deciduous trees (birch and later alder) became much more prominent. This was the result of a warmer and drier climate, which affected the mesic conditions necessary for spruce growth to occur. Although birch trees are also a mesic species, they are less so than spruce, as is indicated by the fact that today the Northern Mixedwood Subregion of Alberta's Boreal Forest Natural Region is dominated by white and black spruce, whereas birch tends to be more common in the drier (but still mesic) Central and Dry Mixedwood Natural subregions (Natural Regions Committee 2006, 131–140 and 157–161). Lack of peat moss and higher values of greasewood and cattail are also suggestive of warmer and drier conditions at this time. As indicated by higher NAP percentages, this deciduous-dominated northern woodland was more open than its predecessor, which probably contributed to the absence of charcoal in the early Holocene interval.

Starting around 7,580 ^{14}C yr BP (about 8,390 cal yr BP), jack pine appeared in the area, as it did in other nearby locales. The spread of jack pine occurred in response to a greater frequency of fires near the end of the Hypsithermal dry period, prompted by an increased coniferous presence as conditions started to cool again. Spruce, birch, and alder trees became more abundant, suggesting a return to cooler and wetter conditions. Some forbs and herbs were now scarcer, and the forests became more closed. Peat moss became more abundant, suggesting peatland formation. In terms of species composition, these shifts essentially represent the establishment of the modern boreal forest.

These data indicate that the pollen record of Kearl Lake is in general agreement with those of the relatively nearby sites of Eaglenest Lake and Mariana Lake. However, this study has also demonstrated the existence of some important differences in the vegetation composition and the timing of various changes. Whereas at Eaglenest Lake, with its relatively high altitude, cooler and wetter conditions persisted throughout the Hypsithermal, the oil sands region became warmer and drier by about 9,800 ^{14}C yr BP (about 11,210 cal yr BP). Also, the change to more open, deciduous forests with greater forb and herb representation occurred some 800 years earlier in the Kearl Lake area than at Mariana Lake, 150 kilometres further south. Likewise, the onset of muskeg formation by about 7,100 ^{14}C yr BP (about 7,940 cal yr BP) in the oil sands area predates that of the Mariana Lake area by approximately a millennium. These differences may prove significant for the interpretation of the archaeological record of the oil sands region.

The Kearl Lake pollen record has some important implications for the understanding of human occupation in the region. In particular, the record indicates that upland areas were well vegetated in the terminal late Pleistocene and earliest Holocene. This suggests that, far from being a harsh and inhospitable landscape, the region would in all likelihood have offered a range of plant and animal resources, as does the northern boreal forest margin today.

Somewhat later in the early Holocene, the lowland areas supported a mix of deciduous trees and open herbaceous vegetation. This terrain might well have been attractive to groups of Early Prehistoric plains hunter-gatherers as well as to big game animals, who were perhaps finding the plains to the south increasingly inhospitable owing to the desiccating effects of the Hypsithermal. It is probably no coincidence that the number of sites in the oil sands area declines dramatically following the end of the Hypsithermal. It is against this background that Saxberg and Reeves's hypothesis (2003) regarding the remarkable abundance of Early Prehistoric archaeological sites in the oil sands region must be considered.

ACKNOWLEDGEMENTS

We thank Brian Ronaghan for his invitation to contribute to this volume. We are grateful to Komex Canada for providing the core, to Brandon Beierle for the radiocarbon dates, to Robin Woywitka, Amandah van Merlin, and Britta Jensen for drafting the maps, and to Timothy Fisher for clarifying the location of the Firebag Moraine. Luc Bouchet thanks Charles Schweger and Harvey Friebe for laboratory space and technical guidance, James White for comments on the pollen work, and the University of Calgary for funding support during his graduate studies.

REFERENCES

Bayrock, L. A.
  1971 *Surficial Geology, Bitumount, NTS 74E.* Map 34, Alberta Research Council, Edmonton. Available as Map 140, Alberta Geological Survey, Edmonton, http://ags.aer.ca/publications/MAP_140.html.
Beaudoin, Alwynne B.
  1993 A Compendium and Evaluation of Postglacial Pollen Records in Alberta. *Canadian Journal of Archaeology* 17: 92–112.
Beierle, Brandon D.
  1996 Holocene Environments of Kearl Lake, N.E. Alberta. Unpublished report prepared for Syncrude Canada Ltd. Copy on file, Syncrude Canada Ltd., Edmonton.
Birks, H. John B., and Sylvia M. Peglar
  1980 Identification of *Picea* Pollen of Late Quaternary Age in Eastern North America: A Numerical Approach. *Canadian Journal of Botany* 58: 2042–2058.
Bouchet-Bert, Luc
  2002 When Humans Entered the Northern Forests: An Archaeological and Palaeoenvironmental Perspective. MA thesis. Department of Archaeology, University of Calgary.

Brookes, D., and K. W. Thomas

    1967   The Distribution of Pollen Grains in Microscope Slides. Part I: The Non-randomness of the Distribution. *Pollen et Spores* 9: 621–630.

Dyke, Arthur S., Andrew Moore, and Louis Robertson

    2003   *Deglaciation of North America*. Geological Survey of Canada, Open File 1574, **doi: 10.4095/214399.**

Dyke, Arthur S., and Victor K. Prest

    1987   *Late Wisconsinan and Holocene Retreat of the Laurentide Ice Sheet*. Map 1702A, Geological Survey of Canada, Ottawa.

Farrar, John L.

    1995   *Trees in Canada*. Fitzhenry and Whiteside, Markham, Ontario, and the Canadian Forest Service, Ottawa.

Faegri, Knut, Peter E. Kaland, and Knut Krzywinski

    1989   *Textbook of Pollen Analysis*. 4th ed., revised. John Wiley, New York.

Fenton, Mark M., E. Joan Waters, Steven M. Pawley, Nigel Atkinson, Dan J. Utting, and Kirk Mckay

    2013   *Surficial Geology of Alberta*. Map 601. Alberta Geological Survey. http://ags.aer.ca/publications/MAP_601.html.

Fiedel, Stuart J.

    1999   Older Than We Thought: Implications of Corrected Dates for Paleoindians. *American Antiquity* 64(1): 95–115.

Fisher, Timothy G.

    2007   Abandonment Chronology of Glacial Lake Agassiz's Northwestern Outlet. *Palaeogeography, Palaeoclimatology, Palaeoecology* 246: 31–44.

Fisher, Timothy G., Nicholas Waterson, Thomas V. Lowell, and Irka Hajdas

    2009   Deglaciation Ages and Meltwater Routing in the Fort McMurray Region, Northeastern Alberta and Northwestern Saskatchewan, Canada. *Quaternary Science Reviews* 28: 1608–1624.

Habgood, Thelma, and E. P. Simons

    1985   *A Key to Pollen and Spores from Alberta*. 3rd ed. Palaeoenvironmental Studies Laboratory, Anthropology Department, University of Alberta, Edmonton.

Hickman, Michael, and Charles E. Schweger

    1996   The Late Quaternary Palaeoenvironmental History of a Presently Deep Freshwater Lake in East-central Alberta, Canada, and Palaeoclimate Implications. *Palaeogeography, Palaeoclimatology, Palaeoecology* 123: 161–178.

Hutton, Michael J., Glen M. MacDonald, and Robert J. Mott

    1994   Postglacial Vegetation History of the Mariana Lakes Region, Alberta. *Canadian Journal of Earth Sciences* 31: 418–425.

Ives, John W.

    1977   Pollen Separation of Three North American Birches. *Arctic and Alpine Research* 9(1): 73–80.

    1993   The Ten Thousand Years Before the Fur Trade in Northeastern Alberta. In *The Uncovered Past: Roots of Northern Alberta Societies*, edited by Patricia A. McCormack and R. Geoffrey Ironside, pp. 5–31. Circumpolar Research Series no. 3. Canadian Circumpolar Institute, University of Alberta, Edmonton.

Jacobson, George L., and Richard H. W. Bradshaw

    1981   The Selection of Sites for Paleovegetational Studies. *Quaternary Research* 16: 80–96.

Johnson, Derek, Linda Kershaw, Andy MacKinnon, and Jim Pojar

    1995   *Plants of the Western Boreal Forest and Aspen Parkland*. Lone Pine Publishing, Edmonton.

Lichti-Federovich, Sigrid

    1970   The Pollen Stratigraphy of a Dated Section of Late Pleistocene Lake Sediment Core from Central Alberta. *Canadian Journal of Earth Sciences* 7: 938–945.

Lowell, Thomas V., Timothy G. Fisher, Gary C. Comer, Irka Haidas, Nicholas Waterson, Katherine Glover, Henry M. Loope, Joerg M. Schaefer, Vincent Rinterknecht, Wallace Broecker, George Denton, and James T. Teller

    2005   Testing the Lake Agassiz Meltwater Trigger for the Younger Dryas. *Eos, Transactions, American Geophysical Union* 86(40): 365–373.

MacDonald, Glen M.

    1993   Methodological Falsification and the Interpretation of Palaeoecological Records: The Cause of the Early Holocene Birch Decline in Western Canada. *Review of Palaeobotany and Palynology* 79: 83–97.

    1987   Postglacial Vegetation History of the Mackenzie River Basin. *Quaternary Research* 28: 245–262.

    1989   Postglacial Palaeoecology of the Subalpine Forest-Grassland Ecotone of Southwestern Alberta: New Insights on Vegetation and Climate Change in the Canadian Rocky Mountains and Adjacent Foothills. *Palaeogeography, Palaeoclimatology, Palaeoecology* 73: 155–173.

MacDonald, Glen M., Chris P. S. Larsen, Julian M. Szeicz, and Katrina A. Moser

    1991   The Reconstruction of Boreal Forest Fire History from Lake Sediments: A Comparison of Charcoal, Pollen, Sedimentological, and Geochemical Indices. *Quaternary Science Reviews* 10: 53–71.

Maher, Lou J., Jr.

    1981   Statistics for Microfossil Concentration Measurements Employing Samples Spiked with Marker Grains. *Review of Palaeobotany and Palynology* 32: 153–191.

McAndrews, John H., Albert A. Berti, and Geoffrey Norris

    1973   *Key to the Quaternary Pollen and Spores of the Great Lakes Region.* Life Sciences Miscellaneous Publication. Royal Ontario Museum, Toronto.

McLeod, T. Katherine, and Glen M. MacDonald

    1997   Postglacial Range Expansion and Population Growth of *Picea mariana, Picea glauca* and *Pinus banksiana* in the Western Interior of Canada. *Journal of Biogeography* 24: 865–881.

McPherson, R. A., and C. P. Kathol

    1977   *Surficial Geology of Potential Mining Areas in the Athabasca Oil Sands Region.* Open File Report 1977-04. Alberta Research Council, Edmonton.

Moore, Peter D., J. A. Webb, and Margaret E. Collinson

    1991   *Pollen Analysis.* 2nd ed. Blackwell Scientific Publications, Oxford.

Moss, Ezra H., and John G. Packer

    1983   *Flora of Alberta.* 2nd ed. University of Toronto Press, Toronto.

Natural Regions Committee

    2006   *Natural Regions and Subregions of Alberta.* Compiled by David J. Downing and Wayne W. Pettapiece. Government of Alberta Publication no. T/852. https://www.albertaparks.ca/media/2942026/nrsrcomplete_may_06.pdf.

Prather, Colleen, and Michael Hickman

    2000   History of a Presently Slightly Acidic Lake in Northeastern Alberta, Canada, as Determined Through Analysis of the Diatom Record. *Journal of Paleolimnology* 24(2): 183–198.

Prior, G. J., B. Hathway, P. M. Glombick, D. I. Pană, C. J. Banks, D. C. Hay, C. L. Schneider, M. Grobe, R. Elgr, and J. A. Weiss

    2013   *Bedrock Geology of Alberta.* Map 600, Alberta Geological Survey. http://ags.aer.ca/publications/MAP_600.html.

Ritchie, James C.

    1984   *Past and Present Vegetation of the Far Northwest of Canada.* University of Toronto Press, Toronto.

Ritchie, James C., and Glen M. MacDonald

    1986   The Patterns and Post-glacial Spread of White Spruce. *Journal of Biogeography* 13: 527–546.

Rutter, Nathaniel W., Andrew J. Weaver, Dean Rokosh, Augustus F. Fanning, and Daniel G. Wright

    2000   Data Model Comparison of the Younger Dryas Event. *Canadian Journal of Earth Sciences* 37: 811–830.

Saxberg, Nancy, and Brian O. K. Reeves

    2003   The First Two Thousand Years of Oil Sands History: Ancient Hunters at the Northwest Outlet of Glacial Lake Agassiz. In *Archaeology in Alberta: A View from the New Millennium,* edited by Jack W. Brink and John F. Dormaar, pp. 290–322. Archaeological Society of Alberta, Medicine Hat.

Schweger, Charles E., and Michael Hickman

    1989   Holocene Paleohydrology, Central Alberta: Testing the General-Circulation-Model Climate Simulations. *Canadian Journal of Earth Sciences* 26: 1826–1833.

Smith, Derald G., and Timothy G. Fisher

    1993   Glacial Lake Agassiz: The Northwestern Outlet and Paleoflood. *Geology* 21(1): 9–12.

Stockman, Jens

    1971   Tablets with Spores Used in Absolute Pollen Analysis. *Pollen et spores* 13: 615–621.

Vance, Robert E.

    1986   Pollen Stratigraphy of Eaglenest Lake, Northeastern Alberta. *Canadian Journal of Earth Sciences* 23: 11–20.

Vance, Robert E., Alwynne B. Beaudoin, and Brian H. Luckman

    1995   The Paleoecological Record of 6 ka BP Climate in the Canadian Prairie Provinces. *Géographie physique et Quaternaire* 49: 81–98.

Weninger, Bernhard, Olaf Jöris, and Uwe Danzeglocke

    2005   CalPal: University of Cologne Radiocarbon Calibration Program Package. http://www.calpal-online.de/.

White, James M., and Rolf W. Mathewes

    1982   Holocene Vegetation and Climatic Change in the Peace River District, Canada. *Canadian Journal of Earth Sciences* 19: 555–570.

    1986   Postglacial Vegetation and Climatic Change in the Upper Peace River District, Alberta. *Canadian Journal of Botany* 64: 2305–2318.

Yansa, Catherine H.

    2006   The Timing and Nature of Late Quaternary Vegetation Changes in the Northern Great Plains, USA and Canada: A Re-assessment of the Spruce Phase. *Quaternary Science Reviews* 25: 263–281.

Zoltai, Stephen C., and Dale H. Vitt

    1990   Holocene Climatic Change and the Distribution of Peatlands in Western Interior Canada. *Quaternary Research* 33: 231–240.

2 Human History

5 The Early Prehistoric Use of a Flood-Scoured Landscape in Northeastern Alberta

GRANT M. CLARKE, BRIAN M. RONAGHAN, AND LUC BOUCHET

Archaeological investigation in Alberta's oil sands region began in the early 1970s but proceeded intermittently until the late 1990s, when the price of synthetic crude made the widespread development of one of the world's largest reserves of petroleum economically attractive. From the outset, a small area within the larger Athabasca oil sands region exhibited a relatively dense distribution of prehistoric sites in comparison to other areas in Alberta's forested north. Of note in the results of the earliest studies (see, for example, Syncrude Canada Ltd. 1973, 1974; McCullough and Reeves 1978) was the occurrence of considerable numbers of prehistoric sites situated in forested terrain at some distance from major river systems and lakeshores, contrary to expectations for boreal forest environments. As investigations increased in their extent in concert with proposed development projects, and as intensive shovel testing began to be employed as a site discovery technique, a rich pattern of prehistoric use of a particular series of terrestrial landforms started to emerge. In this chapter, we will review the available geological and palaeoenvironmental information in an effort to understand the basis for, and the chronology of, this remarkable prehistoric record and to present an alternative to previously advanced interpretations of the human use of the area.

Alberta's oil sands region lies fully within the boreal forest and, over most of its breadth, exhibits archaeological site distributions that are characteristic of similar areas throughout western Canada's middle latitudes. As in other boreal forest regions, archaeological investigations in this area face many challenges.

The dense forest and muskeg cover inhibit site discovery, the acidic nature of regional soils frequently removes most or all of the organic materials that often accompany human occupations, and the long-term stability of the forested landscape inhibits sediment buildup, so that a clear stratigraphic separation of archaeological deposits is a rare circumstance. All of these factors place limitations on the ability of archaeologists to discover evidence of prehistoric human occupation and to develop detailed interpretations of the landscape use reflected by this occupation.

Despite these limitations, archaeological fieldwork, which has been conducted in the oil sands region primarily in connection with development, has been relatively successful in identifying evidence of prehistoric human use of the area. Of the three formally defined oil sands administrative areas (see fig 13.1 in this volume), the Athabasca Oil Sands Area has received the most intensive study, given that it is the area within which surface-minable bitumen deposits occur. Although portions of the area have yet to receive detailed study, archaeological investigations have now been undertaken across most of it, and, to date, more than 2,500 archaeological sites have been recorded. As figure 5.1 illustrates, these sites form an unusually dense cluster in the central portion of the surface-minable area.[1]

Archaeological sites identified in studies conducted outside the relatively confined surface-minable zone (see, for example, Balcom 1996; Bouchet-Bert 2003, 2005, 2007; Clarke 1998, 2000; Green 2000, 2001, 2003; McCullough and Wilson 1982; Meyer 2000; and Unfreed and Blower 2005) generally appear to represent localized, short-term, task-specific occupations by small, dispersed populations, a pattern initially discussed by Clarke (2002). For the most part, the results of these studies align with what we would expect for areas throughout the Canadian boreal forest, in which patterns of prehistoric occupation reflect a focus on the use of lakeshores and drainage systems. These studies have also covered an extremely broad range of territory, as they were designed to assess the impact of extensive exploratory activities (needed to determine the depth and content of bitumen-bearing formations) and widely dispersed production facilities.

In view of these studies, we are convinced that had concentrated site distributions comparable to those within the surface-minable zone been present beyond this area, they would have been recognized. In all likelihood, the extraordinary number of archaeological sites in that zone represents the densest concentration of sites yet identified in the Canadian boreal forest region. Moreover, as development proceeds, bringing with it the need to extract detailed excavation samples from sites that lie within in the Muskeg River basin, to the east of the Athabasca River, as well as to its west, along the eastern flanks of the Birch

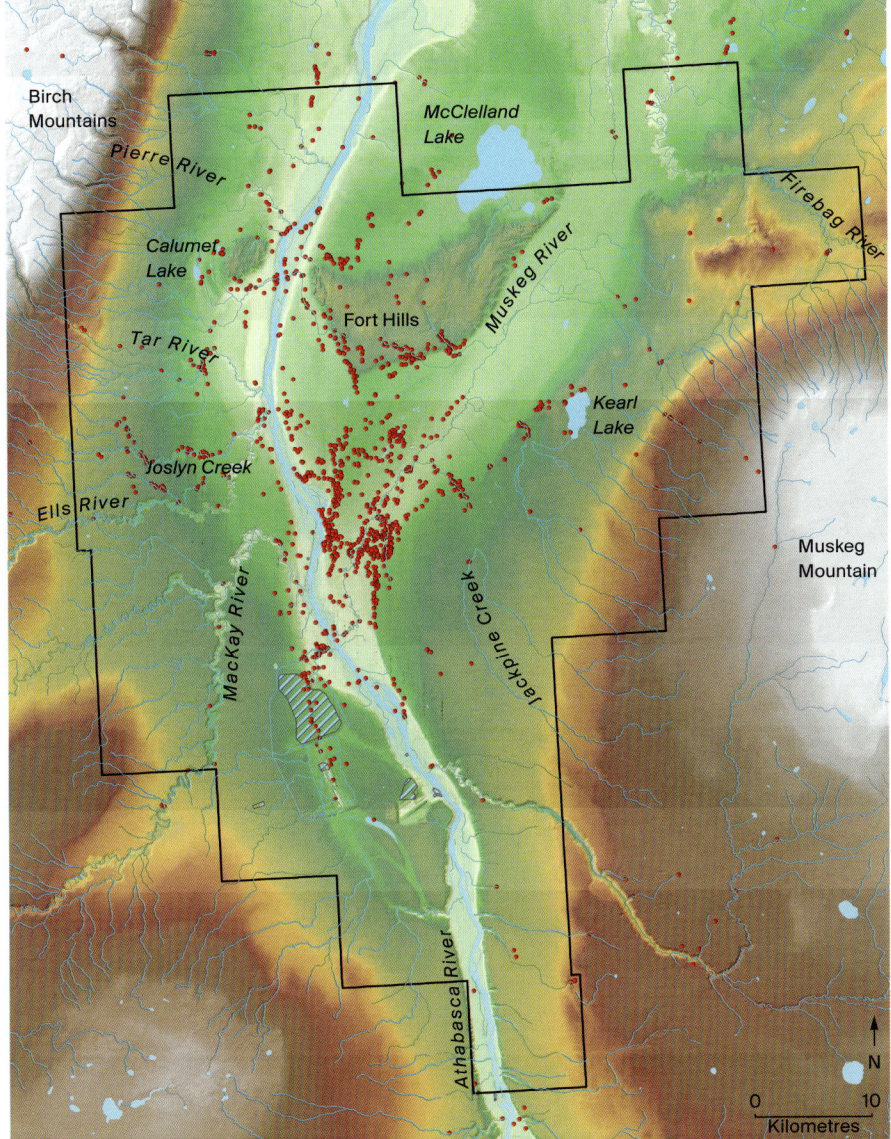

Figure 5.1. Archaeological sites recorded to date in the surface-minable oil sands area

Mountains, information is emerging that suggests that the central portion of surface-minable zone may be unique in western Canada with respect to its record of environmental and human history.

Key to understanding this emerging picture are the events surrounding the retreat of the Wisconsinan ice sheets from Canada's western interior. Aspects of these events and their environmental and cultural implications are discussed in several chapters in this volume, and a preliminary interpretation has already been presented by Saxberg and Reeves (2003). Earlier research (Clarke and Ronaghan 2001) suggested, however, that the time-transgressive land use model they proposed had significant limitations. The present chapter follows up on that earlier research, which was directed toward developing an alternative model. In what follows, we hope to clarify the basis for our views on the factors that might explain this exceptional prehistoric pattern of land use.

LANDFORM GENESIS

Archaeologists and anthropologists have long recognized the intimate relationships between hunter-gatherer societies and the environments in which these peoples live.[2] Landscape structure, climate, and the seasonal distribution of resources profoundly condition the character of land use and the distribution of hunter-gatherer populations within a given region. Although temporal variations in each of these factors have played a role in the archaeological record of the oil sands region, most influential in our view were the remarkable geological events of the early postglacial period, which created a landscape that stood apart from the surrounding region and was for millennia the locus of intense prehistoric human land use.

Geologically, the oil sands region lies at the northeastern extent of the Western Canadian Sedimentary Basin, where it abuts the Canadian Shield (Carrigy 1973). Here, the Precambrian basement is overlain by Devonian-age limestone and then by the Cretaceous sediments of the McMurray Formation (Carrigy and Green 1965). In addition to bitumen-impregnated sands, these latter sedimentary layers include metamorphosed sandstones suitable for the manufacture of stone tools and have thus been of key importance not only as a modern source of energy but also for prehistory. A constituent of the McMurray Formation known as Beaver River Sandstone is the source of the vast majority of the artifacts found in the region, and the availability of this material, as well as the nature and distribution of artifacts manufactured of it, is crucial to tracing the evolution of the pattern of prehistoric land use visible in the region.

As chapters 1 and 2 in this volume demonstrate, regional glacial events, especially during their terminal stages, have had a critical influence on both previous and current land use. As is the case in much of Canada, most of the region is blanketed by till consisting of unsorted gravels silts and clays left in the wake of retreating ice sheets (Bayrock 1971), creating a moderately undulating surface. Subsequent modification of these deposits by slope wash and alluvial action has tended to reduce surface variation and to increase the proportional surface expression of the finer elements of the tills, thus reducing drainage capacity. The fact that fine glaciolacustrine silts are not widespread within the area (Bayrock 1971) suggests that long-lived impoundments of outwash water were not a prominent feature of glacial retreat in this area.

Outwash events, relating to the drainage of waters that had accumulated along the margins of retreating ice sheets, did, however, modify surficial deposits within the minable oil sands area, creating a landscape that differs significantly from the surrounding region. In chapter 1 of this volume, Burns and Young, following Rains et al. (1993), describe the catastrophic drainage of pressurized meltwater beneath the Laurentide Ice Sheet in the early stages of its retreat, prior to the flooding of Glacial Lake Agassiz. This water is said to have coursed southward along the existing Athabasca River valley, eroding a wide channel and stripping some of the Pleistocene ice age deposits from the area (see fig 1.2). A mapping of existing bedrock topography (Carrigy and Green 1965) also reveals an absence of later Cretaceous-age deposits along this path, depicting instead a near-surface expression of McMurray Formation oil-impregnated sandstone— which, as Burns and Young note, has benefitted modern bitumen recovery by making surface mining financially viable. We would further note the benefits to prehistoric peoples of the near-surface occurrence of Beaver River Sandstone, which is a component of the McMurray Formation. (For further discussion of Beaver River Sandstone, see chapters 9 and 10 in this volume.)

As the Laurentide Ice Sheet retreated to the east of the oil sands region, its blockage of natural drainage channels resulted in the accumulation of extremely large lakes along its front, the largest of which is now known as Glacial Lake Agassiz. Research by Fisher and Smith (Fisher 1993; Smith and Fisher 1993; Fisher and Smith 1994) has provided concrete evidence of a massive flood event at one stage in the history of this lake, an event that fundamentally altered the oil sands landscape and provided the basis for subsequent development of the unique prehistoric land use pattern that is the subject of this chapter. This flood occurred as a moraine or ice dam in northwestern Saskatchewan was breached and vast quantities of water sped westward down the current Clearwater River valley and then northward along the Athabasca valley, where flood waters overtopped the valley walls before entering another large lake (Glacial Lake McConnell), north of the Fort Hills (fig 5.2).

As Fisher and Lowell report in chapter 2, this event took place somewhere between 9,800 and 9,600 ^{14}C yr BP and was followed by steady-state flow for a few hundred years, with abandonment of the outlet occurring in the period between 9,590 and 9,070 ^{14}C yr BP (Fisher 2007). The magnitude of this flood is

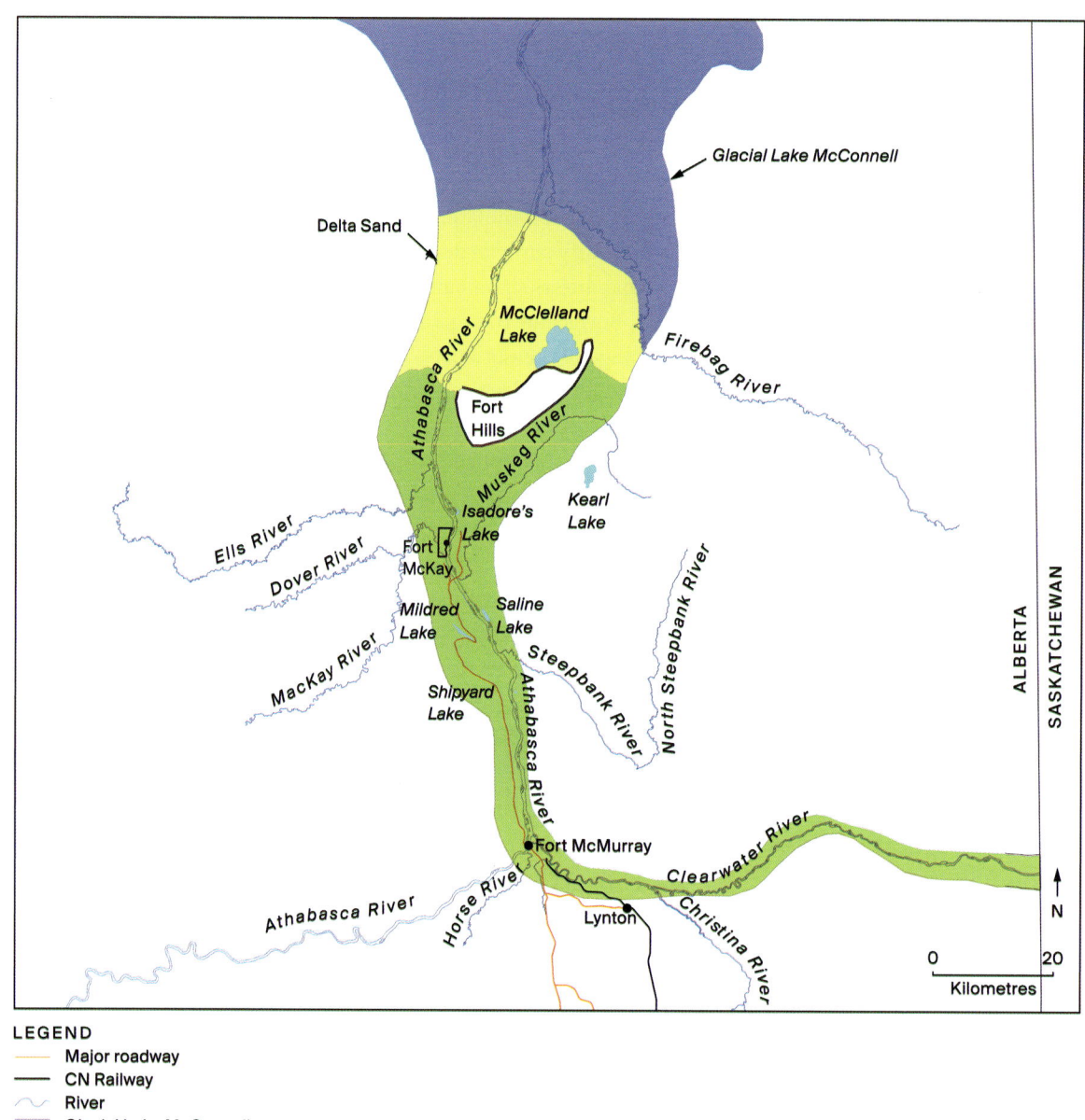

Figure 5.2. The Glacial Lake Agassiz flood zone. Adapted from Smith and Fisher 1993, figure 1.

startling in modern terms. Smith (1989) has estimated that, during a 78-day period, more than 1,000 cubic kilometres of water passed through the study area each day, lowering Glacial Lake Agassiz by 46 metres. During this period, the Clearwater River valley was incised to a width of 2.5 to 30 kilometres and deepened by 200 metres (Smith 1989). The Athabasca River valley in the vicinity and downstream of Fort McMurray also experienced significant scouring.

The initial stage of the flood was so erosional that all of the glacial sediments, and perhaps remnants of unconsolidated late Cretaceous sediments, would have been scoured from within its path. North of Fort McMurray the flood waters overtopped the Athabasca valley, bifurcating around the Fort Hills and spilling over the lowlands of the Muskeg River valley to the east as well as the lower portions of the rivers draining the Birch Mountains (fig 5.3). Flood waters slowed as they backed up, entering Glacial Lake McConnell, north of the Fort Hills, which occupied the current Lake Athabasca basin and surrounding terrain.

As the waters slowed, a massive delta formed, known as the Late Pleistocene Athabasca braid delta (Rhine and Smith 1988). The deposits released in this process show distinctive fining upward from south to north and over time. Large boulders and gravel were deposited along the margins of the Athabasca River valley north of its junction with the Clearwater River, while a series of gravel ridges, or bars, and intervening channels, representing a large-scale version of the braided channels seen in mountain streams, formed where flows diverged around the Fort Hills in the Muskeg River valley (figs. 5.4 and 5.5). These latter deposits are of critical importance for the prehistoric land use patterns discussed in this chapter. During the later stages of inundation, these braided channel and point bar deposits, as well as ones located further south, were draped by an upward-fining series of sands. North of the Fort Hills, a massive deltaic sand deposit filled the previously scoured valley, forming a level plain that stretches to the Richardson Lake area.

Within the central portion of the minable oil sands area, which includes the Muskeg River valley, to the east of the Athabasca River, and portions of the lowest slopes of the Birch Mountains, west of the Athabasca, the post-flood landscape can be characterized as one formed initially by the scouring action of a massive outwash flood, then partially filled with outwash gravels in the form of linear-oriented ridge and channel features, and finally draped with sand. As the glacial outwash waters receded, this landscape would have stood in dramatic contrast to the level, forested till plain that surrounded it. The minable oil sands area, as defined by the Alberta government (see fig 5.1), extends somewhat beyond the Glacial Lake Agassiz flood zone but for the most part is coincident with it. Portions that extend outside the flood zone to the south and east may

Figure 5.3. The maximum extent of the flood and associated scouring in the Athabasca lowlands

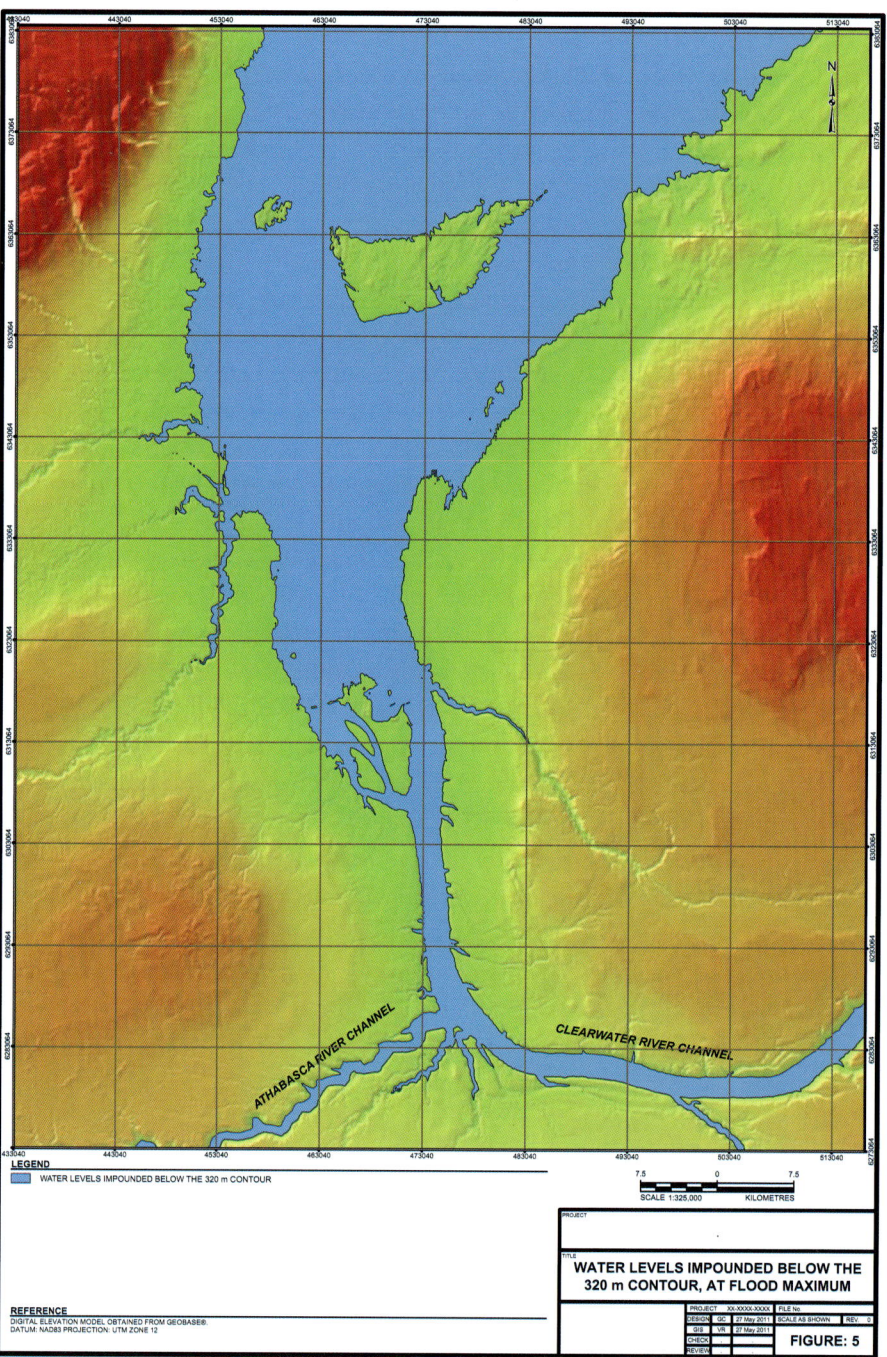

Figure 5.4. Aerial view of present-day braided channel deposits in the Muskeg River basin

have been scoured by earlier subglacial outwash activity that Burns and Young discuss in chapter 1, but they remain relatively featureless and lack the granular fill deposited by the Lake Agassiz event.

The scouring action of flood waters would have served to decrease the depth of overburden on the Cretaceous McMurray Formation bitumen-bearing deposits in the region, thereby increasing their accessibility for surface mining. But it also served to create greater exposure of one of the McMurray Formation's lower members, which contains Beaver River Sandstone, the principal source of workable stone for tool manufacture by prehistoric hunting populations. Much has been written about the stratigraphic origin of this material and its near-surface expression, which will not be recounted here. Suffice to say that this material occurs in situ in variable quality, typically along the eroding margins of the Athabasca River and the lower courses of the tributaries that feed it. Elsewhere in the flood zone, as testimony to the intensity of the flood, Beaver River Sandstone has been noted to occur on the surface as dispersed boulders of varying size (Saxberg and Reeves 2003).

Perhaps the most readily available exposure of high-quality workable stone is located inland of the Athabasca in the Muskeg River basin along the eastern margin of the flood zone, where Devonian limestone rises to a near-surface expression. Here, Beaver River Sandstone has been recognized on the now forested ridge tops across an area of over 100 hectares now known as the Quarry of the Ancestors. Given this area's presence within the flood margins, we believe it likely that prehistoric people using this landscape would have easily recognized

Figure 5.5. Aerial view of modern ridge and channel topography

the exposure of this material and integrated it into their use of the surrounding region.

EARLY ECOLOGY OF THE FLOOD ZONE

Climate and Vegetation

The characteristics of past environments can be inferred through the application of a variety of analytical techniques. Most commonly used are analyses that focus on the sediments that accumulate at the bottoms of lakes and ponds. Microscopic and macroscopic remains of surrounding environments—many of which provide indications of the character of local and regional vegetation communities and, inferentially, the climatic conditions that produced them—can remain preserved in such sediments, and radiocarbon dating of the organic elements in these sediments can yield chronologically ordered sequences within which to situate further interpretation. Among these forms of evidence, one of the most useful is fossil pollen, some varieties of which are extremely resistant to degradation and provide a fine-grained record of past vegetation communities surrounding a collection basin. One of the challenges involved in pollen analysis centres on variations in the preservation rates of specific taxa, which may cause certain taxa to be under-represented in the pollen record. Another stems from the possibility that air-borne pollen may have been carried long distances from its source (see Lichti-Federovich and Ritchie 1968; Ritchie and Lichti-Federovich 1967). Naturally, palaeoenvironmental interpretations of the pollen record must take such factors into account.

In chapter 4, Bouchet and Beaudoin provide a comprehensive discussion of the palaeoenvironmental record that emerges from a sediment core recovered from Kearl Lake, which lies a few kilometres to the east of the flood zone, in the Muskeg River basin. In determining the relevance of this evidence for the flood zone, one must remember that the Kearl Lake data reflect elements of both the immediate environs of the lake and the surrounding region (anywhere from 100 metres to more than 1,500 kilometres away), including forested terrain unaffected by the flood. Although an exact correspondence with the flood zone is perhaps lacking, the palaeoenvironmental record revealed in the Kearl Lake core is inferentially valuable to any discussion involving the environment of the study area at the time of its most intense human occupation, as climatic conditions would be directly comparable and many elements of flood zone vegetation communities would be represented in the Kearl Lake core.

One of the single most important of the factors that serve to explain the rich archaeological finds in the study area is the fact that the flood event coincided with the end of the cooler and wetter Younger Dryas interval and the beginning of the warmer and dryer Hypsithermal, about 10,000 14C yr BP. By the time of the flood, around 9,800 14C yr BP, stands of spruce forest, which replaced the original sparse herb and shrub community that followed the retreating ice sheets, had in turn largely disappeared. As Zone K-2 of the Kearl Lake core indicates, roughly between 9,820 and 7,580 14C yr BP, increasingly xeric growing conditions resulted in a great reduction of spruce cover and ushered in a new vegetation regime consisting largely of deciduous trees with an increased presence of various herbaceous species. The warmer and drier climate caused lakes to shrink, producing higher concentrations of cattails around shallow water basins, and allowed greasewood (which today grows around saline flats in the grasslands of southern Alberta) to extend its range into the study area. These conditions were likely more prevalent at lower elevations, such as those within the flood zone, than in nearby upland areas such as the Birch Mountains (see Vance 1986).

While the surrounding region's poorly drained clay soils would presumably have supported open-forest mixtures of birch, alder, and willow, with some spruce cover, the areas of well-drained sandy soils left behind by the flood probably supported vegetation communities that reflected these more xeric microenvironments, which may not be accurately represented in the Kearl Lake core.

The Athabasca Sand Dunes of northern Alberta and Saskatchewan, not very far from Alberta's oil sands region, provide an example of landform and drainage characteristics that may not be very different from those left in the wake of the Lake Agassiz flood. Although describing relatively modern conditions, ecological baseline studies conducted in this environment (Raup and Argus 1982) may provide a useful analogy for the Hypsithermal-era Lake Agassiz flood zone, between approximately 9,900 and 7,500 14C yr BP. These studies show that while open jack pine forests, which feature limited undergrowth and a floor dominated by lichen, today commonly grow on the stabilized dune surfaces, areas of actively blowing sand support grasses as the most common vegetation (Raup and Argus 1982, 53–54). Taxa identified include brome, fescue, wheat, and reed grass, with the largest number of taxa recognized in areas between dunes known as "slacks," particularly those with bottoms at or near the water table.

The jack pine forest that currently dominates well-drained ridges throughout the region was not present during the early postglacial Hypsithermal interval. In fact, it is thought that jack pine migration into the region did not occur until the end of this period, roughly around 7,580 14C yr BP (see chapter 4), and then it is believed to have begun along fluvial systems, with interfluves such as the flood zone being colonized later (Raup and Argus 1982). The Kearl Lake data suggest that, during the Hypsithermal, the surrounding landscape would have supported birch, alder, and aspen in relatively open communities. Within the flood zone, however, the highly porous substrate, coupled with higher temperatures and lower moisture regimes, may have favoured a largely grassland environment. These factors, as well as the susceptibility to wind erosion of the sand that blankets these features, probably contributed to a delay in the establishment of forest communities until later in the period, and even then an open aspect would have been favoured.

Intervening channels, equivalent to slacks in dune areas, are likely to have contained meadow-like vegetation, initially grasses such as those identified above and herbaceous plants such as greasewood, sand heather in dry areas, and cattails and other sedges where surface water was present. However, as indicated above, during this period surface water may have been limited largely to flowing streams.

As Hypsithermal conditions decreased in intensity, a trend toward moister, cooler conditions set in between 7,700 and 7,200 14C yr BP, eventually resulting in ecosystems more typical of the modern boreal forest. As this trend continued, coniferous arboreal species became more prevalent components of forested communities. Although perhaps delayed to some degree by the flood zone's interfluvial aspect, it is likely that jack pine and lichen communities began replacing the open deciduous and herbaceous communities on the well-drained bar features, perhaps in response to an increased frequency of fire as the Hypsithermal waned and forests became more closed (see chapter 4).

At Kearl Lake during this period, sphagnum, which occurred only in trace amounts earlier, began to occur in greater abundance, indicating that muskegs may have started forming in the channels between bars. As well, it is likely that spruce, tamarack, and other coniferous taxa, indicative of moister conditions, began to establish themselves in less well-drained locations such as bar slopes.

These developments were probably gradual, with muskegs forming in channels as increased levels of moisture caused water tables to rise. This process was accretive, as revealed in the presence of artifact deposits that extend off the sides of bar features, under peat deposits that have subsequently advanced into these areas. Despite the present lack of data, it seems reasonable to assume that peat formation began slowly, accelerated as modern moisture regimes were achieved, and may then have stabilized during the latter part of the Holocene. Modern development planning research indicates that muskegs reach depths of 1.8 metres in the channels between bar features within the main part of the lower Muskeg River basin (Golder Associates 1997). As well as creating distinctly different ecosystems, this growth would have reduced the topographic variation initially present in the post-flood landscape.

In such environments, shrubs such as bog willow, dwarf birch, leatherleaf, and Labrador tea, as well as a wide range of other vascular plants supported by sphagnum, would have replaced the grasses and herbaceous taxa characteristic of the Hypsithermal era. Further, in areas close to the water table, the small ponds currently present in various parts of the basin would have started forming. Here, grass and sedge meadows were likely to have developed around the margins of the open water, characterized by taxa such as horsetail, bur reed, rough bent grass, reed grass, spike rush, water sedge, and other types of sedges.

In summary, while evidence from some upland sites, such as the Birch Mountains, suggests that postglacial ecosystems reflect a relatively continuous development of the coniferous-dominated boreal forest, lower elevation locations, such as Kearl Lake, provide evidence of warmer, drier conditions more typical of areas further south during the Hypsithermal climatic maximum between 9,800 and 7,580 ^{14}C yr BP.[3] We can infer that relatively more open forest existed at lower elevations, consisting largely of deciduous arboreal taxa and various herbs and forbs (see chapter 4). We have also considered the differences between the gravel-cored, sand-draped landforms of the Lake Agassiz outwash braid channel and the surrounding till-based landscape. We conclude from this comparison that the enhanced drainage characteristics of the braid channel are likely to have resulted in recolonization during this period by vegetation communities that, while they may have few modern analogues, would reflect the relatively xeric conditions present in this microenvironment. We have also used descriptions of the vegetation community in the Athabasca Sand Dunes north of the oil sands region as a potential analogue for what we expect would have been an open grass-and-herb-dominated landscape with deciduous arboreal species present in open communities in more stable topographic situations and possibly delayed in their appearance.

As the climate began its trend toward cooler, moister conditions more typical of modern environments, coniferous taxa were favoured and began to occur in greater proportions, forests became increasingly more closed, and muskegs began to develop in lower topographic locations. Although this trend began around 7,500 ^{14}C yr BP, its effects would have developed gradually and may have been restricted to some degree by an increased frequency of fire. As a result, the predicted contrast between the flood zone ecosystems and the surrounding forest may have been maintained for some time after the initiation of more modern climatic regimes.

Animal Species

Owing to the acidic nature of regional soils, direct evidence of the animals that would have been available to prehistoric hunter-gatherers is essentially absent. Bone does not preserve well in the near-surface circumstances that contain the remains of prehistoric occupations in the region. This problem, which is pervasive throughout the boreal forest, poses a significant obstacle to our understanding of prehistoric lifeways in the north. For the most part, archaeologists are forced to draw inferences from existing or predicted environmental parameters, modern wildlife information, and the character of the stone tool assemblages recovered in archaeological investigations. However, other interpretive avenues do exist. For example, with the help of forensic techniques, some archaeologists have been able to identify and classify blood residue remaining on the edges of stone tools. The accuracy of the taxonomic identifications and the strength of the association between residues and the tools on which they occur have, however, been questioned in the archaeological literature (see, for example, Fiedel 1996). As result, inferences about the role of animals in prehistoric lifeways must be made with caution.

As of 2009, multiple artifacts recovered from twenty-one sites throughout the flood zone had provided positive reactions to blood antisera. Species identified in order of frequency include deer, bovid (bison), caribou, rabbit (local hare), rodent (probably beaver or muskrat) moose, bear, cat (lynx, cougar, or bobcat), sheep, chicken (most likely grouse), dog (wolf, coyote, or fox), striped bass (probably walleye or pike), and elephant.

The occurrence of elephant in the area is unexpected and, if it is not a misidentification, may represent a late survival of mastodon in the region (see chapter 6, note 10, and Saxberg 2007), on the assumption that, with the exception of the flood zone, the postglacial regional environment was largely forested and thus would have favoured browsers rather than grazers. The latest date for other elephant remains in Alberta is 10,240 ± 325 ^{14}C yr BP, for a mammoth from an area in the central foothills of the Rocky Mountains, not far north of present-day Calgary (see chapter 1 in this volume), one that would have been deglaciated several hundred, perhaps a thousand, years earlier. The residue evidence from the oil sands comes from within the flood zone and so probably postdates 9,600 ^{14}C yr BP. Given the relatively open forest that we would predict to have existed at the time, it is not inconceivable that remnants of the ice age megafauna populations could have survived for a short time after the outwash event. DNA evidence emerging from Alaska supports the survival of Pleistocene species such as mammoth as recently as 7,600 years ago (Haile et al. 2009).

As noted in chapter 6, a knife found at a site not far from the Birch Mountains reacted positively to sheep antisera. The presence of sheep is intriguing, as the area lies well outside the current range of bighorn sheep. Dall, Stone, and bighorn sheep are all acknowledged to be adapted to the rugged terrain along the Rocky Mountain chain from Alaska to Mexico (Geist 1971; Krausman and Bowyer 2003). The precipitous topography provides the necessary escape ground needed especially by females during breeding season. Consequently, one would not expect bighorn sheep to represent a common species in the oil sands region, except perhaps in immediate postglacial times. Moreover, an isolated identification of sheep should not be granted a great deal of significance, given the uncertainty associated with the analysis techniques employed.

Some of the surviving residues may be the product of the use of sinew to tie a stone tool to a wooden or bone handle. Such an interpretation has been advanced to explain the high frequency of deer, rabbit, and canids (see chapter 6). It is evident, however, that the deer family, which includes moose and elk as well as mule and whitetail deer, formed a major food source for prehistoric hunting groups throughout North America. Modern wildlife surveys (see, for example, AXYS Environmental Consulting Ltd. 2005) count deer, both whitetail and mule, among the most common ungulate species recorded in predevelopment assessment in the minable oil sands area. However, ethnographic information and traditional ecological knowledge suggest that when other sources of meat are available, Aboriginal populations do not favour whitetail or mule deer as a food source.[4] One hesitates to ascribe such preferences to prehistoric populations, but we would predict that deer and possibly elk would have been continually present among the ungulate species available to prehistoric hunters within the flood zone.

It is interesting to note that elk are not mentioned as a traditional food resource by local Aboriginal communities today (Fort McKay Environment Services 1997), perhaps because elk have not been seen in the region for many decades (Fort McKay Environment Services 1996a, 1996b). However, given that elk prefer areas of woodland mixed with open grassland, such as forest edges and mountain meadows, and were once extremely widespread, the predicted habitat throughout the flood zone during much of the Early Prehistoric record would in all likelihood have supported this species. Their herding tendencies and large body size would have made them attractive for group hunting activities. The blood residue evidence for the oil sands region is distinguished minimally with regard to family, but some genus identifications have been made, and although moose, caribou, and deer have been recognized, so far elk has not. As the recovery of an elk skull dating to approximately 5,550 ^{14}C yr BP from the Agassiz flood zone indicates, however (see chapter 1), this species was regionally present toward the end of the Hypsithermal climatic maximum.

Moose is relatively commonly identified in the results of residue analysis, although not to the extent that would be expected if modern Aboriginal consumptive patterns applied in prehistory. The Fort McKay study within the oil sands region (Fort McKay Environment Services 1997) indicates that the consumption of moose exceeds that of all other species. Given its extensive northern range and tolerance for extremes in climate, moose was most likely present in the minable oil sands area throughout the full range of prehistory. Moose are browsers that prefer boreal habitats, which include lakes and ponds, and subsist largely on tender leaves and twigs and aquatic plants. Although their large body weight provides considerable return for the hunter, their solitary habits and widely dispersed populations make their exploitation a less likely explanation for the intense patterns of human use reflected by the vast quantities of stone tool manufacturing debris found in the minable oil sands area. This may be especially true for the predicted Hypsithermal ecology of the flood zone, with its reduced forest cover and grassland ecosystems. However, because of its desirability, moose would presumably have been hunted whenever encountered.

Caribou is represented in several of the blood residue results and is also mentioned in the Fort McKay traditional food consumption study as a desirable species. Genetically, two subspecies, barren-ground and woodland, are distinguished (Miller 2003). Although this nomenclature reflects distributional variation and consequential behavioural differences, some overlap between the two subspecies has been noted (Geist 1982). As we saw earlier, open jack pine forest, with its lichen-dominated sub-storey, is likely to have become established in the flood zone only toward the end of the Hypsithermal, after it was already widespread in the surrounding regions. Because of their preference for and ability to subsist on lichens, caribou may therefore have been one of the most prevalent seasonally available big game in the later vegetational expression of the predicted Hypsithermal ecosystem of the flood zone.

Caribou tendencies toward herding behaviour and seasonal movement may have also been attractive to prehistoric hunters, providing opportunities for high-return communal intercept-and-kill strategies. Although we cannot confirm that use of the linear ridges in the abandoned braid delta of the flood zone figured in such strategies, it is considered highly probable. These opportunities would probably have been reduced if local herds adopted the more sedentary habits of the modern woodland species (Stuart-Smith et al. 1997). However, an Aboriginal informant reported to Preble just after the turn of the twentieth century that barren-ground caribou were once known to extend their winter migration range to the area of Fort McMurray (Preble 1908, cited in Soper 1964, 359). The distribution of the species during the early Holocene is, however, a matter of speculation.

The blood residue results reported for flood zone archaeological sites include frequent reactions to bovid antisera, indicating the presence of bison (the sole regional representative of the bovid family in prehistoric times). Following the almost total extirpation of bison in the late nineteenth century, survivors were placed under protection in Wood Buffalo National Park, north of the oil sands region. Accordingly, the only mention of bison in the Fort McKay consumption study relates to the distribution of meat after the death of an escaped member of the Syncrude experimental herd. Nevertheless, elsewhere in Alberta, where faunal material is preserved along with archaeological artifacts, bison appear to be the overwhelming preference of many prehistoric cultures. In fact, many cultural expressions on the Plains in historic and, inferentially, prehistoric times were focused around exploitation of bison as the principal food resource.

For example, the Cody Complex—a widespread early postglacial cultural entity (see, for example, Forbis 1968; Frison and Todd 1987) that is strongly represented in flood zone archaeological sites (Ronaghan 2005)—is commonly associated with large-scale communal bison killing throughout the Plains region. Although bison, as grazing animals, are typically found in Plains environments, their tolerance for temperature extremes and their ability to subsist on sedges as well as grass, as demonstrated in Wood Buffalo National Park, indicates that bison would have been present throughout the subarctic regions of northwestern Canada through to historic times (Gates, Chowns, and Reynolds 1992; see also Guthrie 1980).

Although boreal forest sites have produced only blood residue evidence for bison exploitation, historical records suggest that, prior to their near extermination, bison were relatively plentiful in the area around Fort McMurray and were hunted extensively by regional Aboriginal groups. For example, Roe (1951, 304) cites several accounts, the earliest provided by Dr. Richardson of Franklin's first expedition (1819–22), in which the Lower Athabasca valley between Lake Athabasca and the Clearwater junction at Fort McMurray is described as "well wooded and frequented by buffaloes." Similarly, in 1841–42, George Simpson, governor of the Hudson's Bay Company, reported of the area around Fort McKay that it "abounds in Buffalo and Deer" and that "the Indians about there desired to devote themselves to buffalo hunting." However, as early as 1833, John McLean, also of the Hudson's Bay Company, noted for the Clearwater region just upstream of Fort McMurray that "in former times these hills were covered with herds of buffaloes but not one is to be seen now" (both quoted in Roe 1951, 304). In short, it seems reasonably well established that bison constituted a major (if gradually dwindling) element of the regional big game population even in relatively recent times.

The predicted postglacial and Hypsithermal flood zone ecosystem, with its more open character and its higher frequency of herbaceous elements, would have been especially appealing to bison. Even as the channels between ridges began to accumulate muskeg or developed fen-related vegetation, the sedges therein would have been attractive as graze for small bison herds. The linear ridge and channel topography of the flood zone, especially in the early post-glacial period, before the accumulation of muskeg reduced initial variations in elevation, may have been particularly suited to the communal drive-and-entrap techniques known to have been used by Palaeoindian groups to harvest bison in more southerly landscapes.

Along with caribou, bison are the big game species with the greatest tendency to aggregate. This tendency to form herds, coupled with their large body weight and predictable seasonal patterns of movement, further enhances their suitability for communal hunting techniques. Given the predicted contrast in the vegetation community between the flood zone and the surrounding higher-elevation forest during the Hypsithermal interval and perhaps later, it may be that the flood zone became an integral component of bison seasonal migration patterns, such as is recognized for the herds in Wood Buffalo Park to the north (Carbyn, Oosenbrug, and Anions 1993). This contrast and the attendant attraction of this area for grazing ungulates, particularly bison, may be a principal explanation for the high density of Palaeoindian and Middle Prehistoric sites.

As these differences diminished with the advent of cooler, wetter climates, the appearance of relatively closed, less productive forests, and the accumulation of muskeg, we infer that the flood zone would have been less attractive for grazing herd ungulates such as bison and caribou, favouring instead browsing species more typical of the boreal forest. These more dispersed, solitary species are less suitable for communal hunting, which may explain the lower site density and use patterns more typical of the boreal forest that characterized the flood zone in Late Prehistoric times.

Other animal species represented in the blood residue results indicate that not all hunting and processing activities were dedicated to big game. Smaller game species, such as hare, appear in the residue results. Similarly, in the Fort McKay consumption study, snowshoe hare is identified as the second most commonly harvested food species, which probably reflects their widespread distribution and seasonal availability. Hare is generally considered a lower-value food than either fish or moose, but prehistoric groups undoubtedly hunted rabbits, typically capturing them by means of snares. Their meat, hide, and, perhaps, sinew are likely to have found use in the past, as they do today.

Reactions to chicken antiserum probably reflect the presence of a species of grouse, either spruce or ruffed grouse, which occur throughout the region today and are a preferred source of food. As with hare, these would have been hunted in an opportunistic fashion, but the harvest of grouse would not account for the vast quantities of archaeological materials present in the flood zone.

Reactions to dog and cat probably reflect the presence of species such as wolf, coyote, and fox, and lynx, cougar, or bobcat. It seems most likely that these species were used in connection with non-food-related activities, such as the manufacture of clothing and decorative items, as well as for products such as sinew. However, we would not expect the harvest of these species to be a major component of the activities that produced such large numbers of prehistoric assemblages in the flood zone. In contrast, bear, which is represented in two of the blood residue identifications, may have been harvested as food (although none of these species is mentioned by members of the Fort McKay community as a traditional food source). In addition, the use of bear hide, fur, teeth, and claws in clothing, blankets, and rugs, as well as for decoration, is well documented and would have warranted the hunting of these animals well beyond their possible use as food.

Other taxa represented in the blood residue results include "rat" and "guinea pig," which probably represent members of the rodent family such as muskrat and beaver. The Fort McKay study identifies muskrat as a traditional food, but we would expect that these species were hunted primarily for their pelts. In prehistoric times, however, demands for fur would have been limited to the local community, given that these species were widespread and thus readily available and no commercial fur market yet existed. But even if these species were hunted for both food and fur, this cannot account for the density of archaeological occupation within the flood zone as opposed to any other part of the boreal forest.

Finally, only one reaction to fish antiserum has been identified in the residue results on archaeological materials. While this might suggest that fish did not figure very prominently in the diet of prehistoric peoples, it may simply reflect the fact that relatively few archaeological specimens from assemblages adjacent to fish-bearing streams within the flood zone have been submitted for analysis. The Fort McKay study mentions the consumption of a wide array of fish species in considerable quantity, which agrees with our understanding of subsistence patterns for most Aboriginal communities throughout the Canadian boreal forest. However, without placing too much weight on the blood residue results, the facts that the majority of the assemblages within the flood zone occur inland, away from any fish-bearing stream, and that only one reaction to fish antiserum has been noted might also suggest that glacial lake outwash waters, which may have provided a potential source of harvestable fish populations, had receded before the principal prehistoric land use took place. Evidence of the early use of such fish populations might have occurred along retreating shorelines well distant from current water bodies, had these glacial outwash waters remained when prehistoric people moved in to occupy the post-flood landscape.

One possibly surprising outcome of the blood residue results currently available is the absence of any indication of waterfowl. The major flyway along the Athabasca River to the Peace-Athabasca Delta (Peace-Athabasca Delta Project Group 1972) would presumably have been used annually by large populations of ducks, geese, swans, and other birds throughout much of prehistory, and this resource would hardly have been overlooked by prehistoric occupants of the region. Moreover, waterfowl are a significant component of the traditional Aboriginal diet (Fort McKay Environmental Services 1997). The fact that the flood zone contains little by way of open water may have restricted the seasonal availability of large quantities of waterfowl. At the same time, the lack of waterfowl residue may relate more to the type of tools tested, most of which are projectile points and other shaped tools. Such tools were probably not frequently used to hunt waterfowl, with the result that existing residue samples may under-represent the extent to which prehistoric people relied on waterfowl as a food source.

In summary, we must reiterate that only limited, indirect evidence is available from which to develop predictions about the food resources exploited by prehistoric groups—patterns of use that might then help to explain the vast quantities of archaeological materials found within the minable oil sands region. Nevertheless, on the basis of landscape form and predicted vegetation, we believe that the dense pattern of occupation reflected in the archaeological record, which includes a high concentration of stone reduction and tool production activities, relates to the harvest of food resources, probably big game animals. In this regard, we would offer several suggestions.

Information about landforms and predicted vegetation indicates that the abandoned flood zone would have made a significant contrast with the surrounding forest. The area would have had superior drainage capacity, and the landscape would have been recolonized with the open vegetation communities favoured in well-drained areas and in the warmer, drier conditions of the postglacial Hypsithermal climatic maximum. The parkland-like environment—consisting initially of grassland communities and, later, of open deciduous forest along ridges with grass or sedge meadows in intervening channels—would have attracted grazing ungulates. Of these, bison and barren-ground caribou (if seasonally present) are the species with the greatest tendency toward herding behaviour and therefore those with the most predictable movement patterns and the highest potential return for co-operative hunting efforts. Elk may also have favoured this environment. Both bison and caribou are capable of coping with climatic extremes and are known to have been present under the glacial conditions that would have preceded the outwash event.

In their preliminary interpretation of the settlement patterns in the Agassiz flood zone, Saxberg and Reeves predicted that outwash waters had been impounded for perhaps 2,500 years after the initial flood. Sequential occupation by prehistoric people was presumed to have focused on retreating lakeshores during a period from about 10,000 to 7,000 years ago, with exploitation of lake resources and littoral zones as a significant focus of subsistence strategies (Saxberg and Reeves 2003). In view of the geologically accepted interpretation of relatively rapid flood water drainage, however, together with the absence of lake sediments in the area, the imprecise correlation between chronologically ordered occupations and specific shoreline elevations, and the lack of evidence for the use of aquatic resources, we consider it most likely that the site distribution points to a more traditional, terrestrial-based use of resources, as influenced by landscape and by Hypsithermal conditions.

We would argue that the complex braided-channel topography created by the Lake Agassiz outwash event in the Muskeg River basin and areas to the west of the Athabasca River could have been used to drive, corral, or otherwise entrap bison, possibly caribou, and perhaps also elk, as part of the seasonal rounds of Palaeoindian and early Middle Prehistoric occupants of the region. These activities seem to us the most likely explanation for the distinctive pattern of archaeological resources seen in the region.

As the effects of the Hypsithermal maximum wore off, jack pine and lichen communities became established on ridge tops, producing a habitat more favourable for woodland or possibly barren-ground caribou. Like bison, barren-ground caribou have significant herding tendencies and would be suited to communal hunting strategies, especially by groups familiar with the species and its migratory patterns. However, bison would have remained in the region, albeit perhaps in smaller numbers, taking advantage of the sedge meadow habitat still present in channels. Middle Prehistoric occupation is thought likely to relate to harvest of these species.

As postglacial climates evolved toward more modern conditions, with forests becoming more closed and muskegs developing in the channels between braid bars, subsistence strategies would require adjustment. Without open woodland or grassland environments, animal populations presumably shifted, with herd animals giving way to animals that would spend much of the year in isolation or in very small groups, whether because existing species adapted their behaviour to the new conditions or because new species occupied the area. The subsistence focus for human populations would likewise have shifted toward browsing ungulates such as moose and, to a lesser extent, deer. Wood bison and woodland caribou would have remained important, but in small groups, probably no longer hunted communally. A more broadly based subsistence strategy that would have included fish and other boreal forest food resources is likely to have resulted in a less intense and more widespread settlement pattern more typical of what would be expected throughout the Canadian boreal forest.

Lithic Resources

As mentioned above, archaeological assemblages in the minable oil sands area and surrounding regions are dominated by a single type of stone material, most commonly called Beaver River Sandstone (BRS). Well in excess of a million specimens of the stone have so far been recovered, and the extraction, processing, and use of this material are key to the prehistoric site distribution pattern seen in the area. Discussions concerning the potential sources of this material have been an ongoing feature of oil sands archaeological study since its outset (see Fenton and Ives 1982; Ives and Fenton 1983).

Recognition of potential sources of this material emerged early in the archaeological work conducted in advance of oil sands development (Syncrude Canada Ltd. 1974), with the discovery of the Beaver River Quarry. Fenton and Ives subsequently identified the lower member of the McMurray Formation as the geological origin of BRS, just below the occurrence of bitumen-saturated sands (Fenton and Ives 1984). The Beaver River Quarry (HgOv-29) is situated on the west side of the Athabasca River in the Lake Agassiz flood zone, and the outwash event probably facilitated access to the formations containing this material. However, the material present at that site is coarser in grade than the material from which most regional artifacts are made, and the site is now considered to be a secondary source.

Quarries may also be present in the vicinity of the Cree Burn Lake site (HhOv-16), on the east side of the Athabasca River, north of its junction with the Muskeg River, in the direct path of the Lake Agassiz outwash event and at the head of the braid channel. The extreme density of archaeological sites in this region, and the steep walls of the valley along this section of the river, suggested to several researchers that formations containing BRS must be available somewhere along the now forested and slumping valley walls. Ives and Fenton (1983) subsequently reported the discovery of an in situ unit of BRS at HhOv-55, a site to the south of HhOv-16, where blocks of the stone also occurred in the area of a slump.[5] To date, however, no extensive source of BRS suitable for stone tool manufacture has been identified in the vicinity of Cree Burn Lake.

The possibility of secondary (that is, redeposited) sources of BRS was investigated by Ives and Fenton (1985) through an inspection of gravel pits, road cuts, and natural exposures in the hinterlands beyond the main river valley, but they encountered only limited evidence of this source material. Reeves (1996) identified large boulders of Beaver River Sandstone perched on ridges and table lands in various locations throughout the Muskeg River basin. His theory was that these boulders had been displaced by the Lake Agassiz outwash event and could have provided the source for stone tool manufacture in the area. However, in almost all of the instances where material available in these boulders has been examined, it has proved to be too coarse to have been the source of the fine-grained material from which artifacts are typically fashioned. Further, it is considered doubtful that surficial boulder sources could explain the vast quantities of artifacts associated with prehistoric use of this landscape. The virtual absence of water-altered cortex in flood zone prehistoric assemblages indicates that in situ bedrock formations were the ultimate source of this material.

The differences between the fine-grained varieties of Beaver River Sandstone observed in artifact samples throughout the oil sands region and the coarser grade observed at the Beaver River quarry, as well as in virtually all other recorded surface and near-surface exposures of the lower members of the McMurray Formation (see Tsang 1998), has puzzled archaeologists for decades. The possibility that heat treatment may have been used to upgrade the quality of the coarser stone, as Gryba suggests in chapter 9 of this volume, could provide a partial basis for understanding the widespread use of this material.

In considering the questions surrounding the variation in quality, Saxberg and Reeves (2004) cite geological studies that suggest that silicification may have occurred in specific locations within the McMurray Formation. The silicification would have been caused by an upwelling of high-temperature siliceous fluids along faults in underlying formations. The notion that fine-grained Beaver River Sandstone originates only with the presence of mineral-rich or chemically active fluids is an attractive explanation for the origin and the limited distribution of fine-grained material within a formation that is characterized primarily by coarse-grained material. It would seem reasonable to conceive of fluids dispersing from points of issuance along the top of the largely impermeable Devonian Waterways Formation through the porous lower members of the McMurray Formation, resulting in high levels of silicification near the point of issuance and diminishing levels as the distance increases. This process could have resulted in localized in situ occurrences of high-quality Beaver River Sandstone. Such a scenario is, however, speculative and would require geological confirmation. Furthermore, to be accessible to prehistoric hunting groups, such stone would have had to be available in near-surface locations.

Many of the uncertainties surrounding the principal source of Beaver River Sandstone were potentially resolved in 2003 with the identification of a site named the Quarry of the Ancestors (Saxberg and Reeves 2004; see also chapters 10 and 12 in this volume). This site is actually a complex of related sites that centre on two near-surface exposures of fine-grained Beaver River Sandstone, which are generally assumed to be outcrops of the McMurray Formation. In this area, even Devonian-age limestone, which underlies the McMurray Formation, occurs sufficiently near the surface to make commercial mining of this material economically feasible, and the Quarry of the Ancestors was discovered in the course of an archaeological impact assessment required in connection with a proposed limestone quarry. The Quarry of the Ancestors lies approximately 6 kilometres inland of the current Athabasca River valley but is within the path of the Agassiz outwash event, near the eastern margin of the flood zone. Although it has yet to be conclusively determined whether these near-surface exposures of Beaver River Sandstone are outcrops of bedrock, in places the formation that contains it occurs on ridge tops and would have been relatively easy to recognize as flood waters retreated. The quality of the material and the considerable extent of its occurrence suggest that this was probably at least one of the principal sources of the Beaver River Sandstone used throughout the oil sands region, and possibly *the* principal source.

Although typically present in small quantities, other material types occur within oil sands archaeological assemblages. Some of these include quartzites and cherts that are likely to have regional fluvial sources, but materials from far-away sources, such as obsidian from northeastern British Columbia and Swan River chert from the Saskatchewan parklands, also occasionally occur. The presence of less common types of stone is of interest, as it can reveal something of a prehistoric group's seasonal patterns of movement and/or trade networks and interactions with other cultural groups.

THE CHRONOLOGY OF PREHISTORIC OCCUPATION

For a variety of environmental reasons, we have suggested that, for roughly three millennia, the Lake Agassiz outwash braid channel provided a microenvironment significantly different from the surrounding forest ecosystems, one that was seasonally attractive to game animals and their prehistoric hunters. We have further suggested that this environment, along with the occurrence of near-surface deposits of bedrock especially well-suited for tool manufacture and use, resulted in an intense pattern of prehistoric activity in the area. In addition, we have suggested that, as climatic conditions deteriorated, and as the forests began closing in and wetlands developed in former channels, human occupation became less focused on this microenvironment and instead became adapted to the more widely distributed resources characteristic of the greater boreal forest region.

Is such a scenario supported by the chronological information recovered to date from the numerous archaeological excavations that have been conducted throughout the minable oil sands region? Accurate dating of archaeological assemblages in the Canadian boreal forest is fraught with significant challenges. Perhaps most problematic is the acidic nature of the soils of the region, which degrades organic residues that could otherwise be employed for radiocarbon dating. Often only more recently deposited materials remain preserved. Several radiocarbon dates are available from archaeological sites in the region, but, as we will see, it cannot be said that they make a significant contribution to our understanding of regional prehistory.

Aggravating this problem is the general lack of a stratigraphic separation of sequential occupations. Except in major river valleys, landforms in the boreal forest have been relatively stable since glacial retreat, with sediments held in place by vegetation that quickly recolonized exposed areas and has persisted in one form or another ever since. As a result, there has been little sedimentary accumulation over most of the region, with the same surfaces available for repeated reoccupation throughout the full span of prehistory. When this factor is coupled with the disruption of near-surface sediment caused by the continuous growth and renewal of forest vegetation communities, separation of chronologically isolated cultural activities becomes very difficult using standard archaeological methods.

These issues constrain archaeological endeavour throughout the forested regions of Canada and present challenges that may never be totally resolved. However, stylistic variations in specific artifact types, particularly projectile points, can potentially allow us to establish relative cultural chronologies for the sites and regions in which they have been recovered, on the basis of radiocarbon estimates obtained for these styles elsewhere. In chapter 6, Reeves, Blakey, and Lobb present a comprehensive model of a postglacial cultural chronology for the oil sands region founded on a direct examination of most of the specimens they discuss (see also Saxberg and Reeves 2003). Although we recount that sequence here, we also note that, as with any cultural construct based on style, there will be differences of opinion as to how variations should be organized, a fact to which some of the chapters in this volume attest (see, for example, chapter 8).

Reeves, Blakey, and Lobb adopt a tripartite scheme, as is typical in discussions of the prehistory of the plains, parklands, and eastern slopes regions of Alberta and has also been employed for Alberta's forested regions (Ives 1993). In their scheme, the Early Prehistoric period is represented by three sequential cultural constructs, or complexes, believed to have occupied the region in the period from about 9,800 to 7,750 BP, during early postglacial and early Hypsithermal times: the Fort Creek Fen, Nezu, and Cree Burn Lake complexes. All three employed technologies that featured the production of lanceolate projectile points, that is, long, relatively narrow points without barbs or notches, presumably intended for use on spears that were either thrown or thrust. The stylistic variation in these specimens and in some associated tool types are used to define the three cultural complexes, and inferences can be drawn about the relationships of these cultures with more southerly cultural constructs.

The Middle Prehistoric period, which Reeves, Blakey, and Lobb date from roughly 7,750 to 2,650 BP, extends from the latter part of the Hypsithermal interval through to the period of climatic decline leading to modern conditions. This period has been divided into two sequential complexes, the Beaver River and the Firebag Hills. Again, these complexes can be distinguished and certain external cultural relationships inferred on the basis of stylistic variations in projectile point types, as well as other tool types presumed to be chronologically diagnostic. The technology employed during this period involved the use of notched points, which probably tipped long dart shafts. These shafts were propelled by a throwing device that, although generally referred to by the Aztec term *atlatl*, is believed to have been in use throughout North America during this period. The Beaver River Complex (7,750 to 4,000 BP) is thought to represent local groups that had adopted cultural traits characteristic of a number of surrounding areas, first to the south and then to the west and north. The subsequent Firebag Hills Complex, which seems to have existed for only about 1,400 years (4,000 to 2,650 BP), is thought to exhibit closer affinities to groups defined on the northern forest edge and the Barrenlands.

The Late Prehistoric period covers the final stages of in the development of the modern boreal forest environment. The single cultural complex defined for this period, the Chartier, is believed to represent the prehistoric Dene occupation of the region. It features a technology that, although first identified in the boreal and tundra-transitional regions of the central Mackenzie district (Noble 1971), is widely recognized throughout western boreal forest regions as far south as north-central Saskatchewan (Noble 1971) and in the Peace River district of Alberta (Bryan and Conaty 1975). Diagnostic specimens include a range of notched and unnotched projectile point styles, some of which may represent the adoption of bow and arrow technology, which becomes a mainstay of hunting methods in areas to the south in the latter part of this period.

As noted above, while often our only recourse in the absence of absolute radiocarbon dates and solid stratigraphic evidence, the chronological ordering of archaeological assemblages on the basis of stylistic variation in diagnostic tool types is imprecise and is frequently the subject of disagreement among experts. Nevertheless, we undertook a review of the proposed chronological affiliation of all the diagnostic projectile points recovered from the minable oil sands region (an area roughly equivalent to the Glacial Lake Agassiz flood zone) as a rough means of assessing whether the archaeological evidence tends to corroborate a comparatively intense use of the post-flood landscape and Hypsithermal ecosystems and resources as we have outlined earlier. For these purposes, we considered only the assignment of projectile points to the three major temporal periods rather than to their constituent complexes, as the former roughly correspond with the environmental variations discussed earlier.

As of December 2009, sites within the minable oil sands region had yielded 105 projectile points deemed sufficiently diagnostic to permit chronological assignment. Of these, 71 (67.6%) have been classified as belonging to complexes within the Early Prehistoric period (9,800 to 7,750 BP), 23 (21.9%) have been assigned to the Middle Prehistoric period (7,750 to 2,650 BP), and 11 (10.5%) have been assigned to groups present during the Late Prehistoric period (2,650 to 300 BP). The fact that two-thirds of these artifacts appear to date to the Early Prehistoric period strongly supports the contention that early post-flood landscapes, vegetation communities, and resources proved very attractive to prehistoric peoples and supported intense human use. The proportional prominence of Early Prehistoric occupations in regional assemblages becomes even more striking when one considers that the period represents little more than 2,000 years of the 9,500-year time span of regional prehistoric human use.

Although we have accepted the cultural classifications applied by the original researchers to their diagnostic projectile points, we are aware that controversy surrounds some of these decisions. Principally this revolves around the assignment to Early Prehistoric occupations of specimens the features of which bear a resemblance to a range of styles that have been grouped with a much later cultural construct known as the Taltheilei Tradition. This cultural assemblage was originally termed the Taltheilei Shale Tradition (Noble 1971), but Gordon (1977) renamed it simply the Taltheilei Tradition because commonly used raw materials included quartzite as well as shale.

The Taltheilei, from whom the Chipewyan and related Dene groups are believed to be descended, were an Athapaskan people who lived primarily by hunting caribou. The Taltheilei Tradition appeared rather suddenly, around 2,600 BP, in the southeastern portion of the Mackenzie district (Gordon 1977) and the southern half of the Keewatin district (Wright 1972). The tradition is characterized largely by variants of lanceolate and stemmed points that lack lateral grinding, as well as by notched points, and is defined by ten successive complexes (Noble 1977). On the basis of radiocarbon dates obtained from intact stratified contexts, supplemented by dates inferred from beach strand chronologies and from stylistic similarities with assemblages from sites in areas east of Great Slave Lake, the ten complexes that constitute the Taltheilei Tradition have been grouped into four developmental phases, Earliest, Early, Middle, and Late, that span the period from roughly 2,650 to 300 BP (Gordon 1996).

The most productive Early Taltheilei Tradition occupation areas are located just inside what is now the northern margin of the boreal forest, reflecting the presumption that, as caribou hunters, the Taltheilei summered only briefly on the Barrenlands (Noble 1977, 65). But Taltheilei sites occur deeper inside the forest as well. The majority of diagnostic materials at Black Lake, in northern Saskatchewan, are, for example, considered to reflect the Early Taltheilei Tradition (Minni 1976, 53); the same is true of sites along the eastern margin of Lake Athabasca (Wright 1975). Middle and Late Taltheilei sites have also been reported in the Lake Athabasca region (Gordon 1977, 74–75; Wright 1975), as well as in the Peace River district (Bryan and Conaty 1975). Taltheilei occupation in the oil sands region, as expressed in the Chartier Complex, has been reasonably well established (see the discussion in chapter 6).

Disagreements have, however, arisen surrounding the age of three particular styles of lanceolate spear point. The first is a variety of unstemmed, parallel-sided point that is sometimes described as "waisted" because the lateral margins of the point constrict as they near its base, forming a "waist" along its lateral margins and resulting in a slight flair before its typically straight base develops (see Reeves and Saxberg 2003, figure 9A). This style, which is represented in the oil sands in the Early Prehistoric Fort Creek Fen Complex (ca. 9,800 to 9,600 BP), in the form of the Fort Creek Fen lanceolate (see plate 6.1 in this volume), has been estimated to date to around 9,500 BP, on the basis of similarities to firmly dated specimens in Montana and Wyoming. There, examples of the style predate Cody Complex material, which has also been recovered in considerable quantity in the flood zone in association with the Nezu Complex (ca. 9,600 to 8,600 BP), suggesting the same cultural-chronological sequence. However, formal similarities also exist with specimens grouped with Late Taltheilei assemblages dated to no earlier than 1,400 BP (see Gordon 1977, figure 4 and plate 8; see also Bryan and Conaty 1975, plate 3).

A similar situation occurs in relation to a second style, namely, unstemmed lanceolate points, which represent perhaps the simplest of style of spear point. This style is characteristic of well-dated Early Prehistoric occupations throughout the northern Plains but also occurs in later assemblages. Its most distinctive example was defined at the Agate Basin site in Wyoming, where forms that range from parallel-sided to leaf-shaped and display specific manufacturing characteristics have been firmly dated to 10,500 to 10,000 ^{14}C yr BP (Frison and Stanford 1982). Agate Basin–style points found along the southern margin of the boreal forest in Canada have been dated to 8,500 to 7,500 ^{14}C yr BP (Buchner 1981), but similar styles are known to occur as early as 10,770 ^{14}C yr BP at the Mesa site in the Alaska interior (Kuntz and Reanier 1995). In Alberta contexts, unstemmed lanceolate specimens are attested in early occupations that postdate the Fluted Point Complex (11,050 to 10,200 ^{14}C yr BP: see Peck 2011, 24, 55; see also Vickers 1986), as well as somewhat later occupations, during the terminal stages of the Early Prehistoric period, between 9,000 and 7,500 BP (Driver 1978).

Within the oil sands region, Agate Basin–style points have been recovered from the Gardiner Lake Narrows site in the Birch Mountains and the Beaver River Quarry (HgOv-29; see Syncrude Canada Ltd. 1974) and have been assigned to a period between 10,000 and 8,000 BP (Ives 1993). Somewhat similar forms have been also grouped with the Cree Burn Lake Complex, which is estimated to date from about 8,600 to 7,750 BP, during the terminal stages of the Early Prehistoric period (Saxberg and Reeves 2004; Reeves, Blakey, and Lobb, chapter 6 in this volume). However, unstemmed lanceolate points, sometimes leaf-shaped in form, have also been identified in both Early and Middle Taltheilei assemblages (Gordon 1996, figures 5.2 and 5.3) that are estimated to date between 2,600 and 1,450 BP.

Finally, stemmed lanceolate points are noted throughout the North American Plains in a distinctive and well-defined cultural complex known as the Cody Complex (Frison and Todd 1987). These points represent some of the most technically well-made and aesthetically pleasing specimens in the North American prehistoric record. In the United States, the Cody Complex has been dated to the period between 9,200 and 8,800 BP (Frison 1991). The complex was initially thought to represent the first intensive occupation of Alberta (Wormington and Forbis 1965, 185). More recent studies have recognized a richer and more complicated record of Early Prehistoric use of Alberta landscapes than was originally envisaged (Dawe 2013), but the Cody occupation remains one of the most distinctive facets of Alberta Plains prehistory, where it is estimated to date to a period between 9,600 and 8,600 BP (see Peck 2011, 67–93). In the oil sands region, a distinctive Cody Complex occupation has been recognized. Termed the Nezu Complex (Saxberg and Reeves 2003; see also chapter 6 in this volume), it occurs in relatively high frequency and is likewise thought to date in the range of 9,600 to 8,600 BP. As in the case of the unstemmed lanceolate forms, however, stemmed spear points of roughly similar formal outline have been included in the Early and Middle Taltheilei assemblages mentioned above (see Gordon 1996, figures 5.2, 5.3, and 6.3), which, as noted, are thought to date between 2,600 and 1,450 BP.

The suggestion that some stemmed Early Taltheilei points have been mistaken by archaeologists for Plains Cody Complex points arose shortly after the initial recognition of Tatheilei occupations in areas south of the Northwest Territories (Noble 1971, 111). Stewart (1991) has shown how Taltheilei points may occasionally be confused with Agate Basin types, while Wilson and Burns (1999, 228–231) have gone so far as to argue, in connection with the lanceolate points of the Early Prehistoric Northern Plano Tradition found in the Canadian subarctic, that these points are not related to those found in Agate Basin cultures largely because "unstemmed, leaf-shaped points are the simplest form to make and the easiest to reinvent" (228). In their view, "unlike biological species, projectile-point types can recur without necessary historical linkages" (Wilson and Burns 1999, 228; see also Wright 1975, 86).

To evaluate the suggestion that Late Prehistoric Taltheilei points have been misclassified as Early Prehistoric Agate Basin or Cody Complex–related specimens, one needs to look beyond the formal outline of the styles and consider attributes relating to their manufacture, as well as other characteristics of the assemblages in which they occur. Whereas individual stylistic features might occur independently, Agate Basin points are characterized by several such features that, in combination, are less likely to have recurred in other cultural contexts. Both Agate Basin and Cody Complex styles exemplify features of a technological tradition that Bradley (1991) calls the Collateral Point Complex. Flake removal is generally well controlled, proceeding perpendicularly from the margin toward the midline. This results in a profile that ranges from flat to lenticular (lens-shaped) in longitudinal section, with a cross-section featuring a slightly low to medium ridge either on one face or on both faces, forming a diamond-like lozenge-shape. These profiles are typical for Agate Basin points (Roberts 1961; Ebell 1980). Lateral edges are ground from the base toward the tip for a distance of one quarter to one half of the total length of the blade. Somewhat similar manufacturing patterns are exhibited in the typically square-based Cody Complex types. These include relatively well-controlled flake removal that is perpendicular to the midline but may either overlap the midline or meet evenly at it, which produces lenticular or distinctly diamond-shaped cross-sections (Bradley 1991, 390). The regularity of the flake removal often results in sinuous edges (Frison and Todd 1987).

Notions of what a tool should look like and how it should be made are learned and passed on within a cultural context that is situated in a particular time and place. Recognition of technological characteristics consistent with the Collateral Point Complex allows archaeologists working in the oil sands to assign unstemmed and stemmed lanceolate points to Early Prehistoric cultural complexes that either derive from or share relationships with the Agate Basin and Cody complexes, respectively. While some Taltheilei points (especially their bases) mimic the *shape* of Agate Basin points, these Taltheilei points exhibit little or no grinding along the lateral basal edges (Noble 1971, 111–113). Also, in contrast to Agate Basin points, Taltheilei points generally display either wide transverse flaking patterns (Gordon 1977, 74) or simply uncontrolled flaking (Noble 1971, 111–113). In other words, while some Taltheilei points may exhibit one or more Agate Basin attributes, it is unlikely that they share most of them. Likewise, the well-controlled flaking patterns typical of Cody Complex specimens, coupled with their parallel sides and square bases, distinguish them from similar-looking Taltheilei specimens.

Along with these differences in projectile points, other parts of the tool kit may also serve as diagnostic indicators. For example, Frison and Stanford (1982) note that gravers, notched tools, blades, end scrapers on blades, wedges, and burin spalls are commonly found in the Agate Basin occupation at the Agate Basin site. The Cree Burn Lake Complex, which is associated with a range of unstemmed lanceolate point styles, is also characterized by blades, burins, gravers, and end scrapers on blades (Saxberg and Reeves 1998). In contrast, in his discussion of the Hennessy, Taltheilei, and Windy Point complexes of the Taltheilei Tradition—whose point styles include unstemmed and stemmed lanceolate forms that resemble Agate Basin and Cody Complex types—Noble (1971) notes that tool kits *do not include* gravers, burins, or scrapers on the ends of blades.

Together with the points, these data suggest that most Early Prehistoric points and affiliated assemblages identified in the oil sands region have been correctly dated as such. We should also bear in mind that the issues raised by the similarities between certain Early Prehistoric point styles and certain Late Prehistoric Taltheilei forms have for the most part been recognized since the 1970s. Archaeologists are well aware of these issues and have taken them into consideration in making their judgments. There is no question that considerable uncertainty surrounds chronologies based principally on style, without the assistance of a comprehensive series of radiocarbon dates obtained from stratified contexts. However, on the basis of the information available to date, we would argue that the disproportionately large percentage of diagnostic artifacts that date to the Early Prehistoric period, along with the material assemblages with which these artifacts are associated, confirms the conclusion that the greatest prehistoric use of the minable oil sands area occurred during this period. For the reasons explained above, we would argue that it was the exceptional circumstances of the flood-scoured landscape, in conjunction with Hypsithermal climatic conditions, that provided the basis for this unparalleled occurrence.

To further assess this hypothesis, we also considered the limited information available from efforts to obtain radiocarbon dates for organic materials recovered within the oil sands region. We have already mentioned the generally acidic nature of boreal forest soils and the deleterious effects of such acidity on organic remains. Perhaps more critical is the fact that the landforms exhibiting the highest concentration of prehistoric occupation (the elevated tops of the braided bars) are not environments that were subject to much, if any, sedimentary deposition, on which stratigraphic separations can then be based. Multiple revisits to site locations could easily take place, but the lack of deposition would tend to result in an indistinguishable mass of occupational materials from vastly different periods. The problem is compounded by the fact that forest regrowth over the millennia is likely to have contributed to the mixing of evidence from different occupations. In such circumstances, one is left with questions that need to be addressed on a case-by-case basis and may not, in the end, be resolvable.

With these cautions in mind, we considered the radiocarbon dates available from twenty-four sites in the oil sands region (see table 5.1), which range from 5,250 ± 40 ^{14}C yr BP to essentially modern times. Several have been obtained from sites that lie outside the Glacial Lake Agassiz flood zone, and none appear to relate to the early post-flood occupations that we have been discussing here. This situation is largely reflective of the reasons outlined above.

Clearly, dated specimens from sites that lie outside the Lake Agassiz flood zone do not directly relate to the early human use of this area. Remains that are reported to be less than four hundred years old can be considered modern. The fact that prehistoric assemblages can occur in locations that also exhibit modern use, which often leaves behind bone and/or charcoal, illustrates some the difficulties in drawing direct, firm relationships between any of the recovered organic material and the remains of prehistoric occupation. At HhOv-184, for example, where a clearly Early Prehistoric Cody Complex occupation (the Nezu) was identified, the charcoal that produced a date of 1,640 ± 80 ^{14}C yr BP was interpreted as the remains of an ancient forest fire. A similar case could be made for the 1,240 ± 60 ^{14}C yr BP date from the Cree Burn Lake site (HhOv-16), where a sample from a shallow 20- to 25-centimetre sediment horizon could as easily relate to a forest fire as a cultural occupation.

We are further confounded by the results from sites within the flood zone where no cultural diagnostics have been recovered. This situation applies to the two oldest dates in the sample, one of which (4,740 ± 40 ^{14}C yr BP) was obtained from charcoal found in what appeared to be a hearth at HhOv-256 and the other (5,250 ± 40 ^{14}C yr BP) from dispersed burned fragments of bone at HhOv-520. The lack of chronologically sensitive artifacts recovered from these sites limits what can be said about cultural relationships of the prehistoric assemblage recovered. In both cases, however, calibration of these radiocarbon dates to arrive at more accurate calendar dates would place both in the terminal stages of the Hypsithermal climatic interval.

Perhaps the most relevant radiocarbon estimate remains the one of 3,990 ± 170 ^{14}C yr BP obtained in 1983 from the Bezya site (HhOv-73), where a distinctive microcore and blade assemblage was recovered that has appropriately been assigned to the Middle Prehistoric Northwest Microblade Tradition (Le Blanc and Ives 1986). A date of 3,970 ± 30 ^{14}C yr BP obtained on burned fragments of bone from HhOv-156 could not be related to any chronologically sensitive artifacts, however, and provides little information about cultural relationships.

TABLE 5.1 Radiocarbon dates from sites in and around the flood zone

Site	Location	Material	Date (¹⁴C yr BP)
HhOu-70	Flood zone (Muskeg River)	Composite bone	1,650 ± 40
HhOv-16 (Cree Burn Lake site)	Flood zone (Athabasca River)	Sediment	1,240 ± 60
HhOv-73 (Bezya site)	Flood zone	Composite charcoal	3,990 ± 170
HhOv-156	Flood zone	Calcined bone fragments	3,970 ± 30
HhOv-184	Flood zone	Composite charcoal thought to relate to forest fire	1,640 ± 80
HhOv-245	Flood zone	Calcined bone fragments	520 ± 40
HhOv-256	Flood zone	Charcoal from hearth	4,740 ± 40
HhOv-351	Flood zone	Composite, widely spread bone sample	1,910 ± 30
HhOv-384	Flood zone	AMS from bone carbonate	2,930 ± 40
HhOv-387	Flood zone (Muskeg River)	AMS from bone carbonate	1,900 ± 40
HhOv-449	Flood zone	Composite charcoal above cultural layer	650 ± 40 BP
HhOv-520	Flood zone	Calcined bone fragments	5,250 ± 40
HhOw-20	Joslyn Creek, outside flood zone	Calcined bone fragments	1,670 ± 40
HhOw-30	Tar River, outside flood zone	Calcined bone fragments	Modern
HhOw-37	Calumet River tributary outside flood zone	Charcoal sample from hearth	1,300 ± 40
HhOw-45	Lower flanks of Birch Mountains, outside flood zone	Calcined bone	2,320 ± 40
HhOw-46	Ells River valley, outside flood zone	Calcined bone	1,980 ± 40
HhOw-55	Lower flanks of Birch Mountains, outside flood zone	Bone	280 ± 100 and 100 ± 40
HhOx-9	Lower flanks of Birch Mountains, outside flood zone	Bone	Modern (< 50 years)
HhOx-18	Lower flanks of Birch Mountains, outside flood zone	Calcined bone fragments	2,080 ± 40
HiOu-8	Delta deposits, north of flood zone	Calcined bone fragments	130 ± 40
HiOv-46	Fort Hills uplands, northern edge of flood zone	Calcined bone	2,270 ± 40
HiOv-70	Fort Hills uplands, northern edge of flood zone	Calcined bone	1,710 ± 40
HiOv-126	Fort Hills uplands, delta deposit not braid channel	Calcined bone	Modern (< 50 years)

Association	Reference (with ASA permit no.)
Middle Taltheilei, but no direct association	Roskowski, Landals, and Blower 2008 (07-219)
No diagnostics	Head and Van Dyke 1990 (88-032)
Northwest Coast Microblade Tradition (Firebag Hills Complex) (Middle Prehistoric)	Le Blanc and Ives 1986 (83-53)
No diagnostics	Roskowski and Netzel 2012 (11-167)
Nezu Complex (Early Prehistoric)	Clarke and Ronaghan 2000 (99-073)
No diagnostics	Wickham and Graham 2009 (06-376)
No diagnostics	Wickham and Graham 2009 (06-376)
No diagnostics	Roskowski and Netzel 2011 (10-148)
No diagnostics	Woywitka et al. 2008 (07-280)
No diagnostics	Woywitka et al. 2008 (07-280)
No associated diagnostics	Wickham and Graham 2009 (06-376)
No diagnostics	Roskowski and Netzel 2012 (11-167)
Side-notched projectile point, probably Late Prehistoric	Youell et al. 2009 (07-393)
Possible Early Prehistoric point, but no direct association	Bryant 2004 (03-269)
No diagnostics	Bryant 2004 (03-269)
No diagnostics	Boland, Brenner, and Tischer 2009 (08-208)
No diagnostics	Boland, Brenner, and Tischer 2009 (08-208)
Mixed assemblage: Mummy Cave Complex (Middle Prehistoric); Taltheilei; dates considered recent	Kjorlien, Mann, and Tischer 2009 (08-166)
Associated with historic component of site	Graham and Tischer 2009 (09-298)
No associated diagnostics	Kjorlien, Mann, and Tischer 2009 (08-166)
No associated diagnostics	Woywitka et al. 2009 (08-163)
No associated diagnostics	Woywitka et al. 2009 (08-163)
No associated diagnostics	Woywitka et al. 2009 (08-163)
Undetermined	Woywitka et al. 2009 (08-163)

The 1,650 ± 40 ^{14}C yr BP date obtained on bone from another site, HhOu-70, tends to corroborate the stylistic identification of the site as a Middle Taltheilei occupation, although the association between the bone dated and the diagnostic specimen is not a direct one. The site's location along the Muskeg River could be expected, given the predicted settlement pattern during this period, with a focus on use of riparian habitats along drainages. A comparable date of 1,910 ± 30 ^{14}C yr BP, at HhOv-351, belongs to the same time frame, but it was obtained on scattered fragments of bone from a small rodent that are probably of natural rather than cultural origin.

Two Late Prehistoric period dates are similarly uninformative about cultural occupations. At HhOv-245, a date of 520 ± 40 ^{14}C yr BP was obtained on scattered bone with which no chronologically sensitive artifacts were associated, and, at HhOv-449, a date of 650 ± 40 ^{14}C yr BP was obtained on charcoal found above the cultural occupation and probably relates to forest fire. In sum, the dates that can be said to relate to the prehistoric use of the Lake Agassiz flood zone are, at present, too few and too uncertain to allow us to make firm statements about whether the hypothetical use patterns discussed here for the postflood environment can be validated by the use of absolute dating techniques.

With regard to dating, one final characteristic of flood zone use warrants mention. During some of the earliest sample recovery programs undertaken in advance of modern industrial development within the flood zone, it was noted that evidence of prehistoric occupation occasionally extended off the crests of the elevated bar deposits, the place where occupation typically occurs, down their slopes and under muskeg deposits at the base of these landforms (Saxberg and Reeves 1998). Archaeological components within these depositional contexts are beginning to be observed in greater frequency within the flood zone. While the extent of this phenomenon has not yet been established, nor have occurrences of it been reliably dated, it provides clear indication that the initial use of the flood zone landscape precedes muskeg development in the area. It has not been ascertained whether the shift to the pattern of smaller and more dispersed populations typical of boreal forest occupations had already occurred prior to muskeg development. As the chronology of muskeg development is not yet fully understood, little more can be concluded at this time.

SUMMARY AND CONCLUSION

The numerous archaeological studies conducted within the minable oil sands region over roughly the past two decades have revealed a striking pattern of intense prehistoric land use. Because these studies have been prompted not by a systematic program of research but rather by the need to assess the impact of specific development projects, they have been scattered over a fairly wide area. As a result, our understanding of the diverse elements that make up this pattern of human use and the reasons underlying the pattern has been long in developing. However, sufficient information has emerged to allow some level of synthesis to take place. In the foregoing discussion, we have attempted to draw together geological, environmental, and cultural information to provide a time-transgressive model of landscape and resource use that we feel best explains what may be one of the most fascinating and significant chapters of northern Canadian prehistory.

The basis of the story is a massive postglacial flood that scoured a broad area within the Lower Athabasca basin, exposing near-surface bedrock containing material suitable for stone tool manufacture and laying down gravel and sand deposits. Subsequently, during a period of warm, dry climate, revegetation of this landscape took place, which in all likelihood produced a distinct ecology that we suggest was preferentially attractive to big game animals that grazed there on a regular and seasonal basis. These ecological resources, coupled with a landscape in which the availability of lithic resources suitable for making stone tools coincided with diverse geomorphological features conducive to the communal hunting of ungulate species, supported an intense pattern of human use that persisted for perhaps 5,000 years—throughout the entire Early Prehistoric period and much of the Middle Prehistoric. As modern climates and ecosystems developed, this use decreased in intensity. But it never ceased, instead evolving into a pattern more typical of the boreal forest. The use made of the area today by Aboriginal communities represents a continuation of practices whose legacy stretches back to the end of the Pleistocene ice age.

We are fully cognizant of the speculative nature of the conclusions we have drawn from this evidence. As new information emerges from the ongoing efforts to manage the effects of development and from the research currently proceeding in various quarters, no doubt aspects of the scenario we have presented here will require revision. Nevertheless, we are confident that some of the ideas we have outlined here can be tested, perhaps with existing evidence and certainly as more is acquired. We look forward to interpretations that are firmly grounded in such evidence.

NOTES

1   As initially defined by Alberta's Energy Resources Conservation Board (now the Alberta Energy Regulator), the surface-minable area covered an area of roughly 3,500 square kilometres. In 2009, in response to discoveries of potentially surface-minable bitumen reserves beyond the original boundaries of the surface-minable area, those boundaries were officially extended (see ERCB 2009, figure 2.4), increasing the total area to about 4,750 square kilometres. Figure 5.1 shows the area as it was defined prior to 2009, when the archaeological studies in question were carried out.

2   Cultural ecological theory, as developed by Steward (1955), understands cultural evolution as a series of adaptive human responses to environmental factors. Outgrowths of this theory have formed the basis for archaeological interpretations of the structure and evolution of past cultures.

3   For the Mariana Lakes region, which lies in the Athabasca River basin about 100 kilometres southwest of Fort McMurray, see Hutton, MacDonald, and Mott (1994), who provide
a record from an elevation of 688 metres above sea level.

4   For example, a survey of the consumption patterns of the Aboriginal residents of Fort McKay (Fort McKay Environmental Services 1997, 5), who consider the minable oil sands region their traditional land, indicates that deer "fall far down the list of preference as a staple."

5   For discussions of HhOv-55, see Fenton and Ives (1983, 1984) and Unfreed and Fedirchuk (2001); see also Ives and Fenton (1983, figure 7) for a map of areas of potential exposure of BRS. As Gryba notes in chapter 9, artifacts manufactured of BRS also occur in significant proportions in sites along the Clearwater River, leading researchers to speculate that a source (or sources) of the stone may exist in that area. Thus far, however, no such source has been confirmed.

REFERENCES

AXYS Environmental Consulting Ltd
     2005   *Albian Sands Energy Inc., Muskeg River Mine Expansion Project: Terrestrial Environmental Setting Report*. Report prepared for Shell Canada Limited, Calgary. http://www.ceaa. gc.ca/050/documents_staticpost/cearref_16259/MK-PRO-0018.pdf.

Balcom, Rebecca J.
     1996   *Historical Resources Impact Assessment, Steepbank Mine Project (ASA Permit 95-083)*. Copy on file, Archaeological Survey, Historic Resources Management Branch, Alberta Culture, Edmonton.

Bayrock, L. A.
     1971   *Surficial Geology, Bitumount, NTS 74E*. Map 34, Alberta Research Council, Edmonton. Available as Map 140, Alberta Geological Survey, Edmonton. http://ags.aer.ca/publications/MAP_140.html.

Boland, Dale E., Bonnie Brenner, and Jennifer C. Tischer
     2009   *Historical Resources Impact Mitigation, Total E&P Joslyn Limited Joslyn North Mine Project, 2008 Mitigation Studies (HhOw-18, HhOw-29, HhOw-32, HhOv-42, HhOw-43, HhOw-45, HhOw-46): Final Report (ASA Permit 08-208)*. Copy on file, Archaeological Survey, Historic Resources Management Branch, Alberta Culture, Edmonton.

Bouchet-Bert, Luc
     2003   *Historical Resources Impact Assessment for the Suncor South Tailings Pond Project: Final Report (ASA Permit 03-244)*. Copy on file, Archaeological Survey, Historic Resources Management Branch, Alberta Culture, Edmonton.

2005   *Historical Resources Impact Assessment for the Suncor Voyageur Project: Final Report (ASA Permit 04-272)*. Copy on file, Archaeological Survey, Historic Resources Management Branch, Alberta Culture, Edmonton.

2007   *Historical Resources Impact Assessment for the Suncor Voyager South Project: Final Report (ASA Permit 06-545)*. Copy on file, Archaeological Survey, Historic Resources Management Branch, Alberta Culture, Edmonton.

Bradley, Bruce A.

1991   Lithic Technology. In *Prehistoric Hunters of the High Plains,* 2nd ed., edited by George C. Frison, pp. 369–396. Academic Press, San Diego.

Bryan, Alan L., and Gerald Conaty

1975   A Prehistoric Athapaskan Campsite in Northeastern Alberta. *Western Canadian Journal of Anthropology* 5(3–4): 64–91.

Bryant, Laureen

2004   *Historical Resources Impact Assessment and Mitigation, Fall 2003, Canadian Natural Resources Limited Horizon Oil Sands Project: Final Report (ASA Permit 03-269)*. Copy on file, Archaeological Survey, Historic Resources Management Branch, Alberta Culture, Edmonton.

Buchner, Anthony P.

1981   *Sinnock: A Paleolithic Camp and Kill in Manitoba*. Papers in Manitoba Archaeology Final Report no. 10. Historic Resources Branch, Department of Cultural Affairs and Historical Resouces, Winnipeg.

Carbyn L. N., S. M. Oosenbrug, and D. W. Anions

1992   *Wolves, Bison and the Dynamics Related to the Peace-Athabasca Delta in Canada's Wood Buffalo National Park*. Circumpolar Research Series no. 4. Canadian Circumpolar Institute, University of Alberta, Edmonton.

Carrigy, M. A.

1973   Introduction and General Geology. In *Guide to the Athabasca Oil Sands Area,* edited by M. A. Carrigy, pp. 1–14. Information Series no. 65, Alberta Research, Edmonton.

Carrigy, M. A., and R. Green

1965   *Bedrock Geology*. Map reproduced as an insert in L. A. Bayrock (1971), *Surficial Geology, Bitumount, NTS 74E*. http://ags.aer.ca/publications/MAP_140.html.

Clarke, Grant M.

1998   *Historical Resources Impact Assessment, Mobil Lease 36 (ASA Permit 98-145)*. Copy on file, Archaeological Survey, Historic Resources Management Branch, Alberta Culture, Edmonton.

2000   *Historical Resources Impact Assessment, Suncor Firebag Project: Final Report (ASA Permit 00-042)*. Copy on file, Archaeological Survey, Historic Resources Management Branch, Alberta Culture, Edmonton.

2002   *Historical Resources Impact Assessment, Canadian Natural Resources Limited Project Horizon: Final Report (ASA Permit 01-248)*. Copy on file, Archaeological Survey, Historic Resources Management Branch, Alberta Culture, Edmonton.

Clarke, Grant M., and Brian M. Ronaghan

2000   *Historical Resources Impact Mitigation, Muskeg River Mine Project: Final Report (ASA Permit 99-073)*. Copy on file, Archaeological Survey, Historic Resources Management Branch, Alberta Culture, Edmonton.

2001   Early Prehistoric Use of a Flood-Scoured Landscape in Northeastern Alberta. Paper presented at the 34th annual meeting of the Canadian Archaeological Association, Banff, Alberta, 18–21 May.

Dawe, Robert J.

2013   A Review of the Cody Complex in Alberta. In *Paleoindian Lifeways of the Cody Complex,* edited by Edward J. Knell and Mark P. Muñiz, pp. 144–187. University of Utah Press, Salt Lake City.

Driver, Jonathan C.

    1978   Holocene Man and Environments in the Crowsnest Pass, Alberta. PhD dissertation, Department of Archaeology, University of Calgary, Calgary.

Ebell, S. Biron

    1980   *The Parkhill Site: An Agate Basin Surface Collection in South-Central Saskatchewan.* Past Log Series no. 4. Saskatchewan Culture and Youth, Regina.

ERCB (Energy Resources Conservation Board)

    2009   *Alberta's Energy Reserves 2008 and Supply/Demand Outlook, 2009-2018.* Report no. ST98-2009. June. Calgary: Energy Resources Conservation Board. https://www.aer.ca/documents/sts/ST98/st98-2009.pdf.

Fenton, Mark M., and John W. Ives

    1982   Preliminary Observations on the Origins of Beaver River Sandstone. In *Archaeology in Alberta, 1981,* compiled by Jack Brink, pp. 166-189. Archaeological Survey of Alberta Occasional Paper no. 19. Historic Resources Management Branch, Alberta Culture, Edmonton.

    1984   The Stratigraphic Position of Beaver River Sandstone. In *Archaeology in Alberta, 1983,* compiled by David Burley, pp. 128-136. Archaeological Survey of Alberta Occasional Paper no. 23. Historic Resources Management Branch, Alberta Culture, Edmonton.

Fiedel, Stuart J.

    1996   Blood from Stones? Some Methodological and Interpretive Problems in Blood Residue Analysis. *Journal of Archaeological Science* 23: 139-147.

Fisher, Timothy G.

    1993   Glacial Lake Agassiz: The Northwest Outlet and Paleoflood Spillway, N.W. Saskatchewan and N.E. Alberta. PhD dissertation, Department of Archaeology, University of Calgary, Calgary.

    2007   Abandonment Chronology of Glacial Lake Agassiz's Northwestern Outlet. *Palaeogeography, Palaeoclimatology, Palaeoecology* 246: 31-44.

Fisher, Timothy G., and Derald G. Smith

    1994   Glacial Lake Agassiz: Its Northwest Maximum Extent and Outlet in Saskatchewan (Emerson Phase). *Quaternary Science Reviews* 13: 845-858.

Forbis, Richard G.

    1968   Fletcher: A Paleo-Indian Site in Alberta. *American Antiquity* 33(1): 1-10.

Fort McKay Environment Services Ltd.

    1996a   *A Fort McKay Community Document Traditional Uses of the Renewable Resources on the Proposed Suncor Steepbank Mine Site.* Report prepared for Suncor Energy Inc., Oil Sands Group. Fort McKay, Alberta.

    1996b   *The Community of Fort McKay Traditional Uses of the Renewable Resources on the Proposed Syncrude Aurora Mine Local Study Area.* Report prepared for Syncrude Canada Ltd. Fort McKay, Alberta.

    1997   *A Survey of the Consumptive Use of Traditional Resources in the Community of Fort McKay.* Report prepared for Syncrude Canada Ltd. Fort McKay, Alberta.

Frison, George C.

    1991   *Prehistoric Hunters of the High Plains.* 2nd ed. Academic Press, San Diego.

Frison, George C., and Dennis J. Stanford

    1982   *The Agate Basin Site: A Record of Paleoindian Occupation of the Northwestern High Plains.* Academic Press, New York.

Frison, George C., and Lawrence C. Todd

    1987   *The Horner Site: The Type Site of the Cody Cultural Complex.* Academic Press, New York.

Gates, C., T. Chowns, and H. Reynolds

    1992   Wood Buffalo at the Crossroads. In *Buffalo,* ed. John Foster, Dick Harrison, and I. S. MacLaren, pp. 139-165. University of Alberta Press, Edmonton.

Geist, Valerius

    1971   *Mountain Sheep: A Study in Behavior and Evolution.* University of Chicago Press, Chicago.

    1982   Taxonomy: On an Objective Definition of Subspecies, Taxa as Legal Entities, and Its Application to *Rangifer tarandus* Lin. 1758. *Proceedings of the North American Caribou Workshop* 4: 1–36.

Golder Associates

    1997   Terrain and Soil Baseline for the Muskeg River Mine Project. Unpublished report prepared for Shell Canada Limited, Calgary.

Gordon, Bryan H. C. A.

    1977   Chipewyan Prehistory. In *Problems in the Prehistory of the North American Subarctic: The Athapaskan Question*, edited by James W. Helmer, Stanley Van Dyke, and François J. Kense, pp. 72–76. Chacmool Archaeological Association, Department of Archaeology, University of Calgary, Calgary.

    1996   *People of Sunlight, People of Starlight: Barrenland Archaeology in the Northwest Territories of Canada.* Mercury Series no. 154. Archaeological Survey of Canada, Canadian Museum of Civilization, Gatineau, Québec.

Graham, James W., and Jennifer C. Tischer

    2009   *Historical Resources Impact Mitigation, Total E&P Joslyn Limited Joslyn North Mine Project, 2008 Mitigation Studies (HhOw-22, HhOw-30, HhOw-38, HhOx-9, HhOx-13): Final Report (ASA Permit 08-298).* Copy on file, Archaeological Survey, Historic Resources Management Branch, Alberta Culture, Edmonton.

Green, D'Arcy

    2000   *Historical Resources Impact Assessment, OPTI Long Lake Project (ASA Permit 00-136).* Copy on file, Archaeological Survey, Historic Resources Management Branch, Alberta Culture, Edmonton.

    2001   *Historical Resources Impact Assessment, Petro-Canada Meadow Creek Project (ASA Permit 01-202).* Copy on file, Archaeological Survey, Historic Resources Management Branch, Alberta Culture, Edmonton.

    2003   *Historical Resources Impact Assessment, Suncor Millennium Mine Acceleration Project (ASA Permit 02-179).* Copy on file, Archaeological Survey, Historic Resources Management Branch, Alberta Culture, Edmonton.

Guthrie, R. Dale

    1980   Bison and Man in North America. *Canadian Journal of Anthropology* 1: 55–73.

Haile James, Duane G. Froese, Ross D. E. MacPhee, Richard G. Roberts, Lee J. Arnold, Alberto V. Reyes, Morten Rasmussen, Rasmus Nielsen, Barry W. Brook, Simon Robinson, Martina Demuro, M. Thomas P. Gilbert, Kasper Munch, Jeremy J. Austin, Alan Cooper, Ian Barnes, Per Möller, and Eske Willerslev

    2009   Ancient DNA Reveals Late Survival of Mammoth and Horse in Interior of Alaska. *Proceedings of the National Academy of Sciences* 106(52): 22352–22357.

Head, Thomas H., and Stanley Van Dyke

    1990   *Historical Resources Impact Assessment and Mitigation, Cree Burn Lake Site (HhOv-16) Jct. S.R. 963 to Gravel Pit Source "A" in NW 29-95-10-4 (ASA Permit 88-032).* Copy on file, Archaeological Survey, Historic Resources Management Branch, Alberta Culture, Edmonton.

Hutton, Michael J., Glen M. MacDonald, and Robert J. Mott

    1994   Postglacial Vegetation History of the Mariana Lakes Region, Alberta. *Canadian Journal of Earth Sciences* 31: 418–425.

Ives, John W.

    1993   The Ten Thousand Years Before the Fur Trade in Northeastern Alberta. In *The Uncovered Past: Roots of Northern Alberta Societies,* edited by Patricia A. McCormack and R. Geoffrey Ironside, pp. 5–31. Circumpolar Research Series no. 3. Canadian Circumpolar Institute, University of Alberta, Edmonton.

Ives, John W., and Mark M. Fenton

1983   Continued Research on Geological Sources of Beaver River Sandstone. In *Archaeology in Alberta 1981,* compiled by Jack Brink, pp. 78–88. Archaeological Survey of Alberta Occasional Paper no. 19. Historic Resources Management Branch, Alberta Culture, Edmonton.

1985   *Progress Report for the Beaver River Sandstone Geological Source Study (ASA Permit 83-054).* Copy on file, Archaeological Survey, Historic Resources Management Branch, Alberta Culture, Edmonton.

Krausman, Paul R., and R. Terry Bowyer

2003   Mountain Sheep (*Ovis canadensis* and *O. dalli*). In *Wild Mammals of North America: Biology, Management, and Conservation,* 2nd ed., edited by George A. Feldhamer, Bruce C. Thompson, and Joseph A. Chapman, pp. 1095–1115. Johns Hopkins University Press, Baltimore.

Kjorlien, Yvonne P., Leah Mann, and Jennifer C. Tischer

2009   *Historical Resources Impact Mitigation, Total E&P Joslyn Limited Joslyn North Mine Project, 2008 Mitigation Studies (HhOw-49, HhOw-54, HhOw-55, HhOx-10, HhOx-11, HhOx-15, HhOx-17, HhOx-18): Final Report (ASA Permit 08-166).* Copy on file, Archaeological Survey, Historic Resources Management Branch, Alberta Culture, Edmonton.

Kuntz, Michael L., and Richard E. Reanier

1995   The Mesa Site: A Paleo-Indian Hunting Lookout in Arctic Alaska. *Arctic Anthropology* 32(1): 5–30.

Le Blanc, Raymond J., and John W. Ives

1986   The Bezya Site: A Wedge-Shaped Core Assemblage from Northeastern Alberta. *Canadian Journal of Archaeology* 10: 59–98.

Lichti-Federovich, Sigrid, and James C. Ritchie

1968   Recent Pollen Assemblages in the Western Interior of Canada. *Review of Palaeobotany and Palynology* 7: 297–344.

McCullough, Edward J., and Brian O. K. Reeves

1978   *Historical Resources Impact Assessment Syncrude Canada Ltd. Western Portion of Lease 17 (ASA Permit 77-087).* Copy on file, Archaeological Survey, Historic Resources Management Branch, Alberta Culture, Edmonton.

McCullough, Edward J., and M. C. Wilson

1982   *A Prehistoric Settlement-Subsistence Model for Northeastern Alberta. Canstar Oil Sands Ltd. Bituminous Sands Leases 33, 92, and 95: A Preliminary Statement.* Environmental Research Monograph 1982-1. Canstar Oil Sands Ltd., Calgary.

Meyer, Daniel A.

2000   *Historical Resources Impact Assessment, Petro-Canada Oil and Gas, Mackay River Oil Sands Development (ASA Permit 00-118).* Copy on file, Archaeological Survey, Historic Resources Management Branch, Alberta Culture, Edmonton.

Miller, Frank L.

2003   Caribou (*Rangifer tarandus*). In *Wild Mammals of North America: Biology, Management, and Conservation,* 2nd ed., edited by George A. Feldhamer, Bruce C. Thompson, and Joseph A. Chapman, pp. 965–997. Johns Hopkins University Press, Baltimore.

Minni, Sheila J.

1976   *The Prehistoric Occupations of Black Lake, Northern Saskatchewan.* Mercury Series no. 53. Archaeological Survey of Canada, National Museum of Man, Ottawa.

Noble, William C.

1971   Archaeological Surveys and Sequences in Central District of Mackenzie, N.W.T. *Arctic Anthropology* 8(1): 102–135.

1977   The Taltheilei Shale Tradition: An Update. In *Problems in the Prehistory of the North American Subarctic: The Athapaskan Question,* edited by James W. Helmer, Stanley

Van Dyke, and François J. Kense, pp. 72–76. Chacmool Archaeological Association, Department of Archaeology, University of Calgary, Calgary.

Peace-Athabasca Delta Project Group

1972 *The Peace-Athabasca Delta: A Canadian Resource.* Summary report. Governments of Canada, Alberta, and Saskatchewan, Information Canada, Ottawa.

Peck, Trevor R.

2011 *Light from Ancient Campfires: Archaeological Evidence for Native Lifeways on the Northern Plains.* Edmonton: Athabasca University Press.

Preble, Edward A.

1908 *A Biological Investigation of the Athabasca-Mackenzie Region.* North American Fauna no. 27. U.S. Department of Agriculture, Bureau of Biological Survey, Washington, D.C.

Rains, R. Bruce, John Shaw, K. Robert Skoye, Darren B. Sjogren, and Donald R. Kvill

1993 Late Wisconsin Subglacial Megaflood Paths in Alberta. *Geology* 21: 323–326.

Raup, Hugh M., and George W. Argus

1982 *The Lake Athabasca Sand Dunes of Northern Saskatchewan and Alberta, Canada.* Publications in Botany no. 12. National Museums of Canada, Ottawa.

Reeves, Brian O. K.

1996 *Aurora Mine Project, Historical Resources Baseline Study.* Copy on file, Archaeological Survey, Historic Resources Management Branch, Alberta Culture, Edmonton.

Rhine, Janet L., and Derald G. Smith

1988 The Late Pleistocene Athabasca Braid Delta of Northeastern Alberta, Canada: A Paraglacial Drainage System Affected by Aeolian Sand Supply. In *Fan Deltas: Sedimentology and Tectonic Settings,* edited by W. Nemec and R. J. Steel, pp. 158–169. Blackie and Son, Glasgow and London.

Ritchie, James C., and Sigrid Lichti-Federovich

1967 Pollen Dispersal Phenomena in Arctic-Subarctic Canada. *Review of Palaeobotany and Palynology* 3: 235–266.

Roberts, Frank H. H.

1961 The Agate Basin Complex. In *Homenaje a Pablo Martínez del Río en el vigésimoquinto aniversario de la primera edición de Los orígenes americanos,* pp. 125–132. Instituto Nacional de Antropología e Historia, Mexico City.

Roe, Frank G.

1951 *The North American Buffalo.* University of Toronto Press, Toronto.

Ronaghan, Brian M.

2005 Cody Complex Occupation of the Boreal Forest Region Northeastern Alberta. Paper presented at the 63rd annual meeting of the Plains Anthropological Society, Edmonton, Alberta, 19–23 October.

Roskowski, Laura, and Morgan Netzel

2011 *Historical Resources Impact Mitigation, Shell Canada Energy, Stage II Mitigative Excavation of HhOv-351: Final Report (ASA Permit 10-148).* Copy on file, Archaeological Survey, Historic Resources Management Branch, Alberta Culture, Edmonton.

2012 *Historical Resources Impact Mitigation, Shell Canada Energy Muskeg Rive Mine Expansion, Area 6, Stage I Mitigation of Sites HhOv-156 and HhOv-520: Final Report (ASA Permit 11-167).* Copy on file, Archaeological Survey, Historic Resources Management Branch, Alberta Culture, Edmonton.

Roskowski, Laura, Alison J. Landals, and Morgan Blower

2008 *Historical Resources Impact Mitigation, Shell Canada Limited Albian Sands Muskeg River Mine Expansion Project, Mitigation for Sites HhOu-68, HhOu-69, HhOu-70, HhOu-94, HhOu-95, HhOv-378, HhOv-379, HhOv-380, HhOv-381, and HhOv-383: Final Report (ASA Permit 07-219).* Report on file, Archaeological Survey, Historic Resources Management Branch, Alberta Culture, Edmonton.

Saxberg, Nancy

2007   *Birch Mountain Resources Ltd. Muskeg Valley Quarry, Historical Resources Mitigation, 2004 Field Studies: Final Report (ASA Permit 05-118).* 2 vols. Copy on file, Archaeological Survey, Historic Resources Management Branch, Alberta Culture, Edmonton.

Saxberg, Nancy, and Brian O. K. Reeves

1998   *Aurora Mine North Utility Corridor, Historical Resources Impact Assessment, 1998 Field Studies: Interim Report (ASA Permit 98-039).* Copy on file, Archaeological Survey, Historic Resources Management Branch, Alberta Culture, Edmonton.

2003   The First Two Thousand Years of Oil Sands History: Ancient Hunters at the Northwest Outlet of Glacial Lake Agassiz. In *Archaeology in Alberta: A View from the New Millennium,* edited by Jack W. Brink and John F. Dormaar, pp. 290–322. Archaeological Society of Alberta, Medicine Hat.

2004   *Birch Mountain Resources Ltd. Muskeg Valley Quarry, Historical Resources Impact Assessment, 2003 Field Studies: Final Report (ASA Permit 03-249).* Copy on file, Archaeological Survey, Historic Resources Management Branch, Alberta Culture, Edmonton.

Smith, Derald G.

1989   Catastrophic Paleoflood from Glacial Lake Agassiz 9,900 Years Ago in the Fort McMurray Region, Northeast Alberta. Paper presented at the biennial meeting of the Canadian Quaternary Association (Late Glacial and Post-glacial Processes and Environments in Montane and Adjacent Areas), Edmonton, Alberta, 25–27 August. Abstract available in *Program and Abstracts,* p. 47.

Smith, Derald G., and Timothy G. Fisher

1993   Glacial Lake Agassiz: The Northwestern Outlet and Paleoflood. *Geology* 21(1): 9–12.

Soper, J. Dewey.

1964   *The Mammals of Alberta.* Hamley Press, Edmonton.

Steward, Julian H.

1955   *Theory of Culture Change: The Methodology of Multilinear Evolution.* Champaign: University of Illinois Press.

Stewart, Andrew M.

1991   Recognition of Northern Plano in the Context of Settlement in the Central Northwest Territories: Developing a Technological Approach. *Canadian Journal of Archaeology* 15: 179–191.

Stuart-Smith, A. Kari, Corey J. A. Bradshaw, Stan Boutin, Daryll M. Hebert, and A. Blair Rippen

1997   Woodland Caribou Relative to Landscape Patterns in Northeastern Alberta. *Journal of Wildlife Management* 61: 622–633.

Syncrude Canada Ltd.

1973   *Syncrude Lease No. 17: An Archaeological Survey.* Environmental Research Monograph 1973-4. Syncrude Canada Ltd., Edmonton.

1974   *The Beaver Creek Site: A Prehistoric Stone Quarry on Syncrude Lease No. 22.* Environmental Research Monograph 1974-2. Syncrude Canada Ltd., Edmonton.

Tsang, Brian W. B.

1998   The Origin of the Enigmatic Beaver River Sandstone. MSc thesis, Department of Geology and Geophysics, University of Calgary, Calgary.

Unfreed, Wendy J., and David Blower

2005   *Historical Resources Impact Assessment, Imperial Oil Resources Ventures Limited Kearl Oil Sands Project, Leases 6, 87, and 88A (Twps. 95 to 97, Rges. 7 to 8, W4M): Final Report (ASA Permit 04-375).* Copy on file, Archaeological Survey, Historic Resources Management Branch, Alberta Culture, Edmonton.

Unfreed, Wendy J., and Gloria J. Fedirchuk

    2001   *Historical Resources Impact Assessment, ATCO Pipelines Limited, Muskeg River Pipeline Project (Section 29-92-20-W4M to Section 23-95-10-W4M): Final Report (ASA Permit 00-064).* Copy on file, Archaeological Survey, Historic Resources Management Branch, Alberta Culture, Edmonton.

Vance, Robert E.

    1986   Pollen Stratigraphy of Eaglenest Lake, Northeastern Alberta. *Canadian Journal of Earth Sciences* 23: 11–20.

Vickers, J. Roderick

    1986   *Alberta Plains Prehistory: A Review.* Archaeological Survey of Alberta Occasional Paper no. 27. Historic Resources Management Branch, Alberta Culture, Edmonton.

Wickham, Michelle D., and Taylor Graham

    2009   *Historical Resources Impact Mitigation of the TransCanada Pipelines Ltd. Fort McKay Mainline Expansion: Final Report and Post-construction Audit (ASA Permits 06-376 and 07-266).* Copy on file, Archaeological Survey, Historic Resources Management Branch, Alberta Culture, Edmonton.

Wilson, Michael C., and James A. Burns

    1999   Searching for the Earliest Canadians: Wide Corridors, Narrow Doorways, Small Windows. In *Ice Age Peoples of North America: Environments, Origins, and Adaptations of the First Americans*, edited by Robson Bonnichsen and Karen L. Turnmire, pp. 213–248. Oregon State University Press, Corvallis.

Wormington, H. M., and Richard G. Forbis

    1965   *An Introduction to the Archaeology of Alberta, Canada.* Denver Museum of Natural History, Denver.

Woywitka, Robin J., Jennifer C. Tischer, Laura Roskowski, and Angela M. Younie

    2009   *Historical Resources Impact Mitigation, Fort Hills Energy Corporation, Fort Hills Oil Sands Project, 2008 Mitigation Studies: Final Report (ASA Permit 08-163).* Copy on file, Archaeological Survey, Historic Resources Management Branch, Alberta Culture, Edmonton.

Woywitka, Robin J., Angela M. Younie, Morgan Blower, and Alison J. Landals

    2008   *Historical Resources Impact Mitigation, Shell Canada Limited Albian Sands Muskeg River Mine Expansion Project, Mitigation for Sites HhOv-384, HhOv-385, HhOv-387, HhOv-431, and HhOv-432: Final Report (ASA Permit 07-280).* Copy on file, Archaeological Survey, Historic Resources Management Branch, Alberta Culture, Edmonton.

Wright, James V.

    1972   *The Shield Archaic.* Publications in Archaeology no. 3. National Museum of Man, Ottawa.

    1975   *The Prehistory of Lake Athabasca: An Initial Statement.* Mercury Series no. 29. Archaeological Survey of Canada, National Museum of Man, Ottawa.

Youell, A. J., Jennifer C. Tischer, Morgan Blower, and Lauren Copithorne

    2009   *Historical Resources Impact Mitigation, Total E&P Joslyn Limited, Joslyn North Mine Project, 2007 Mitigation Studies (HhOw-20, HhOw-21, HhOw-22, HhOw-24, HhOw-26, HhOw-27, HhOw-30, HhOw-38, HhOw-39, HhOw-40, HhOx-7): Final Report (ASA Permit 07-393).* Copy on file, Archaeological Survey, Historic Resources Management Branch, Alberta Culture, Edmonton.

6 A Chronological Outline for the Athabasca Lowlands and Adjacent Areas

BRIAN O. K. REEVES, JANET BLAKEY, AND MURRAY LOBB

Since the early 1970s, archaeological studies in the Athabasca lowlands and the Birch Mountains have recorded a large number of sites that contain significant numbers of stone tools, including diagnostic artifacts, among them projectile points (fig 6.1). While some of the earlier studies originated in academic research, most of them have taken place in connection with proposed oil sands development or local or regional infrastructure projects. Our intent in this chapter is to outline, refine, and update the cultural chronology of the Lower Athabasca region in the light of existing studies and on the basis of the provincial database of archaeological sites.

PAST CHRONOLOGICAL STUDIES

The first generalized chronologies of the Athabasca lowlands and adjacent areas date to the mid-1970s and were based on reconnaissance surveys, test excavations, and detailed excavations carried out by Archaeological Survey of Alberta staff archaeologists Paul Donahue (1976), John Ives (1977), and John Pollock (1978b). The results of these studies were later summarized by Ives (1981). Detailed discussions of chronology also appeared in the Historical Resource Impact Assessment reports pertaining to two major oil sands leases, Shell's Alsands Lease 13 (Ronaghan 1981a, 1981b) and Syncrude Lease 22 (Van Dyke and Reeves 1984). Ives (1993) subsequently provided an in-depth discussion of the

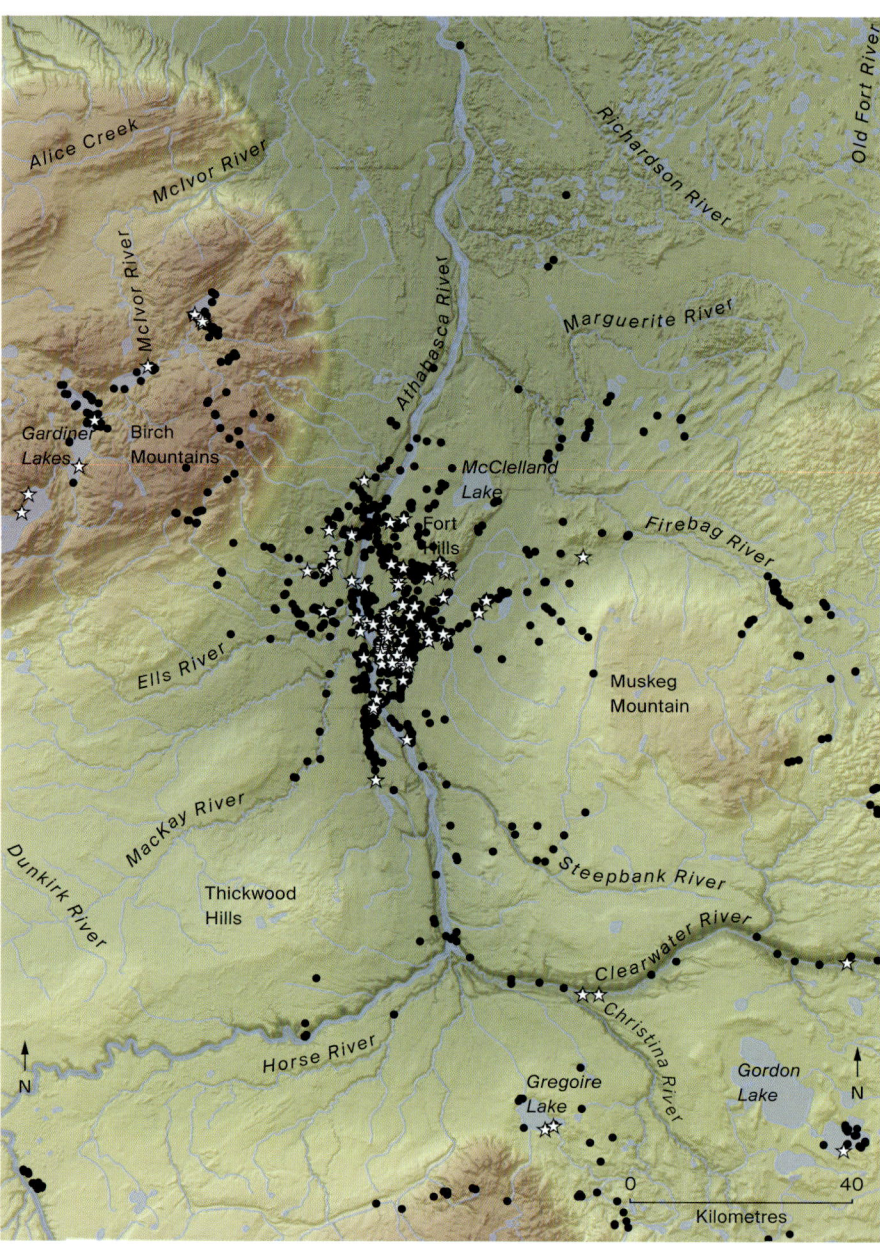

Figure 6.1. Lower Athabasca study area. Chronologically assignable sites are marked with a star.

cultural chronology of the oil sands area in the larger context of the precontact history of northeastern Alberta.

Ives (1993) did not, however, attempt to develop a local cultural sequence, given that only limited chronological data were available at the time and were derived primarily from the Birch Mountains. Saxberg and Reeves (2003) were the first to develop a localized cultural sequence. Their proposed chronology was based upon the archaeological assemblages found in the 1998 and 1999 mitigative excavations carried out in connection with Syncrude's Aurora North mine project. These assemblages were recovered from the Lake Agassiz flood zone and, by means of a comparative analysis of projectile points, were dated to roughly 9,900 to 7,000 BP. The cultural chronology presented in this chapter is a refined and considerably expanded version of that proposed by Saxberg and Reeves.

METHODS

In researching this study, our primary goals were to locate and review published papers and unpublished archaeological permit reports that discuss and illustrate projectile points and other chronologically sensitive tools from the region and then to group these artifacts into a series of cultural complexes (see figs. 6.2 and 6.3). To organize the data, we constructed a database of the relevant sites and the tools associated with them and then generated a series of maps for the cultural complexes. We identified a total of 156 sites, site components, or isolated occurrences of artifacts that could be assigned to a specific cultural complex.

Once this review of the literature was completed, we conducted a search of site inventory forms, which identified sites associated with permits for which final reports had not yet been submitted but that we considered pertinent to our research. This search turned up an additional six sites at which projectile points had reportedly been recovered. We then contacted the consulting companies or permit holders to request further information and, if possible, images of the points, although little was forthcoming. In addition, a number of projectile points were recovered in mitigative archaeological studies carried out by Altamira in the early 2000s in connection with the expansion of Highway 63. Although the reports on this work were not available, we were able to conduct a detailed examination of the points, which had been archived at the Royal Alberta Museum (RAM).

The principal published works consulted were Ives 1981 and 1993, as well as Saxberg and Reeves's 2003 review paper. Ives's articles are particularly useful,

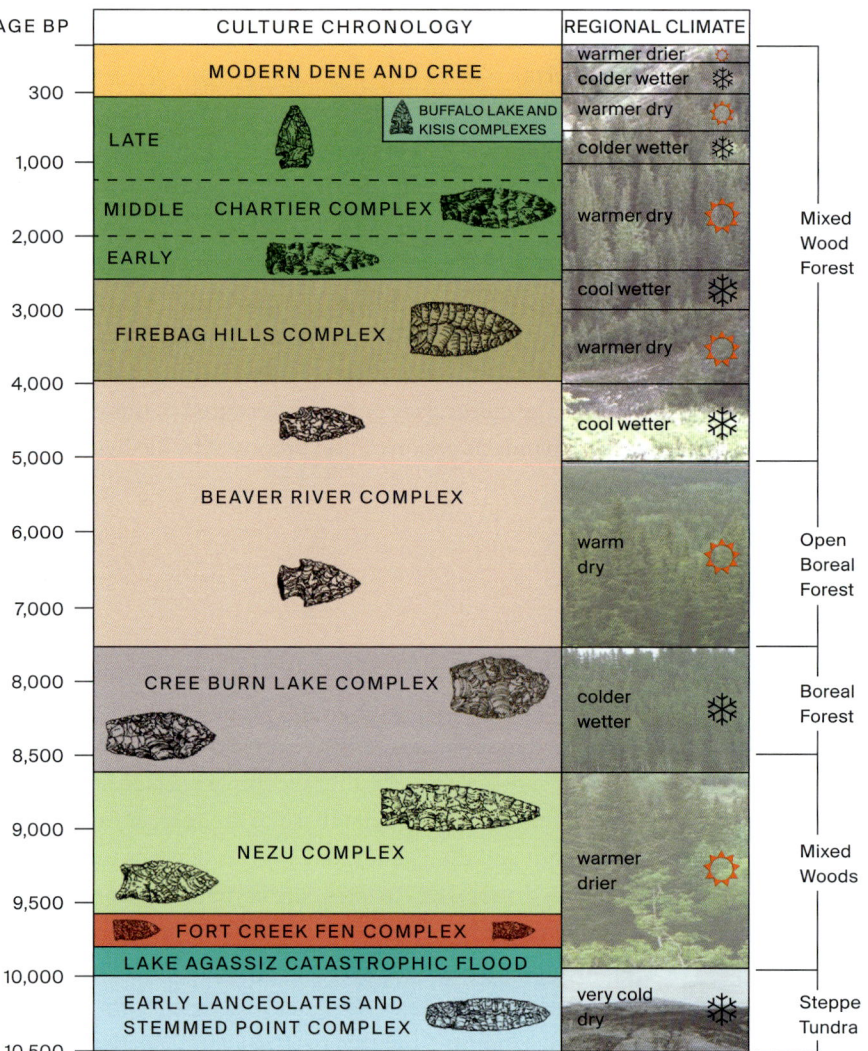

Figure 6.2. Cultural chronology of the Lower Athabasca region

AGE BP	CULTURE CHRONOLOGY	REGIONAL CLIMATE	
	MODERN DENE AND CREE	warmer drier / colder wetter	Mixed Wood Forest
300			
	LATE — BUFFALO LAKE AND KISIS COMPLEXES	warmer dry / colder wetter	
1,000			
	MIDDLE — CHARTIER COMPLEX	warmer dry	
2,000			
	EARLY		
3,000		cool wetter	
	FIREBAG HILLS COMPLEX	warmer dry	
4,000			
		cool wetter	
5,000	BEAVER RIVER COMPLEX		
6,000		warm dry	Open Boreal Forest
7,000			
8,000	CREE BURN LAKE COMPLEX	colder wetter	Boreal Forest
8,500			
9,000	NEZU COMPLEX	warmer drier	Mixed Woods
9,500			
	FORT CREEK FEN COMPLEX		
10,000	LAKE AGASSIZ CATASTROPHIC FLOOD	very cold dry	Steppe Tundra
	EARLY LANCEOLATES AND STEMMED POINT COMPLEX		
10,500			

for they contain information on and illustrations of projectile points recovered by Cort Sims during excavations in the mid-1970s at the Gardiner Lake Narrows site (HjPd-1) in the Birch Mountains.[1] HjPd-1 is a stratified site of key significance to the region, and yet no final report on it exists. Other relevant regional reconnaissance-level studies from the mid-1970s include Donahue's 1975 survey of the Clearwater and Athabasca rivers and the Birch Mountains (Donahue 1976), as well as Pollock's 1976 survey of the Clearwater and Christina drainages (Pollock 1978b). Another important but unpublished document from this time is Ives's

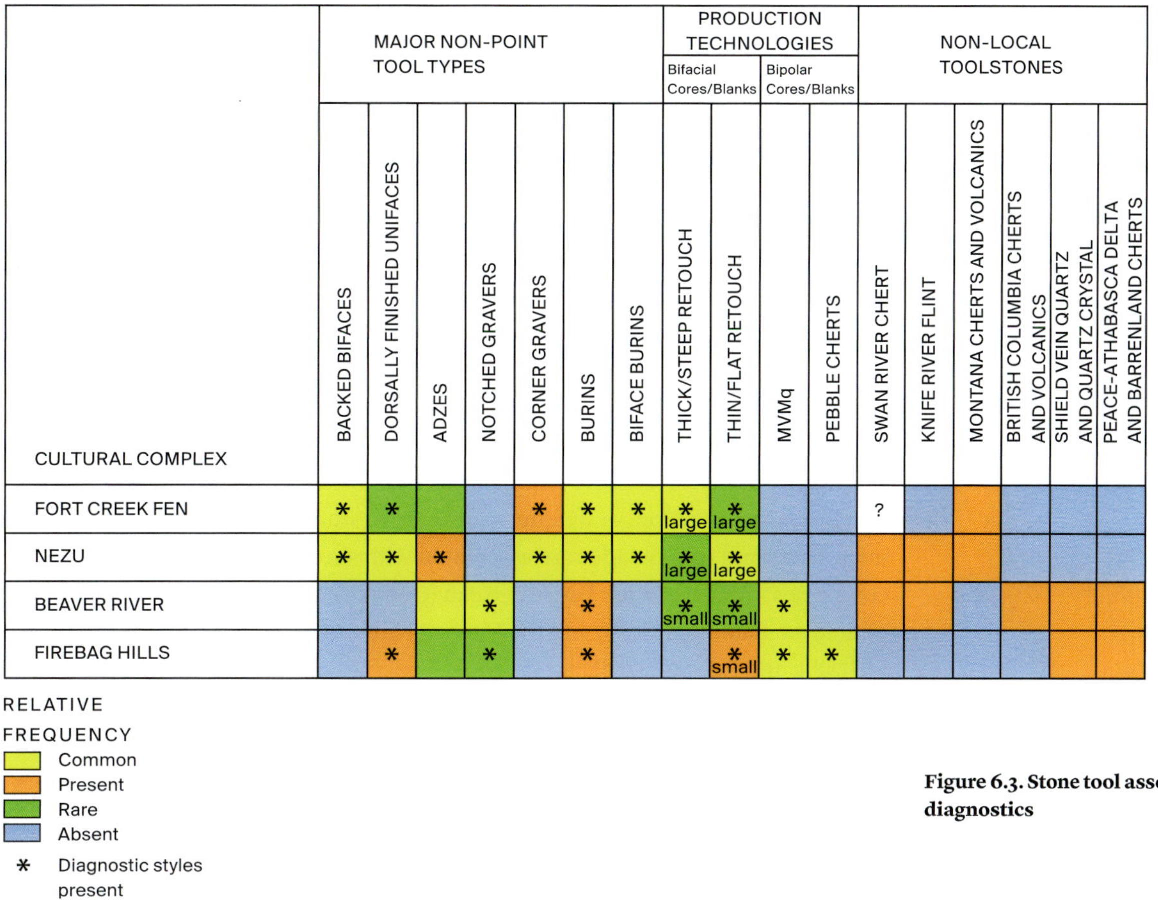

| CULTURAL COMPLEX | MAJOR NON-POINT TOOL TYPES | | | | | | | PRODUCTION TECHNOLOGIES | | | | NON-LOCAL TOOLSTONES | | | | | |
| | Backed Bifaces | Dorsally Finished Unifaces | Adzes | Notched Gravers | Corner Gravers | Burins | Biface Burins | Bifacial Cores/Blanks | | Bipolar Cores/Blanks | | Swan River Chert | Knife River Flint | Montana Cherts and Volcanics | British Columbia Cherts and Volcanics | Shield Vein Quartz and Quartz Crystal | Peace-Athabasca Delta and Barrenland Cherts |
								Thick/Steep Retouch	Thin/Flat Retouch	MVMq	Pebble Cherts						
FORT CREEK FEN	*	*			*	*	*	* large	* large			?					
NEZU	*	*	*		*	*	*	* large	* large								
BEAVER RIVER				*		*		* small	* small	*							
FIREBAG HILLS		*		*	*	*			* small	*	*						

RELATIVE
FREQUENCY
- Common
- Present
- Rare
- Absent

* Diagnostic styles present

Figure 6.3. Stone tool assemblage diagnostics

(1977) permit report on his excavations at the Eaglenest Portage site (HkPa-4) in the Birch Mountains.

Over the past few decades, a great many archaeological permits have been issued in connection with Cultural Resource Management (CRM), Historical Resource Impact Assessment (HRIA), and Historical Resource Impact Mitigation (HRIM) studies. In comparison to their number, however, and to the number of shovel prospects excavated, relatively few of these studies have resulted in the recovery of diagnostic projectile points. Prior to the late 1990s, when extensive compliance work resumed, several studies did recover diagnostics. Notable among these were the 1973 HRIA of Syncrude Lease 17 (Syncrude Canada Ltd. 1973), the 1974 Beaver River Quarry studies (Syncrude Canada Ltd. 1974), the 1974 HRIA of Highway 63 (Losey, Freeman, and Priegert 1975), and the 1979

HRIA of the Highway 63 approach to the Peter Lougheed Bridge (Gryba 1980), all on the west side of the Athabasca River. Significant studies on the east side of the Athabasca River included the 1974 surface survey of Shell Lease 13 (Sims and Losey 1975), although this report is unfortunately quite cursory and its illustrations of the artifacts recovered are less than satisfactory.

In the 1980s, the first relevant studies are the Alsands and Fort McMurray Energy Corridor HRIAs (Ronaghan 1981a, 1981b). The Bezya site (HhOv-73), a provincially significant site that was identified during the Alsands HRIA, was the first site at which evidence of microblade technology was discovered (Ronaghan 1981b). Subsequent excavations at the site by the Archaeological Survey of Alberta in the early 1980s (Le Blanc 1985; Le Blanc and Ives 1986) yielded a radiocarbon date but unfortunately did not recover any projectile points, nor did later mitigative excavations by Golder Associates Ltd. (Green and Blower 2005). The Bezya site is no longer unique, as mitigative studies have since indicated that microcores and microblades are associated with sites grouped with four cultural complexes: the Fort Creek Fen, the Nezu (Cody Complex), the Beaver River (Shield Archaic Tradition), and the Firebag Hills (Arctic Small Tool Tradition) (de Mille and Reeves 2010).

Studies over roughly the past fifteen years have considerably augmented the sample for the east side of the Athabasca River. These include studies associated with Shell Albian Sands' Muskeg River and Jackpine mines (Ronaghan 1997; Clarke 2002b; Clarke and Ronaghan 2000; Clarke and Ronaghan 2004; Green et al. 2006; Tischer 2004, 2005; Bouchert-Bert 2007); Birch Mountain Resources' Muskeg Valley Quarry and Hammerstone projects (Saxberg and Reeves 2004, 2006; Saxberg 2007; de Mille and Reeves 2009); Syncrude's Aurora North project (Shortt and Reeves 1997; Shortt, Saxberg, and Reeves 1998; Saxberg and Reeves 1998; Saxberg, Shortt, and Reeves 1998; Reeves, Bourges, and Saxberg 2009; Reeves et al. 2013a, 2013b, 2014a, 2014b, 2014c; Saxberg, Somer, and Reeves 2003; Somer 2005; Somer and Kjar 2007); and TransCanada's Fort McKay Mainline pipeline project (Wickham 2006a, 2006b). For the west side of the Athabasca River, studies associated with the Horizon Oil Sands Project, an undertaking by Canadian Natural Resources Limited (CNRL), have provided the first new chronological information since the pioneering studies of the 1970s (see Bryant 2004, 2005; Clarke 2002a; Tischer 2006).

TOOL STONE TYPES AND TERMINOLOGY

Most readers are already aware that the vast majority of artifacts in the Athabasca lowlands are manufactured from a locally obtained material variously known as Beaver River Sandstone, Beaver River Silicified Sandstone, Fine-Grained Beaver River Silicified Sandstone, and Muskeg Valley Microquartzite (MVMq). The search for the source of the fine-grained variety has been of continuing interest since the beginnings of archaeology in the oil sands region. When the Beaver River quarry was excavated in 1974 (Syncrude Canada Ltd. 1974; Reardon 1976), it was discovered that the bedrock tool stone quarried at the site, a coarse Beaver River Silicified Sandstone, was not in fact the source of the small chipped stone tools found in excavations at the quarry-workshop; rather, these tools had been manufactured from a fine-grained variety of the stone. In the same year, that variety was shown to be common in assemblages from the east side of the Athabasca River in the area of Shell Lease 13 (Sims and Losey 1975).

The search continued over the following two decades for a primary quarry source on the east side of the Athabasca River (see Fenton and Ives 1982, 1984, 1990; Ives 1993). While surface exposures of the coarse-grained variety were occasionally found, as were cobbles and boulder-sized occurrences of similar material (often at sites identified as workshops), a primary source of the fine-grained variety remained elusive. This changed in 2003, when Lifeways of Canada Limited carried out an HRIA in connection with Birch Mountain Resources' Muskeg Valley limestone quarry (Saxberg and Reeves 2004). During this assessment, two bedrock quarries of the fine-grained variety of the tool stone were identified, in an area now collectively known as the Quarry of the Ancestors (Saxberg and Reeves 2006; Saxberg 2007). Subsequent petrographic analysis of samples from the quarry determined that the material is best classified as a micro-quartz-cemented orthoquartzitic siltstone (De Paoli 2005). Saxberg, Reeves, and De Paoli have thus chosen to call it Muskeg Valley Microquartzite (MVMq), rather than Beaver River Sandstone.[2] Given that MVMq dominates the lowlands assemblages, in the discussions to follow we do not usually specify that a particular point or other artifact is manufactured of it. The reader should thus assume that, unless otherwise indicated, the tool stone type is MVMq.

The second most common tool stone from which artifacts are manufactured is quartzite of various kinds, exhibiting a range of colours and granule sizes and apparently recovered in cobble form. In the Lower Athabasca region, many of these varieties fall into Gryba's Northern quartzite category (Gryba 2001), which most commonly consists of high-quality brown and grey and silver-grey-white quartzites, sometimes with orthoquartz inclusions.[3] The specific sources for the cobbles are unknown, although one may lie in the Birch Mountains, where the assemblages are characterized by large amounts of grey quartzite of varying quality.

Salt and pepper quartzite is another variety, one that occurs primarily in sites on the west side of the Athabasca River (see Ives 1993). This material type was first identified during the HRIA of the proposed Canstar Project (McCullough and Wilson 1982), and large boulders of salt and pepper quartzite have since been identified in CRM studies connected with CNRL's Horizon Project, which incorporates portions of the former Canstar Project area. Salt and pepper quartzite is the dominant material type in tools and debitage recovered from these project areas (Clark 2002a; Bryant 2004, 2005; Tischer 2006). Curiously, however, this variety of quartzite is quite infrequent in sites that belong to the same cultural complexes but are located on the east side of the Athabasca River.

In addition, to MVMq and quartzite, a limited variety of silicified siltstones, cherts, and chalcedony tool stone types occur in Lower Athabasca artifact assemblages. Interestingly, they tend to be associated with tools other than projectile points and occur more frequently at sites belonging to post–Early Precontact period complexes, particularly the Firebag Hills Complex. The most common are pebble cherts, often found in Chartier Complex sites (although less so in the Firebag Hills Complex). Their secondary sources could be local proglacial and postglacial gravels, lag gravels, or erosional caps or chert conglomerate stringers in bedrock. In the Plains region to the south, these stringers are found in the Upper Cretaceous marine Bearspaw Formation, which appears to have been actively mined in the Neutral Hills and adjacent to Sullivan Lake in east-central Alberta (Brian Ronaghan, pers. comm., 2006). Our understanding of the Chartier Complex would be enhanced by determining the source(s) of these pebble cherts in the Lower Athabasca region.

One type of chert that appears occasionally, particularly in Beaver River and Chartier Complex sites, is Lake One Dune chert, which is common in collections from sites on Lac Claire (Stevenson 1981). This chert, which can be mistaken for MVMq, ranges from a blend of creamy white and grey to a dirty brown, with inclusions and fossil fragments. It is different both in texture and structure from the cherts at the Peace Point site (IgPc-2; see fig. 6.4), which are nonetheless similar in colour, ranging from white to cream to brown (Stevenson 1986; Reeves, observation of the Lac Claire collections at Parks Canada, Winnipeg, 2004). At the Lake One Dune site (IgPc-9; see fig. 6.4), the stone was obtained in sufficiently large enough pieces to fabricate large-sized (roughly 6 to 10 cm) lanceolate points.

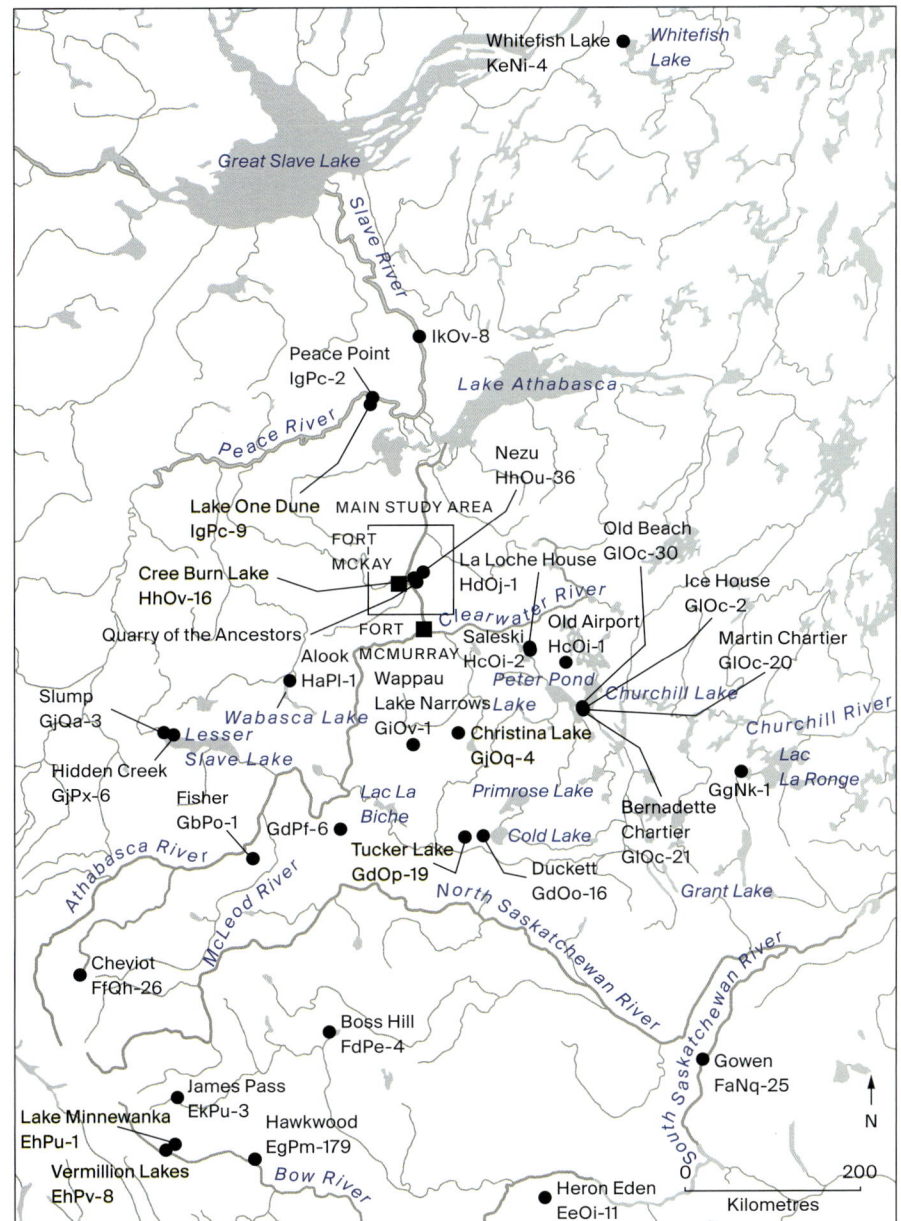

Figure 6.4. Sites in the region surrounding the Lower Athabasca valley. The main study area lies within the square. The broader region encompasses the Peace, Athabasca, and North Saskatchewan river basins and extends from as far north as Whitefish Lake, in the Northwest Territories, all the way south to the Bow River, as well as eastward into Saskatchewan.

We now turn to our outline of the cultural chronology, the primary focus of this chapter.

EARLY PRECONTACT PERIOD COMPLEXES (CA. 10,000 TO 7,750 BP)

The earliest evidence of human occupation in Alberta consists of projectile points typical of the Clovis and Folsom cultural complexes. As expressed in Alberta, these complexes, which are generally termed Early Palaeoindian, have been dated to the period from roughly 11,050 BP to 10,200 BP (see Peck 2011, 24–47). Early Precontact period archaeological complexes, which postdate the Clovis and Folsom complexes, are characterized by lanceolate points that may be broad- or narrow-bladed and stemmed or unstemmed. Generally considered to be Middle or Late Palaeoindian, these complexes span a period from about 10,000 BP to 7,750 BP. Around the end of this period, corner- and side-notched dart points first appear in sites in the Saskatchewan plains and parklands and in the Rocky Mountains to the south of the Athabasca region.

As noted above, in his discussion of the Early Precontact period in the Lower Athabasca region, Ives (1993, 9) did not propose a detailed chronology, given the limited projectile point data available at that time. Rather, he drew primarily on general comparisons of the lanceolate and stemmed points that had thus far been found in the area with points known from the Yukon, Alaska, and the Plains region to the south.

Saxberg and Reeves (2003) subsequently proposed three Early Precontact period complexes for the Athabasca lowlands, all of which postdated the Lake Agassiz flood: the Fort Creek Fen Complex (ca. 9,800–9,600 BP), the Nezu Complex (ca. 9,600–8,600 BP), and the Cree Burn Lake Complex (ca. 8,600–7,750 BP). Their constructs were based on the distinctive projectile points, as well as certain other key elements, in the assemblages recovered from excavations in sites on the east side of the Athabasca River associated with elevated landforms along the Aurora North utility corridor (Saxberg, Shortt, and Reeves 1998; Reeves et al. 2013a, 2013b, 2014a, 2014b, 2014c). Also reflected in this chronology were the results of studies from the 1970s and 1980s that had recovered projectile points from sites in the lowlands, such as the Beaver Creek Quarry excavations (Syncrude Canada Ltd. 1974) and the 1974 survey of Shell Lease 13 (Sims and Losey 1975).

Initial human settlement of the Athabasca lowlands and adjacent uplands, such as the Birch Mountains, was naturally controlled by the timing of deglaciation, the recession of flood waters, and subsequent repopulation of the landscape by plant and animal communities. Palynological data from sites such as Kearl Lake (see chapter 4 in this volume), as well the radiocarbon-dated spruce trees recovered in the late 1970s from the gravel and sand pit area at Syncrude's Mildred Lake facility (Van Dyke and Reeves 1984), indicate that pioneering spruce-dominated boreal forest communities were well established in the lowlands by about 10,200 to 10,000 BP. As we will see, the evidence is that humans were also present at this time.

Pre-flood (?) and Early Post-flood Lanceolate and Stemmed-Point Complexes

The earliest identifiable cultural complexes in Alberta are characterized by stemmed lanceolate projectile points, including the Agate Basin and Hell Gap styles, as well as by fluted points and points exhibiting basal thinning, that generally date in the range of about 11,000 to 9,500 BP. Basal thinning is a technique that results in attributes functionally similar to those produced by classic fluting, and such points have been grouped with those exhibiting true fluting to describe an Alberta variant of the Fluted Point Tradition (Gryba 2001). This tradition is attested at sites scattered throughout the central and southern regions of the province.

In assessing the evidence for the Athabasca lowlands, we must remember that the prime settlement areas, which would have lined the existing banks of the Athabasca and Muskeg rivers, were destroyed by the Lake Agassiz flood—which, at its maximum extent, submerged the Athabasca valley beneath water up to an elevation of roughly 300 metres above sea level. The earliest indications of human occupation within the Athabasca lowlands consist of water-rolled artifacts found both within and above the Lake Agassiz flood zone. In addition, a number of projectile points have been recovered from sites at or above the flood line, as well as in the Birch Mountains, that exhibit features typical of point styles that predate the flood.[4]

Water-rolled artifacts have been recovered in excavations at three sites located within the flood zone (HhOv-4, HhOv-163, and HhOv-173) and one (HhOu-52) in the Fort Hills (table 6.1 and fig 6.5). The three artifacts from within the flood zone might have been found elsewhere by later occupants of the area, who then transported them to these sites, or they might have been deposited at these sites by fluvial processes associated with the flood. The finds consist of three examples of water-rolled quartzite bifaces—from HhOv-4, a Nezu Complex campsite and workshop; from HhOv-173, an isolated find; and from HhOu-52, a Beaver River Complex campsite on Stanley Creek, on the south edge

TABLE 6.1 Sites yielding pre-flood water-rolled artifacts and early lanceolate points

Borden no.	Elevation (masl)	Projectile points and or other diagnostic tools	Reference(s)
Water-rolled artifacts			
HhOv-4	280.71	Ovate quartzite biface	Mallory 1980, figure 4: a
HhOv-163	295.93	Athabasca quartzite Hell Gap stemmed point base	Saxberg, Shortt, and Reeves 1998; Reeves et al. 2013a
HhOv-173	303.00	Quartzite biface	Saxberg, Shortt, and Reeves 1998, plate 53: 1
HhOu-52	287.78	Quartzite biface	Somer and Kjar 2007, plate 53: d
Lanceolate points			
HiOu-34	300.54	Stemmed lanceolate point (fire-fractured)	Saxberg, Somer, and Reeves 2003, plate 74: a
HiOu-56	310.00	Megaquartzite broad-bladed stemmed lanceolate point	Somer 2005, plate 51: d
HhOv-132	295.35	Lanceolate midsection	Site inventory form
HhOv-174	301.33	2 lanceolate point fragments	Shortt, Saxberg, and Reeves 1998, plate 35: 6
HhOv-455	297.97	Black silicified siltstone Agate Basin lanceolate point	Tischer 2005, plate 72
HhOu-1	285.00	Hell Gap point stem	Sims and Losey 1975; Ives 1993, figure 2: c
HjPd-1	683.00	Tertiary Hills clinker Agate Basin point	Ives 1981; Ives 1993, figure 2: a
HkPa-4	714.00	Lanceolate point	Ives 1993, figure 2: g

of the Fort Hills—as well as the stem of a quartzite Hell Gap point snapped at the shoulders (plate 6.6: 6) from HhOv-163, another Beaver River Complex camp-site. The only other Hell Gap point identified to date is a specimen also snapped across the shoulders that was recovered by Sims in 1974 from the surface at HhOu-1, the Shell Airstrip site on the Muskeg River (Ives 1993, figure 2: c; Sims and Losey 1975, figure 3), but it is unclear from existing descriptions whether this artifact is water-rolled.

In addition, six finds of lanceolate points that appear, from their stylistic features, to predate the flood have been recovered from sites at and just above the maximum flood level (300 metres above sea level) in various locales east of the Athabasca River.[5] Five of these six are point fragments that were discovered along the southern edge of the Fort Hills. These consist of:

- a fire-fractured stemmed lanceolate point, broken at mid-blade, with a damaged base from HiOu-34 (Saxberg, Somer, and Reeves 2003, plate 74: a)
- a snapped, brown megaquartzite stemmed point with excurvate blade edges from HiOu-56 (Somer 2005, 65 and plate 51: d)
- the mid-blade section, with excurvate lateral edges, of a finely made subparallel flaked lanceolate point from HhOv-132, a lithic scatter site located on a beach ridge (Reeves, personal examination of the collections at RAM, 2006)

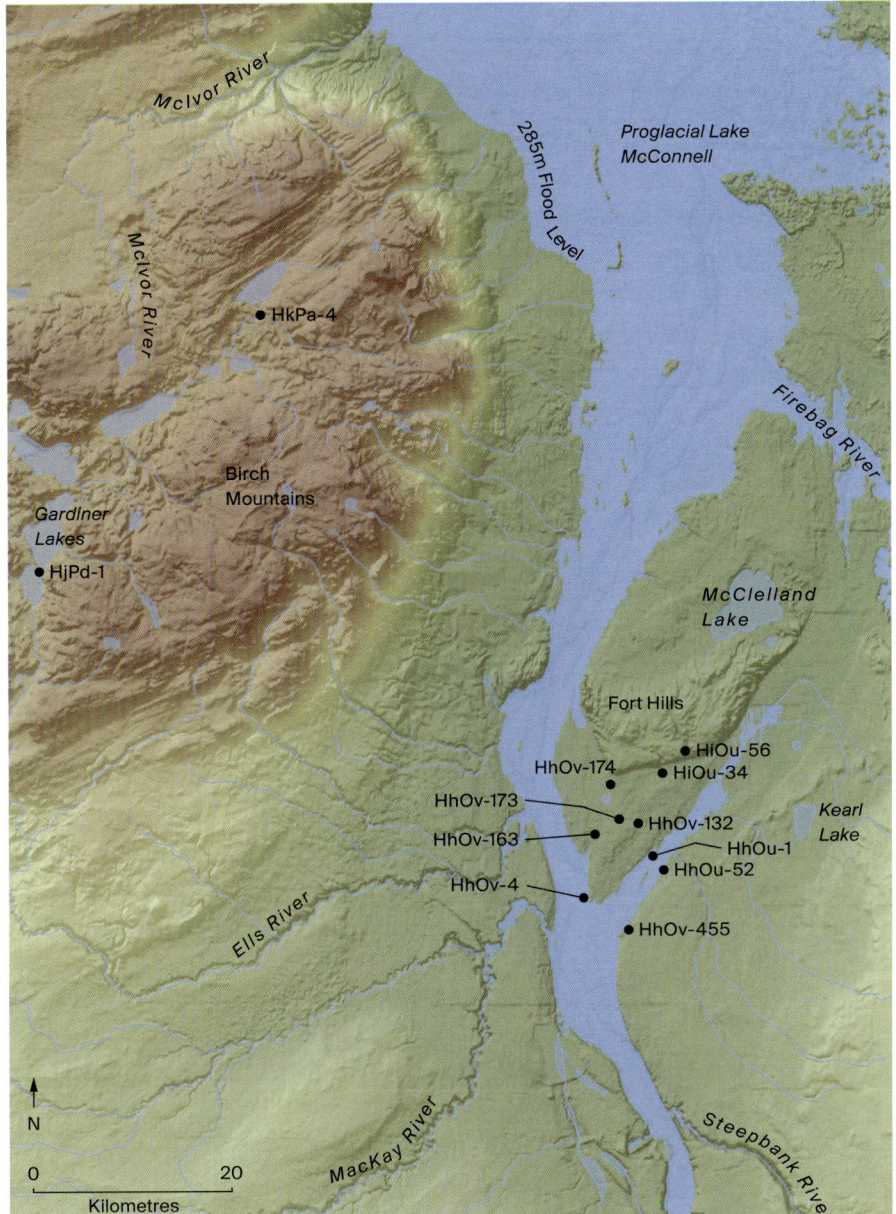

Figure 6.5. Sites yielding water-rolled artifacts or early lanceolate points (ca. 10,000 to 9,600 BP)

- portions of two different lanceolate projectile points that were unearthed in a single shovel test at HhOv-174 (Shortt, Saxberg, and Reeves 1998, 10 and plate 35: 6)

The sixth find was made a little to the south, on the lower Muskeg River. In the course of compliance studies (Tischer 2004, 2005), during which a number of sites were identified at or above the flood line, a point fragment was recovered from HhOv-455, a small lithic scatter site (Tischer 2005, 153 and plate 72). It is a black silicified siltstone lanceolate point, 8.1 centimetres long. Tischer typed it as Agate Basin, as it compares most favourably with some of the "classic" specimens illustrated by Frison and Stanford (1982) from the Agate Basin type site. The probable source of the tool stone is the Palaeozoic-age black silicified siltstones and cherts found in the Rocky Mountains (Landals 2008; Reeves, Bourges, and Saxberg 2009).

In the Birch Mountains, early lanceolates are represented by a complete, finely finished Agate Basin point made of Tertiary Hills clinker (then known as Tertiary Hills welded tuff) that was excavated by Sims from the Gardiner Lake Narrows site, HjPd-1 (Ives 1981, figure 7: first row, first point on left; Ives 1993, figure 2a). These early points also include obliquely flaked lanceolates recovered from this site and from Eaglenest Portage (HkPa-4). (These are discussed below, under the Cree Burn Lake Complex.)

The above evidence clearly suggests an initial occupancy of the Lower Athabasca region possibly prior to, almost certainly coincident with, and clearly shortly after the Lake Agassiz flood of about 9,900 BP.[6] Lanceolate and stemmed points that fall within the range of variation found in the Agate Basin and Hell Gap types have been recovered in both primary and secondary contexts. Although the data are scarce, we suggest that the early occupants made only limited use of MVMq, given the probable impoundment of glacial outwash in the lowlands and the consequent submergence of sources of the stone, and that, should large sites be discovered above the flood zone, it is likely that the assemblages will be dominated by non-lowland quartzites. Exotic tool stones present are Rocky Mountain Palaeozoic black silicified siltstone and Tertiary Hills clinker. These are indicative of early distant trade, exchange, and movement networks of the first peoples to enter the region.

The Fort Creek Fen Complex (ca. 9,800 to 9,600 BP)
The Fort Creek Fen Complex was defined on the basis of 1998 excavations at two sites, HhOv-87 and HhOv-164, both located in the Aurora North utility corridor northeast of Cree Burn Lake (fig 6.6). The most common type of point

Alberta's Lower Athabasca Basin

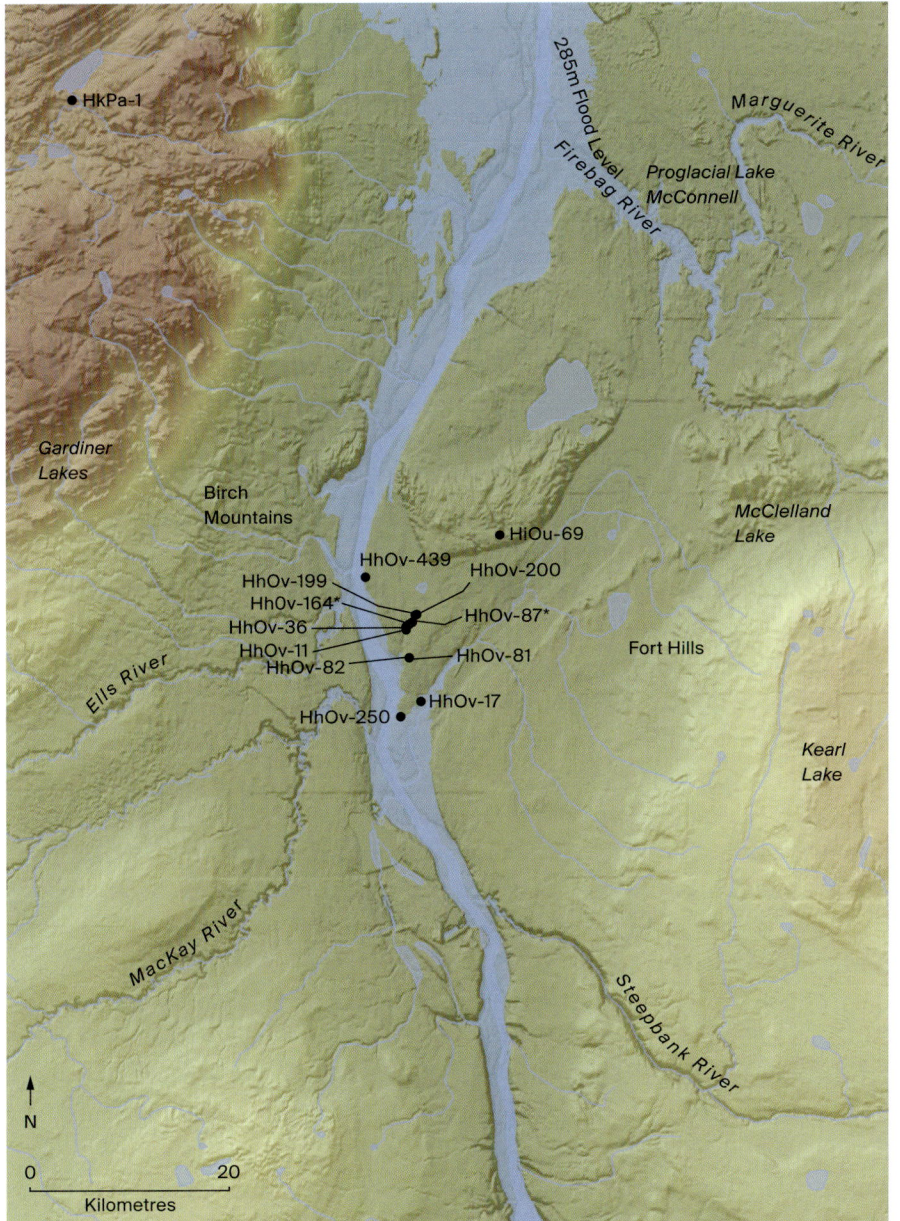

Figure 6.6. Fort Creek Fen Complex sites (ca. 9,800 to 9,600 BP). An asterisk indicates that a Scottsbluff point and/or Cody knife occurred on the same landform.

found—the Fort Creek Fen lanceolate—was a thin, broad-bladed, waisted form with constricted lateral edges and a straight to slightly concave base, sometimes with incipient side notches set within the lateral constrictions near the base (plate 6.1; Saxberg and Reeves 2003, 307–308). These points are comparable to points recovered from excavations at sites in the greater Yellowstone region that date to approximately 9,500 to 9,400 BP, including Barton Gulch (Davis et al. 1989; the Alder point type), Mummy Cave (Husted and Edgar 2002), and Medicine Lodge Creek (Frison 1991).

The Fort Creek Fen lanceolates are also similar in form and manufacture to points found in the region of the glacial Great Lakes. These include points associated with the Chesrow Complex in Wisconsin (Overstreet 1993) and the Hi-Lo Complex as expressed in southwestern Ontario, along the north shore of Lake Ontario, as well as elsewhere in the Great Lakes area (Ellis 2004; Jackson 2004; Stewart 2004). The Chesrow and the Hi-Lo are related complexes (Ellis 2004) and are considered to date to roughly 10,000 to 9,500 BP. The most recent point variants in these complexes include forms similar to the oil sands specimens, with shallow side notches set within the lateral constrictions above the base, that have been radiocarbon dated to as early as 9,600 ^{14}C yr BP (Ellis 2004, 64).

In addition to projectile points, Fort Creek Fen sites have yielded a number of other tools (table 6.2). Bifaces include a distinctive backed lanceolate-semi-lunar form that is also part of the Hi-Lo assemblage (Ellis 2004, 65 and figure 3-4: C). Fort Creek Fen end scrapers also share similarities with Hi-Lo end scrapers in their overall size and shape, and both assemblages are characterized by the absence of the finely finished dorsally retouched end scrapers that are found in later Nezu (Cody Complex) sites. Convex-edge side scrapers and marginally utilized and/or retouched flakes, generally manufactured from thick core or flake fragments and often backed, are again characteristic of both Fort Creek Fen and Hi-Lo assemblages. In addition, notches and denticulates occur in Fort Creek Fen assemblages (but not in Hi-Lo ones, although they may not have been recognized as such). Fort Creek Fen burins are produced on intentionally snapped or radially fractured biface blanks as well as flake blanks. Forms include dihedral, angle, and transverse burins; notched "Donnelly Ridge"–style flake burins are rare.

There is also some evidence of microblade technology. Small numbers of microblade cores and microblades were recovered from both HhOv-87 and HhOv-164. The blanks produced for microblade preforms are either wedge- or boat-shaped or else hemi-conical and can be further modified and used as formal wedge- or boat-shaped microblade cores (see plate 6.10: 1, 2) or as less formal face-faceted and pillar microblade cores. These blanks are derived from

TABLE 6.2 Fort Creek Fen Complex sites and diagnostic artifacts

Borden no.	Elevation (masl)	Projectile points or other diagnostic tools	Reference(s)
HhOv-11	294.00	backed bifaces	Reeves et al. 2014a
HhOv-17	276.13	Fort Creek Fen lanceolate point	Green et al. 2006, plate I: 23
HhOv-36	295.00	microblade core	Reeves et al. 2014a
HhOv-82	296.26	Stubby Fort Creek Fen lanceolate point	Clarke and Ronaghan 2004, plate HhOv-82-1
HhOv-87	297.90	2 Fort Creek Fen lanceolate points	Reeves et al. 2014a
HhOv-164	294.23	3 Fort Creek Fen lanceolate points, 2 Scottsbluff points, 1 Cody knife	Reeves et al. 2014a
HhOv-199	295.00	backed bifaces	Reeves et al. 2014a
HhOv-200	295.00	backed bifaces	Reeves et al. 2014a
HhOv-250	280.91	Fort Creek Fen lanceolate point	Reeves, examination at RAM
HhOv-439	284.77	Fort Creek Fen lanceolate point	Somer and Kjar 2007: plate 52: q
HiOu-69	313.00	Scottsbluff Type I point	Somer 2005, plate 51: e
HjPd-1	683.00	Fort Creek Fen point?	Reeves, examination at RAM
HkPa-1	716.84	Fort Creek Fen point?	Ives 1977, plate 21: j

thick bifacial cores that are either transversely or longitudinally snapped or burinated. This technology is very comparable to that described by Fladmark (1985) for the Ice Mountain Microblade Tradition (Smith 1971, 1974) of the Mount Edziza–Telegraph Creek region in northwestern British Columbia.[7] Similar cores of Edziza obsidian have recently been recovered from the Finlay Reach of Lake Williston (Eldridge et al. 2008), as have points resembling Fort Creek Fen lanceolates (Reeves, personal examination of the Lake Williston collections; Eldridge et al. 2010).

Non-local tool stones are relatively uncommon in Fort Creek Fen assemblages. Out of the 51,594 pieces of debitage initially recovered at HhOv-87, only 26 were non-local, consisting of twenty chert and six Northern quartzite flakes.[8] The cherts here and at HhOv-164 are primarily black to brownish-grey varieties, often with a weathered cortex and most of unknown provenance. Montana cherts are very rare. In addition to Northern quartzites, the HhOv-164 assemblage contains examples of Athabasca quartzite (Landals 2008; Meyer, Roe, and Dow 2007; Reeves, Bourges, and Saxberg 2009).

In addition to HhOv-87 and HhOv-164, Fort Creek Fen Complex artifacts have been recovered from seven other sites in Syncrude's Aurora North project area (HhOv-11, HhOv-36, HhOv-82, HhOv-199, HhOv-200, HhOv-250, HhOv-439), as well as at HhOv-17, on the Muskeg River, and at HiOu-69, in the Fort Hills (see fig 6.6). In addition, there are two possible occurrences of Fort Creek Fen points at sites in the Birch Mountains (see table 6.2). At HhOv-439, a finely worked bifacial meat-filleting knife reacted positively to sheep antisera (Somer 2005).

In summary, we view the Fort Creek Fen Complex as representing the first post-flood archaeological reoccupation of the Lake Agassiz flood zone in the Athabasca lowlands. While the number of identified sites is small in comparison to the following Nezu Complex, suggesting a shorter temporal span for the occupation than for Nezu, some sites, such as HhOv-87 and HhOv-164, are large workshops or campsites with dense accumulations of MVMq debris. This may indicate that Fort Creek Fen Complex groups had access to large quantities of high-quality MVMq tool stone, which could have been obtained from MVMq blocks deposited by the Agassiz flood along the Athabasca River, or from as-yet-undiscovered primary outcrops along the river, or from the Quarry of the Ancestors, the higher portions of which would have been accessible when the water levels dropped below 290 metres above sea level.

Scottsbluff points were also found, although not necessarily in the same excavation loci as the Fort Creek Fen points, at HhOv-87 and HhOv-164, the latter of which also yielded a Cody knife. This suggests that these particular sites are transitional and that, by extension, the majority, if not all, of the Fort Creek Fen sites date within a few hundred years of each other, since they are relatively few in comparison to those of the Nezu Complex. We also note in passing that a Scottsbluff point and a Cody knife were recovered at Medicine Lodge Creek from site components containing waisted points similar to Fort Creek Fen forms (Frison 1991, figure 2.3; Frison and Walker 2007, figures 3.6 and 3.8).

The Fort Creek Fen Complex originated, in our view, in the Glacial Great Lakes region of northeastern North America, moving westward as early peoples travelled in watercraft across Glacial Lake Agassiz, probably sometime after 11,000 BP, during the lake's Moorhead Phase, when it drained into the Glacial Great Lakes and its water levels were relatively low. The fact that Fort Creek Fen lanceolates occur in collections from Agassiz beaches in the Swan River Valley of west-central Manitoba and adjacent Saskatchewan (David Meyer, pers. comm., 2008) indicates that these peoples were present in areas to the west of the lake.

Fort Creek Fen groups moved into the Athabasca lowlands shortly after the Lake Agassiz flood, during a period when Lake Agassiz continued to drain via its northwestern outlet into the Athabasca River and then into Glacial Lake McConnell, its waters entering McConnell through a broad delta that extended to south of the Fort Hills. At that time, an arm of Glacial Lake McConnell extended far up the Peace River valley to just downstream of the modern community of High Level. This may be the route by which the Fort Creek Fen groups reached the Upper Peace and the Rocky Mountain Trench, where they came into contact with northern British Columbian microblade users, thereby adding the Ice Mountain Microblade Tradition, the oldest microblade technology in northwestern North America, to their technological repertoire. This technology was, in turn, carried at least as far east as the Lower Athabasca, where its occurrence in the Fort Creek Fen Complex represents the first appearance of microblade technology in what is today northeastern Alberta.

Nezu Complex (ca. 9,600 to 8,600 BP)

The Nezu Complex—the regional expression of the Cody Complex in northeastern Alberta—was defined by Saxberg and Reeves (2003) and by Reeves, Bourges, and Saxberg (2009) primarily on the basis of the results of 1997 and 1998 excavations at the Nezu site (HhOu-36), now located on the middle reaches of the Muskeg River in the Aurora North tailings pond (fig 6.7). At the time of occupation, however, the site was situated on the shores of a lake known as Lake Nezu, which formed in the Muskeg valley as a result of the catastrophic Lake Agassiz flood of roughly 9,900 BP (but see Fisher and Lowell, chapter 2 in this volume, for a slightly younger estimate for the timing of this event). A total of 139 square metres, representing a little over 90% of the main site area, was excavated. Three discrete activity loci were defined, each measuring about 2 to 3 by 3 to 4 metres, which are thought to represent two tent-frame locales and an outside activity area. The large number of tools associated with these loci in comparison to other single-component Nezu Complex campsites in the Fort Hills suggests that these areas of the Nezu site were seasonally occupied a number of times. Evidence of tool caching and repeated reuse was found at Locus 2.

Blood antisera and faunal analysis indicate that a variety of big game and fur-bearing animals were hunted. At the Nezu site, for example, caribou antisera reactions occurred on seven artifacts, moose on two artifacts, deer on eight artifacts, bovid (bison) on four artifacts, rabbit on five artifacts, and bear on three artifacts. Fifteen of 579 small calcined bone fragments recovered were identifiable to species bear (6 specimens), beaver (1 specimen), canid (3 specimens), and bison (5 specimens).[9] This analysis suggests that the Nezu occupants used sinews or tissues from canids, lagomorphs, and cervids as hafting materials. The results support environmental reconstructions based on other lines of evidence (Bouchert-Bert 2007) of an open mixed forest interspersed with grasslands and wetlands. Multiple lines of evidence lead to the conclusion that the Nezu site was a fall caribou hunting camp reoccupied a number of

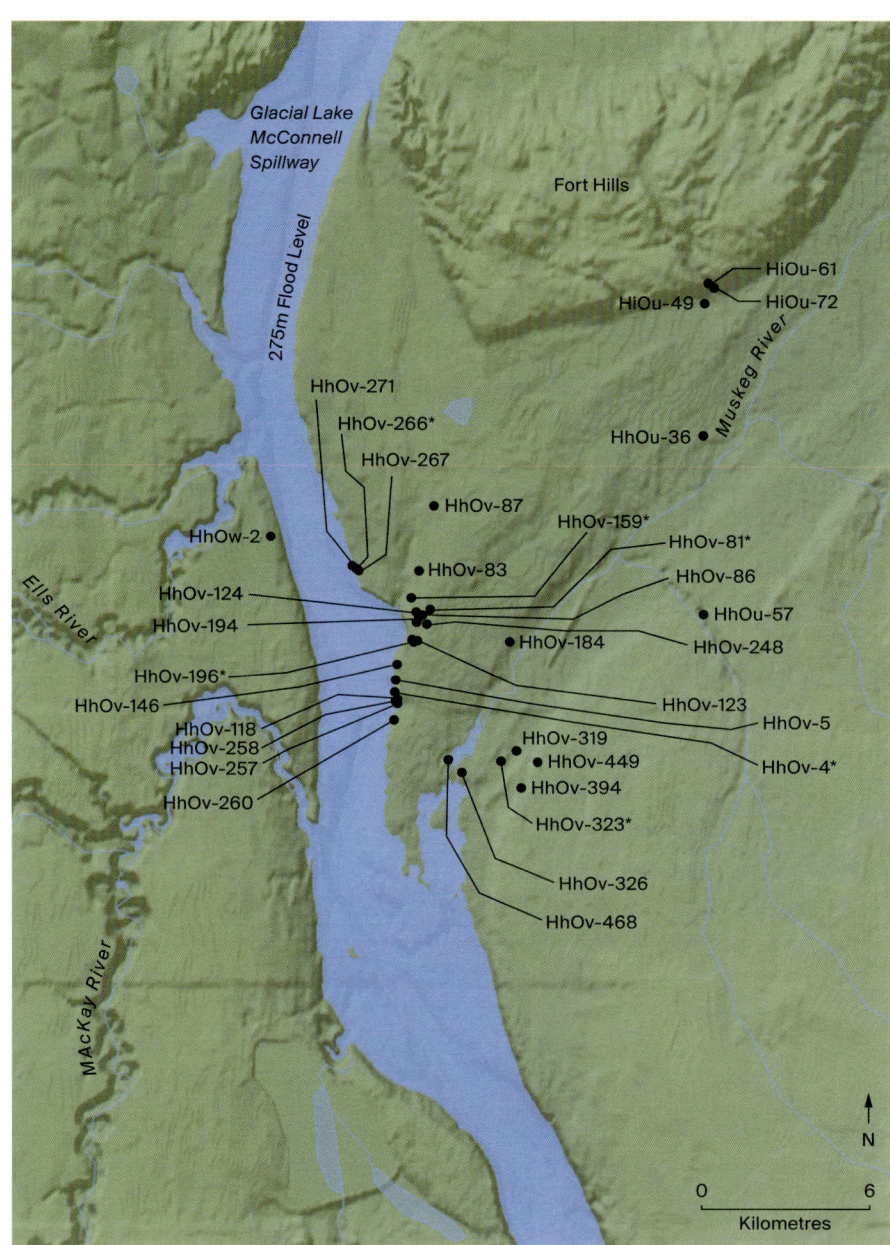

Figure 6.7. Nezu Complex sites (ca. 9,600 to 8,600 BP). An asterisk indicates the presence of Eden or other collaterally flaked points.

times, possibly by a small extended family group (Reeves, Bourges, and Saxberg 2009).

To define the Nezu Complex, Reeves, Bourges, and Saxberg (2009) carried out a comparative analysis of the Nezu site artifact assemblage and other excavated Nezu Complex sites along the Athabasca escarpment and the Muskeg River valley (table 6.3), as well as making broader comparisons to Cody Complex assemblages to the south. This work is briefly summarized below.

Nezu Complex stemmed projectile points are classifiable as Scottsbluff Type I or Type II points (plate 6.2: 4–10) and narrow and broad-bladed Eden points (plate 6.3: 10–12). These styles may or may not occur at the same site. Point preforms are rare. "Fish-tailed" and waisted lanceolate points also occur infrequently at the Nezu site (plate 6.2: 1–3), as is the case at other Cody Complex sites to the south, such as the Horner site (see Bradley and Frison 1987, 226 and figure 6: 14). A number of snapped Scottsbluff points have been recycled as Cody knives. Cody knives, although uncommon, include both single- and double-shouldered types (plate 6.3: 6–9). A reworked convex-based Agate Basin point (see plate 6.6: 1) was recovered from HhOv-148.

Two types of drills (or perforators or awls) are present: T-butt (plate 6.3: 1) and Niska (plate 6.3: 3–5). Niska drills, named after the Niska site in southwestern Saskatchewan (Meyer and Liboiron 1990), are characterized by a defined subrectangular haft or butt, generally with hafting modifications.

The bifacial knife assemblage is characterized by thin, flat-flaked meat-filleting knives of specific shapes, including Nezu knives, as well as knives of an asymmetric ovate form and narrow- and broad-bladed subrectangular knives. Lanceolate and oval-shaped knives are rare. Heavy-duty tools include backed bifaces and notched axes of specific type (plates 6.4 and 6.5: 10 and 11). More expedient heavy-duty tools include bifacial choppers and quartzite cortical spall tools of specific type and manufacture that distinguish them from the chithos often associated with the Taltheilei Tradition.

There are two types of Nezu Complex adzes. The first type is made on prismatic or truncated flakes with distal end and lateral-edge retouch. The second type, which may be recycled microblade cores (plate 6.10: 3), is characterized by retouched dorsal surfaces. The Nezu adzes are smaller and more regularly shaped than the adzes in later cultural complexes.

Formed flake tools include dorsally finished ovate or teardrop-shaped "humped-back" end scrapers and thin-to-thick tabular dorsally unretouched end scrapers of specific type (plate 6.5: 1–9), as well as specifically formed single- or double-edged side scrapers. A variety of marginally utilized or retouched flakes occur in Nezu assemblages, some of which feature characteristic

TABLE 6.3 Nezu Complex sites and diagnostic artifacts

Borden no.	Elevation (masl)	Projectile points or other diagnostic tools	Reference(s)
HhOu-36	285.45	Scottsbluff Type I points, Scottsbluff Type II points (1 of Northern quartzite)	Shortt, Saxberg, and Reeves 1998; Reeves, Bourges, and Saxberg 2009; Reeves et al.,2014b
HhOu-57	302.00	Scottsbluff base	Green et al. 2006, plate I-15
HhOv-4	280.71	Eden blade	Reeves et al. 2014b
HhOv-5	280.58	Possible Nezu hafted knife	Sims and Losey 1975, figure 3: f
HhOv-81	297.28	Broad-bladed Eden point	Clarke and Ronaghan 2004, plate HhOv-81-3
HhOv-83	293.93	Nezu tools	Reeves et al. 2014b
HhOv-86	294.32	Nezu tools	Reeves et al. 2014b; Clarke and Ronaghan 2004
HhOv-87	297.90	2 Scottsbluff points (type unspecified)	Clarke and Ronaghan 2000, plate I-8
HhOv-118	278.17	Nezu tools	Reeves et al. 2014b
HhOv-123	292.33	Scottsbluff base, 3 preform tips	Saxberg, Shortt, and Reeves 1998, plate 51: 3; Reeves et al. 2014b; Green et al. 2006
HhOv-124	293.09	Strangulated prismatic blade end scraper, Nezu tools	Shortt and Reeves 1997, plate 17: 7; Reeves et al. 2014b
HhOv-146	283.19	Scottsbluff Type I point, broad-bladed Eden point, Nezu adze	Reeves, examination at RAM
HhOv-148	293.00	Nezu tools	Reeves et al. 2014b
HhOv-159	292.02	Narrow-bladed Eden point	Reeves et al. 2014b
HhOv-184	283.11	Scottsbluff point, pink quartzite Scottsbluff stem, Nezu tools	Clarke and Ronaghan 2000, plates I-17 and I-18
HhOv-194	292.93	1 Scottsbluff point, 2 snapped Scottsbluff preforms	Saxberg, Shortt, and Reeves 1998; Reeves et al. 2014b
HhOv-196	290.53	Narrow-bladed Eden point tip	Saxberg, Shortt, and Reeves 1998; Reeves et al. 2014b
HhOv-198	285.00	1 Cody subrectangular knife, 1 Nezu end scraper	Reeves et al. 2014b
HhOv-248	295.55	Narrow-bladed collaterally flaked point tip	Clarke and Ronaghan 2004, plate HhOv-248-2
HhOv-257	273.00	Broad-bladed Eden point	Reeves, examination at RAM
HhOv-258	275.03	Scottsbluff Type II point stem, lanceolate point, Nezu tools	Reeves, examination at RAM
HhOv-260	276.00	Truncated microblade, burin spall	Reeves, examination at RAM
HhOv-266	278.95	2 Scottsbluff points, 1 broad-bladed Eden point	Reeves, examination at RAM
HhOv-267	280.35	Scottsbluff point, Nezu drill	Reeves, examination at RAM
HhOv-271	277.18	Stemmed obsidian point fragment	Reeves, examination at RAM
HhOv-319	281.91	Miniature Scottsbluff point, drill tip, Nezu tools	Saxberg 2007, plate B.19: e and f
HhOv-323	281.22	1 broad-bladed Eden point (positive test for proboscidian antisera), 1 Montana chert Scottsbluff point	Saxberg and Reeves 2004, plate 65: f; Saxberg 2007, plate B.22: a
HhOv-326	279.00	Nezu bifacial quarry blank	Saxberg 2007, plate B.27: a
HhOv-394	293.60	Nezu bifacial knife	Tischer 2004
HhOv-449	298.06	Scottsbluff point blade fragment	Wickham 2006b, figure 12
HhOv-468	278.00	2 Scottsbluff points (1 of red chert)	Wickham 2006b, figures 14 and 15
HhOw-2	291.31	Scottsbluff Type II point	Losey, Freeman, and Priegert 1975, 34; Ives 1993, figure 2: d
HiOu-49	294.00	2 Cody knives	Somer 2005, plate 51: f and g
HiOu-61	304.91	Nezu waisted point, Nezu tools	Somer and Kjar 2007, plate 53: i
HiOu-72	306.00	Cody knife	Somer and Kjar 2007, plate 53: c
HkPa-4	714.00	Cody knife?	Ives 1977, plate 2: k; Ives 1981, figure 7: row 1, fourth from left

"fingertip"-shaped employable units (Knudson 1983) and include beaked and carinated gravers, notches, and denticulates. Wedges are extremely rare.

Marginally utilized or retouched flakes and intentionally snapped bifaces have been used or modified to serve as end gravers, corner gravers, pseudo-burins, and burins. Burins are quite common in some Nezu sites, particularly in Locus 2 of the Nezu site, and are thought to represent intensive processing of caribou antler. Flake burins include transverse, angle, and dihedral burins, often manufactured on specific secondary flake blanks with prepared platforms, and are backed. Notched Donnelly Ridge–style burins do occur. The range of variation in the Nezu burin assemblage is comparable to that found in Denali Complex sites in Alaska.

Nezu Complex tool production technology is based on the fabrication of large biface preforms or cores and a core reduction strategy designed to create bifacial rough-outs that, through the various stages of bifacial reduction, are shaped into knives, points, or drills. Secondary flake blanks derived from these cores appear to have been used or modified to serve as end and side scrapers, adzes, and a variety of marginally utilized or retouched flakes, including gravers and burins. Large primary and secondary decortication quarry flakes were also recovered, indicating a specific Levallois-style reduction technology similar to that identified by Knudson (1983) at the Cody Complex occupations at the McHaffie site. Along with bifacial reduction, these decortication flakes are a key part of the Nezu tool reduction and manufacturing strategies.

Nezu Complex microblade technology was first identified in 1997 in the Aurora North utility corridor HRIA (Shortt, Saxberg, and Reeves 1998; Reeves et al., 2014b). This technology is based on the transverse or longitudinal snapping or burination of biface cores to produce bifacially edged, boat-shaped or wedge-shaped microblade core preforms (see plate 6.10: 3 and 4 for illustrations of the former). The snap fracture or burination served as the core preform platform, and the thickest end usually as the core face, which could be further trimmed before microblade removal (plate 6.10: 4). Similar boat-shaped cores occur in the Ice Mountain Microblade Tradition (Fladmark 1985; Smith 1974), as well as in Denali assemblages, such as that at Dry Creek, in Alaska (see Powers, Guthrie, and Hoffecker 1983, figures 4.8: A and 4.15). The majority of Denali wedge-shaped cores are based on bifacially worked blanks, however, rather than on snapped or burinated biface blanks.

The frequency of the cores and microblades varies considerably among different Nezu Complex sites and seems to be inversely correlated with the frequency of stone dart points. This pattern perhaps reflects the use of microblades as side blades in dart or spear points made of caribou antler, rather than as the tip of a traditional stone projectile, as may have been the case for the Denali

Complex in Alaska (Ackerman 2007, 168–170; Larsen 1968). Ackerman (2007) concluded that these composite side blade points may have been used at Denali sites, which are summer occupations, to hunt dispersed caribou during the summer, after the spring migration and before the herd reassembled in the fall. Perhaps the same was true of some of the Nezu Complex summer hunting practices in the Lower Athabasca region.

Nezu Complex sites contain a small percentage of non-local tool stones, usually found in exhausted and recycled or discarded tools. In addition to Northern quartzite, these stones include Swan River chert, Knife River flint, Bear Gulch obsidian, various Montana cherts from the South Everson (Bonnichsen et al.1992; Douglas 1991), Doggett (Roll 2003), and Helena (Knudson 1983) quarries, metamorphosed green argillite from the Waterton-Glacier region (Reeves 2003), and tourmaline chert from the central Canadian Rockies (Reeves 2003). Basalt is also present.

Nezu Complex artifacts have been recognized in a total of thirty-six sites in the Athabasca lowlands and Birch Mountains (table 6.3). Unusual or unique finds of particular interest include a probable bola stone recovered from HhOv-184 (Clarke and Ronaghan 2000, plate I-22). Another is an isolated Eden lanceolate point found at the Quarry of the Ancestors site HhOv-323 (Saxberg and Reeves 2004, plate 65: f) that reacted positively to elephant antisera (see plate 6.3: 12).[10] Nezu Complex artifacts have also been identified at four sites in the Aurora North Mine in the Stanley Creek area.

Only one Nezu Complex occurrence—HhOw-2, a small lithic scatter—has been identified on the west side of the Athabasca River (Losey, Freeman, and Priegert 1975, 28). A Scottsbluff Type II specimen was collected from the site (Losey, Freeman, and Priegert 1975, 34; see also Ives 1993, figure 2: d). The dearth of Nezu Complex sites on the west side of the river probably reflects the fact that, at the time, the shoreline was some distance west of today's river escarpment (compare figs. 6.7 and 6.8).

Nezu Complex artifacts are poorly represented in collections from the Birch Mountains. Diagnostic Nezu Complex points from the Gardiner Lake Narrows site, HjPd-1, were either not recovered or no longer remain in the collections housed at the RAM. However, Cody Complex materials are represented in the collections from HkPa-4. In addition to a probable oolitic Northern quartzite Cody knife observed by Reeves in the RAM collection (see Ives 1977, plate 2: k; see also Ives 1981, figure 7, row 1, fourth from left), these include a large, thin, finely flaked flake end scraper with a graver spur on the left lateral edge (see Ives 1977, plate 5, row C, first on left) and a flake knife manufactured of a thin slab of oxidized MVMq (see Ives 1977, plate 4: f).

The Nezu Complex is the best represented of the all archaeological complexes in the Athabasca lowlands. It was during roughly the first half of Nezu Complex times (ca. 9,600 to 9,000 BP) that the waters of the Athabasca embayment lowered sufficiently to expose the main quarry areas at the Quarry of the Ancestors. The major mining and workshop activity at the Quarry of the Ancestors evidently also dates to the period of the Nezu Complex, and archaeological studies in the area of the quarry have further enhanced the numbers and the visibility of Nezu Complex sites in this region.

We suggest that the greater frequency of Nezu Complex sites reflects a longer period of occupancy in comparison to the other two Early Precontact period complexes, particularly Fort Creek Fen. The greater number of Nezu Complex sites could also mean that a more open, warmer, dryer environment was present during this time than was the case later, during Cree Burn Lake Complex occupations. In large part, this latter period correlates temporally with the negative climatic impact that the discharge of Glacial Lake Agassiz into Hudson Bay had on the northern hemisphere (see Alley and Ágústsdóttir 2005). A more productive subsistence environment, both terrestrial and aquatic, during Nezu Complex times could have resulted in a higher frequency of occupancy of the Lower Athabasca region during the warm season.

Nezu Complex territory extended into the Firebag Hills and to the headwaters of the Descharme River. Field studies in this region related to oil sands development have recorded three Nezu Complex isolated finds or artifact scatters and four other sites that are most probably Nezu or earlier in age (Reeves, Cummins, and Lobb 2008). To the southeast, Nezu occupation extends into the northwest precincts of Lake Agassiz. Cody Complex artifacts—some of which, in view of the colour and texture of the stone, are probably made of MVMq—are present in the collections from the Old Beach site (GlOc-30) at Buffalo Narrows (see fig 6.4).[11] Approximately 460 kilometres southwest of Buffalo Narrows lies the Cody Complex Heron Eden Bison Kill (EeOi-11), located near Kindersley, Saskatchewan (Linnamae and Corbeil 1993; Corbeil 1995). An MVMq Scottsbluff point and retouched MVMq flakes were among the artifacts recovered at the Heron Eden site, which dates to about 9,000 BP. Points fashioned from Knife River flint were also associated with the site.

A similar association was present in surface collections from locations near the towns of Boyle and Barrhead, 300 kilometres up the Athabasca River. GdPf-6, near Boyle (see fig 6.4), contained an MVMq Scottsbluff point, Alberta and Scottsbluff points made of Knife River flint, a Scottsbluff point of Peace River chert, and another of a Montana chert (GdPf-6 site inventory form). The Fisher site (GbPo-1; see fig 6.4), on an old shore of Shoal Lake near Barrhead, produced an MVMq Eden point, two Knife River flint Scottsbluff points, and obsidian artifacts. The obsidian would most probably have originated from either the Bear Gulch or the Obsidian Cliff quarries (see also Ives 1993, 26; Fenton and Ives 1982, 1990).[12]

Limited as it is, the tool stone evidence suggests that Nezu Complex peoples were seasonal residents of the Lower Athabasca region, which was situated at one extreme of a much larger trading network that characterizes the Cody Complex throughout the northwestern Plains. In Nezu Complex sites in the Lower Athabasca region, the presence of Swan River chert—a Cody Complex tool stone type found in the Alberta-Saskatchewan parklands (Bob Dawe, pers. comm., 2008)—as well as cherts that most probably originate in central and/or southern Montana indicates that the Nezu Complex occupants of the Lower Athabasca interacted seasonally with other Cody Complex groups in the parklands or lakelands of central Alberta and Saskatchewan. Such interaction is further suggested by the occurrence of Scottsbluff points manufactured of MVMq in Cody Complex sites in that region, such as Heron Eden.

As previously mentioned, the Nezu Complex is the regional expression of the Cody Complex, a culture that is clearly attested in areas of what is today the boreal forest ranging from northeastern British Columbia to northwestern Saskatchewan, as well as southward through the Northern Rockies to the greater Yellowstone region (Johnson and Reeves 2013). The patterns of occupancy are similar in Cody sites throughout the region, and evidence indicates that a broad spectrum of migratory and non-migratory game animals and smaller fur-bearers was harvested. Lakes were important seasonal settlement locales, and although no direct evidence exists, it is likely that fish were caught and fowl hunted, and food and medicinal plants were gathered. Particularly favoured tool stones were procured and traded over thousands of kilometres. That these early Cody Complex–related people in the Lower Athabasca region used watercraft as a means of transportation seems only logical given the generally forested nature of the landscape at that time and the extensive system of navigable lakes and rivers. Evidence for the use of watercraft by other Late Palaeoindian groups around the Great Lakes, for example, and in Late Palaeoindian sites in the Rocky Mountains, further supports this conclusion, as do Cody Complex occupations in the area of Yellowstone Lake.

On the basis of the distribution of MVMq Scottsbluff and Eden points, we suggest that Nezu Complex peoples spent the winter season along the edge of what were then parklands and lakelands and the northwestern precincts of Lake Agassiz to the south. Once the ice went out in the spring, they journeyed downstream along the Athabasca and Clearwater rivers in skin-covered boats to the Athabasca lowlands, where they spent the warm-weather months, returning to their wintering grounds in the south before freeze-up in the fall. While in the Athabasca region, they quarried MVMq, hunted a variety of big game (bison, deer, moose, caribou, and bear), and took smaller fur-bearers such as beaver. They travelled up into highlands such as the Birch Mountains and the Firebag Hills to hunt and probably fish. The ice cap was not far to the east and, while winters would have been extremely harsh, summers were probably warmer than they are today.

In their wintering grounds, Nezu Complex peoples resided with other Cody Complex groups who, as indicated by the occasional presence of Knife River flint and small quantities of tool stones from central Montana and the greater Yellowstone region, were in contact with other Cody groups along the western fringe of the Plains to the south during the warm-weather months. We presume that these groups interacted with more southerly groups, obtaining and exchanging tool stones and finished objects, and that some of this economic and cultural exchange subsequently reached Nezu peoples. Nezu hunters and groups may have joined in communal bison hunts, such as that at Heron Eden in southwestern Saskatchewan, which at this time was a mesic parkland landscape, rather than xeric grasslands (Beaudoin and Oetelaar 2003), and not far removed from the South Saskatchewan River or Lake Agassiz. Somewhere around 8,600 BP, however, the Nezu occupancy of the Athabasca lowlands came to an abrupt end, perhaps because of adverse climatic change. It was succeeded by the Cree Burn Lake Complex.

Cree Burn Lake Complex (ca. 8,600 to 7,750 BP)

The Cree Burn Lake Complex takes its name from Cree Burn Lake (Ronaghan 1981a, 1981b; Head and Van Dyke 1990; Shortt and Reeves 1997), an abandoned oxbow of the Athabasca River north of Fort McKay. The complex was initially defined by Saxberg and Reeves (2003) on the basis of the 1998 and 1999 excavations in the Aurora North utility corridor. The characteristic points are obliquely parallel-flaked lanceolates of the Lusk and Frederick types, as defined in the archaeological literature on the northwestern Plains and Rocky Mountains (Frison 1991; Reeves 1972; Driver 1978; Langemann and Perry 2002). Perpendicular parallel-flaked Agate Basin points occasionally co-occur. In the northern boreal forest area and the adjacent Barrenlands, obliquely parallel-flaked lanceolates are generally referred to as Northern Plano (Arnold 1985; Gordon 1975, 1996; Wright 1972a, 1972b, 1976).

Saxberg and Reeves (2003) initially proposed that the Cree Burn Lake Complex dated back to about 9,500 BP and thus overlapped with the Nezu

Complex, co-existing with it in the Athabasca lowlands in the wake of the Lake Agassiz flood. They identified two sites in the immediate vicinity of Cree Burn Lake, HhOv-148 and HhOv-194, that they considered to contain early Cree Burn Lake components (Saxberg and Reeves 2003, 301). On the basis of technological analysis, however, both sites have subsequently been reclassified as Nezu Complex (Reeves et al. 2014b). The later sites grouped with the Cree Burn Lake Complex are characterized by a more opportunistic lithic technology, bipolar reduction, and a high degree of expediency in tool manufacture and use. Although some of the sites that we assign to the Cree Burn Lake Complex may be technologically earlier, as Saxberg and Reeves (2003) suggested, we are restricting the dating of the Cree Burn Lake Complex to post-Nezu occurrences of obliquely flaked lanceolate points.

Studies have identified a total of thirteen Cree Burn Lake Complex sites and other occurrences of obliquely flaked lanceolate points in the region (see table 6.4 and fig 6.8). The artifact assemblage and technological characteristics of the Cree Burn Lake Complex are not well defined regionally, however, as most of the points either occur in mixed stratigraphic contexts (as, for example, is the case for the three sites in the Birch Mountains, HjPd-1, HkPa-4 and HkPb-1) or are obliquely flaked lanceolates collected and recycled or discarded in sites along the Athabasca escarpment by later Beaver River Complex occupants. One of these—a classic Jimmy Allen point made of an opaque mottled grey and brown chert with black-coloured linear inclusions and light-coloured fibrous inclusions with heavily worn arrises (plate 6.6: 5)—was recovered from HhOv-193, a site otherwise assigned to the Beaver River Complex. An obliquely flaked lanceolate from HhOv-256, a site that was originally part of the HhOv-55 quarry (Wickham 2006a), has also been included in the Cree Burn Lake Complex. Excavations at HhOv-256 also recovered small, pitted anvil "nutting stones" and bipolar cores and wedges like those found at Beaver River Complex sites, however, suggesting that this site may have been dominated by a Beaver River Complex occupation.

Other sites of interest on the eastern Athabasca escarpment include the Cree Burn Lake site itself, HhOv-16, from which a snapped Early Precontact period lanceolate biface (Shortt and Reeves 1997, plate 17: 1) and snapped biface gravers were recovered. (The former reacted positively to caribou antisera, and one of the latter to deer antisera.) These snapped tools suggest a potential association with lanceolate point complexes, possibly the Cree Burn Lake Complex. At HhOv-167, not far south of HhOv-16, a lanceolate point (see plate 6.6: 7) was recovered during shovel tests (Saxberg, Shortt, and Reeves 1998, 28), as were a microblade core, microblades, and other tools and tool fragments. The lanceolate point, manufactured of what may be heat-treated Swan River chert, has been identified as a Mesa point on the basis of hand comparisons with casts of Mesa points (Kunz, Bever, and Adkins 2003).

Lanceolates are recorded from five sites on the western Athabasca escarpment (see table 6.4). Finds from early studies include a point from the Beaver River Quarry, HgOv-29 (Syncrude Canada Ltd. 1974, figure 6a: A; Ives 1993, figure 2b). Beaver River Complex artifacts were also found at this site, suggesting that the occurrence of the Cree Burn Lake Complex at HgOv-29 is similar to its occurrence at sites on the east side of the river. At least on the basis of the described artifacts, the other four sites (HhOv-2, HiOw-38, HiOw-52, and HiOw-30) appear to be part of the Cree Burn Lake occupations.

In the Birch Mountains, obliquely parallel-flaked lanceolate points have been recovered from three stratigraphically compressed sites. These include three manufactured of grey quartzite: two broken, then retipped, specimens from HjPd-1, the Gardiner Lake Narrows site (Ives 1981, figure 7, first row, second and third from the left; Ives 1993, figure 2: e, f), and one broken retipped specimen from HkPa-4 (Ives 1993, figure 2g). A retipped ground convex-based lanceolate, which Reeves observed in the RAM collections, came from HkPb-1 on Eaglenest Lake. It is manufactured from a dirty yellow brown chert.

Ives (1993, 9) draws comparisons between the three obliquely flaked quartzite lanceolates from the Birch Mountains and oblanceolate points from the Yukon and Alaska. While the Birch Mountain points do share certain similarities with these northern oblanceolate forms, they also generally fit within the range of variation for obliquely flaked lanceolates (excepting, perhaps, one of the Gardiner Lake Narrows specimens: see Ives 1993, figure 2: f) found in the both pre- and post-Cody lanceolate point complexes of the northwestern Plains and Rocky Mountains (Frison 1991; Reeves 1972; Husted and Edgar 2002; Johnson, Reeves, and Shortt 2004; Langemann and Perry 2002).

As noted above, the Cree Burn Lake Complex is not as well represented in the Athabasca lowlands and nearby Fort Hills as the temporally adjacent Nezu and Beaver River complexes (see tables 6.3 and 6.5). While the numeric differences may simply reflect differences in the length of occupancies, there also appear to be differences in site patterning between the Nezu Complex and the Cree Burn Lake Complex. Quite a few Cree Burn Lake Complex sites occur on the west side of the Athabasca River, in comparison to only one Nezu Complex site. On the eastern escarpment, the Cree Burn Lake Complex is poorly represented by primary sites (at least in the present sample): most occurrences appear to be secondary in nature and are associated with early Beaver River Complex sites. Nor does the Cree Burn Lake Complex appear to have an expression in the Quarry of the Ancestors. (There, very small lanceolate bases recovered from HhOv-305 are thought to be

TABLE 6.4 Cree Burn Lake Complex sites and diagnostic artifacts

Borden no.	Elevation (masl)	Projectile points and other diagnostic tools	Reference(s)
HgOv-29*	256.00	1 lanceolate point, 1 point tip, 1 Agate Basin point	Syncrude Canada Ltd. 1974, figure 6: a
HhOv-2	297.00	Agate Basin point	Losey Freeman, and Priegert 1975
HhOv-16	292.98	Early period biface, snapped biface gravers	Shortt and Reeves 1997
HhOV-167**	273.67	1 Mesa point, microblades and/or microcores	Saxberg, Shortt, and Reeves 1998, plate 51: 1
HhOv-193*	277.00	Obliquely flaked Jimmy Allen point, of mottled grey and brown chert	Reeves et al. 2014c
HhOv-256**	274.54	Agate Basin point	Wickham 2006a
HhOv-445**	284.44	Blade midsection	Tischer 2005
HiOw-30	295.00	1 dark grey siltstone lanceolate point tip, 1 bipoint	Bryant 2004, plate III-5
HiOw-38	287.00	1 quartzite Agate Basin base, 1 Agate Basin point tip	Clarke 2002a, plate III-4
HiOw-52	318.00	Salt and pepper quartzite lanceolate point base	Bryant 2004, plate III-13
HjPd-1**	683.00	2 grey quartzite obliquely flaked lanceolate points, retipped	Ives 1981
HkPa-4**	714.00	Grey quartzite obliquely flaked lanceolate point, retipped	Ives 1993, figure 29
HkPb-1**	715.65	Dirty yellow chert lanceolate point	Reeves, examination at RAM

NOTE: The two sites marked with an asterisk (HgOv-29 and HhOv-193) are Beaver River Complex sites at which obliquely flaked lanceolate points were recovered. At sites marked with two asterisks, such points occurred in mixed stratigraphic contexts.

associated with the Firebag Hills Complex: see table 6.6.) In contrast, Cree Burn Lake Complex is seemingly better represented in the Birch Mountains than the Nezu Complex, with three occurrences as opposed to one. Possibly these differences in site patterning reflect a short-term variation in climate.

Neither Agate Basin nor Lusk points manufactured of MVMq have been reported south of the Lower Athabasca region. Obliquely flaked lanceolates of undescribed tool stone types occur at the Old Beach site (GlOc-30) on Buffalo Narrows at Peter Pond Lake, in nearby Saskatchewan (Millar 1983), and in dune field sites on the south shore of Lake Athabasca (Wilson 1981; Reeves, observation of collections at the Canadian Museum of Civilization, 1986). None of the points found in the Athabasca dune fields are manufactured of MVMq, however. Similarly, in his study of Northern Plano points from sites in the Mackenzie River area and Barrenlands and other point assemblages (such as that at Acasta Lake), Arnold (1985) did not identify any MVMq points. These data suggest that trade involving MVMq did not extend into these more northerly areas during Cree Burn Lake times. Perhaps the short-term but fast-acting adverse climatic events that took place at that time (see Alley and Ágústsdóttir 2005) marginalized seasonal settlement in the Athabasca lowlands, as similar adverse conditions may have done during the Little Ice Age.

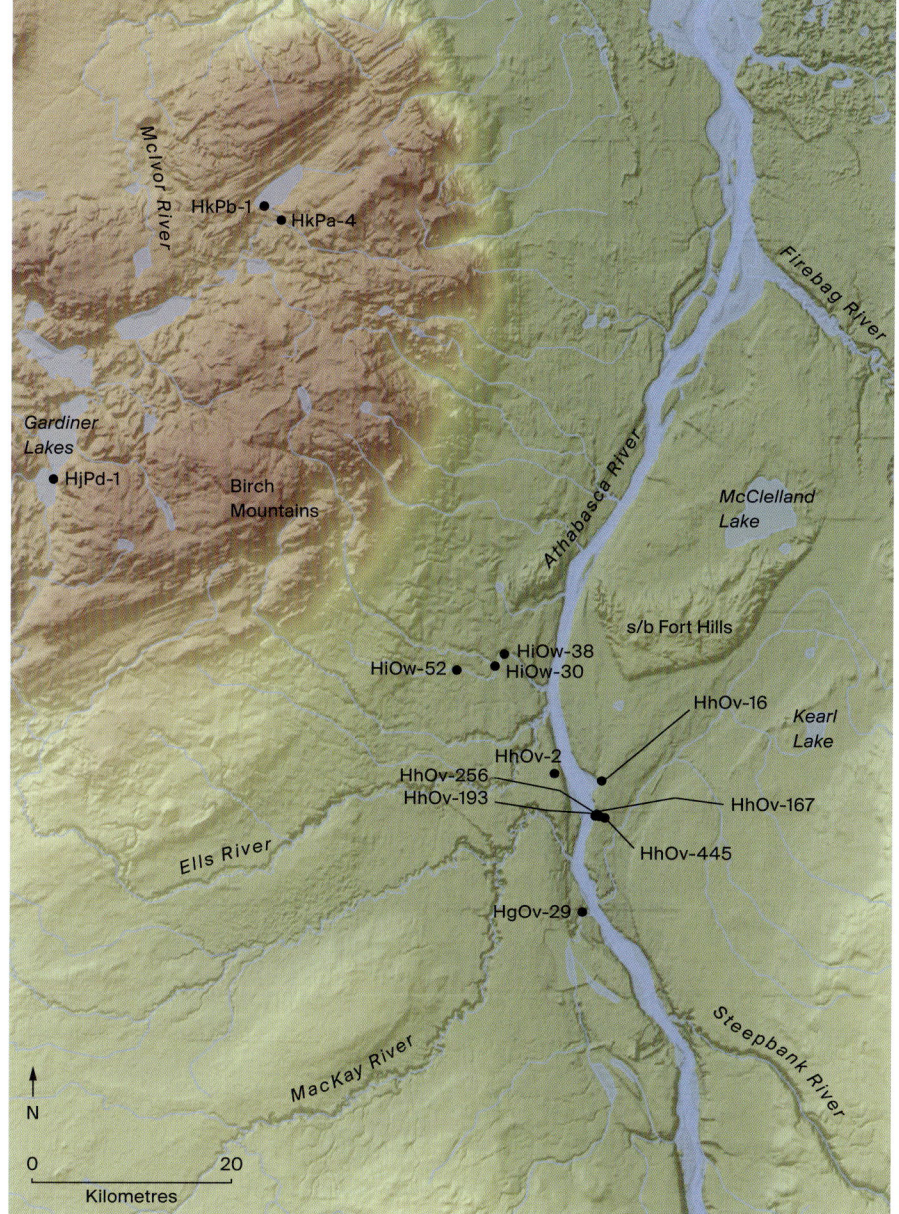

Figure 6.8. Cree Burn Lake Complex sites (ca. 8,600 to 7,750 BP)

MIDDLE PRECONTACT PERIOD (PRE-TALTHEILEI) COMPLEXES (CA. 7,750 TO 2,650 BP)

During the Middle Precontact period, the Lower Athabasca region was dominated by two consecutive cultural complexes. The first of these was the Beaver River Complex, which endured for well over three millennia, from approximately 7,750 to 4,000 BP. The second was the Firebag Hills Complex, which existed from roughly 4,000 to 2,650 BP and immediately predated the appearance of the Taltheilei Tradition in the region.

Beaver River Complex (ca. 7,750 BP to 4,000 BP)

An early expression of the Beaver River Complex (ca. 7,750 to 7,000 BP) was proposed by Saxberg and Reeves (2003) to encompass the early side-notched dart point sites found during the 1998 and 1999 excavations in the Aurora North utility corridor (Reeves et al. 2014c). They took the name of the complex from the Beaver River Quarry, where excavations in 1974 first identified this style of point (Syncrude Canada Ltd. 1974). Later site components, containing side-notched points and/or Oxbow points, were not identified at the time of the Aurora North studies. Saxberg and Reeves (2003) chose a beginning date of about 7,750 BP for the Beaver River Complex. This date was based on the earliest dates for the appearance of large side-notched dart points at sites in the northwestern Plains and Rocky Mountains to the south, including the Hawkwood site in Calgary (Van Dyke and Stewart 1984), the James Pass site (Ronaghan 1993), and the Boss Hill site (Doll 1982) (see fig 6.4).

Although, as others have pointed out (see Green et al. 2006, for example), many of the side-notched points in the Beaver River Complex are comparable to side-notched specimens from the Gowen site at Saskatoon (Walker 1992) and other sites in the Upper Saskatchewan and Missouri basins, they also compare well to side-notched points illustrated by Wright (1972a) and Gordon (1996) from Shield Archaic sites to the northeast that date from about 6,450 to 3,500 BP. Lanceolate points also occur in early Beaver River sites (such as HhOv-112, plate 6.6: 2–4), as they do in early Mummy Cave sites in the Northern Rockies (Reeves 1972).

Oxbow and stemmed indented-base points (similar to Duncan points) occur in terminal Shield Archaic sites. Gordon (1996) has good radiocarbon control on his terminal Shield Archaic components. Stemmed indented-base points have been dated to 4,040 ± 125 ^{14}C yr BP (S-1435) at KeNi-4, on Whitefish Lake (Gordon 1996, table 9.1). We have therefore considered sites that contain Oxbow and/or stemmed indented-base points to represent the late expression of the Beaver River Complex and have selected 4,000 BP as the end date for the complex.

The Beaver River projectile point assemblage (table 6.5 and plate 6.7) includes side-notched dart points and occasional broad corner-notched dart points, as well as the Oxbow and stemmed indented-base points found in later components. The side-notched dart points include both square and rounded or pointed basal edge variations. In the context of the Northern Rockies of Montana and Alberta, these variations are known as Bitterroot and Salmon River Side-Notched points. Also present are side-notched points with convex bases, a variant found in Shield Archaic and age-equivalent sites in northern British Columbia and the Yukon. Square and convex-based narrow-bladed lanceolates may also occur, particularly in early components, some of which may be point preforms.

A number of biface types are present, most of which are produced on bipolar cores and blanks. Finished biface forms are the result of bifacial reduction and generally include:

- small-sized, thick asymmetric subrectangular, ovate, and trianguloid forms, which are sometimes backed
- subrectangular to subovate forms with one or more strongly convex-shaped edges
- backed bifacially worked bipolar cores with lateral edges and ends that resemble oversized side blades
- bifacially worked edge pieces.

The first category is less common than the others.

End scrapers include some that are manufactured on recycled Nezu Complex microblade cores, as well as small flake end scrapers that are sometimes notched and have graver spurs with retouch confined to the distal end and lateral edges. Dorsally finished forms are rare or absent. Flake side scrapers include a distinctive backed form with one worked lateral edge combined with an obliquely set distal working edge.

Also found are adzes and gouges that are relatively large in size. Frequently they are manufactured on bipolar core edge flakes and often appear to have been utilized as gravers or pseudo-burin blanks. Flake burins are present, particularly in early components, and include angle, transverse, and carinated forms. Notched burins are absent, and backed burins uncommon. One particular manufacturing technique for gravers and burins involves radially fracturing thin pieces

TABLE 6.5 Beaver River Complex sites and diagnostic artifacts

Borden no.	Elevation (masl)	Projectile points and other diagnostic tools	Reference(s)
HcOs-3	478.00	1 Oxbow point, 1 quartzite dart point	Pollock 1978b, figure 43: 15 and 16
HgOv-29	256.00	Side-notched point	Syncrude Canada Ltd. 1974, figure 6: b
HgOv-31	251.00	Oxbow (?) point	Sims 1976, figure 12: c
HgOv-32	256.00	Side-notched point	Sims 1976, figure 12: b
HgOv-50	245.00	1 grey siltstone Duncan-like point, 1 quartzite side-notched dart point	Gryba 1980; Reeves et al. 2014c
HhOv-4	280.00	Side-notched point	Ronaghan 1981a
HhOv-17	276.13	Quartzite notched point base	Green et al. 2006, plate I-22
HhOv-16	292.98	3 Oxbow points, 1 small quartzite stemmed point	Ives 1993, plate I-28; Head 1979; Clarke and Ronaghan 2000
HhOv-55	271.00	Notched point tip	Wickham 2006b
HhOv-87	297.90	Northern quartzite side-notched point	Clarke and Ronaghan 2000, plate I-8
HhOv-112	280.25	Side-notched point, bipolar technologies	Reeves et al. 2014c
HhOv-113	282.35	3 side-notched points, bifacial knife, bipolar technology	Green et al. 2006; Unfreed 2001, plate 22
HhOv-146	283.19	Bedded volcanic side-notched point	Reeves, examination at RAM
HhOv-163	295.93	5 quartzite notched points, 1 black chert notched point	Reeves et al. 2014c
HhOv-191	283.26	Fishtail point	Green et al. 2006, plate I-55
HhOv-193	276.82	White orthoquartzite side-notched dart point, bipolar technology	Reeves et al. 2014c
HhOv-212	277.09	Side-notched point	Green et al. 2006, plate I-65
HhOv-265	280.00	Swan River chert side-notched point	Reeves, examination at RAM
HhOv-282	291.00	Quartzite side-notched point	Clarke 2002a, plate III-I
HhOv-302	280.59	Early Beaver River point base	Saxberg 2007, plate B.36
HhOv-305	282.48	2 side-notched points	Saxberg 2007, plate B.8: a and b
HhOv-308	280.54	Quartzite side-notched point	Saxberg and Reeves 2004, plate 65: d
HhOv-319	281.91	Oxbow point	Saxberg and Reeves 2004, plate 65: e
HhOv-332	282.24	Grey chert side- or corner-notched point	Saxberg 2007, plate B.29: a

of MVMq (about 12 mm thick) and using these pieces as gravers or burin blanks. Corner and end gravers as well as notches are common in the assemblages. One diagnostic type is notched flake corner gravers. Denticulates may occur.

Wedges are extremely common. Most are manufactured on bipolar cores, but some are made on thin, transversely snapped, recycled Nezu biface and point blade fragments, which were then worked on all four edges to produce a thin, discoidal-shaped tool. Sometimes the distinctive Nezu thin, flat flaking remains visible.

Quartzite, granodiorite, and diorite cobbles were used as hammer and anvil stones in bipolar reduction. Semicircular pitted "nutting stones," of roughly the same shape and size as cobbles and made of the same materials, sometimes with what appears to be red ochre staining, were also found in early Beaver River sites. One of these "nutting stones," from HhOv-112, tested positive for *Chenopodium*.

Bipolar reduction characterizes the core, flake, and tool blank manufacturing trajectory, as is clearly evident from the large numbers of bipolar cores and pre-forms, core edge fragments and flakes, finished tools such as backed bifaces, and flakes with bifacial edges and wedges, as well as cores and core fragments, found in many Beaver River sites in which tool fabrication and modification were major activities. This bipolar technology was designed to maximize the use of quarry pieces of a specific shape and size (including earlier Fort Creek Fen and Nezu Complex cores and tools), which were obtained either by mining or by collection from outcrops and/or earlier archaeological sites, to produce a wide set of tool blanks and tools. Much of the MVMq used is of low quality, suggesting that the high-quality, fine-grained MVMq that characterizes the earlier Nezu and Fort Creek Fen assemblages may have been largely mined out by Beaver River times.

Multifaceted microblade cores and numbers of microblades do occur in some sites. Primarily they are face-faceted and pillar, tabular, or, on occasion, conical in form (de Mille and Reeves 2009, plate 20:5 and 6). They can be difficult to dis-criminate from bipolar cores and related debitage, as the latter often have micro-blade-like facets (as, for example, at HhOv-113; Green et al. 2006, plate I-40).

Beaver River Complex tool stones other than MVMq include Swan River chert and Northern quartzites. The tool stone pattern exhibits some significant differences from those of the earlier Nezu Complex, one of which is the small, but variable, amounts of various quartzes and vein quartzes of Canadian Shield provenance. These range from milky-white opaque to translucent varieties, and some contain muscovite flecks. Most of the material probably came from the Grandfather Quarry complex on Granville Lake in northern Manitoba (Brownlee and Sitchon 2010), as the Athabasca lowlands specimens are visually identical to specimens from these quarries (Kevin Brownlee, pers. comm. regarding HhOv-212 specimens, 2010). These tool stones were particularly common at HhOv-212.

Similar frequencies of these specific quartz varieties occur in the Beaver River Complex component at the Old Airport site (HcOi-1) and at Saleski Creek (HcOi-2), just east of Lac La Loche (Hanna 1982; Reeves, personal examination of the University of Saskatchewan collections, 2006). The high quality of the quartz tool stone at these sites and the lack of a pebble- or cobble-weathered cortex sug-gest to us that primary outcrops or large erratic boulders were accessed by Beaver River Complex peoples. The material may represent an exchange pattern between groups that wintered together in the Lac La Loche–Peter Pond Lake area, with some groups migrating to the northeast or down the Churchill River in summer, where they obtained the quartz, and others travelling to the Athabasca lowlands in summer, where they obtained MVMq. Then, upon their return to the wintering range, they could have exchanged tool stones in nearby encampments.

Other non-local tool stones appear in Beaver River components at HhOv-112, which is part of a very large, repeatedly occupied early Beaver River site complex that includes HhOv-17, HhOv-113, HhOv-193, and HhOv-212. Among these tool stones are Lake One Dune chert, black Peace River chert, specific varieties of copper-enriched chalcedonies and green cherts, and fine-grained volcanics (basalt, andesite, and rhyolite). Two small obsidian retouching flakes of visually different varieties recovered from HhOv-112 were sourced, one green in colour to Mount Edziza and the other to Batza Tena, suggesting that these non-local tool stones probably originated in northern British Columbia. The presence of these northern British Columbia tool stones suggests seasonal movement of watercraft-born groups who wintered in the upper reaches of the Peace River and, in the spring, travelled downstream to the Peace-Athabasca delta and then up the Athabasca, where they co-occupied areas with groups that had wintered in the Lac La Loche–Peter Pond Lake area and at the headwaters of the Churchill.

The Beaver River Complex is well represented in the Lower Athabasca region, with thirty-eight sites or site components so far identified, including a number at the Quarry of the Ancestors. One site of note is HiOu-55, located on Stanley Creek, on the south side of the Fort Hills, from which an Oxbow-like point and two convex-based, side-notched points were recovered, along with a possible McKean lanceolate (Somer and Kjar 2007, plate 52: d, f, g and e, respectively). The McKean lanceolate, which shows traces of fire burning, is the only such find that we are aware of from the Athabasca lowlands.

On the west side of the Athabasca River, the Beaver River Complex is represented at the Beaver River Quarry site (HgOv-29) by a typical early Beaver River side-notched point (Syncrude Canada Ltd. 1974, figure 6: B). It is further represented at three other nearby sites, including HgOv-50, at the Peter Lougheed Bridge, which also has a Firebag Hills component, and two other sites downstream, in the area of CNRL's Horizon Project.

A late Beaver River Complex occupation is represented at HcOs-3, located on Gregoire Lake, to the south of Fort McMurray (Pollock 1978b, 116–118). Excavations at the site yielded fine-grained, white quartzite Oxbow point, along with an MVMq dart point of the style found in Beaver River Complex sites (Pollock 1978b, figure 43: 15 and 16).

The Beaver River Complex is well attested to the north, in the Birch Mountains area, by finds from five sites (see fig 6.9). Most of the points from Beaver River Complex sites are manufactured of grey quartzite. The Gardiner

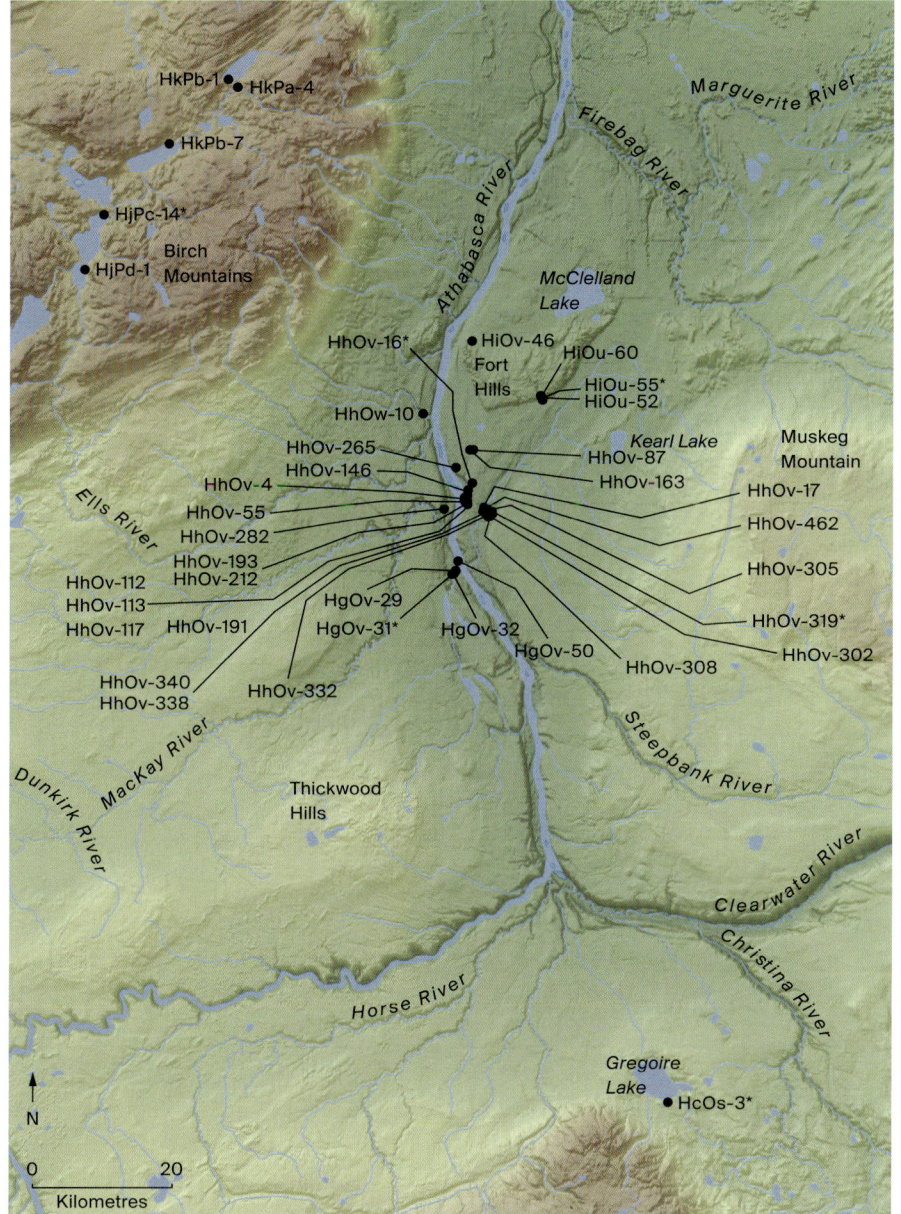

Figure 6.9. Beaver River Complex sites (ca. 7,750 to 4,000 BP). An asterisk indicates the presence of Oxbow or stemmed, indented-base points.

Marguerite River

Firebag River

HkPb-1 • • HkPa-4

• HkPb-7

• HjPc-14*

Birch
Mountains

• HjPd-1

Athabasca River

McClelland
Lake

HhOv-16* • • HiOv-46   HiOu-60
Fort
Hills   HiOu-55*
• HiOu-52

HhOw-10 •

HhOv-265   Kearl Lake   Muskeg
HhOv-146   • HhOv-87   Mountain
HhOv-4   HhOv-163
HhOv-55   HhOv-17
HhOv-282   HhOv-462
Ells River

HhOv-193
HhOv-212   HhOv-305
HhOv-112
HhOv-113   HgOv-29   HhOv-319*
HhOv-117 HhOv-191 HgOv-31*   HgOv-32
HgOv-50   HhOv-302
HhOv-340   HhOv-308
HhOv-338   HhOv-332

MacKay River

Steepbank River

Dunkirk River

Thickwood
Hills

Clearwater River

Christina River

Horse River

Gregoire
Lake
• HcOs-3*

N

0   20
Kilometres

Lake Narrows site (HjPd-1) contained both early and late side- and corner-notched dart points, including examples of the early form with a convex base (Ives 1993, figure 6, row 2, third and fourth from the left, and row 3, first and second from the left). In addition, the complex is represented to the east, in the Firebag Hills–Descharme River headwaters, by two isolated side-notched point finds, one manufactured of MVMq (Reeves, Cummins, and Lobb 2008).

The Beaver River Complex is thus well represented in sites in various areas of the Lower Athabasca region, both in the lowlands and in local uplands such as the Fort Hills, and the Birch Mountains. Quarrying and workshop activities continued at the Quarry of the Ancestors during the Beaver River period, which spanned over three millennia. During this time, there was probably a significant shift in settlement to the occupation of Athabasca River valley terraces and hospitable locales along the escarpment, such as Cree Burn Lake.

We view Beaver River peoples as a part of a larger, regionally focused group whose traditional territory extended southeastward at least as far as Peter Pond Lake. As had earlier inhabitants of the Lower Athabasca region, Beaver River peoples continued to make seasonal use of the Clearwater–Methye Portage–Lac La Loche route between Peter Pond Lake and the Lower Athabasca. MVMq is common in sites on Lac La Loche (see Fenton and Ives 1982, 1990) and Peter Pond Lake, and at Buffalo Narrows, and some of the MVMq side-notched points collected from surface surveys or test excavations date to the period of the Beaver River Complex. These include a side-notched point found in excavations beneath the main room of La Loche House (HdOj-1) (Steer 1977, figure 35: f), as well as points from Saleski Lake collected by David Meyer in 1978 (Reeves, personal examination of the University of Saskatchewan collections, 2006), from the Old Airport site (HcOi-1) on Saleski Creek, test-excavated by Hanna (1982), and from the Old Beach site (GlOc-30) at Buffalo Narrows (Millar 1983) (see fig 6.4).

MVMq side-notched points have not been reported for the Shield Archaic sites discussed by Gordon (1996), although the collections at the Canadian Museum of Civilization, both from these sites and from other sites of equivalent age in the Mackenzie region, have not yet been examined specifically for the occurrence of MVMq. As the presence of quartz tools demonstrates, however, Beaver River Complex peoples who resided in the Lower Athabasca area participated in the Shield Archaic Tradition of the northern boreal forest and the Barrenlands, interacting with culturally related groups situated farther north and west in the forests of the Upper Athabasca and Peace rivers. In addition, Beaver River peoples had contact with groups in the parklands and plains to the south. This is indicated not only by the early side-notched point styles and specific biface types but also by the presence of bipolar reduction technology (Kasstan 2004; Low 1997; Walker 1992), which is associated in those regions with the working of quartzite and cobbles and pebbles of chert.

As noted above, the appearance of Oxbow points at a number of sites in the Lower Athabasca basin, as well as at other locations such as the Alook site (Sims 1981), La Loche House (Steer 1977), the Old Airport site, Saleski Creek (Hanna 1982), and the Old Beach site (Millar 1983; see fig 6.4), is more likely to represent the adoption of this point style by regionally resident groups, as it does in the Northern Rockies and on the northern British Columbia plateau, than the appearance of hunters from the Plains. Of interest, however, is the apparent absence in these northern forests of McKean lanceolates or other features of the McKean Complex of the northwestern Plains and Rocky Mountains. As noted above, one fire-burned point from HiOu-55 in the Fort Hills is stylistically similar to a McKean lanceolate. In addition, a stemmed, indented-base point recovered from HgOv-50 bears some resemblance to a Duncan point, another characteristic style of the McKean Complex. These artifacts may best be regarded as evidence of possible contact between the cultural traditions represented by the Beaver River Complex and Plains cultures.

Firebag Hills Complex (ca. 4,000 to 2,650 BP)

The Beaver River Complex was succeeded by a technologically, and presumably linguistically and culturally, unrelated complex that we call the Firebag Hills Complex. It is part of the larger Pre-Dorset Arctic Small Tool Tradition of the northern edge of the boreal forest and the Barrenlands, which has been dated to 3,450 to 2,650 BP in these areas (Gordon 1996). Ives (1981) was the first to recognize the stylistic similarities between points from the Birch Mountains and Arctic Small Tool Tradition lanceolates (see also Van Dyke and Reeves 1984). We take the name of this complex from the Firebag Hills, an area that straddles the Alberta-Saskatchewan border north of the Clearwater River. Archaeological studies in that region (Reeves, Cummins, and Lobb 2008) have identified a number of the tools distinctive of the Arctic Small Tool Tradition as discussed by Gordon (1975, 1996).

The Firebag Hills tool assemblage (table 6.6 and plate 6.8) includes thin, triangular-shaped points or end blades, lateral blade and side-blade insets, and small, notched points. Other tools include finely made spurred end scrapers, microgravers (sometimes notched and often manufactured on prismatic microblades), transverse and mitt burins, microblades, and microcores (plate 6.10: 7–9). The non-MVMq tool stone assemblage includes a high-quality,

TABLE 6.6 Firebag Hills Complex sites and diagnostic artifacts

Borden no.	Elevation (masl)	Projectile points and other diagnostic tools	Reference(s)
HdOs-2	254.00	Graver point, scrapers (black chert)	Donahue 1976
HgOv-29	256.00	Side-notched point	Syncrude Canada Ltd. 1974, figure 6: b
HgOv-50	245.00	Peace River chert end blade; FBH assemblage	Reeves et al. 2013b
HgOv-85	240.00	FBH assemblage	Reeves et al. 2013b
HiOu-14	293.00	Microblades, microcores, gravers	Saxberg, Somer, and Reeves 2004
HiOv-89	303.19	Microblades, microcores	Unfreed 2001, HiOv-89 site inventory form (05-328)
HiOw-44	293.00	End scrapers (some quartzite)	Clarke 2002a
HiOw-37	291.00	Bifaces, end scraper (quartzite)	Bryant 2004, 2005
HiOw-39	272.00	End scraper, micrograver (salt and pepper quartzite)	Tischer 2006
HiOw-43	265.00	Chalcedony side-notched point	Tischer 2006, plate 16
HhOt-32	313.00	Brown quartzite side-notched point base	Bouchet-Bert 2007, plate 33
HhOu-56	308.00	Small notched point, 3 bifaces or point tips	Green et al. 2006, plate I-9
HhOv-3	274.99	Notched burin	Sims 1977
HhOv-18	280.00	2 side-notched points	Ronaghan 1981b, plate 23: 1 and 2
HhOv-73	294.32	Multiple tools (cherts and quartzites)	Ronaghan 1981b; Le Blanc 1985; Le Blanc and Ives 1986; Green and Blower 2005
HhOv-78	285.79	Small notched point	Clarke and Ronaghan 2004, plate HhOv-78-1
HhOv-87	298.00	Small notched point, of chert (isolated find)	Reeves et al. 2014a
HhOv-304	278.00	2 side-notched points, 2 lanceolate side blades, microtools, and other tools (quartzite)	Saxberg 2007; de Mille and Reeves 2009
HhOv-305	282.48	Lanceolates	Saxberg and Reeves 2004, plate 65: a and b
HhOv-307	281.89	Lanceolate point	Saxberg 2007, plate B.11: a
HhOv-324	277.00	Notched graver, end scraper	Saxberg 2007
HhOv-364	277.00	Brown quartzite lanceolate point	Tischer 2004, plate 155
HjPc-14	646.00	Two ovate points, side blades (quartzite)	Donahue 1976, plate XII: g and h
HjPd-1	683.00	Thin lanceolate point (quartzite)	Ives 1993, figure 5
HkPa-4	714.00	Thin lanceolate point (quartzite)	Ives 1993, figure 5
HkPb-1	715.65	Black chert side-notched point	Ives 1993; Reeves, examination at RAM

honey-to-brown-coloured Northern quartzite, as well as other varieties of that stone; pebble cherts, including black and olive green; and a distinctive banded cream-and-white chert that appears to be the same as that described by Gordon (1996) as a diagnostic Pre-Dorset tool stone type. Bipolar technology associated with the reduction of Northern quartzite or chert pebbles dominates. The frequency of MVMq varies, with sites on the Athabasca River, such as HgOv-50 and HgOv-85, containing small amounts of poor-quality MVMq. Both sites also contained some fire-cracked rock, suggesting that they might represent early-spring occupations by groups returning to the lowlands who had not yet accessed, or could not yet access, the Quarry of the Ancestors (Reeves et al. 2013b).

The Firebag Hills Complex is represented in the Lower Athabasca region by twenty-six sites (see fig 6.10). At this time, the tree line was both latitudinally and altitudinally depressed relative to its present position. The region may therefore have been either within or adjacent to the main wintering range of the Beverly caribou herd and thus seasonally occupied by Pre-Dorset Tradition peoples who followed the migrations of this herd between its wintering grounds at the edge of the boreal forest and its summer habitat in the Barrenlands to the north (Gordon 1975, 1996).

The Bezya site (HhOv-73) is assigned to the Firebag Hills Complex. Located more or less midway between the Athabasca and Muskeg rivers some 4 kilometres northeast of Cree Burn Lake, the site was discovered in 1980 during shovel prospecting on an elevated knoll located in Alsands Lease 13 (Ronaghan 1981b: 88–99). Ronaghan's initial studies, as well as those of Le Blanc in 1982 and 1983 (Le Blanc 1985; Le Blanc and Ives 1986), recovered microblades and microcores, core tablets, and a notched transverse burin and other tools but no projectile points. Twenty-five faunal fragments were found and were radiocarbon dated to $3,990 \pm 170$ [14]C yr BP (Beta-7839). More recently, in 2000 and 2003, Golder Associates Ltd. (Green and Blower 2005) carried out mitigative excavations, as the site would be destroyed in the development of Shell's Albian Sands Muskeg River Mine open pit. While they did not find any projectile points, they did recover a number of other tools, which include a typical Pre-Dorset knife that, as illustrated by Green and Blower (2005, plate III-1), compares favourably with one illustrated by Gordon (1996, figure 8.7, KeNi-4: 125 and KeNi-4: 263).

To date, the Firebag Hills Complex is very poorly represented at sites along the escarpment on the east side of the Athabasca River, generally by isolated finds such as a small, side-notched dart point made of brown-and-white chert from HhOv-87 (plate 6.8: 11). This weak representation appears to reflect a transitory occupation of the escarpment and a preference for lower terraces within the river valley. In contrast, the Firebag Hills Complex is well represented at the Quarry of the Ancestors site complex on the Muskeg River, where it was identified at five sites. Among the sites of interest is HhOv-304, located on a ridge on the northwest margin of the Quarry of the Ancestors (Saxberg 2007; de Mille and Reeves 2009). Excavations at the site—a large workshop and campsite characterized by a number of distinct occupational loci—yielded a number of points. Two lanceolate points were recovered from Locus 6. One is a thin, finely crafted lanceolate point, over 5 centimetres in length but broken in manufacture (plate 6.8: 8). Lanceolate points of this nature are common in Pre-Dorset occupations in the Barrenlands (Gordon 1975, 1996), although Arctic archaeologists tend to call them "end blades." The other was also broken in manufacture (plate 6.8: 7; see

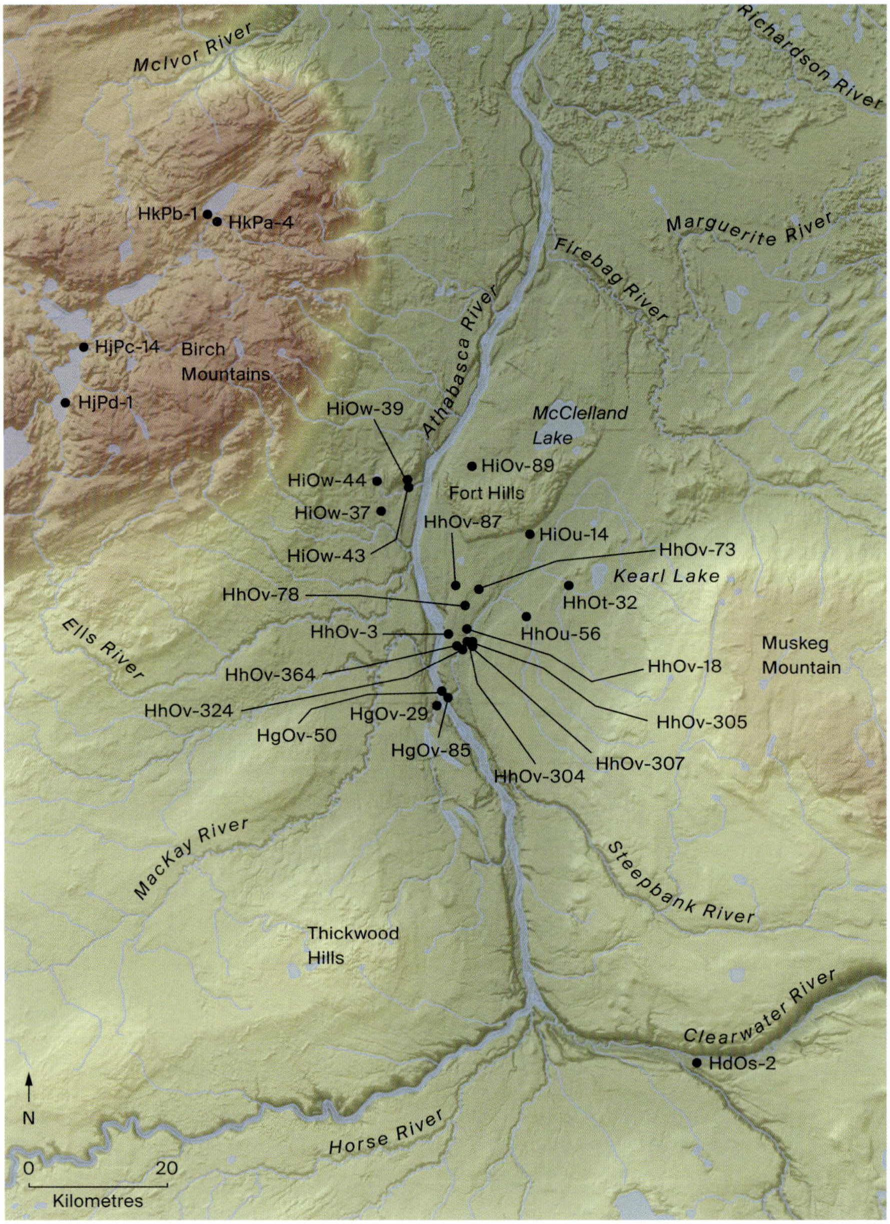

Figure 6.10. Fire Bag Hills Complex sites (ca. 4,000 to 2,650 BP)

also de Mille and Reeves 2009, plate B.1: a and c). In addition, a small, stemmed point was recovered (plate 6.8: 13), along with a side blade (de Mille and Reeves 2009, plate B.1: e and g, respectively).

Side-notched points from Locus 6 are represented by a complete side-notched quartzite point (plate 6.8: 14; see also de Mille and Reeves 2009, plate B.1: b), as well as a side-notched point snapped diagonally across the shoulders and a large side-notched convex point made of black chert that appears to have been recycled and reworked as a knife (de Mille and Reeves 2009, plate B.1: d and f). A variety of other tools, including microblades and microgravers and notched microblade gravers, were also found in Locus 6. Although MVMq over-whelmingly dominated the Locus 6 assemblage (99.78%), 202 non-MVMq arti-facts were recovered, a number of which were tools.

Two small, side-notched point bases (plate 6.8: 12; Saxberg 2007, plate B.5: a and b), one of which was manufactured of orange-brown quartzite, were recovered from Locus 1 and Locus 2 at HhOv-304. Locus 1 also yielded a number of notched graving tools manufactured on microblade and microblade-like flakes (plate 6.8: 9 and 15–18; Saxberg 2007, plate B.7), which Saxberg (2007, 33) notes are "formally and functionally similar to notched 'burin-like' tools found at Dorset sites in the Canadian Arctic." These tools also occur in the Pre-Dorset sites (Gordon 1996, figure 8.6: KjNb-7: 11-95). HhOv-304 also contained a small, finely crafted spurred end scraper of brown megaquartzite similar to styles found in Gordon's Pre-Dorset sites. Positive reactions were obtained to deer antisera on an awl fragment and to cat antisera on a scraper.

Other sites of interest in the Muskeg drainage are two campsites, HhOv-18 (Ronaghan 1981b), situated not far north of the Quarry of the Ancestors, and HhOu-56, located to the east, on Jackpine Creek. Excavations at the latter (Green et al. 2006) recovered a blade fragment of a small, corner-notched point (Green et al. 2006, plates I-8 and I-9) and the tips of three small, thin projectile points or bifaces, which may be lateral blade inserts (Green et al. 2006, plate I-9). These specimens are comparable to those illustrated by Donahue for HjPc-14 (Donahue 1976, plate XII: l and m). Other tools include end scrapers, a micro-graver manufactured on a microblade (Green et al. 2006, plate I-11), wedges, and a bipolar microblade core (Green et al. 2006, plate I-14).

The Firebag Hills Complex is represented by two sites in the Fort Hills, HiOu-14 and HiOv-89. Of interest is an obsidian flake sourced to Mount Edziza, which was recovered from HiOu-14, located at the mouth of Stanley Creek. HiOu-14 also contained microblades and microcores (plate 6.10: 9), wedges, a micro-graver, and bipolar split pebbles of black chert (Saxberg, Somer, and Reeves 2004, 35–37; Somer 2005). Another probable Firebag Hills Complex site is

HiOv-89 (Unfreed, Fedirchuk, and Gryba 2001; Younie 2008). Excavations at this site recovered a number of microblade cores, microblades, burins, and other tools. Pebble cherts were dominant.

Six Firebag Hills components have been identified on the west side of the Athabasca River (see fig 6.10), including HgOv-50, located on the 10-metre-high river terrace at the Peter Lougheed Bridge. Further to the south, the Firebag Hills Complex is also represented on the Clearwater River in the artifact assemblage collected by Donahue in 1975 from HdOs-2, located on an 8-metre-high terrace (Donahue 1976, 49–50). Two of the unifaces that he illustrates are dorsally unretouched triangular to subrectangular end scrapers, possibly with graver points (Donahue 1976, plate IV: e and f). According to Donahue (1976, 51), another uniface had a burin-like scar.

The Firebag Hills Complex is represented at four sites in the Birch Mountains. Thin, lanceolate-shaped points of grey quartzite recovered from Gardiner Lake Narrows (HjPd-1) and Eaglenest Portage (HkPa-4) were recognized as being "similar to Arctic Small Tool tradition artifacts from the central Northwest Territories" (Ives 1993, 12 and figure 5). Other sites of interest include HjPc-14, on Big Island Lake, test-excavated by Donahue, where a charcoal concentration yielded an age of 3,610 ±120 ^{14}C yr BP (RL-5333; Donahue 1976). Among the artifacts Donahue recovered were two ovate points (Donahue 1976, plate XII: g and h), both manufactured of grey quartzite, which fit well into the Firebag Hills Complex. From Donahue's description (1976, 102), side blades are also present, although they are illustrated in his thin biface category.

Another site in the Birch Mountains is Satsi (HkPb-1), on Eaglenest Lake, where a small black chert side-notched point was collected that fits within the range of variation for side-notched points found in the Firebag Hills Complex (Reeves, personal examination of the collections at RAM). A smudge pit was also found in test excavations at the site and was radiocarbon dated to 2,795 ± 85 ^{14}C yr BP (S-2174; Ives 1993, 11). Microblades and microcores have not been identified at sites in the Birch Mountains, however, in contrast to Firebag Hills Complex sites in the lowlands, including those in the Fort Hills. Given that evidence of the manufacturing and use of microblades was also present in the earlier Beaver River and Nezu complexes in the lowlands, the absence of such evidence from the Birch Mountains suggests to us a difference in hunting strategies between these areas.

As noted earlier, the Firebag Hills Complex also occurs in the area of the Firebag Hills and Descharme River headwaters, where it is represented in ten sites, as reflected by its distinctive tool kit (Reeves, Cummins, and Lobb 2008). Millar (1983) illustrates three unnotched triangular points from sites at Buffalo

Narrows that fit comfortably with those associated with the Firebag
Hills Complex.

In view of the relatively short temporal duration of the Firebag Hills
Complex (roughly 1,500 radiocarbon years), its presence in the Lower
Athabasca region appears to have been fairly intense in comparison to the following Chartier Complex, which endured for well over two millennia. The contrast may reflect a local intensification of occupancy and resource harvesting
during the Firebag Hills period in response to a climatic event that resulted in
the southward shift in the tree line and thus in the seasonal range of the Beverly
caribou herd (Gordon 1975, 32–56).

LATE PRECONTACT PERIOD: THE CHARTIER COMPLEX
(CA. 2,650 TO 300 BP)

The Chartier Complex is the regional representative of the Taltheilei Tradition in
the Lower Athabasca–Lac La Loche–Peter Pond Lake region (see fig 6.4). This
complex, first proposed by Millar (1983), takes its name from the Martin Chartier
(GlOc-20) and Bernadette Chartier (GlOc-21) sites on the Kisis Channel at
Buffalo Narrows (Millar 1983; Millar and Ross 1982). Excavations at these and
nearby sites in the early 1980s indicated the presence of major Taltheilei occupations spanning the entire duration of the Taltheilei Tradition. Some 10% to 15%
of the tool assemblage is manufactured of MVMq, indicating a significant relationship between Taltheilei Tradition occupations there and those in the Lower
Athabasca (Reeves, personal examination of the University of Saskatchewan collections, 2006).

In the boreal forest and Barrenlands, the Taltheilei Tradition follows the Pre-
Dorset Tradition and is generally understood to represent the precontact Dene
occupation of the region. On the basis of changes in artifact types and styles, the
tradition can be divided into three successive phases: the Early (2,650 to 1,800
BP), the Middle (1,800 to 1,300 BP), and the Late (1,300 to 200 BP) (Gordon
1996).[13] The Late Taltheilei is characterized by both spear or dart points and
arrow points.

In general, the Taltheilei stone tool assemblages from the Lower Athabasca
region fit well with those described by Gordon (1996) and Le Blanc (2005), and
the reader is referred to these, as well as to Ives's reports on site HkPa-4, in the
Birch Mountains (Ives 1977, 1985), for descriptions and illustrations. Among the
tools and technologies useful in identifying Taltheilei sites in the Lower
Athabasca region is the bipolar split chert pebble technology (discussed below

for HgOv-107) and the presence of elongated, bi-pointed, leaf- or lanceolate-shaped bifacially worked adze-like tools, which are characterized by plano-convex transverse sections. Other typical tools include chithos. Care must be taken in discriminating these tools from the cortical spall tools found in the Nezu Complex (Reeves, Bourges, and Saxberg 2009).

The Chartier Complex is on the whole well represented in the Lower Athabasca region, with a total of twenty-five sites, site components, or isolated occurrences assigned to this complex (table 6.7 and fig 6.11). However, it does not appear to be represented along the eastern escarpment of the Athabasca River, which probably reflects the focus of the great majority of studies to date on the higher terraces and the river escarpment benchlands above these terraces rather than on the lower river terraces. That the latter were occupied, however, has been known since 1975, when, during the course of his canoe-based reconnaissance of the Athabasca River, Donahue (1976, 55) located a site (HhOv-29) on a low terrace of the river, where he recovered split black chert pebbles, a black chert uniface, and some MVMq flake and core fragments. Donahue also recorded a site (HgOv-33) on a low terrace near Saline Lake that had "metal" in it and an MVMq biface (Donahue 1976, 51 and plate V: g). (Unfortunately, the "metal" is not described in the report's artifact analysis.) The only other find is an isolated lanceolate bipoint at HgOv-92, a site situated on the edge of a highway borrow source 2.3 kilometres north-northeast of the Peter Lougheed Bridge (site inventory form, Archaeological Research Permit 00-175). The point, examined by Reeves at the RAM, compares favourably with Middle Taltheilei specimens illustrated by Gordon (1996, figure 5.2: see, for example, KjNb-7: 2-38).

The Chartier Complex may be somewhat better represented in the Muskeg River drainage by two isolated finds (HhOt-14, HiOs-2) in the vicinity of Kearl Lake, as well as farther downstream at HhOu-50, located at the mouth of Jackpine Creek, and possibly southwest of the Quarry of the Ancestors, at HhOv-391 and HgOv-107. HhOu-50 artifacts include a small arrow point of the general Prairie Side-Notched style (plate 6.9: 2; see also Clarke and Ronaghan 2000, plate I-4), typical of the forms found in Late Taltheilei sites on Peter Pond Lake (Reeves, personal examination of the University of Saskatchewan collections, 2006).

The HgOv-107 assemblage (Saxberg and Reeves 2006, 35–36) lacks projectile points but contains a number of bipolar split black chert pebbles and bipolar flakes worked and/or used as tools (Saxberg and Reeves 2006, plate B.3). This site could be an earlier Firebag Hills Complex occupation, as some burins are present (although no microcores or microblades). However, the bipolar split black chert pebble industry resembles that represented at Late Taltheilei sites in the Birch Mountains, such as Eaglenest Portage (HkPa-4; Ives 1977, plate 13),

TABLE 6.7 Chartier Complex sites and diagnostic artifacts

Borden no.	Elevation (masl)	Projectile points and other diagnostic tools	Reference(s)
HeOn-1[a]	283.87	Dart point, split black chert scrapers	Pollock 1978b, figure 38: 1
HdOr-1	256.00	Dart point, split black chert scrapers	Pollock 1978b
HcOn-3[b]	478.00	Split black chert pebbles, Avonlea dart point	Pollock 1978b
HcOs-1	477.00	Late Taltheilei arrow point, Avonlea arrow point, point tip, point base, split black chert pebbles	Pollock 1978b
HgOv-22	297.00	Side-notched dart point, barbed bone point, chitho, split pebbles	Syncrude Canada Ltd. 1973
HgOv-33	232.69	Biface	Donahue 1976, plate V: g
HgOv-92	274.63	Lanceolate bipoint	HgOv-92 site inventory form (00-175)
HgOv-107	301.00	Split black chert pebbles	Saxberg and Reeves 2006
HhOt-14	318.00	Late Taltheilei dart point	Clarke 1998, figure IV-4
HhOu-50	282.94	Late Taltheilei arrow point	Clarke and Ronaghan 2000
HhOu-73	303.02	Corner-notched arrow point	Somer and Kjar 2007, plate 52: p
HhOv-29	241.00	Split black chert pebbles	Donahue 1976
HhOv-294	302.93	Split black chert pebbles	Saxberg, Somer, and Reeves 2004
HhOv-391	291.00	Dart point	Tischer 2004, plate 176
HhOw-20	299.00	Side-notched point	HhOw-20 site inventory form (04-094)
HiOs-2	360.60	Bipointed leaf-shaped biface, adze	HiOs-2 site inventory form (04-375)
HiOv-57	331.00	3 lanceolate points	Saxberg, Somer, and Reeves 2004, plate 21: b, c, and d
HiOv-140	321.21	Adzes	Saxberg, Somer, and Reeves 2004
HiOw-50	295.56	Slightly stemmed lanceolate point	Bryant 2005, plate II-4
HiPd-2	741.28	Quartzite Late Taltheilei arrow point	Brian Ronaghan, pers. comm.
HiPe-2	723.66	Quartzite Late Taltheilei arrow point	Brian Ronaghan, pers. comm.
HjPd-1	683.00	Early to Middle Taltheilei points, bipointed side scraper or adze	Ives 1981, 1993; Reeves, examination at RAM
HkPa-4	714.00	Late Taltheilei dart points, bipointed side scraper, adze (quartzite)	Donahue 1976, plate X; Ives 1977, 1981, 1993
HkPa-11	714.00	Dart points (quartzite)	Donahue 1976, Plate IX: h
HkPb-1	715.65	Dart points (quartzite)	Donahue 1976, Plate X: h

[a] Date of 1735 ± 35 ^{14}C yr BP (NMS-1275) on charcoal from hearth (Pollock 1978b, 44)

[b] Date of 570 ± 115 ^{14}C yr BP (NMS-1274) on calcined bone fragments (Pollock 1978b, 80)

and at the Slump and Hidden Creek sites on Lesser Slave Lake (Le Blanc 2005; see fig 6.4), as well as at Late Taltheilei sites at Peter Pond Lake, in Saskatchewan (Reeves, personal examination of the University of Saskatchewan collections, 2006). We consider the bipolar split black chert pebble industry to be a Late Taltheilei diagnostic, which was probably introduced through contact with Late Avonlea and Old Women's groups to the south, whose technologies

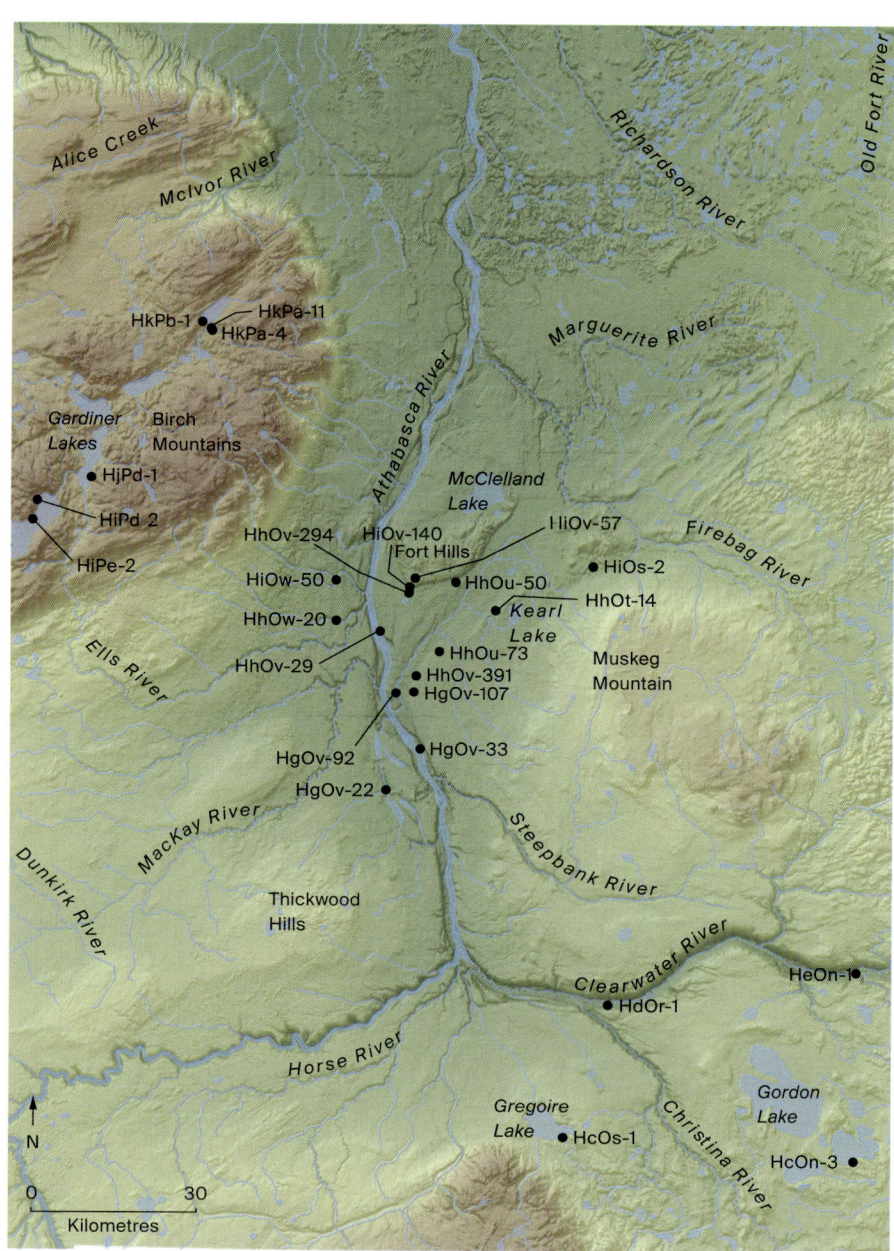

Figure 6.11. Chartier Complex sites (ca. 2,650 to 300 BP)

include a significant bipolar split black chert pebble industry known as the Rundle Technology (Reeves 1969).

The Chartier Complex is represented at four sites along the south edge of the Fort Hills (HiOv-57, HiOv-140, HhOu-73, and HhOv-294). Three lanceolate points were recovered from HiOv-57 (plate 6.9: 3–5; see also Somer, Saxberg, and Reeves 2004, 3941), one of which reacted positively to striped bass antisera, suggesting the use of a member of the *Perciformes* order of fish (such as perch, walleye, pike).These points compare favourably with Middle and Late Taltheilei forms illustrated by Ives for the Birch Mountains and by Gordon for the Barrenlands (see Ives 1993, figure 6; Gordon 1996, figures 4.4 and 5.4). A Late Taltheilei small, well-finished grey quartzite arrow point (plate 6.9: 1) was recovered from excavations at HhOu-73, which Somer and Kjar (2007, 71–73) rightly argue is best described as an Avonlea type. A number of Taltheilei arrow points from sites in northeastern British Columbia (see, for example, Driver et al. 1996, figure 9k; Eldridge et al. 2008. figure 80: a–e; Spurling 1980, 284, figure 40d) and in the Spence River Complex of the Mackenzie drainage (Morrison 1984, figure 2: a and b, for example) generally conform to the template characteristic of the early Avonlea horizon, as was the case for arrow point styles in other cultural groups peripheral to the northwestern Plains when first adopting bow and arrow technology (Reeves 2003).

On the west side of the Athabasca River, the first probable Taltheilei point recovered in the region was found at HgOv-22 during the 1973 HRIA of Syncrude's Lease 17 (Syncrude Canada Ltd. 1973, 98). Other Taltheilei tools identified during this study included a chitho and a number of split black chert pebbles. Losey also collected a fine example of a complete unilaterally barbed bone point (Syncrude Canada Ltd. 1973, 92). Other Taltheilei finds include a Middle Taltheilei point at HiOw-50 (Bryant 2005, plate II-4) and, at HhOw-20 (a campsite), a "small side-notched point of grey northern quartzite," to which Gryba and Tischer (2005) assigned a Late Prehistoric age.

The Chartier Complex is represented at two sites on the Clearwater River, HeOn-1 and HdOr-1. Test excavations at the Gros Roche site (HeOn-1), located on a portage, uncovered a basin-shaped hearth lined with limestone cobbles from which a sample of charcoal was dated to 1,735 ^{14}C yr BP ± 105 (NMS-1275; Pollock 1978b, 44). An MVMq dart point body fragment was also recovered at the site (Pollock 1978b, figure 38: 1). The faunal remains were dominated by mallard duck, representing 13.0%, followed by beaver (4.9%), muskrat (2.2%), red squirrel (1.1%), and canid, moose, and northern pike (each 0.5%). Most large mammal remains were unidentifiable, suggesting that they were being intensively processed for marrow and bone grease and/or that only the meat was being brought back from the kills.

Pollock also found a Late Chartier Complex site on Gypsy Lake. Testing at HcOn-3 recovered sixty-one pieces of carbonized and calcined bone fragments (unidentifiable), which yielded a date of 570 ± 115 ^{14}C yr BP (NMS-1274; Pollock 1978b, 80). Five of the artifacts recovered were bipolar split black chert pebble tools (Pollock 1978b, figure 40: 1–5). Prager (in Pollock 1978b, 157) suggests that the site may have been a short-term fishing camp, as the only identifiable remains were those of northern pike.

HcOs-1, located on Gregoire Lake, also contains Middle and Late Taltheilei occupations. Tools recovered included a Late Taltheilei–style arrow point and a small, poorly made Avonlea-style arrow point (Pollock 1978b, figure 43: 1 and 2), the latter manufactured of a semi-translucent grey-white chert. The site also yielded various dart point fragments, one of which was the convex base of a dart made of Lake One Dune chert (Pollock 1978b, figure 43: 5; Pollock misidentifies it as MVMq). Black chert pebble wedges, end scrapers, spalls, and cores were also found.

The Chartier Complex is well represented in the Birch Mountains, particularly in the excavations at the Gardiner Lake Narrows (HjPd-1) and Eaglenest Portage (HkPa-4) sites (see fig 6.11). A number of Early to Middle Taltheilei dart points from HjPd-1 are among the stemmed, side- and corner-notched points illustrated by Ives (Ives 1981, figure 7; Ives 1993, figure 6). Also present in the HjPd-1 collections at the RAM is a typical Taltheilei-style bipointed side scraper or adze. Middle to Late Taltheilei sites in the Birch Mountains include HkPa-11, on Clear Lake, and HkPb-1, on Eaglenest Lake, from which quartzite dart points with broad corner notches and convex bases were recovered (Donahue 1976, plates IX: h and XI: h). Two additional sites, HiPd-2 and HiPe-2, were identified more recently by Ronaghan and Ives during a post-fire inventory of Namur Lake and proved to contain two small, typical Late Taltheilei arrow points of grey quartzite and salt and pepper quartzite, respectively (Brian Ronaghan, pers. comm., 2007).

Ives's excavations at Eaglenest Portage (HkPa-4) indicate that it contains a major Middle to Late Taltheilei occupation. He obtained a radiocarbon date of 1,030 ± 110 ^{14}C yr BP (DIC-720) from charcoal collected from the surface of a buried organic horizon (Ives 1985, 32–33). Ives illustrates a range of dart points from the site, a number of which are quartzite (see Ives 1981, figure 7; Ives 1993, figure 6, top row); those illustrated include specimens that we assign to the Beaver River Complex. Late Taltheilei dart and possible arrow points were also recovered from Eaglenest Portage by Donahue, who discovered and tested the site in 1975. These included both stemmed and corner-notched forms, including a specimen with the typical Late Taltheilei rounded basal edge–straight base configuration (Donahue 1976, plate X: i; see also Ives 1981, figure 7, second row, second from the right). Specimens like this have been found elsewhere in the Athabasca lowlands, as well as at Buffalo Narrows and other sites in northwestern Saskatchewan and at Lesser Slave Lake, in Level 2 of the Slump site (Le Blanc 2005, figure 48: d).

Ives illustrates four smaller projectile points from Eaglenest Portage that he considers to be arrow points (Ives 1993, figure 7; see also Ives 2003, figure 6). The points include one manufactured of black chert, one of brown quartzite, one of Lake One Dune chert, and one of salt and pepper quartzite (see Ives 1993, figure 7: a–d). We would classify these as Late Taltheilei dart points rather than arrow points. The basal edge–base configurations of these points fit better within the range of variation illustrated by Gordon (1996) for Late Taltheilei dart points than with the Late Taltheilei arrow points recovered from the Lower Athabasca basin, including sites on Namur Lake, the Buffalo Narrows sites at Peter Pond Lake, and the Slump and Hidden Creek sites at Lesser Slave Lake (Le Blanc 2005, figures 48 and 68). Ives does illustrate a small arrow point (Ives 1981, figure 7, row 2, fourth from left). The site provenance is not given, but the point bears a close resemblance to the Avonlea-style arrow point (plate 6.9: 1) recovered from HhOu-73 in the Fort Hills, discussed above.

Ives recovered a number of other typical Taltheilei tools from the HkPa-4 excavations, including a bipointed side scraper or adze made of MVMq, a well-formed subrectangular cortical-backed quartzite cortical spall tool, and a considerable number of bipolar split black chert pebbles (see Ives 1977, plates 4b, 16b, and 13, respectively). Ives also examined the tool stones present in the HkPa-4 assemblage (Ives 1977, 18–20; see also Ives 1985, 33–35). Interestingly, MVMq—represented in 12 out of 91 (13.2%) of the formed tools—constituted only 4.5% of the total assemblage. Salt and pepper quartzite represented 2.0%, other chert 6.0%, and black chert 2.8% of the assemblage, which was dominated by quartzite (69.6%). In considering these percentages, however, the reader should bear in mind that, in our opinion, all of the earlier cultural complexes, except for the Fort Creek Fen Complex and possibly the Nezu Complex, are represented in the HkPa-4 artifact assemblage.

In summary, in terms of site distribution, the presence of the Chartier Complex in the Lower Athabasca region roughly corresponds to that of the preceding Firebag Hills Complex, with good representation in the Fort Hills and Birch Mountains, possibly somewhat poorer representation in the Muskeg River area, and quite limited representation along the eastern escarpment of the Athabasca River (see fig 6.11). As we suggest for the Firebag Hills Complex, the groups who occupied the area during the period of the Chartier Complex appear to have focused on the lower terraces within the river valley, as indicated by Donahue's finds of over thirty years ago. It is to these potential locales, many of which are eroding out along the river, that we need to direct archaeological attention if we are to characterize the full extent and nature of a seasonal Chartier Complex occupation of the Lower Athabasca River.

Temporal trends in Chartier Complex occupations, if any exist, between the Early and Middle Taltheilei and the Late Taltheilei in the Lower Athabasca basin are not yet discernible. At least to date, Late Taltheilei arrow points seem less well represented than dart points. However, if not merely the result of the sampling, this may simply reflect the fact that the use of unilaterally barbed bone and antler points, which do not preserve well, became more common during the Late Taltheilei in the Athabasca lowlands, as it did among Dene peoples in the southwest Yukon (Hare et al. 2004). In addition to the specimen recovered from HgOv-22 noted above, portions of these bone tools have also been found at the Alook site (HaPl-1) at Wabasca Lake (Sims 1981, figure 4: l) and in Level 2 at the Hidden Creek site on Lesser Slave Lake (Le Blanc 2005, figure 65).

The peoples of the Chartier Complex clearly used MVMq for stone tool manufacture, in some locales quite extensively, as is evident from the sites in the Fort Hills area, where MVMq entirely dominates the Chartier Complex assemblages. This suggests access to large amounts of the material. However, we have yet to find any conclusive evidence that these peoples occupied the Quarry of the Ancestors. There, the most recent evidence of use, in the form of projectile points and other tools, is associated with the earlier Firebag Hills Complex. In contrast to the Fort Hills area, MVMq is not the major tool stone at Chartier Complex sites in the Birch Mountains. There, assemblages are dominated by grey quartzites, reflecting a local, but as yet unidentified, source for this tool stone, which also occurs extensively in earlier cultural complexes in the Birch Mountains.

At Chartier Complex sites on Lac La Loche and Peter Pond Lake, MVMq represents 10% to 15% of the formed tool assemblage (Reeves, personal examination of the University of Saskatchewan collections, 2006). This suggests that the precontact Dene peoples of the region continued to rely significantly on this material and thus had direct access to its source. Evidently, then, Chartier Complex groups travelled back and forth between the Birch Mountains–Lower Athabasca area and the Lac La Loche–Peter Pond Lake area, presumably via the Methye Portage and the Clearwater River. We suggest that this continues a basic regional land use pattern dating back 9,000 years or more. How geographically extensive or chronologically varied this pattern is in the Churchill and adjacent drainages in northwestern Saskatchewan remains to be determined. However, some 200 kilometres southeast of Buffalo Narrows, a blocky, split-pebble wedge of MVMq and two MVMq flakes were found at GgNk-1, a campsite located on a point of land where the Montreal River exits Sikachu Lake (Hanna 2004). This occurrence suggests that the pattern of travel and trade extended well downriver from Buffalo Narrows during Taltheilei times.

The geographic extent of MVMq usage also remains to be determined. There has been no systematic archaeological survey of precontact sites along the Athabasca River upstream of the Clearwater River. As Sims (1981) noted, MVMq is present at the Alook site (HaPl-1), a stratified site located at the head of North Wabasca Lake west of the Lower Athabasca in the Peace River basin (see fig 6.4). In addition to preserved organic faunal material, including copious amounts of fish bones, the site contains a major Middle to Late Taltheilei occupation. Interestingly, according to Sims (1981), MVMq represents less than 1% of the tool stone assemblage. The Alook site has not been revisited in some forty years, and, if it still exists, a major excavation should be undertaken at it.

Southward into the headwaters of the Christina River and tributaries of the Beaver River, MVMq has been reported in association with a Late Taltheilei point recovered from GjOq-4, a site located on a relic shoreline at the east end of Christina Lake (site inventory form, Archaeological Research Permit 07-186) (see fig 6.4). In areas north of the oil sands, however, the distribution of MVMq at Taltheilei sites appears to be very limited. It is absent at Peace Point and Lake One Dunes in Wood Buffalo National Park (Reeves, personal examination of the collections at Parks Canada, Winnipeg).[14] Archaeological studies south of Fort Smith associated with the once-proposed Slave River Hydro Project recovered two secondary and four flake fragments of MVMq from IkOv-8 (McCullough 1984; see fig 6.4), which may or may not be associated with the Taltheilei Tradition, as no diagnostics were found. These data suggest that Chartier Complex peoples had little interaction with those precontact Dene groups who traditionally occupied the Peace-Athabasca delta, Lake Athabasca, or the Slave River areas and that trade involving MVMq did not extend that far north.

The yearly round of the peoples of the Chartier Complex and other Taltheilei Tradition regional complexes was clearly tied to fish-bearing lakes. We suggest that these groups occupied the Athabasca lowlands and the Birch Mountains primarily during the frost- and snow-free seasons and generally wintered not in the Lower Athabasca basin but on Lac La Loche, Peter Pond Lake, and adjacent lakes and locales. We further suggest that, once spring arrived, Chartier Complex groups would travel via the Methye Portage to the Birch Mountains and Athabasca lowlands to hunt and fish and to quarry MVMq, returning via the Methye Portage before freeze-up.

At no time in the past would MVMq quarries have been accessible in the winter, in view of frozen ground and snow cover. Consequently, before moving to winter quarters, precontact groups who resided in the Lower Athabasca region would have had to stockpile MVMq blanks for the manufacture of small formed tools or smaller tool blanks over the coming winter. It may even be, then, that the Chartier Complex sites located on the south slopes of the Fort Hills represent the wintering sites of a few families who had stockpiled MVMq and remained in the area to manufacture tools over the winter. The evidence we recovered indicates that ample supplies of MVMq were available and little conservation of the material took place, as is reflected by the high proportion of waste material present at sites. In contrast, the Chartier Complex sites on Peter Pond Lake have very low frequencies of MVMq core fragments, shatter, and debitage, of the sort relating to primary tool production, and a high frequency of small-sized debitage, of the sort produced by resharpening and retouching tools. This suggests that finished tools and final-stage tool preforms were prepared before groups left the Athabasca lowlands.

The Chartier Complex appears to have drawn to a close sometime before the establishment of the inland fur trade, given that, to the best of our knowledge, no trade goods dating to the fur trade era have been found either in Chartier Complex sites or as isolates in the Lower Athabasca region. Perhaps the absence of such goods is simply the result of sampling, but it does seem to indicate that precontact Dene peoples had abandoned the Lower Athabasca region by the time the fur trade arrived there. Alternatively, it is possible that local and regional bands were decimated by the first smallpox epidemic recorded in oral history, which spread through the tribes of the Upper Missouri and Saskatchewan Plains and Rocky Mountains in the early 1730s (Reeves and Peacock 2001), or by earlier epidemics that, on the basis of archaeological evidence and radiocarbon dating, are thought to have swept through northern North America beginning in the sixteenth century (Reeves 2009). The onset of Little Ice Age conditions around roughly the same time may well have been another significant factor in the marginalization of the Lower Athabasca region toward the end of the Late Precontact period.

Thus far, the Late Taltheilei Tradition is the only Late Precontact occupant of the Lower Athabasca region to be firmly identified. In the Peter Pond Lake–Buffalo Narrows area of Saskatchewan, however, two Late Precontact–period ceramic complexes have been identified, which are distinct from Late Taltheilei occupations in the same area: the Buffalo Lake Complex, radiocarbon dated in the range of about 730 to 480 BP, or AD 1200 to 1500 (Young 2006, 218), and the Kisis Complex (Paquin 1995), which is related to the Selkirk Composite (ca.

AD 1300 to 1700) and represents the ancestral culture of the Western Woods Cree (Meyer 1987; Meyer and Russell 1987). Whereas, as noted above, in the Chartier Complex components at Peter Pond Lake, MVMq constitutes 10% to 15% of the tool stone, MVMq is absent from these two complexes. Moreover, in contrast to the typical Taltheilei arrow points, with their low, pointed or rounded basal edges, the small, side-notched arrow points associated with these two ceramic assemblages are characterized by well-defined high, rectangular-shaped basal edges. Many of these points are manufactured of salt and pepper quartzite, which dominates the tool stone assemblage at these ceramic sites, leading Young to conclude that a local source had been discovered and was used extensively by the occupants of these ceramic complexes (Patrick Young, pers. comm., 2005). Evidence thus suggests that groups who produced ceramics were present in this region of Saskatchewan at roughly the same time as the Dene peoples of the Late Taltheilei occupations, who did not.

In Alberta, ceramics have been found at the Wappau Lake Narrows site (GiOv-1; see fig 6.4), 65 kilometres north of Lac La Biche, which Pollock (1978a, 53–54) suggests may represent precontact or protohistoric occupations by Cree-speaking groups. The probability is reasonably high that ceramics will eventually be discovered at locales such as Gypsy Lake to the north, where Pollock (1978b) found sites that appeared to be relatively recent, although he did not recover any time-diagnostic artifacts. Targeted research at these locales is required before we can say with any certainty whether groups of Cree speakers were present in the Lower Athabasca region prior to contact.

CONCLUDING REMARKS

In the preceding sections we have presented a revised chronological outline for the precontact archaeological record of the Lower Athabasca region. Our constructs build on the work of previous researchers, notably Ives (1981, 1993), Van Dyke and Reeves (1984), and Saxberg and Reeves (2003), who have undertaken the task of synthesizing the accumulating archaeological record, or parts thereof, for the oil sands region. The current round of oil sands regulatory approvals that began in the late 1990s has increased by many orders of magnitude the archaeological database for the oil sands region, particularly within the area formed by the Agassiz flood. In years to come, as additional sites are excavated, no doubt there will be challenges to our chronology. Hopefully, these will be based on solid comparative archaeological analysis rather than on generalized comparisons of single specimens.

The 10,000-year archaeological record thus far recovered has its inherent problems, most notably the almost universally shallow surface archaeological deposits, the general lack of stratified sites with well-separated components, and the paucity of organics or clearly defined, charcoal-rich archaeological features that could provide radiocarbon dates, coupled with the suspect nature of some of the radiocarbon dates obtained to date. Fortunately, it appears that, in the Athabasca lowlands (as elsewhere), precontact cultural complexes rarely reoccupied precisely the same spot on the escarpments overlooking the river valley or the same elevated feature inland, with the result that there is a relatively limited mixing of the archaeological record. In contrast, in the Birch Mountains, where the pattern is typical of that often found in the Rocky Mountains to the southwest, focal settlement sites occur with shallow deposits, and the archaeological remains within them span some 9,000 years.

Some of the limitations of the oil sands record may eventually be overcome with the development of new chronometric techniques capable of dating artifacts manufactured of siliceous materials and with both continued and refined application of blood-trace analysis to artifacts, including ancient DNA analysis. We expect that detailed technological analysis of the recovered assemblages, which goes beyond the level of analysis currently available, will certainly reveal significant new information.

Except for those set aside as Provincial Historic Resources (Cree Burn Lake, Beaver River Quarry, and the Quarry of the Ancestors), the principal archaeological sites in the oil sands area have been lost or will be lost during this latest round of resource development. While mitigative excavation studies have recovered samples, of varying sizes, from these sites, archaeologists will not be able to return to these sites to collect additional samples or to make use of new methodologies and techniques that are sure to develop in the coming decades.

Except for the protected sites, twenty years from now, or perhaps in as little as a decade, the primary sources of archaeological data regarding the Athabasca lowlands region will be gone. Given that this resource will vanish, should we not be taking a larger and longer view and striving to ensure that, for those sites that remain within the oil sands region, recovery and interpretation is maximized? What industry has proposed, for example, by way of environmental compensation for wetlands and wildlife habitat lost to oil sands development, including the protection or restoration of significant wetlands and wildlife habitat outside development areas, is a worthy beginning. The same can and should be applied to archaeological sites, which are a non-renewable resource. In return for their loss, industry should commit to protecting other significant sites in the boreal forest and make funds available for research and study.

ACKNOWLEDGEMENTS

The authors would like to thank David Meyer, of the Department of Archaeology, University of Saskatchewan, for giving us access to the university collections, as well as for information on the Buffalo Narrows collections and sites. Patrick Young provided important information regarding the distributions of MVMq across northwestern Saskatchewan. Thanks must also be given to Bob Dawe and Jack Brink, of the Royal Alberta Museum, for arranging access to museum collections as well as for information on MVMq point distributions. Additionally, we are grateful to Golder Associates Ltd., specifically David Blower, for making copies of their reports available to us and allowing us to use images from them. We would especially like to thank Syncrude Canada Ltd., specifically Bill Hunter, as well as Birch Mountain Resources Ltd., specifically Don Dabbs and Ken Foster. Finally, we extend our thanks to Alberta Sustainable Resource Development and to Natural Resources Canada's Centre for Topographic Information for furnishing the base data for the maps in this chapter.

Plate 6.1. Fort Creek Fen Complex projectile points

1. HhOv-87: 3983 Fort Creek Fen lanceolate
2. HhOv-87: 3984 Fort Creek Fen lanceolate (positive reaction to rabbit antisera)
3. HhOv-17 Fort Creek Fen lanceolate
4. HiOu-69 Hi-Lo point
5. HhOv-439 Fort Creek Fen lanceolate
6. HhOv-164: 45 Fort Creek Fen lanceolate, grey quartzite
7. HhOv-164: 47 Fort Creek Fen lanceolate, Northern quartzite (positive reaction to deer [base] and moose [blade] antisera)
8. HhOv-164: 14135 Fort Creek Fen lanceolate, oolitic Northern quartzite

Plate 6.2. Nezu Complex projectile points

1. HhOu-36:126 broad-bladed lanceolate with hafting modifications (positive reaction to moose and deer antisera)
2. HhOu-36:113 fish-tailed point (positive reaction to caribou antisera)
3. HhOu-36:6 concave-base lanceolate (positive reaction to deer and caribou antisera)
4. HhOv-164:46 Scottsbluff Type I point
5. HhOv-323:1561/3959 Scottsbluff Type II point, Montana chert
6. HhOu-36:140 Scottsbluff Type II point, Northern quartzite
7. HhOu-36:150 Scottsbluff Type II point (positive reaction to bovid and rabbit antisera)
8. HhOu-36:159 Scottsbluff Type I point (positive reaction to caribou antisera)
9. HhOu-36:2 Scottsbluff Type I point (positive reaction to rabbit antisera on haft)
10. HhOu-36:1 Scottsbluff Type II point

0   3
Centimetres

Plate 6.3. Nezu Complex drills, Cody knives, and Eden points

1   HhOu-36:151 "T-butt" drill (positive reaction to rabbit antisera)

2   HhOu-36:8 drill stem (positive reaction to bear antisera)

3   HhOu-36:9, 10 Niska narrow broken lanceolate drill refit (positive reactions to caribou antisera on both fragments)

4   HhOu-36:11 Niska lanceolate square-based drill

5   HhOu-36:88 Niska lanceolate square-based drill (positive reactions to deer and rabbit antisera)

6   HiOu-49 Cody knife

7   HiOu-49 Cody knife

8   HiOu-72 Cody knife (positive reaction to deer antisera)

9   HhOv-164:14134 Cody knife

10   HhOv-4:166 narrow-bladed Eden blade fragment (positive reaction to canid antisera)

11   HhOv-81:3 broad-bladed Eden point

12   HhOv-323 broad-bladed Eden point (positive reaction to elephant antisera), an isolated find at the Quarry of the Ancestors

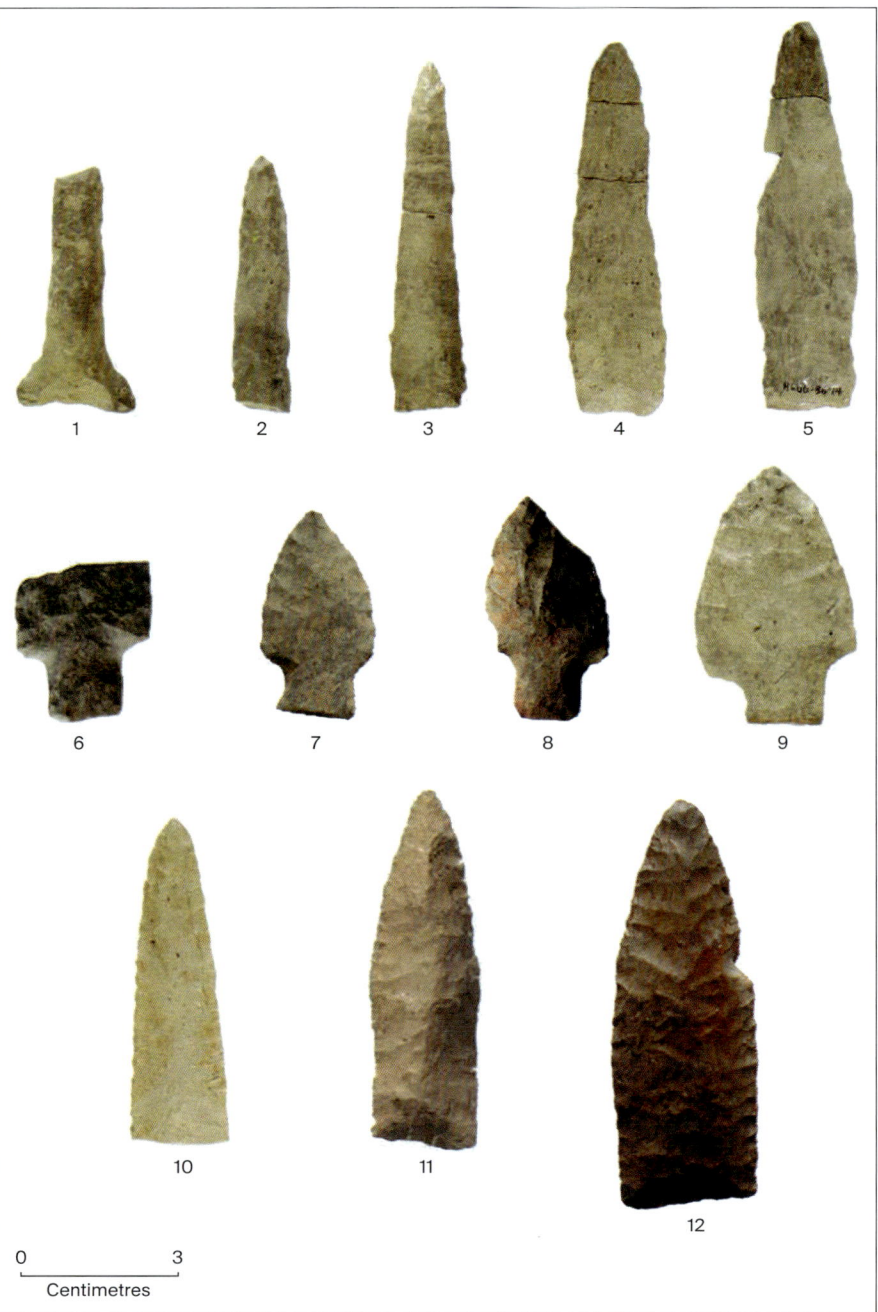

1

2

3

0   3
Centimetres

4

5

Plate 6.4. Nezu Site (HhOu-36)
knives and bifaces

1   HhOu-36:17 Nezu knife
     (positive reaction to bovid
     antisera)
2   HhOu-36:110, 111 Nezu knife
     refit
3   HhOu-36:18, 19 Nezu knife
     refit (positive reaction to bovid
     antisera on both fragments)
4   HhOu-36:62, 149 ovate biface
     refit
5   HhOu-36:52, 2091 axe refit

Plate 6.5. Nezu Site (HhOu-36) end scrapers and knives

1   HhOu-36: 125 small end scraper (positive reaction to cat antisera)
2   HhOu-36: 163 small end scraper
3   HhOu-36: 141 small end scraper
4   HhOu-36: 198 small end scraper
5   HhOu-36: 107 large end scraper
6   HhOu-36: 93 large end scraper
7   HhOu-36: 25 large end scraper (positive reaction to bear antisera)
8   HhOu-36: 97 dorsally finished "teardrop" elongated end scraper
9   HhOu-37: 96 dorsally finished "teardrop" elongated end scraper
10   HhOu-36: 158, 37 Nezu rectangular knife refit (positive reaction to bear antisera)
11   HhOu-36: 39, 40, 41 backed Nezu knife refit

Plate 6.6. Lanceolate projectile points

1 HhOv-148: 23 reworked Agate Basin point
2 HhOv-112: 14000, 13999, 13998 lanceolate point, snapped during finishing
3 HhOv-112: 4, 3190 convex-based lanceolate point
4 HhOv-112: 3 lanceolate point snapped during the thinning process
5 HhOv-193: 2 Jimmy Allen point, mottled grey and brown chert
6 HhOv-163: 26 water-rolled Hell Gap stem, Athabasca quartzite
7 HhOv-167: 714 Mesa point, possibly manufactured of heat-treated Swan River chert

Plate 6.7. Beaver River Complex dart points

1. HhOv-163: 27 Northern quartzite dart point
2. HhOv-163: 21 Northern quartzite dart point (positive reaction to bear antisera)
3. HhOv-163: 16 dart point
4. HhOv-163: 19 dart point
5. HhOv-163: 25 banded brown and black orthoquartzite dart point
6. HhOv-332: 2031 dart point
7. HhOv-349: 8712 grey chert dart point (positive reaction to deer antisera)
8. HhOv-340: 8713 dart point
9. HhOv-340: 8715 dart point
10. HhOv-340: 8714 dart point
11. HhOv-319 quartzite Oxbow point
12. HhOv-212: 4867 Northern quartzite point
13. HhOv-113: 9166 dart point
14. HhOv-113: 61308 dart point
15. HhOv-113: 68597, 68049 dart point refit
16. HhOv-305: 9544 side-notched point
17. HhOv-191: 14, 159 fish-tailed point

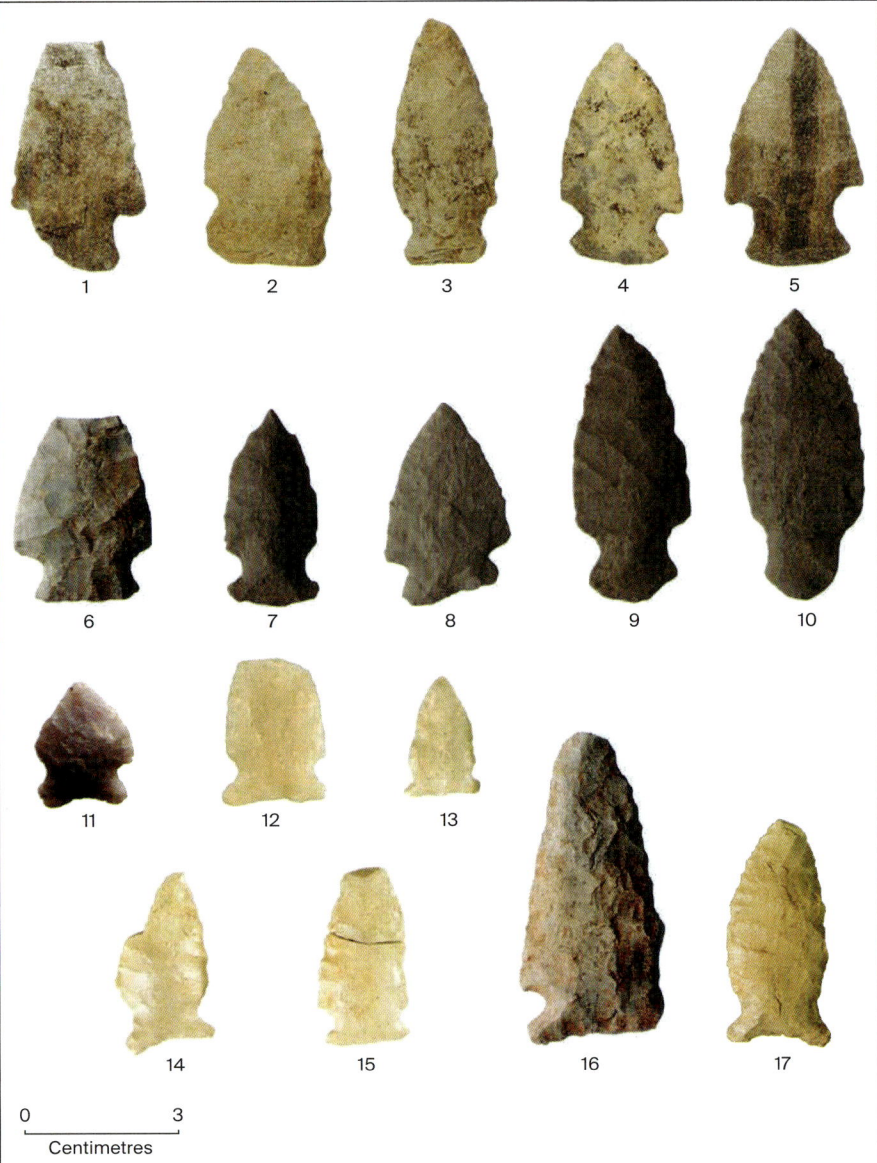

Plate 6.8. Firebag Hills Complex artifacts

1 HgOl-8 quartzite side blade
2 HhOl-32 side blade
3 HhOk-11 quartzite end blade
4 HhOv-305 end blade
5 HhOv-305 end blade
6 HgOv-50:141 Peace River black chert end blade
7 HhOv-304 end blade
8 HhOv-304 end blade
9 HhOv-304 notched graver
10 HhOv-324:1628 notched graver
11 HhOv-87:3985 brown and white chert point
12 HhOv-304:2759 point
13 HhOv-304 point
14 HhOv-304 quartzite point
15 HhOv-304 notched graver
16 HhOv-304 notched graver
17 HhOv-304 notched graver
18 HhOv-304 notched graver
19 HhOl-25 chert mitt burin
20 HgOl-10 grey chert burin
21 HhOk-28 honey brown quartzite micrograver on prismatic microblade
22 HhOk-20 grey and white banded chert micrograver

Plate 6.9. Chartier Complex projectile points

1 HhOu-73 grey quartzite Avonlea-style point
2 HhOu-50 Prairie Side-Notched-style point
3 HhOv-57 stemmed Middle Taltheilei dart point
4 HhOv-57 stemmed Middle Taltheilei dart point
5 HhOv-57 stemmed Middle Taltheilei dart point

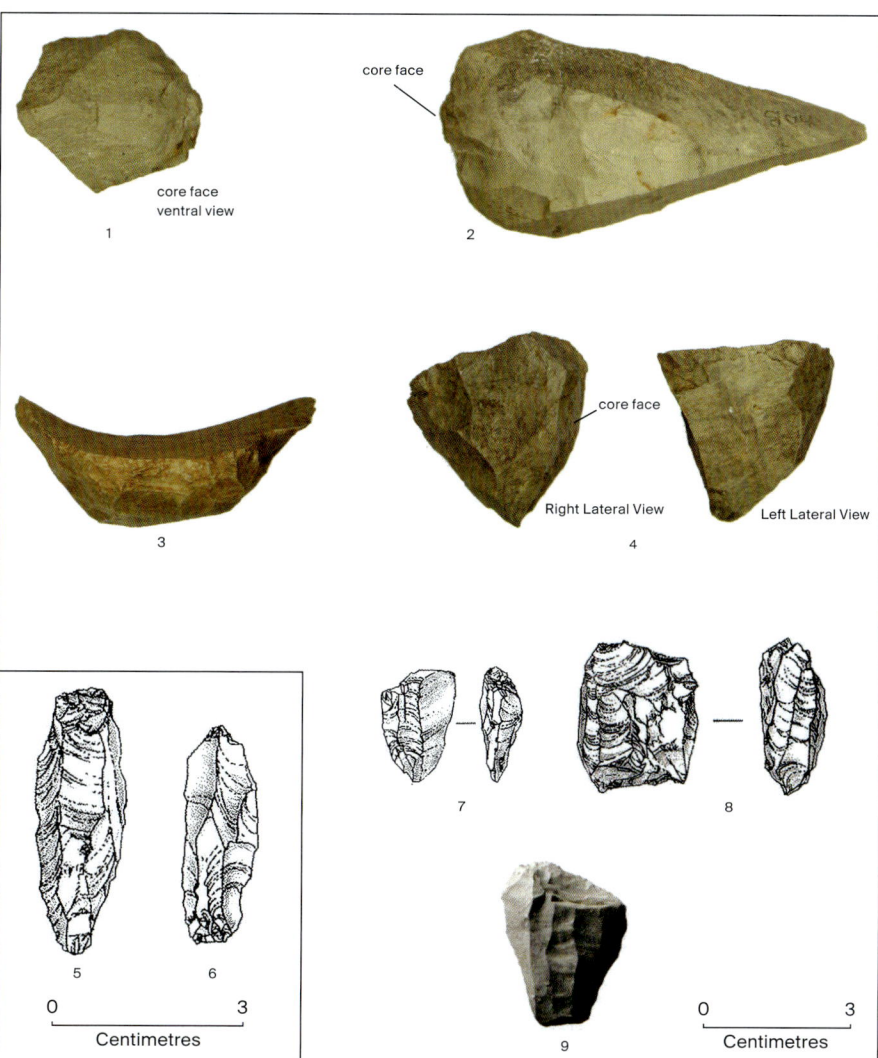

Plate 6.10. Microblade cores

Fort Creek Fen Complex

1 HhOv-164:11576 boat-shaped microblade core; transversely snapped core face, ventral view, manufactured on a thick burinated or snapped biface blank

2 HhOv-164:804 wedge-shaped microblade core, ventral view, manufactured on a thick burinated biface blank

Nezu Complex

3 HhOv-83:2392 exhausted boat-shaped microblade core recycled as an adze, left lateral view

4 HhOv-194:235 transversely snapped boat-shaped preform, manufactured on a thick snapped or burinated biface blank

Beaver River Complex

5 HhOv-121:887 pillar microblade core, bipolar manufacture

6 HhOv-332:5242 pillar microblade core, bipolar manufacture

Firebag Hills Complex

7 HhOv-462:1270 wedge-shaped microblade core

8 HhOv-462:12789 wedge-shaped microblade core

9 HiOu-14:897 microblade core fragment or possibly a rejuvenation tablet

NOTES

1   These excavations were sponsored by the University of Alberta's Boreal Institute for Northern Studies, later the Canadian Circumpolar Institute. Unfortunately, during the years that the collections were housed at the University of Alberta, they became impoverished. These collections now reside at the Royal Alberta Museum.

2   Saxberg and Reeves (2004) initially called this material Muskeg Valley Silicified Limestone, given the close proximity of one of the quarries to outcrops of Waterways Formation limestone and the apparent silicification within that formation. However, further study indicates that the MVMq in fact occurs in the base of the McMurray Formation. It originated as detrital sediment and can be considered a silty facies of Beaver River Sandstone (De Paoli 2006). Tsang (1998) mapped the distribution of Beaver River Sandstone in the area.

3   The usefulness of Gryba's Northern quartzite category has unfortunately been somewhat diluted by researchers using it as a catch-all for grey-white quartzites of varying quality found in northern Alberta.

4   The recovery of water-rolled artifacts offers clear archaeological evidence of the Lake Agassiz flood, which is discussed in some detail in chapters 1 and 2 of this volume. With respect to the date of the flood (or floods), we would point to the lack of evidence for primary occupations within the flood zone prior to the appearance of the Fort Creek Fen Complex around 9,800 BP. The relatively high elevation of the Fort Creek Fen sites— slightly below the maximum flood level of 300 masl—suggests that these occupations occurred very near the beginning of the recession of the flood waters, which in turn lends support to the date of 9,900 BP originally proposed by Smith and Fisher (1993).

5   The elevation of sites is an important consideration in the models proposed by Saxberg and Reeves (2003) and by Clarke and Ronaghan (2004) (see also Green et al. 2006). Site elevations provided in this chapter were determined using the Government of Alberta's Digital Elevation Model (DEM), a three-dimensional, 10 x 10 metre cell-size digital representation of surface topography freely available through the province's spatial data distributor, AltaLIS. The DEM features 5-metre accuracy for 90% of the data set and includes elevation data to the centimetre. This level of precision, which is derived mathematically by the software, is based on the 1:60,000 aerial photography of the province in conjunction with surveyed benchmarks. Using ArcGIS 9.1, elevations were extracted on the basis of the UTM coordinates for each site provided in the site database maintained by the Historic Resources Management Branch of the Alberta government.

6   In addition to the evidence from the Lower Athabasca region itself, two points found further afield suggest that human beings were present in the region prior to the flood. Both are early styles, and both are manufactured of MVMq, a stone that occurs naturally only in the area of the Lower Athabasca. One is a small, basally thinned point from the Duckett Site (GdOo-16) on Ethyl Lake, not far from Cold Lake (see Ives 1993, 25n4; McCullough 1981, figure 17: 1). The other is a fluted point identified in a collection from a site near High Prairie, to the west of Slave Lake (Bob Dawe, pers. comm., 2008). Although radiocarbon dates are not available, these points indicate that local variants of the Fluted Point Tradition were present in the Lower Athabasca basin prior to 10,000 BP.

7   The Ice Mountain Microblade Tradition is the regional expression, in northern British Columbia, of the Dyuktai Microblade Tradition of northeastern Siberia, which appeared in the unglaciated regions of Eastern Beringia some 13,000 years ago (Holmes 2001).

8   Further mitigative studies at HhOv-87 were carried out in 2009 and 2010. In 2009, excavations in both existing and new loci, covering a total of 232 square metres, led to the recovery of 91,171 lithic items and 7,261 faunal remains. In addition, Nezu and Beaver River Complex materials were recovered and a probable Late Taltheilei occupation

identified. This new evidence has significantly expanded our understanding of the sequence of occupations at HhOv-87 and our ability to interpret the activities that took place at the site (Roskowski and Netzel 2011). The 2010 studies investigated two new loci on the southeastern dog legs of the site. These excavations, which covered 44 square metres, identified a possible Chartier Complex occupation as well as Fort Creek Fen and Beaver River Complex occupations (Roskowski and Netzel 2012).

9   Whether the Early Precontact period occupants of the Lower Athabasca region also fished or harvested wildfowl remains an open question. We presume that they did and that this seasonal activity was part of the broad-spectrum pattern of resource harvesting represented at the Nezu site. This pattern is similar to that associated with Cody Complex sites in Yellowstone National Park, which include summer sites along lakeshores as well as on an island in Yellowstone Lake that would have been accessible only by canoe (Johnson and Reeves 2013). Similarly, in eastern Wyoming, the Cody levels at the Hell Gap site, which again reflect summer occupations, contain evidence of a variety of fauna (Knell 2007). A broad-spectrum adaptive strategy that incorporates fishing and fowling is also represented in in Cody Complex sites in Wisconsin (Kuehn 2007), as well as in contemporaneous and earlier sites in Alaska (Yesner 2007). As Knell (2007), Kornfeld (2007), and others point out, although Cody sites do tend to be dominated by bison, there has perhaps been an overemphasis on the role of communal bison hunting in subsistence activities.

10   The point is a collaterally flaked Eden-style point within the Cody Complex. It probably dates to roughly 9,000 BP or, at the outside, to 9,500 BP. If this is the case, and if we accept the blood residue analysis as correct, then we are either left with a late survival most probably of mastodon in the forested lands of the Lower Athabasca region adjacent to the shrinking ice cap, or else we must assume that the point had come into contact with a 500- to 1,000-year-old piece of proboscidian flesh or hide fortuitously preserved in a permafrost deposit. Radiocarbon dates as late as 9,000 ^{14}C yr BP have been obtained on mastodons (see, for example, Dreimanis 1968; Harington 2003), although most of these have been discarded. The possibility does exist, however, that small populations could have survived in the spruce-dominated forests of the Athabasca lowlands and adjacent northwestern precincts of Glacial Lake Agassiz.

     The proboscidian antisera used in this analysis, which was performed by PaleoResearch, was the same as that prepared using a fresh blood sample collected by Lee Bement from the Tulsa Zoo to evaluate the potential for mammoth and other blood antisera on a Clovis point recovered in situ from a site in Oklahoma. A positive reaction was obtained (Lee Bement, pers. comm., 2008). We would also note that in a blood residue study of fluted points from Eastern Beringia, two of the specimens taken from five fluted points found at the Point, Lisburne, and Girls Hill sites reacted positively to elephant antisera (Loy and Dixon 1998, table 2). Other species identified on fluted points from various sites include bison, sheep, bear, caribou, and muskox. Some researchers have questioned the mammoth results: for example, Fiedel (2007, 5) cites his own paper (Fiedel 1996), which predates the completed Loy and Dixon study. However, we see no particular reason to consider the results suspect, given that the earliest dates for human occupation overlap with the latest dates for mammoth (Guthrie 2006; Boulanger and Lyman 2014). Accepting Bement's results from Oklahoma, there is no reason, other than that the date suggested by the style of the point is too late for the generally accepted date for the extinction of mastodons, for the HhOv-323 result to be rejected (Boulanger and Lyman 2014, Froese 2014).

11   Earlier cultural complexes, as reported by Millar (1983), are represented at Buffalo Narrows in private collections from the Old Beach Site. The finds include Early Precontact period lanceolates, which, according to Millar (1983, 39), represent the Plains

Plano Tradition: "Most resemble Agate Basin, Plainview or Frederick varieties. Other types include: small lanceolate points with straight, parallel stems and square bases." The former, illustrated in Millar's plate 1, include obliquely flaked lanceolates characteristic of the Cree Burn Lake Complex and Scottsbluff bases typical of those found in the Nezu Complex. Millar (1983, 66) notes that some were made of "fine-grained chert and Knife River Flint." Millar's plate 1 also includes what may be a Cody knife. What appears to be an Alberta point was collected from the Nordstrom site located in the town of Buffalo Narrows (David Meyer, pers. comm., 2008).

12 To the best of our knowledge, the Scottsbluff point from Boyle is the farthest upstream find of MVMq in the Athabasca drainage to be documented to date. In 1997, studies in connection with the proposed Cheviot mine identified Early Precontact period components as well as flakes of MVMq at site FfQh-27, on Harris Creek, a tributary of the McLeod River near Mountain Park, on the edge of the Front Ranges (Kulle and Neal 1998). Reeves questioned this identification, however, as a local variety of sparkly silicified siltstone that is not uncommon in the foothills of the Upper Athabasca may resemble MVMq. In 2006, Reeves relocated the artifacts in the RAM collections and determined that they were not MVMq but the suspected local silicified siltstone. Cody Complex artifacts, including some manufactured of Knife River flint, have been documented at the Cheviot site (FfQh-26; see fig 6.4), which is also on Harris Creek (Meyer et al. 2007). Nezu knives of Athabasca quartzite have also been recovered in the outer foothills to the east, on southern tributaries of the Athabasca (Meyer, Reeves, and Lobb 2002; Meyer, Roe, and Dow 2007).

13 Gordon divides what we regard as the Early Taltheilei Tradition into two phases: Earliest (2,650 to 2,450 BP) and Early (2,450 to 1,800 BP). However, we are not convinced that these two phases can be clearly distinguished.

14 In 2004, Reeves examined the Peace Point site collections at Parks Canada in Winnipeg with the goal of determining whether there was any MVMq in the collections, which there was not. An additional goal was to examine the microblade core and related artifacts recovered from the site (see Stevenson 1986). These are of some interest as, to the best of our knowledge, they would be the only such artifacts yet found in a Taltheilei Tradition site. Unfortunately, the artifacts were unavailable, on loan to Parks Canada at Wood Buffalo National Park. However, according to unsigned notes left in the artifact number locations, the core was not a microblade core. The Peace Point collections that Reeves examined are dominated by evidence of the bipolar reduction of small pebbles of Peace River chert, resulting in the production of very large numbers of small, parallel-sided flakes, debitage and exhausted bipolar pebble cores, and wedges with parallel-sided flake scars. This technology is the same as that associated with the Chartier Complex bipolar black chert pebbles, which can produce microblade-like cores and flakes. Hence, the notes to the effect that no microblade technology was present at the Peace Point site were probably correct.

REFERENCES

Ackerman, Robert E.

2007   The Microblade Complexes of Alaska and the Yukon: Early Interior and Coastal Adaptations. In *Origin and Spread of Microblade Technology in Northern Asia and North America,* edited by Yaraslov V. Kuzmin, Susan G. Keates and Chen Shen. Archaeology

Press, Department of Archaeology Publication no. 34. Simon Fraser University, Burnaby, British Columbia.

Alley, Richard B., and Anna María Ágústsdóttir

2005   The 8k Event: Cause and Consequences of a Major Holocene Abrupt Climate Change. *Quaternary Science Reviews* 24: 1123–1149.

Arnold, Thomas G.

1985   A Comparison of Northern Plano Complexes. MA thesis, Department of Archaeology, University of Calgary.

Beaudoin, Alwynne B., and Gerald A. Oetelaar

2003   The Changing Ecophysical Landscape of Southern Alberta During the Late Pleistocene and Early Holocene. *Plains Anthropologist* 48(187): 187–207.

Bonnichsen, Robson, Marvin Beatty, Mort D. Turner, and Diane L. Douglas

1992   Paleoindian Lithic Procurement at the South Fork of Everson Creek, Southwestern Montana: A Preliminary Statement. In *Ice Age Hunters of the Rockies,* edited by Dennis J. Stanford and Jane S. Day, pp. 285–321. Denver Museum of Natural History and University Press of Colorado, Denver.

Bouchert-Bert, Luc

2007   *Historical Resources Impact Assessment for Shell Canada Energy's Ten-Year Footprints in the Pierre River, Jackpine Expansion Mining Areas, and Jackpine Mine, Phase 1: Final Report (ASA Permit 06-116).* Copy on file, Archaeological Survey, Historic Resources Management Branch, Alberta Culture, Edmonton.

Boulanger, Matthew T., and R. Lee Lyman

2014   Northeastern North American Pleistocene Megafauna Chronologically Overlapped Minimally with Paleoindians. *Quaternary Science Reviews* 85: 35–46.

Bradley, Bruce A., and George C. Frison

1987   Projectile Points and Specialized Bifaces from the Horner Site. In *The Horner Site: The Type Site of the Cody Cultural Complex,* edited by George C. Frison and Lawrence C. Todd, pp. 199–231. Academic Press, New York.

Brownlee, Kevin, and Myra Sitchon

2010   Research on the Grandfather Quarry: Granville Lake. Paper presented at the 43rd annual meeting of the Canadian Archaeological Association, Calgary, Alberta, 28 April–2 May. Abstract available in *Program and Abstracts,* pp. 46–47.

Bryant, Laureen

2004   *Historical Resources Impact Assessment and Mitigation, Fall 2003, Canadian Natural Resources Limited Horizon Oil Sands Project: Final Report (ASA Permit 03-269).* Copy on file, Archaeological Survey, Historic Resources Management Branch, Alberta Culture, Edmonton.

2005   *Historical Resources Impact Assessment and Mitigation, Summer 2004, Canadian Natural Resources Limited Horizon Oil Sands Project: Final Report (ASA Permit 04-189).* Copy on file, Archaeological Survey, Historic Resources Management Branch, Alberta Culture, Edmonton.

Clarke, Grant M.

1998   *Historical Resources Impact Assessment, Mobil Lease 36 (ASA Permit 98-145).* Copy on file, Archaeological Survey, Historic Resources Management Branch, Alberta Culture, Edmonton.

2002a *Historical Resources Impact Assessment, Canadian Natural Resources Limited Project Horizon: Final Report (ASA Permit 01-248).* Copy on file, Archaeological Survey, Historic Resources Management Branch, Alberta Culture, Edmonton.

2002b *Historical Resources Impact Assessment for Jackpine Mine, Phase 1: Final Report (ASA Permit 01-230).* Copy on file, Archaeological Survey, Historic Resources Management Branch, Alberta Culture, Edmonton.

Clarke, Grant M., and Brian M. Ronaghan

    2000   *Historical Resources Impact Mitigation, Muskeg River Mine Project: Final Report (ASA Permit 99-073)*. Copy on file, Archaeological Survey, Historic Resources Management Branch, Alberta Culture, Edmonton.

    2004   *Historical Resources Impact Assessment and Mitigation Program, Muskeg River Mine Project: Final Report (ASA Permit 00-087)*. 2 vols. Copy on file, Archaeological Survey, Historic Resources Management Branch, Alberta Culture, Edmonton.

Corbeil, Marcel R.

    1995   The Archaeology and Taphonomy of the Heron Eden Site, Southwestern Saskatchewan. MA thesis, Department of Anthropology and Archaeology, University of Saskatchewan. Saskatoon.

Davis, Leslie B., Stephen A. Aaberg, William P. Eckerle, John W. Fisher, Jr., and Sally T. Greiser

    1989   Montana Paleoindian Occupation of the Barton Gulch Site, Ruby Valley, Southwestern Montana. *Current Research in the Pleistocene* 6: 7–9.

de Mille, Christy, and Brian O. K. Reeves

    2009   *Birch Mountain Resources Ltd., Muskeg Valley Quarry Historical Resources Mitigation, 2005 Field Studies: Final Report (ASA Permit 05-230)*. 3 vols. Copy on file, Archaeological Survey, Historic Resources Management Branch, Alberta Culture, Edmonton.

    2010   *Historical Resources Mitigation, HhOv-462, Stage I and Stage II Mitigation Excavations, Hammerstone Corporation North Section (Muskeg Valley Quarry) of the South Haul Road: Final Report (ASA Permit 08-169)*. Copy on file, Archaeological Survey, Historic Resources Management Branch, Alberta Culture, Edmonton.

De Paoli, Glen R.

    2005   Petrographic Examination of the Muskeg Valley Microquartzite (MVMq). Appendix A in Nancy Saxberg and Brian O. K. Reeves (2006), *Birch Mountain Resources Ltd., Hammerstone Project, Historical Resources Impact Assessment, 2004 Field Studies: Final Report (ASA Permit 04-235)*. Copy on file, Archaeological Survey, Historic Resources Management Branch, Alberta Culture, Edmonton.

Doll, Maurice F. V.

    1982   *The Boss Hill Site (FdPe-4) Locality 2: Pre-Archaic Manifestations in the Parkland of Central Alberta, Canada*. Provincial Museum of Alberta, Human History Occasional Paper no. 2. Historic Resources Management Branch, Alberta Culture, Edmonton.

Donahue, Paul F.

    1976   *Archaeological Research in Northern Alberta, 1975*. Archaeological Survey of Alberta Occasional Paper no. 2. Historic Resources Management Branch, Alberta Culture, Edmonton.

Douglas, Diane L.

    1991   Cultural Variability and Lithic Technology: A Detailed Study of Stone Tools from the Mammoth Meadow Locus, Southwestern Montana. MSc thesis, Center for the Study of Early Man, University of Maine, Orono.

Dreimanis, Aleksis

    1968   Extinction of Mastodons in Eastern North America: Testing a New Climatic-Environmental Hypothesis. *Ohio Journal of Science* 68(6): 257–271.

Driver, Jonathan C.

    1978   Holocene Man and Environments in the Crowsnest Pass, Alberta. PhD dissertation, Department of Archaeology, University of Calgary, Calgary.

Driver, Jonathan C., Martin Handly, Knut R. Fladmark, D. Erle Nelson, Gregg M. Sullivan, and Randall Preston

    1996   Stratigraphy, Radiocarbon Dating, and Culture History of Charlie Lake Cave, British Columbia. *Arctic* 9(3): 265–277.

Eldridge, Morley, Jo Brunsden, Roger Eldridge, and Alyssa Parker

  2008   *BC Hydro 2008 Williston Dust Abatement Project, Archaeological Impact Assessment 2008: Final Report.* Millennia Research Limited for BC Hydro. Copy on file, Archaeology Branch, Ministry of Forests, Lands and Natural Resource Operations, Victoria.

Eldridge, Morley, Vashti Theissen, Alyssa Parker, and Roger Eldridge

  2010   *BC Hydro 2010 Williston Dust Abatement Project, Archaeological Impact Assessment 2010: Final Report.* Millennia Research Limited for BC Hydro. Copy on file, Archaeology Branch, Ministry of Forests, Lands and Natural Resource Operations, Victoria.

Ellis, Christopher J.

  2004   Hi-Lo: An Early Lithic Complex in Southern Ontario. In *The Late Palaeo-Indian Great Lakes: Geological and Archaeological Investigations of Late Pleistocene and Early Holocene Environments,* edited by Lawrence J. Jackson and Andrew Hinshelwood, pp. 57–83. Mercury Series no. 165. Archaeological Survey of Canada, Canadian Museum of Civilization, Gatineau, Québec.

Fenton, Mark M., and John W. Ives

  1982   Preliminary Observations on the Geological Origins of Beaver River Sandstone. In *Archaeology in Alberta, 1981,* compiled by Jack Brink, pp. 166–189. Archaeological Survey of Alberta Occasional Paper no. 19. Historic Resources Management Branch, Alberta Culture, Edmonton.

  1984   The Stratigraphic Position of Beaver River Sandstone. In *Archaeology in Alberta, 1983,* compiled by David Burley, pp. 128–136. Archaeological Survey of Alberta Occasional Paper no. 23. Historic Resources Management Branch, Alberta Culture, Edmonton.

  1990   Geoarchaeological Studies of the Beaver River Sandstone, Northeastern Alberta. In *Archaeological Geology of North America,* edited by Norman P. Lasca and Jack Donahue, pp. 123–135. Centennial Special Volume 4. Geological Society of America, Boulder, Colorado.

Fiedel, Stuart J.

  1996   Blood from Stones? Some Methodological and Interpretive Problems in Blood Residue Analysis. *Journal of Archaeological Science* 23: 139–147.

  2007   Quacks in the Ice: Waterfowl, Paleoindians, and the Discovery of America. In *Foragers of the Terminal Pleistocene in North America,* edited by Renee B. Walker and Boyce N. Driskell, pp. 1–14. University of Nebraska Press, Lincoln.

Fladmark, Knut R.

  1985   *Glass and Ice: The Archaeology of Mt. Edziza.* Archaeology Press, Department of Archaeology Publication no. 14. Simon Fraser University, Burnaby, British Columbia.

Frison, George C.

  1991   *Prehistoric Hunters of the High Plains.* 2nd ed. Academic Press, San Diego.

Frison, George C., and Dennis J. Stanford

  1982   *The Agate Basin Site: A Record of Paleoindian Occupation of the Northwestern High Plains.* Academic Press, New York.

Frison, George C., and Danny N. Walker

  2007   *Medicine Lodge Creek: Holocene Archaeology of the Eastern Big Horn Basin, Wyoming.* Vol. 1. Clovis Press, New York.

Froese, Duane

  2014   The Curious Case of the Arctic Mastodons. *Proceedings of the National Academy of Sciences* 111(52): 18405–18406.

Gordon, Bryan H. C. A.

  1975   *Of Men and Herds in Barrenland Prehistory.* Mercury Series no. 28. Archaeological Survey of Canada, National Museum of Man, Ottawa.

1996   *People of Sunlight, People of Starlight: Barrenland Archaeology in the Northwest Territories of Canada.* Mercury Series no. 154. Archaeological Survey of Canada, Canadian Museum of Civilization, Gatineau, Québec.

Green, D'Arcy, and David Blower

2005   *2003 Historical Resources Impact Mitigation of the Bezya Site (HhOv-73): Final Report (ASA Permit 03-198).* Copy on file, Archaeological Survey, Historic Resources Management Branch, Alberta Culture, Edmonton.

Green, D'Arcy, David Blower, Dana Dalmer, and Luc Bouchet-Bert

2006   *Historical Resources Impact Assessment and Mitigation, Albian Sands Energy's Muskeg River Mine and Shell Canada's Jackpine Mine: Final Report (ASA Permit 05-355).* 2 vols. Copy on file, Archaeological Survey, Historic Resources Management Branch, Alberta Culture, Edmonton.

Gryba, Eugene M.

1980   *Highway Archaeological Salvage Projects in Alberta: Final Report (ASA Permit 79-066).* Copy on file, Archaeological Survey, Historic Resources Management Branch, Alberta Culture, Edmonton.

2001   Evidence of the Fluted Point Tradition in Western Canada. In *On Being First: Cultural Innovation and Environmental Consequences of First Peopling,* edited by Jason Gillespie, Susan Tupakka, and Christy de Mille, pp. 251–284. Proceedings of the 31st Annual Chacmool Conference. Chacmool Archaeological Assocation, Department of Archaeology, University of Calgary, Calgary.

Gryba, Eugene M., and Jennifer C. Tischer

2005   *Historical Resources Impact Assessment, Deer Creek Energy Limited Joslyn North Mine Project: Final Report (ASA Permit 05-094).* Copy on file, Archaeological Survey, Historic Resources Management Branch, Alberta Culture, Edmonton.

Guthrie, R. Dale

2006   New Carbon Dates Link Climatic Change with Human Colonization and Pleistocene Extinctions. *Nature* 441 (11 May): 207–209.

Hanna, Margaret G.

1982   *Report on Reconnaissance Survey of Highway 21 and 155 and Assessment and Mitigation of HcOi-1 Old Airport Site.* Report prepared for the Saskatchewan Department of Highways and Transportation, North Battleford District (Permits 82-13 and 82-40). Saskatchewan Research Council Publication no. C-805-52-E-82.

2004   *Simon Eninew Site, GgNk-1, Report on Reconnaissance, July 2004.* Report prepared for the Heritage Branch, Saskatchewan Culture, Youth and Recreation, Government of Saskatchewan, and Lac La Ronge Indian Band, Hall Lake Reserve.

Hare, P. Gregory, Sheila Greer, Ruth Gotthardt, Richard Farnell, Vandy Bowyer, Charles Schweger, and Diane Strand

2004   Ethnographic and Archaeological Investigations of Alpine Ice Patches in the Southwestern Yukon. *Artic* 57: 260–272.

Harington, Charles R. (editor)

2003   *Annotated Bibliography of Quaternary Vertebrates of Northern North America, with Radiocarbon Dates.* University of Toronto Press, Toronto.

Head, Thomas H.

1979   *Conservation Archaeology, Alberta Transportation Highway Construction Program, Project Number 963 (ASA Permit 78-071).* Copy on file, Archaeological Survey, Historic Resources Management Branch, Alberta Culture, Edmonton.

Head, Thomas H., and Stanley Van Dyke

1990   *Historical Resources Impact Assessment and Mitigation, Cree Burn Lake Site (HhOv-16) Jct. S.R. 963 to Gravel Pit Source "A" in NW 29-95-10-4 (ASA Permit 88-032).* Copy on file, Archaeological Survey, Historic Resources Management Branch, Alberta Culture, Edmonton.

Holmes, Charles E.

2001 Tanana River Valley Archaeology circa 14,000 to 9000 B.P. *Arctic Anthropology* 38(2): 154–170.

Husted, Wilfred M., and Robert Edgar

2002 *The Archeology of Mummy Cave, Wyoming: An Introduction to Shoshonean Prehistory.* Midwest Archeological Center, Special Report no. 4, and Southeast Archeological Center, Technical Reports Series no. 9. National Park Service, U.S. Department of the Interior, Midwest Archeological Center, Lincoln, Nebraska.

Ives, John W.

1977 *The Results of Excavations at HkPa-4 Birch Mountains (ASA Permit Number 76-010).* Copy on file, Archaeological Survey, Historic Resources Management Branch, Alberta Culture, Edmonton.

1981 The Prehistory of Northern Alberta. In *Alberta Archaeology: Retrospect and Prospect,* edited by Terry A. Moore, pp. 39–85. Archaeological Society of Alberta, Lethbridge.

1985 *A Spatial Analysis of Artifact Distribution on a Boreal Forest Archaeological Site.* Archaeological Survey of Alberta Manuscript Series no. 5. Historic Resources Management Branch, Alberta Culture, Edmonton.

1993 The Ten Thousand Years Before the Fur Trade in Northeastern Alberta. In *The Uncovered Past: Roots of Northern Alberta Societies,* edited by Patricia A. McCormack and R. Geoffrey Ironside, pp. 5–31. Circumpolar Research Series no. 3. Canadian Circumpolar Institute, University of Alberta, Edmonton.

2003 Alberta, Athapaskans and Apachean Origins. In *Archaeology in Alberta: A View from the New Millennium,* edited by Jack W. Brink and John F. Dormaar, pp. 256–289. Archaeology Society of Alberta, Medicine Hat.

Jackson, Lawrence J.

2004 Changing Our Views of Late Paleo-Indian in Southern Ontario. In *The Late Palaeo-Indian Great Lakes: Geological and Archaeological Investigations of Late Pleistocene and Early Holocene Environments,* edited by Lawrence J. Jackson and Andrew Hinshelwood, pp. 25–56. Mercury Series no. 165. Archaeological Survey of Canada, Canadian Museum of Civilization, Gatineau, Québec.

Johnson, Ann M., and Brian O. K. Reeves

2013 Summer on Yellowstone Lake 9,300 Years Ago: The Osprey Beach Site. Special issue: Memoir 41. *Plains Anthropologist* 58(227-228): 1–194.

Johnson, Ann M., Brian O. K. Reeves, and Mack W. Shortt

2004 *Osprey Beach: A Cody Complex Camp on Yellowstone Lake.* Report prepared for the U.S. National Park Service. Copy on file, National Park Service, U.S. Department of the Interior, Branch of Cultural Resources, Yellowstone National Park, Mammoth Hot Springs, Wyoming.

Kasstan, Steven C.

2004 Lithic Technology at the Below Forks Site, FhNg-25: Stratagems of Stone Tool Manufacture. MA thesis, Department of Archaeology, University of Saskatchewan.

Knell, Edward J.

2007 The Organization of Late Paleoindian Cody Complex Land-Use on the North American Great Plains. PhD dissertation, Department of Anthropology, Washington State University, Pullman.

Knudson, Ruthann

1983 *Organizational Variability in Late Paleo-Indian Assemblages.* Reports of Investigations no. 60. Laboratory of Anthropology, Washington State University, Pullman, Washington.

Kornfeld, Marcel

2007 Are Paleoindians of the Great Plains and Rockies Subsistence Specialists? In *Foragers of the Terminal Pleistocene in North America,* edited by Renee B. Walker and Boyce N. Driskell, pp. 32–58. University of Nebraska Press, Lincoln.

Kuehn, Steven R.

2007 Late Paleoindian Subsistence Strategies in the Western Great Lakes Region Evidence for Generalized Foraging from Northern Wisconsin. In *Foragers of the Terminal Pleistocene in North America,* edited by Renee B. Walker and Boyce N. Driskell, pp. 88–98. University of Nebraska Press, Lincoln.

Kulle, Barbara J., and Barbara Neal

1998 *Historical Resources Mitigation, Cardinal River Coals Ltd., Cheviot Mine Project Sites FfQh-27 and FfQh-32 (ASA Permit 97-115).* Copy on file, Archaeological Survey, Historic Resources Management Branch, Alberta Culture, Edmonton.

Kunz, Michael, Michael Bever, and Constance Adkins

2003 *The Mesa Site: Paleoindians Above the Arctic Circle.* BLM-Alaska Open File Report 86. U.S. Department of the Interior, Bureau of Land Management, Alaska State Office, Anchorage.

Landals, Alison J.

2008 The Lake Minnewanka Site: Patterns in Late Pleistocene Human Use of the Alberta Rocky Mountains. PhD dissertation, Department of Archaeology, University of Calgary, Calgary.

Langemann, E. Gwyn, and William Perry

2002 *Banff National Park Archaeological Resource Description and Analysis.* Parks Canada Agency, Cultural Resources Services, Western Canada Service Centre, Calgary.

Larsen, Helge

1968 *Trail Creek: Final Report on the Excavation of Two Caves on Seward Peninsula, Alaska.* Acta Arctica no. 15. Ejnar Munksgaard, Copenhagen.

Le Blanc, Raymond J.

1985 *The Bezya Site: A Wedge-Shaped Core Assemblage from Northeastern Alberta: Final Report (ASA Permit 83-053).* Copy on file, Archaeological Survey, Historic Resources Management Branch, Alberta Culture, Edmonton.

2005 *Archaeological Research in the Lesser Slave Lake Region: A Contribution to the Pre-Contact History of the Boreal Forest of Alberta.* Mercury Series no. 166. Archaeological Survey of Canada, Canadian Museum of Civilization, Gatineau, Québec.

Le Blanc, Raymond J., and John W. Ives

1986 The Bezya Site: A Wedge-Shaped Core Assemblage from Northeastern Alberta. *Canadian Journal of Archaeology* 10: 59–98.

Linnamae, Urve, and Marcel R. Corbeil

1993 *Report of the Heron Eden Archaeological Project.* Department of Anthropology and Archaeology, University of Saskatchewan, Saskatoon. Report submitted to the Heritage Branch, Government of Saskatchewan, Regina.

Losey, Timothy C., Randall Freeman, and John Priegert

1975 *Archaeological Reconnaissance, Alberta Highways North, 1974 (ASA Permits 74-010 and 74-014).* Copy on file, Archaeological Survey, Historic Resources Management Branch, Alberta Culture, Edmonton.

Low, Bruce D.

1997 Bipolar Technology and Pebble Stone Artifacts: Experimentation in Stone Tool Manufacture. MA thesis, Department of Anthropology and Archaeology, University of Saskatchewan, Saskatoon.

Loy, Thomas H., and E. James Dixon

1998 Blood Residues on Fluted Points from Eastern Beringia. *American Antiquity* 63(1): 21–46.

Mallory, Oscar L.

1980 *1979 Archaeological Investigation on Highway Project No. 963 (ASA Permit 79-124).* Copy on file, Archaeological Survey, Historic Resources Management Branch, Alberta Culture, Edmonton.

McCullough, Edward J.

1981 *The Duckett Site (GdOo-16): An Evaluative Study (ASA Permit 80-155).* Copy on file, Archaeological Survey, Historic Resources Management Branch, Alberta Culture, Edmonton.

1984 *Archaeological and Historical Resources Assessment, Slave River Hydro Project, Phase I (ASA Permit 83-072).* Vol. 5, *Site Description and Assessment.* Copy on file, Archaeological Survey, Historic Resources Management Branch, Alberta Culture, Edmonton.

McCullough, Edward J., and Michael C. Wilson

1982 *A Prehistoric Settlement-Subsistence Model for Northeastern Alberta, Canstar Oil Sands Ltd. Bituminous Sands Leases 33, 92, and 95: A Preliminary Statement.* Environmental Research Monograph 1982-1, Canstar Oil Sands Ltd., Calgary.

Meyer, Daniel A., Brian O. K. Reeves, and Murray Lobb

2002 *Historical Resources Impact Assessment and Mitigation, Weldwood of Canada Limited, Hinton Division, Embarras 12 Forestry Unit, Haul Roads, Cut Blocks, and Gravel Source, Final Report (ASA Permit 01-296).* Copy on file, Archaeological Survey, Historic Resources Management Branch, Alberta Culture, Edmonton.

Meyer, Daniel A., Jason Roe, and Amanda Dow

2007 *Historical Resources Impact Assessment, Hinton Wood Products, A Division of West Fraser Mills, Hinton Wood Products FMA 2006 Developments: Final Report (ASA Permit 06-264).* Copy on file, Archaeological Survey, Historic Resources Management Branch, Alberta Culture, Edmonton.

Meyer, Daniel A., Nancy Saxberg, Brad Somer, Jason Roe, and Carmen Olson

2007 *Heritage Resources Impact Mitigation, Elk Valley Coal Corporation, Cardinal River Operations, Cheviot Mine 2005 Mitigation Excavations: Final Report (ASA Permit 05-396).* Copy on file, Archaeological Survey, Historic Resources Management Branch, Alberta Culture, Edmonton.

Meyer, David

1987 Time-Depth of the Western Woods Cree Occupation of Northern Ontario, Manitoba, and Saskatchewan. In *Papers of the Eighteenth Algonquian Conference,* edited by William Cowan, pp. 187–200. Carleton University, Ottawa.

Meyer, David, and Henri Liboiron

1990 A Paleoindian Drill from the Niska Site in Southern Saskatchewan. *Plains Anthropologist* 35(129): 299–302.

Meyer, David, and Dale Russell

1987 The Selkirk Composite of Central Canada: A Reconsideration. *Arctic Anthropology* 24(2): 1–31.

Millar, James F. V.

1983 *The Chartier Sites: Two Stratified Campsites on Kisis Channel near Buffalo Narrows, Saskatchewan.* Northern Heritage Limited, Saskatoon. Copy on file, Archaeological Survey of Canada, Canadian Museum of Civilization, Gatineau, Québec.

Millar, James F. V., and S. R. M. Ross

1982 *Heritage Mitigation Study, Site Area 1A and 1B, Kisis Channel, Saskatchewan.* Northern Heritage Limited, Saskatoon. Copy on file, Archaeological Survey of Canada, Canadian Museum of Civilization, Gatineau, Québec.

Morrison, David A.

1984 The Late Prehistoric Period in the MacKenzie Valley. *Arctic* 37(3): 195–209.

Overstreet, David F.

1993 *Chesrow: A Paleoindian Complex in the Southern Lake Michigan Basin.* Case Studies in Great Lakes Archaeology no. 2. Great Lakes Archaeological Press, Milwaukee.

Paquin, Todd A.

    1995   Pottery Styles as Indicators of Cultural Patterns: The Kisis Complex. MA thesis, Department of Anthropology and Archaeology, University of Saskatchewan, Saskatoon.

Peck, Trevor R.

    2011   *Light from Ancient Campfires: Archaeological Evidence for Native Lifeways on the Northern Plains.* Athabasca University Press, Edmonton.

Pollock, John

    1978a   Archaeological Research in Central and Northern Alberta, 1977. In *Archaeology in Alberta, 1977,* edited by W. J. Byrne, pp. 43–45. Archaeological Survey of Alberta Occasional Paper no. 5. Historic Resources Management Branch, Alberta Culture, Edmonton.

    1978b   *Early Cultures of the Clearwater River Area, Northeastern Alberta.* Archaeological Survey of Alberta Occasional Paper no. 6. Historic Resources Management Branch, Alberta Culture, Edmonton.

Powers, W. Roger, R. Dale Guthrie, and John F. Hoffecker

    1983   *Dry Creek: Archeology and Paleocology of a Late Pleistocene Alaskan Hunting Camp.* Report prepared for the National Park Service, U.S. Department of the Interior, Washington, D.C.

Reeves, Brian O. K.

    1969   The Southern Alberta Paleo-cultural–Paleoenvironmental Sequence. In *Post-Pleistocene Man and His Environment on the Northern Plains,* edited by Richard G. Forbis, Leslie B. Davis, Ole A. Christensen, and Gloria Fedirchuk, pp. 6–46. Chacmool Archaeological Association, Department of Archaeology, University of Calgary, Calgary.

    1972   *The Archaeology of Pass Creek Valley, Waterton Lakes National Park.* Manuscript Report Series no. 61. National Historic Parks and Sites Branch, Parks Canada, Ottawa.

    2003   *Mistakis: The Archeology of Waterton-Glacier International Peace Park.* Archeological Inventory and Assessment Program, 1993–1996, Final Technical Report, edited by Leslie B. Davis and Claire Bourges. 2 vols. Report on file, Glacier National Park, West Glacier, Montana.

    2009   *The Old Women's Phase: A Spatial Definition and Toolstone Utilization Patterns in Alberta and Montana.* 2 vols. Report prepared for Tribal Government and External Affairs, Kainai Nation, Standoff, Alberta.

Reeves, Brian O. K., and Sandra Leslie Peacock

    2001   *"Our Mountains Are Our Pillows": An Ethnographic Overview of Glacier National Park.* Report on file, National Park Service, U.S. Department of the Interior, Rocky Mountain Regional Office, Denver.

Reeves, Brian O. K., Claire Bourges, and Nancy Saxberg

    2009   *Historical Resources Impact Assessment, Monitoring, and Mitigation of the Nezu Site (HhOu-36) and Other Minor Sites for Syncrude Canada Ltd.'s Aurora Mine North: Final Report (ASA Permit 98-040).* Copy on file, Archaeological Survey, Historic Resources Management Branch, Alberta Culture, Edmonton.

Reeves, Brian O. K., Dan Cummins, and Murray Lobb

    2008   *Heritage Resources Impact Assessment of Oilsands Quest Axe Lake Discovery Area, Saskatchewan.* Draft report on file, Oil Sands Quest, Calgary.

Reeves, Brian O. K., Claire Bourges, Christy de Mille, Janet Blakey, and Nancy Saxberg

    2013a   *Syncrude Canada Ltd., Aurora North Utility Corridor Historical Resources Impact Assessment and Mitigation Studies (ASA Permits 98-039 and 99-024).* Vol. 1, *All Sites Recorded and Revisited.* Draft report on file, Lifeways of Canada Limited, Calgary.

    2013b   *Syncrude Canada Ltd., Aurora North Utility Corridor Historical Resources Impact Assessment and Mitigation Studies (ASA Permits 98-039 and 99-024).* Vol. 2, *Firebag Hills Complex Sites.* Draft report on file, Lifeways of Canada Limited, Calgary.

2014a   *Syncrude Canada Ltd., Aurora North Utility Corridor Historical Resources Impact Assessment and Mitigation Studies (ASA Permits 98-039 and 99-024). Vol. 3, Fort Creek Fen Complex Sites.* Draft report on file, Lifeways of Canada Limited, Calgary.

2014b   *Syncrude Canada Ltd., Aurora North Utility Corridor Historical Resources Impact Assessment and Mitigation Studies (ASA Permits 98-039 and 99-024). Vol. 4, Nezu Complex Sites.* Draft report on file, Lifeways of Canada Limited, Calgary.

2014c   *Syncrude Canada Ltd., Aurora North Utility Corridor Historical Resources Impact Assessment and Mitigation Studies (ASA Permits 98-039 and 99-024). Vol. 5, Beaver River Complex Sites.* Draft report on file, Lifeways of Canada Limited, Calgary.

Reardon, Gerard V.

1976   A Cognitive Approach to Lithic Analysis. MA thesis. Department of Archaeology, University of Calgary, Calgary.

Rhine, Janet L., and Derald G. Smith

1988   The Late Pleistocene Athabasca Braid Delta of Northeastern Alberta, Canada: A Paraglacial Drainage System Affected by Aeolian Sand Supply. In *Fan Delta: Sedimentology and Tectonic Settings,* edited by W. Nemec and R. J. Steel, pp. 158–169. Blackie and Son, Glasgow and London.

Roll, Tom E.

2003   *The Camp Baker Site (24ME467): 2001.* Copy on file, U.S. Department of the Interior, Bureau of Land Management, Montana State Office, Billings.

Ronaghan, Brian M.

1981a   *Final Report: Historical Resources Impact Assessment, Fort McMurray Energy Corridor, Fort Hills Townsite and Airstrip (ASA Permit 80-091).* Copy on file, Archaeological Survey, Historic Resources Management Branch, Alberta Culture, Edmonton.

1981b   *Final Report: Historical Resources Impact Assessment of Selected Portions of the Alsands Lease 13 (ASA Permit 80-091).* Copy on file, Archaeological Survey, Historic Resources Management Branch, Alberta Culture, Edmonton.

1993   The James Pass Project: Early Holocene Occupation in the Front Ranges of the Rocky Mountains. *Canadian Journal of Archaeology* 17: 85–91.

1997   *Historical Resources Impact Assessment for the Muskeg River Mine Project (ASA Permit 97-107).* Copy on file, Archaeological Survey, Historic Resources Management Branch, Alberta Culture, Edmonton.

Roskowski, Laura, and Morgan Netzel

2011   *Historical Resources Impact Assessment, Shell Canada Energy, Muskeg River Mine Expansion of RMS 10, Mitigation for Sites HhOv-87 and HhOv-200: Final Report (ASA Permit 09 168).* Copy on file, Archaeological Survey, Historic Resources Management Branch, Alberta Culture, Edmonton.

2012   *Historical Resources Impact Assessment of Shell Canada's Muskeg River Mine Expansion, RMS 10 Expansion Area, and Mitigation of Sites HhOv-87, HhOv-487, and HhOv-489: Final Report (ASA Permit 10-068).* Copy on file, Archaeological Survey, Historic Resources Management Branch, Alberta Culture, Edmonton.

Saxberg, Nancy

2007   *Birch Mountain Resources Ltd. Muskeg Valley Quarry, Historical Resources Mitigation, 2004 Field Studies: Final Report (ASA Permit 05-118).* 2 vols. Copy on file, Archaeological Survey, Historic Resources Management Branch, Alberta Culture, Edmonton.

Saxberg, Nancy, and Brian O. K. Reeves

1998   *Aurora Mine North Utility Corridor, Historical Resources Impact Assessment, 1998 Field Studies: Interim Report (ASA Permit 98-0390.* Copy on file, Archaeological Survey, Historic Resources Management Branch, Alberta Culture, Edmonton.

2003   The First Two Thousand Years of Oil Sands History: Ancient Hunters at the Northwest Outlet of Glacial Lake Agassiz. In *Archaeology in Alberta: A View from the New Millennium,* edited by Jack W. Brink and John F. Dormaar, pp. 290–322. Archaeological Society of Alberta, Medicine Hat, Alberta.

2004   *Birch Mountain Resources Ltd. Muskeg Valley Quarry, Historical Resources Impact Assessment, 2003 Field Studies: Final Report (ASA Permit 03-249).* Copy on file, Archaeological Survey, Historic Resources Management Branch, Alberta Culture, Edmonton.

2006   *Birch Mountain Resources Ltd. Hammerstone Quarry, Historical Resources Impact Assessment, 2004 Field Studies: Final Report (ASA Permit 04-235).* Copy on file, Archaeological Survey, Historic Resources Management Branch, Alberta Culture, Edmonton.

Saxberg, Nancy, Mack W. Shortt, and Brian O. K. Reeves

1998   *Historical Resources Impact Assessment, Aurora Mine North Highway and Utility and Access Corridors: Final Report (ASA Permit 97-043).* Copy on file, Archaeological Survey, Historic Resources Management Branch, Alberta Culture, Edmonton.

Saxberg, Nancy, Brad Somer, and Brian O. K. Reeves

2003   *Syncrude Aurora Mine North, Historical Resources Impact Assessment, 2002 Field Studies: Final Report (ASA Permit 02-140).* Copy on file, Archaeological Survey, Historic Resources Management Branch, Alberta Culture, Edmonton.

2004   *Syncrude Aurora Mine North, Historical Resources Impact Assessment and Mitigation Studies, 2003 Field Studies: Final Report (ASA Permit 03-279).* Copy on file, Archaeological Survey, Historic Resources Management Branch, Alberta Culture, Edmonton.

Shortt, Mack W., and Brian O. K. Reeves

1997   *Aurora Mine North, 1996 Archaeological Studies, Cree Burn Lake and East Pit, Tailings (ASA Permit 96-072).* Copy on file, Archaeological Survey, Historic Resources Management Branch, Alberta Culture, Edmonton.

Shortt, Mack W., Nancy Saxberg, and Brian O. K. Reeves

1998   *Aurora Mine North, East Pit Opening, Plant Site, Tailings, and Related Workings, HRIA and Mitigation Studies: Final Report (ASA Permit 97-116).* Copy on file, Archaeological Survey, Historic Resources Management Branch, Alberta Culture, Edmonton.

Sims, Cort

1976   A Preliminary Report Concerning an Archaeological Survey of Certain Boreal Forest Highway Projects in Northeastern Alberta, 1975. In *Archaeology in Alberta, 1975,* compiled by J. Michael Quigg and W. J. Byrne, pp. 20–23. Archaeological Survey of Alberta Occasional Paper no. 1. Historic Resources Management Branch, Alberta Culture, Edmonton.

1977   A Notched Burin from Northern Alberta. *Alberta Archaeological Review* 1 (March): 12–15.

1981   Archaeological Investigations in the North Wabasca Lake Area: The Alook Site. *Alberta Archaeological Review* 3 (Autumn): 12–16.

Sims, Cort, and Timothy C. Losey

1975   *Archaeological Investigations on Athabasca Tar Sands Lease 13 (ASA Permit 74-031).* Copy on file, Archaeological Survey, Historic Resources Management Branch, Alberta Culture, Edmonton.

Smith, Derald G., and Timothy G. Fisher

1993   Glacial Lake Agassiz: The Northwestern Outlet and Paleoflood. *Geology* 21(1): 9–12.

Smith, Jason W.

1971   The Ice Mountain Microblade and Core Industry, Cassiar District, Northern British Columbia, Canada. *Arctic and Alpine Research* 3(3): 199–213.

1974   The Northeast Asian–Northwest American Microblade Tradition and the Ice Mountain Microblade and Core Industry. PhD dissertation, Department of Archaeology, University of Calgary, Calgary.

Somer, Brad

    2005   *Syncrude Aurora Mine North, Historical Resources Impact Assessment and Mitigation Studies, 2004 Field Studies: Final Report (ASA Permit 04-192).* 2 vols. Copy on file, Archaeological Survey, Historic Resources Management Branch, Alberta Culture, Edmonton.

Somer, Brad, and Visti Kjar

    2007   *Syncrude Aurora Mine North Historical Resources Impact Assessment and Mitigation Studies, 2005 Field Studies: Final Report (ASA Permit 05-199).* 2 vols. Copy on file, Archaeological Survey, Historic Resources Management Branch, Alberta Culture, Edmonton.

Spurling, Brian E.

    1980   *The Site C Heritage Resource Inventory and Assessment Final Report: Substantive Contributions.* Report prepared for B.C. Hydro and Power Authority. Copy on file, B.C. Archaeology Branch, Ministry of Forest, Lands and Natural Resource Operations, Government of British Columbia, Victoria.

Steer, Donald N.

    1977   The History and Archaeology of a North West Company Trading Post and a Hudson's Bay Company Transport Depot, Lac La Loche, Saskatchewan. MA thesis, Department of Anthropology and Archaeology, University of Saskatchewan, Saskatoon.

Stevenson, Marc G.

    1981   *Preliminary Archaeological Reconnaissance in Wood Buffalo National Park.* Research Bulletin no. 159. Parks Canada, Ottawa.

    1986   *Window on the Past: Archaeological Assessment of the Peace Point Site, Wood Buffalo National Park, Alberta.* Studies in Archaeology, Architecture and History, National Historic Parks and Sites Branch, Parks Canada, Environment Canada, Ottawa.

Stewart, Andrew M.

    2004   Intensity of Land-Use Around the Holland Marsh: Assessing Temporal Change from Regional Site Distribution. In *The Late Paleo-Indian Great Lakes: Geological and Archaeological Investigations of Late Pleistocene and Early Holocene Environments,* edited by Lawrence J. Jackson and Andrew Hinshelwood, pp. 85–116. Mercury Series no. 65. Archaeological Survey of Canada, Canadian Museum of Civilization, Gatineau, Québec.

Syncrude Canada Ltd.

    1973   *Syncrude Lease No. 17: An Archaeological Survey.* Environmental Research Monograph 1973-4. Syncrude Canada Ltd., Edmonton.

    1974   *The Beaver Creek Site: A Prehistoric Stone Quarry on Syncrude Lease No. 22.* Environmental Research Monograph 1974-2. Syncrude Canada Ltd., Edmonton.

Tischer, Jennifer C.

    2004   *Historical Resources Studies, Final Report: Albian Sands Energy Inc. Muskeg River Mine Expansion (ASA Permit 04-249).* 2 vols. Copy on file, Archaeological Survey, Historic Resources Management Branch, Alberta Culture, Edmonton.

    2005   *Historical Resources Studies, Final Report: Albian Sands Energy Inc.* Part I: *Historical Resources Impact Assessment and Historical Resources Mitigation, Muskeg River Mine Expansion (ASA Permit 05-297).* Copy on file, Archaeological Survey, Historic Resources Management Branch, Alberta Culture, Edmonton.

    2006   *Historical Resources Studies, Final Report: Canadian Natural Resources Limited Horizon Oil Sands Project, Historical Resources Mitigation of HiOw-39, HiOw-42, HiOw-43 and Historical Resources Impact Assessment of Tar River Drainage Diversion (Revised) (ASA Permit 05-225).* Copy on file, Archaeological Survey, Historic Resources Management Branch, Alberta Culture, Edmonton.

Tsang, Brian W. B.

    1998   The Origin of the Enigmatic Beaver River Sandstone. MSc thesis, Department of Geology and Geophysics, University of Calgary, Calgary.

Unfreed, Wendy J., Gloria J. Fedirchuk, and Eugene M. Gryba

    2001   *Historical Resources Impact Assessment, True North Energy L.P. Fort Hills Oil Sands Project: Final Report (ASA Permit 00-130).* 2 vols. Copy on file, Archaeological Survey, Historic Resources Management Branch, Alberta Culture, Edmonton.

Van Dyke, Stanley, and Brian O. K. Reeves

    1984   *Historical Resources Impact Assessment, Syncrude Canada Ltd. Lease No. 22 (ASA Permit 84-053).* Copy on file, Archaeological Survey, Historic Resources Management Branch, Alberta Culture, Edmonton. Published in 1985 as Environmental Research Monograph 1985-4, Syncrude Canada Ltd., Edmonton.

Van Dyke, Stanley, and Sally Stewart

    1984   *Hawkwood Site (EgPm-179): A Multicomponent Prehistoric Campsite on Nose Hill.* Archaeological Survey of Alberta Manuscript Series no. 7. Historic Resources Management Branch, Alberta Culture, Edmonton.

Walker, Ernest G.

    1992   *The Gowen Sites: Cultural Responses to Climatic Warming on the Northern Plains (7500–5000 B.C.).* Mercury Series no. 145. Archaeological Survey of Canada, Canadian Museum of Civilization, Gatineau, Québec.

Wickham, Michelle D.

    2006a   *Interim Report, Historical Resource Impact Mitigation (Stage I): TransCanada Fort McKay Mainline (ASA Permit 06-376).* Copy on file, Archaeological Survey, Historic Resources Management Branch, Alberta Culture, Edmonton.

    2006b   *Interim Report, Historical Resource Impact Mitigation (Stage II): TransCanada Fort McKay Mainline (ASA Permit 06-376).* Copy on file, Archaeological Survey, Historic Resources Management Branch, Alberta Culture, Edmonton.

Wilson, James S.

    1981   Archaeology. In *Athabasca Sand Dunes in Saskatchewan,* edited by Zoheir M. Abouguendia, pp. 277–308. MacKenzie River Basin Study Report, Supplement 7. Environment Canada, Ottawa.

Wright, James V.

    1972a   *The Shield Archaic.* Publications in Archaeology no. 3. National Museum of Man, Ottawa.

    1972b   *The Aberdeen Site, Keewatin District, Northwest Territories.* Mercury Series no. 2. Archaeological Survey of Canada, National Museum of Man, Ottawa.

    1976   *The Grant Lake Site, Keewatin District, Northwest Territories.* Mercury Series no. 47. Archaeological Survey of Canada, National Museum of Man, Ottawa.

Yesner, David R.

    2007   Faunal Extinction, Hunter-Gatherer Foraging Strategies, and Subsistence Diversity Among Eastern Beringian Paleoindians. In *Foragers of the Terminal Pleistocene in North America,* edited by Renee B. Walker and Boyce N. Driskell, pp. 15–31. University of Nebraska Press, Lincoln.

Young, Patrick S.

    2006   An Analysis of Late Woodland Ceramics from Peter Pond Lake, Saskatchewan. MA thesis, Department of Archaeology, University of Saskatchewan, Saskatoon.

Younie, Angela M.

    2008   Prehistoric Microblade Technology in the Oilsands Region of Northeastern Alberta: A Technological Analysis of Microblade Production at Archaeological Site HiOv-89. MA thesis, Department of Anthropology, University of Alberta, Edmonton.

7 Lower Athabasca Archaeology | A View from the Fort Hills

ROBIN J. WOYWITKA

The Fort Hills are located on the east side of the Athabasca River, approximately 20 kilometres northeast of Fort McKay, Alberta. Although the area surrounding the hills has been proposed for oil sands mining several times since the late 1970s, mining operations began only in 2013. This chapter serves as an introduction to the prehistoric archaeological record of the area, focusing mainly on work conducted for the various incarnations of the Fort Hills Oil Sands Project.

The study area is located on the northern periphery of the Lower Athabasca archaeology "heartland," that is, the Quarry of the Ancestors, Cree Burn Lake, and the Muskeg River valley region (fig 7.1). The Fort Hills archaeological record is closely tied to that area but also has several characteristics more akin to boreal forest assemblages recovered throughout the southern Canadian subarctic.

LANDSCAPE

The Fort Hills study area is characterized by five notable geographic features: the Clearwater–Lower Athabasca spillway and Fort Creek Fen; the Late Pleistocene Athabasca braid delta; the Fort Hills upland; the McClelland Lake Wetland Complex; and the Athabasca River valley (see fig 7.2). These areas provide a broad geographic context for the archaeological record. Each zone is transected by and dotted with features such as sinkhole lakes, stream valleys, wetlands, and ponds that were the focus of activity during the prehistoric period. The

Figure 7.1. Archaeological sites in the Fort Hills study area (box) and surrounding region

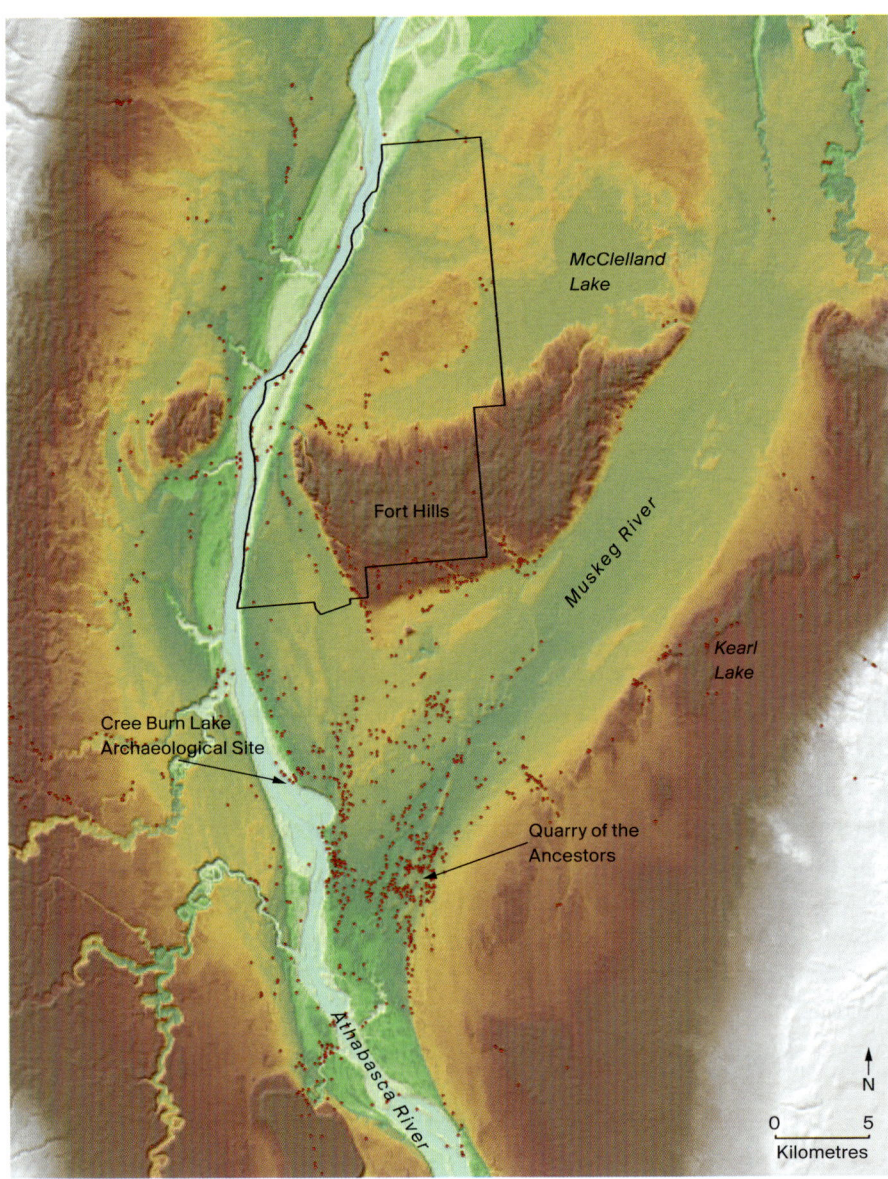

McClelland Lake

Fort Hills

Muskeg River

Kearl Lake

Cree Burn Lake Archaeological Site

Quarry of the Ancestors

Athabasca River

N

0   5
Kilometres

relationships among these landforms, the broader surrounding bio-geomorphic zones, and human occupation throughout the Holocene are key to our understanding of the past in the Fort Hills area.

Clearwater–Lower Athabasca Spillway and Fort Creek Fen

A portion of the terminal end of the Clearwater–Lower Athabasca spillway lies in the southwest corner of the Fort Hills Project lease (see fig 7.2). The Fort Creek Fen covers much of this area, although one large remnant gravel bar is present to the west of this saturated terrain, on the southern border of the lease. This feature rises only 0.5 metres to 1.0 metres above the adjacent wetlands. The east side of the Clearwater–Lower Athabasca spillway is bordered by the edge of the Fort Hills upland, described below. The Clearwater–Lower Athabasca spillway grades into Late Pleistocene Athabasca braid delta deposits north of the upland (see fig 7.3, below). As Fisher and Lowell argue in chapter 2 of this volume, the Clearwater–Lower Athabasca spillway was formed approximately 9,800 to 9,600 ^{14}C yr BP, followed by what they estimate to be a "few hundred years" of steady-state flow.

Late Pleistocene Athabasca Braid Delta

Within the Fort Hills Project area, the surface of the Late Pleistocene Athabasca braid delta is characterized by gently rolling topography consisting of stabilized sand dunes. The dominant forest cover is open pine forest, although large areas of thicker mixed jack pine and aspen also occur. Three intermittent tributaries of the Athabasca River cross the area, the lower reaches of which contain flowing water and are deeply incised. The southernmost of these tributaries (which I call Susan Creek) connects the Athabasca River to the McClelland Lake Wetland Complex via Susan Lake (see fig 7.2). The McClelland Lake Wetland Complex also borders Late Pleistocene Athabasca braid delta deposits in the northeast portion of the Fort Hills Project lease. Aside from these hydrological features, topography and vegetation are remarkably uniform throughout the delta area, with forest density providing the only contrasting landscape element. The Late Pleistocene Athabasca braid delta began to form during deglaciation of northeastern Alberta, when the Athabasca River delivered sediment into Glacial Lake McConnell. Smith and Fisher (1993) propose that most of the estimated 70 cubic kilometres (Rhine and Smith 1988) of sand in the delta was deposited by the Lake Agassiz flood. Following deposition of the Late Pleistocene Athabasca braid delta, the lower reaches of the Athabasca gradually consolidated within the present-day valley and began down cutting through the delta sediments (Smith 1994).

Figure 7.2. Archaeological sites in the Fort Hills area

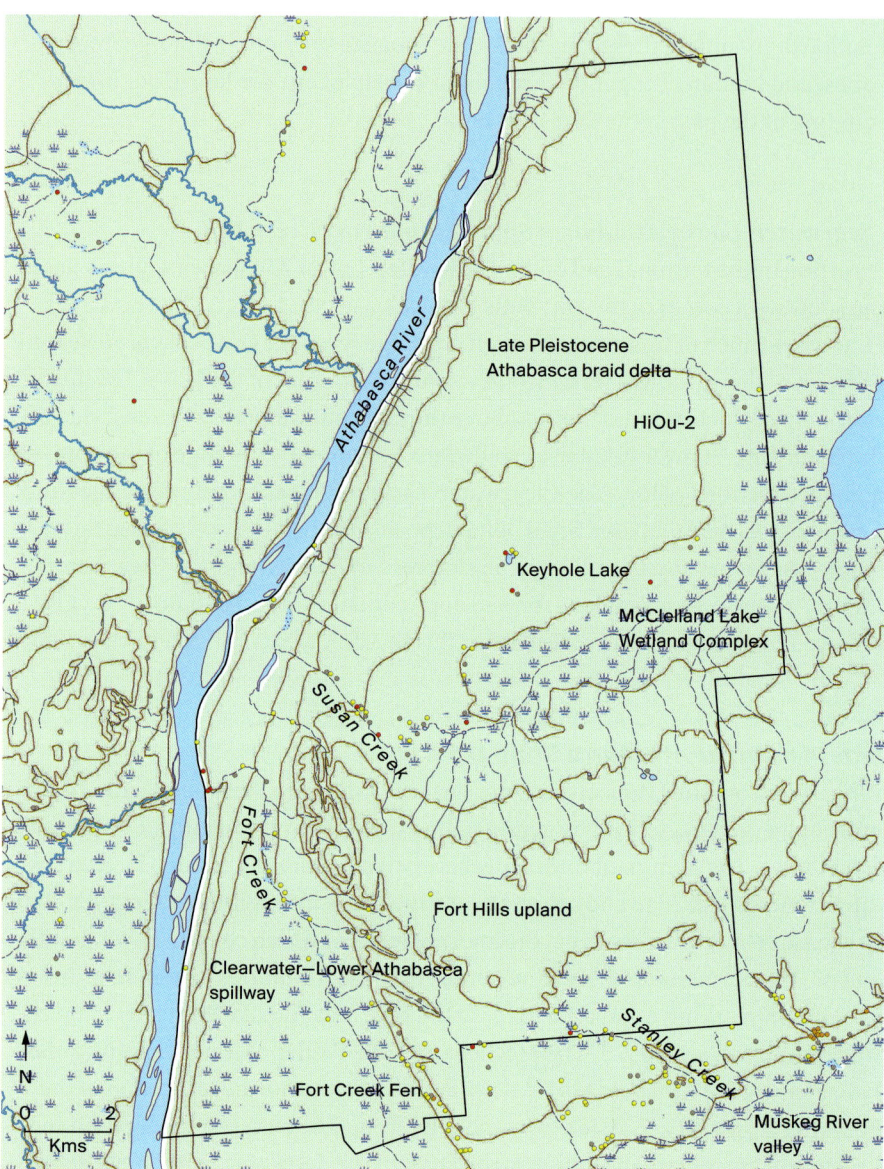

Late Pleistocene
Athabasca braid delta

HiOu-2

Athabasca River

Keyhole Lake

McClelland Lake
Wetland Complex

Susan Creek

Fort Creek

Fort Hills upland

Clearwater–Lower Athabasca
spillway

Stanley Creek

Fort Creek Fen

Muskeg River
valley

N

0   2

Kms

- 🔴 Campsite
- 🟡 Scatter
- 🟠 Scatter/Workshop
- ⚫ Isolated Find
- ☐ Fort Hills Oil
  Sands Project

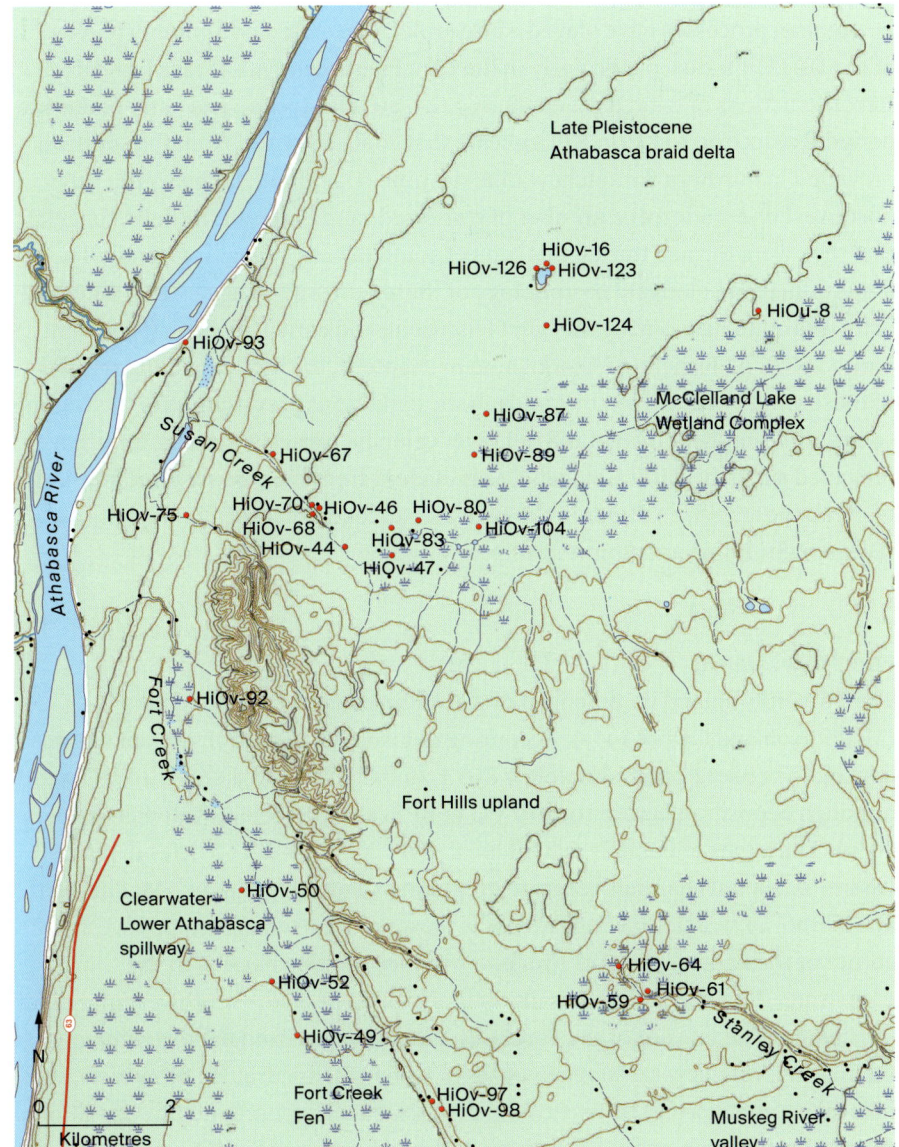

Figure 7.3. Sites excavated for the Fort Hills Oil Sands Project

Aeolian processes have played a large role in the post-flood geomorphology of the Fort Hills area, particularly in the Late Pleistocene Athabasca braid delta. Reconstructions of wind patterns during the late Pleistocene and early Holocene period (ca. 10,000 to 8,800 BP) indicate that cool, dry winds from the southeast prevailed in northern Saskatchewan at the time (David 1981). The southeast-northwest orientation of a number of stabilized sand dunes in the delta suggest that they were formed during this time of southeasterly prevailing winds. The longitudinal and elongate morphology of these dunes also suggest that wind velocity was quite high during the late Pleistocene and early Holocene (Rhine and Smith 1988). As Fisher and Lowell note in chapter 2, the direction and intensity of these winds were driven by atmospheric conditions prevailing over the waning continental ice sheet (David 1981). Lower velocity, westerly wind patterns have prevailed since the early Holocene, as is evident from the east-west orientation of parabolic stabilized dunes in the Late Pleistocene Athabasca braid delta area (David 1981; Rhine and Smith 1988).

Fort Hills Upland

The Fort Hills upland is composed of glacial kame delta or fan deposits (McPherson and Cathol 1977). According to Fisher et al. (2009), the upland was deposited between approximately 9,850 ± 70 ^{14}C yr BP and 9,660 ± 40 ^{14}C yr BP, although deposition before 10,000 ^{14}C yr BP is also possible (Murton et al. 2010). The edges of the hills exhibit local relief in excess of 15 metres in places where "erosional channels separate irregularly shaped highlands" (McPherson and Cathol 1977, 85). This ruggedness is most pronounced in the northwest corner of the zone. A series of irregular benches running parallel to the Clearwater–Lower Athabasca spillway is also present on the southwest margin of the upland. The interior of the Fort Hills upland ranges from flat to gently undulating, with several large expanses of wetlands, dry drainage channels, sinkhole lakes, and discontinuous aeolian deposits superimposed on the depositional surface (McPherson and Cathol 1977). The headwaters of Fort Creek, Stanley Creek, and other tributaries of the Athabasca and Muskeg rivers are located atop the Fort Hills upland.

McClelland Lake Wetland Complex

The McClelland Lake Wetland Complex is a large wetland dominated by wooded, shrubby and open fen, with some permafrost bog and peat plateaus elements (True North Energy 2001). To the north and west, the terrain adjacent to the McClelland Lake Wetland Complex is characterized by the well-drained, stabilized dune topography of the Late Pleistocene Athabasca braid delta. On the south, the complex is bordered by flatter, moister terrain at the base of the Fort Hills upland. The predominant vegetation in this area is mixed forest of aspen and spruce. The McClelland Lake Wetland Complex began accumulating peat approximately 7,000 to 8,000 years ago, and the presence of wood and conifer needles in the basal layers of a peat core from the area suggest that the wetland replaced a forest community around this time (True North Energy 2001).

Athabasca River Valley

The Athabasca River valley began to assume a course similar to its current position sometime after the cessation of flow through the Clearwater–Lower Athabasca spillway. The upper valley edge appears both as a marked break in the terrain and as a gentle slope along the western boundary of the Fort Hills Project area. Intermediate terraces occur along the river and are frequently scarred by slump blocks or dissected by deep tributary stream valleys. Vegetation is primarily mature forest dominated by aspen and spruce. Bogs and fens are located on flatter sections of the Athabasca valley wall (mostly on the heads of slump blocks) and on the modern floodplain. Trees and shrubs grow along the valley wall, while the wetlands on the floodplain are heavily wooded and swampy.

RESEARCH HISTORY

Archaeological studies have been undertaken in the Fort Hills Project area since the mid-1970s, with the greatest activity occurring from 1979 to 1981 and from 2000 to 2008. Although Donahue's Athabasca River survey (Donahue 1976) assessed peripheral parts of the project area, the first targeted archaeological survey was carried out in 1978 in relation to the extension of Highway 63 (then Secondary Highway 963) (Gryba 1980). Gryba's work was carried out around the same time as two studies related to proposed development of a townsite and airport located between McClelland Lake and the Athabasca River, one by Wood (1979) and a second by Ronaghan (1981). The latter study roughly coincided with McCullough's survey of proposed core holes in connection with joint NOVA-PetroCanada mining interests in Leases 5, 8, and 52 (McCullough 1980).

The next survey conducted in the area was undertaken for a proposed oil sands project organized by the SolvEx Corporation (Gorham 1997). This development included a small mine, plant site, and ancillary facilities in Lease 5, as well as improvements to Highway 63. No precontact archaeological sites were identified within the Fort Hills Project area during this study (Gorham 1997).

SolvEx collapsed in the late 1990s, and the project was taken over by True North Energy, a partnership between Koch Exploration Canada and UTS Energy. A suite of Historical Resources Impact Assessments (HRIAs) conducted from 2000 to 2002 for True North constituted the first large-scale archaeological surveys carried out in the project area. Two initial surveys targeted creek and wetland margins, sinkhole lakes, and areas of marked topographic relief (Unfreed, Fedirchuk, and Gryba 2001; Gryba, Unfreed, and Peach 2001), while a third study consisted of post-impact assessment of forestry disturbance within the area of the lease (Tischer 2003). Mitigative excavation was also conducted at one site in 2001, but it was determined that the site had been completely destroyed by access road construction (Gryba and Unfreed 2002).

True North Energy's development plans were abandoned in 2003 owing to unfavourable economic conditions. Two years later, however, economic conditions had improved enough to allow a consortium of companies—Petro-Canada, UTS Energy, and Teck Cominco—to resurrect the project under the Fort Hills Energy Ltd. banner. Between 2005 and 2008, several projects were undertaken within the Fort Hills Project area on behalf of Fort Hills Energy. These included four HRIA surveys (Woywitka 2005, 2007a, 2008; Woywitka and Younie 2007), five Historical Resources Impact Mitigation (HRIM) excavation projects (Woywitka 2007b; Woywitka and Younie 2008a, 2008b, 2010; Woywitka et al. 2009), and two post-impact assessments (Woywitka 2008; Graham, Morton, and Woywitka 2008). Three HRIAs related to ancillary developments for the Fort Hills Project or adjacent projects were also carried out on the lease during this period (Blaikie-Birkigt 2006; Kjorlien and Woywitka 2008; Murphy 2009).

By the end of 2008, nearly the entire Fort Hills Project leasehold had been assessed either in the field or by a desktop review, 111 prehistoric sites had been identified, and 28 precontact archaeological sites had been excavated (Woywitka 2007b; Woywitka and Younie 2008a, 2008b, 2010; Woywitka et al. 2009). The project has since been taken over by Suncor Energy Inc.

SITE LOCATION AND SITE TYPES

Sites in the Fort Hills study area are less abundant and generally less productive than those in the Cree Burn Lake or Quarry of the Ancestors region. The highest site densities occur on the western fringes of the McClelland Lake Wetland

Complex and along Fort Creek and the margins of Fort Creek Fen. Less prominent concentrations are located along the Athabasca River valley near Susan Lake, along the northern perimeter of the McClelland Lake Wetland Complex, at the headwaters of Stanley Creek, and near two sinkhole lakes in the central portion of the study area.

In the Fort Creek drainage, the highest concentration of sites is on the east side of the Clearwater–Lower Athabasca spillway, along the southwestern flanks of the Fort Hills upland (see fig 7.2). This site cluster extends along the margin of Fort Creek Fen to an area outside of the Fort Hills Project lease beyond the southwestern tip of the Fort Hills. The Stanley Creek sites are also an extension of a cluster that lies in part outside the Fort Hills Project boundary, near the intersection of Stanley Creek with the Muskeg River valley.

Most sites to the north of the Fort Hills upland are concentrated in the vicinity of the McClelland Lake Wetland Complex and adjacent to landforms such as sinkhole lakes and intermittent streams. An abundance of sites line the intermittent stream that I call Susan Creek, which connects the Athabasca River, Susan Lake, and the McClelland Lake Wetland Complex. The relative lack of prehistoric sites elsewhere in the Athabasca River valley is not entirely surprising, given the active slumping and generally wet conditions found along its length. The majority of sites in the Fort Hills area are related to hydrological features, although a single site (HiOu-2) does occur in the middle of the sand plain of the Late Pleistocene Athabasca braid delta, away from any prominent topographic feature or water source (Ronaghan 1981). Of the 111 sites identified to date, 16 appear to be campsites, 42 are isolated finds, 51 are scatter sites, and 2 are scatters with workshop components.

LITHICS

Assemblages

To date, approximately 72,049 lithic artifacts have been recovered from twenty-seven archaeological sites (totalling 601.25 square metres) excavated in the area of the Fort Hills lease (see fig 7.3). Average artifact return currently stands at 2,273 items per site.[1] The lithic materials are primarily debitage, although a number of tools have been recovered (Woywitka et al. 2009). These tools account for only 0.2% of the entire excavated assemblage, however, and rarely exceed 1% of the sample at a single site (table 7.1). Just over half (51.5%) of the tools are formed tools, with the remainder consisting of expedient tools. Formed tool types include scrapers, bifaces, projectile points, and multi-function tools; expedient

Table 7.1 Summary of lithic assemblages at excavated sites in the Fort Hills

Borden no.	Site type	Excavation area (m²)	Lithic items	BRS	Non-BRS material	% BRS	Debitage	Cores	Formed tools
HiOu-8	Campsite	14	14	0	14	0.00	11	0	0
HiOv-16	Campsite	55	3,274	3,041	233	92.88	3,252	1	19
HiOv-44	Campsite	20	2,915	2,898	12	99.42	2,910	3	1
HiOv-46	Campsite	24	2,782	2,659	123	95.58	2,771	2	7
HiOv-47	Scatter	10	678	658	20	97.05	675	0	3
HiOv-49	Scatter	50	15,596	15,593	3	99.98	15,562	12	12
HiOv-50	Scatter	8	1,161	1,157	4	99.66	1,140	0	4
HiOv-52	Scatter	50	22,528	22,513	15	99.93	22,490	9	6
HiOv-59	Campsite	10	5	4	1	80.00	4	0	1
HiOv-61	Scatter	22.5	2,181	2,181	0	100.00	2,171	0	3
HiOv-64	Scatter	19	1,881	1,735	146	92.24	1,861	2	6
HiOv-67	Campsite	13.25	1,173	1,173	0	100.00	1,163	1	6
HiOv-68	Campsite	15	1,393	1,370	23	98.35	1,386	2	1
HiOv-70	Campsite	21.5	1,477	1,462	15	98.98	1,476	0	0
HiOv-75	Scatter	16	1,175	1,171	4	99.66	1,167	3	1
HiOv-80	Scatter	20	418	336	82	80.38	413	1	3
HiOv-83	Scatter	14	3,682	3,673	9	99.76	3671	0	7
HiOv-87	Campsite	6	305	287	16	94.10	303	0	0
HiOv-92	Scatter	11	1,150	1,148	2	99.83	1,146	0	4
HiOv-93	Scatter	21	1,796	1,484	312	82.63	1,789	2	5
HiOv-97	Scatter	12	172	172	0	100.00	167	3	1
HiOv-98	Scatter	12	1,324	1,324	0	100.00	1,304	15	0
HiOv-104	Campsite	16	185	72	108	38.92	177	0	4
HiOv-123	Campsite	50	4,691	4,661	30	99.36	4,668	11	9
HiOv-124	Campsite	14	9	8	1	88.89	9	0	0
HiOv-126	Campsite	17	129	33	96	25.58	127	0	2
Total		541.25	72,094	70,813	1,269	98	71,813	67	105

Expedient tools	Total tools	Tools	Cores (%)	Debitage-to-tool ratio	Material types	Dominant reduction stage	Fire-broken rock	Faunal	Diagnostics
3	3	21.43	0.00	5:1	2	Secondary	N	Y	0
3	22	0.67	0.03	148:1	12	Secondary	N	Y	1
2	3	0.10	0.10	1455:1	1	Finishing	Y	Y	1
1	8	0.29	0.07	308:1	15	Secondary	Y	Y	0
0	3	0.44	0.00	255:1	2	Secondary/ finishing	N	N	0
10	22	0.14	0.08	707:1	1	All	N	N	0
16	20	1.72	0.00	54:1	1	Finishing	N	N	0
13	19	0.08	0.04	1183:1	2	Secondary/ finishing	N	N	0
0	1	20.00	0.00	4:1	1	Secondary/ finishing	Y	Y	1
7	10	0.46	0.00	217:1	0	Secondary/ finishing	N	N	1
12	18	0.96	0.11	103:1	4	Secondary/ finishing	N	N	0
3	9	0.77	0.09	129:1	0	Secondary	N	Y	0
4	5	0.36	0.14	277:1	5	Secondary	N	Y	0
1	1	0.07	0.00	1476:1	5	Secondary	Y	Y	0
4	5	0.43	0.26	233:1	1	Secondary	N	N	0
2	5	1.20	0.24	83:1	7	Secondary	N	N	1
4	11	0.30	0.00	334:1	4	Secondary	N	N	3
2	2	0.66	0.00	152:1	1	Secondary/ finishing	N	Y	0
0	4	0.35	0.00	287:1	2	Secondary	N	N	1
0	5	0.28	0.11	358:1	8	Secondary	N	N	0
0	1	0.58	1.74	167:1	0	Primary/ secondary	N	N	0
5	5	0.38	1.13	260:1	0	Primary/ secondary	N	N	0
4	8	4.32	0.00	22:1	2	Secondary/ finishing	Y	Y	0
3	12	0.26	0.23	389:1	4	Second-ary/some primary	Y	Y	0
0	0	0.00	0.00	0	1	Secondary/ finishing	Y	Y	0
0	2	1.55	0.00	64:1	7	Secondary	Y	Y	1
99	204	0.28	0.09						10

tools include retouched and utilized flakes. Of particular note are a microblade assemblage from the Little Pond site (HiOv-89) and three scrapers found at HiOv-16 that were fashioned from Mount Edziza obsidian (fig 7.4; Woywitka et al. 2009). The scrapers exhibit signs of use, but they do not appear to be exhausted; rather, they seem to be still functional specimens. The projectile points, microblade assemblage, and obsidian scrapers are discussed in more detail below.

Secondary and tertiary reduction stages are most prevalent in the debitage assemblages, suggesting that little raw material procurement occurred in the Fort Hills area. However, sites HiOv-49, HiOv-97 and HiOv-98 display primary reduction components, an indication that some level of procurement occurred in the area (Woywitka and Younie 2008a, 2008b). Debitage-to-tool ratios range from 4:1 to 1,476:1, although many of the lower ratios occur in assemblages with very small samples (see table 7.1).

Raw Material

Beaver River Sandstone (BRS) is the predominant raw material in most sites, accounting for over 98% of the total sample collected from the sites excavated (see table 7.1). However, ten of the twenty-seven sites have assemblages that contain at least 5% other raw materials. The most common material type other than BRS is a grey to beige quartzite commonly referred to as Northern quartzite (Unfreed, Fedirchuk, and Gryba 2001; Woywitka and Younie 2008a). Other material types present in the Fort Hills sites include obsidian, silicified siltstone, salt and pepper quartzite, chalcedony, and various cherts and other quartzites. Like Northern quartzite, most of these types can be found in local gravels (Unfreed, Fedirchuk, and Gryba 2001). Obsidian and Swan River chert are exotic materials, however, and thus indicate trade or extensive travel from other areas; they will be discussed in detail below.

Diagnostic Tools

Ten projectile points have been recovered from sites in the Fort Hills Project area (fig 7.5). Side-notched points are the most common (n = 6), followed by corner-notched (n = 2), lanceolate (n = 1), and stemmed (n = 1) specimens. Because of the overlapping morphology and time-transgressive nature of many point types recovered in the subarctic, chronologies that are based solely on the typology of projectile points found elsewhere in northwestern North America can be unreliable (for discussion, see Hare, Hammer, and Gotthardt 2008, for example, as well

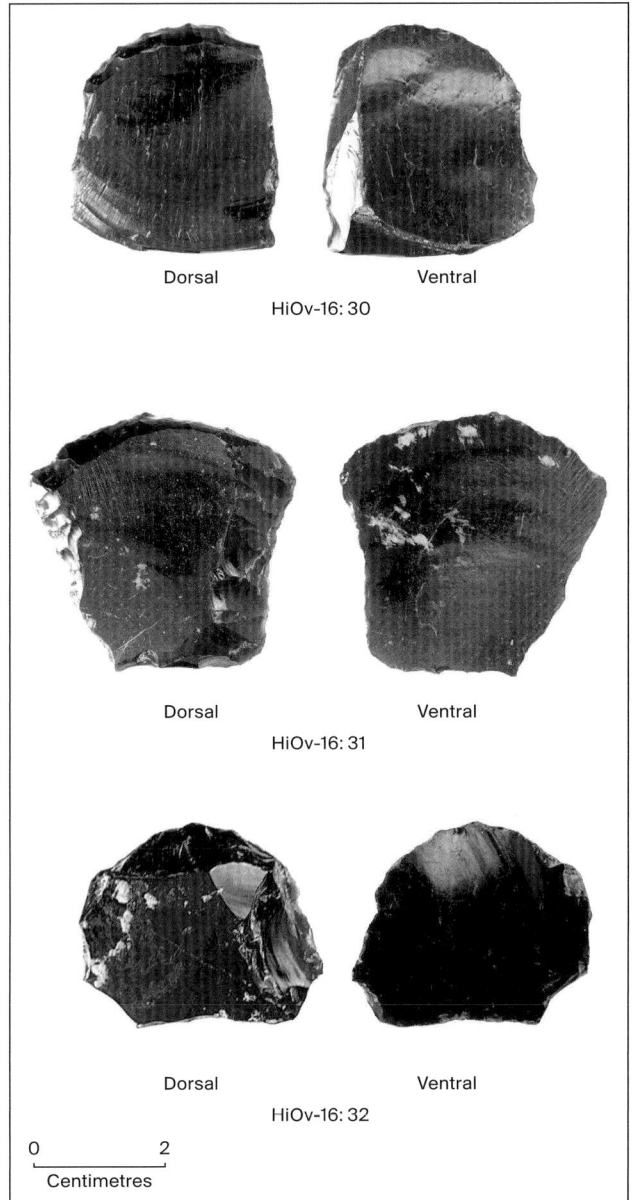

Dorsal Ventral

HiOv-16: 30

Dorsal Ventral

HiOv-16: 31

Dorsal Ventral

HiOv-16: 32

0   2

Centimetres

Figure 7.4. Obsidian scrapers from HiOv-16

as Ives, chapter 8 in this volume). Unfortunately, typological constructs are frequently the best chronological tool currently available to subarctic archaeologists. In lieu of definitive chronological data, table 7.2 is provided to show the closest typological match(es) for each point and the associated estimated age ranges.

Table 7.2   Fort Hills projectile point types

Site	Type	Raw material	Typological match(es)	Estimated age
HiOv-16: 33	Stemmed	Red and grey mottled chert	Birch Mountains/Pointed Mountain	7,500 to 1,600 BP
			Manitoba/Lovell Constricted	8,500 to 8,000 BP
HiOv-44: 2581	Side-notched	Beaver River Sandstone	Late Taltheilei, Frank Channel Phase	650 to 450 BP
			Pelican Lake	3,100 to 1,600 BP
HiOv-59: 11	Side-notched	Beaver River Sandstone	Birch Mountains/Pointed Mountain	Suspected Middle Prehistoric
			Late Taltheilei	1,300 to 200 BP
HiOv-61: 235	Side-notched	Beaver River Sandstone	Middle Prehistoric	7,500 to 1,600 BP
HiOv-83: 20	Corner-notched	Northern quartzite	None	Unknown
HiOv-80: 20	Side-notched	Beaver River Sandstone	Middle Prehistoric	7,500 to 1,600 BP
			Beaver River Complex	7,750 to 3,500 BP
HiOv-80: 21	Side-notched	Quartzite	Late Taltheilei, Frank Channel Phase	650 to 450 BP
			Shield Archaic	6,500 to 3,500 BP
			Beaver River Complex	7,750 to 3,500 BP
HiOv-80: 22	Side-notched	Beaver River Sandstone	Anderson, Taye Lake Phase	ca. 1,500 BP
			Late Taltheilei	1,300 to 200 BP
HiOv-92: 20	Lanceolate	Beaver River Sandstone	Northern Plano	8,000 to 7,000 BP
			Cree Burn Lake Complex	8,600 to 7,750 BP
			Middle Taltheilei	1,800 to 1,300 BP
HiOv-126: 210	Corner-notched	Silicified siltstone	Pelican Lake	3,100 to 1,600 BP

Lanceolate points. A lanceolate point was recovered from two centimetres below the organic-mineral soil contact at HiOv-92 (fig 7.5), a site that lies within the Clearwater–Lower Athabasca spillway, near its point of contact with the Late Pleistocene Athabasca braid delta. The site is situated on a knoll adjacent to wetlands associated with Fort Creek (fig 7.6; Woywitka et al. 2009). The point is fashioned from fine-grained BRS. In cross-section, both lateral edges are convex, and the base is slightly concave. The tip is very sharp, and the body is right-skewed. The flaking ranges from parallel to slightly oblique and is regular on both sides. The overall workmanship is quite fine, and light grinding is evident on the lateral edges, near the base of the point. The base itself is unground. The style of the point is similar to the Agate Basin style defined on the northern Plains, which has been dated to approximately 10,200 to 9,600 BP (Peck 2011) A northern variant of this style, termed Northern Plano, occurs throughout the prairie provinces, as well as in Nunavut and the Northwest Territories (Gordon 1996). These artifacts are usually found in younger

Figure 7.5. Projectile points recovered from Fort Hills excavations

HiOv-16: 33
0 — 2
Centimetres

HiOv-92: 20
0 — 2
Centimetres

HiOv-59: 11
0 — 2
Centimetres

HiOv-61: 235
0 — 2
Centimetres

HiOv-80: 20
0 — 2
Centimetres

HiOv-80: 22
0 — 2
Centimetres

HiOv-80: 163
0 — 2
Centimetres

HiOv-44: 2341
0 — 2
Centimetres

HiOv-126: 210
0 — 2
Centimetres

HiOv-83: 40
0 — 2
Centimetres

contexts than their Plains counterparts (ca. 8,000 to 7,000 BP) and have been interpreted as evidence of Plains hunters moving northward into regenerating postglacial landscapes (Gordon 1996). Reeves, Blakey, and Lobb group these artifacts into their Cree Burn Lake Complex (8,600 to 7,750 BP): see chapter 6 in this volume.

Figure 7.6. Sites that yielded diagnostic tools

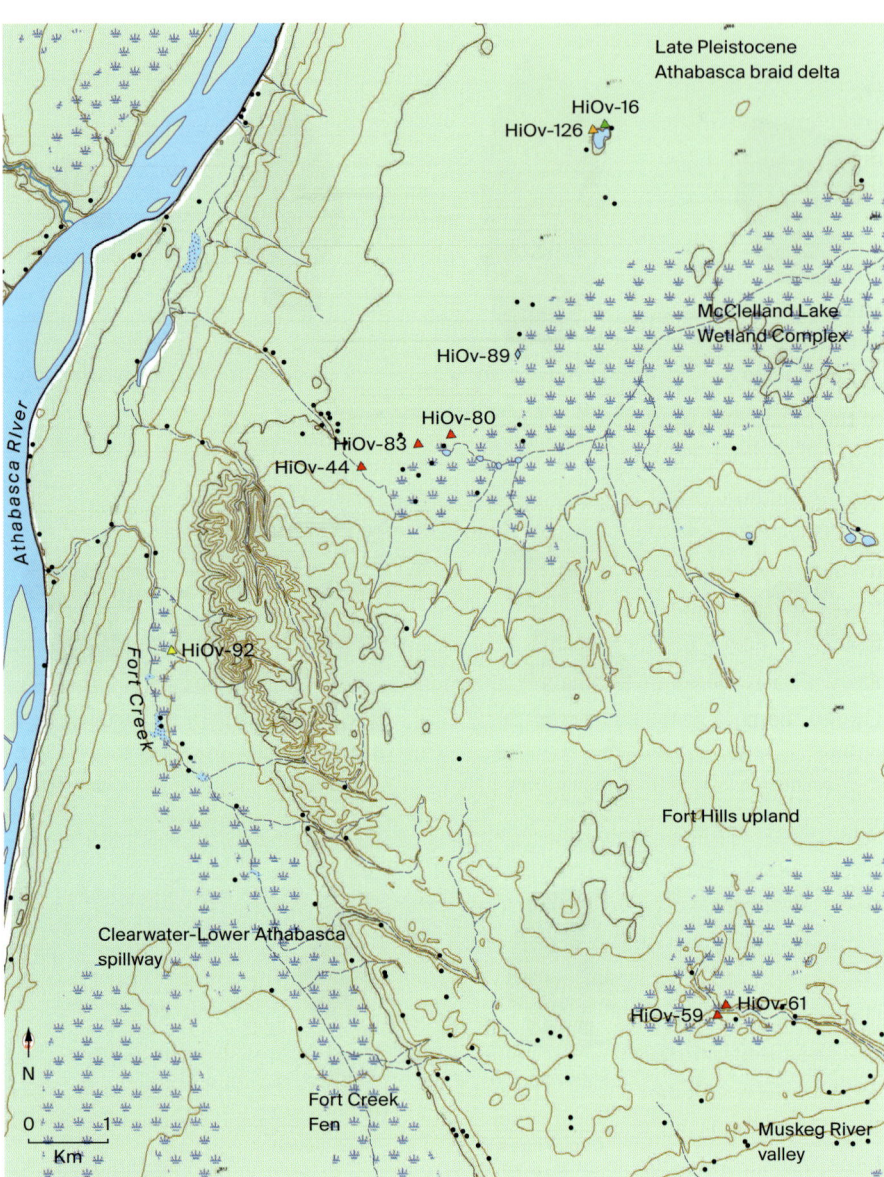

Late Pleistocene
Athabasca braid delta

HiOv-16
HiOv-126

McClelland Lake
Wetland Complex

HiOv-89

HiOv-80
HiOv-83
HiOv-44

Athabasca River

Fort Creek

HiOv-92

Fort Hills upland

Clearwater-Lower Athabasca
spillway

HiOv-59 HiOv-61

N

0   1

Km

Fort Creek
Fen

Muskeg River
valley

△   Lanceolate Point
▲   Stemmed Point
▲   Side Notched
    Point
△   Corner Notched
    Point
◊   Microblades
•   Archaeological
    Site

Lanceolate points of similar shape are also found in Middle Taltheilei (ca. 1,800 to 1,300 BP) assemblages in the Barrenlands of the Northwest Territories and Nunavut (Gordon 1996). The Taltheilei lanceolates tend to be flaked in a more irregular pattern and are somewhat less likely to exhibit basal grinding than Northern Plano points (Gordon 1996).

Stemmed points. A stemmed projectile point fragment was recovered from HiOv-16 (figs. 7.5 and 7.6; Woywitka et al. 2009). The specimen, which was fashioned from a red and grey mottled chert, has a constricting stem and concave base and exhibits an irregular flaking pattern. Two notches are evident on one lateral edge of the base, and the tip is missing. The site is located on the north side of a sink-hole lake (which I will call Keyhole Lake), approximately 1.5 kilometres north of the McClelland Lake Wetland Complex (see fig 7.3).

Projectile points with constricting stems and concave bases are not common in the Lower Athabasca region (Saxberg and Reeves 2003; Somer 2005; Roskowski and Blower 2009; Woywitka et al. 2009), but two have been identified in the Fort Hills area, to the south of the Fort Hills Project lease (Saxberg and Reeves 2003; Somer 2005; Reeves, Blakey, and Lobb, chapter 6 in this volume). One of them—a point recovered from HiOu-56, situated on the banks of Stanley Creek—has been attributed to the Hell Gap type (Somer 2005).[2] This point has a constricting stem, but the base is straight, and the specimen also lacks the lateral-edge notches seen on the HiOv-16 point. Saxberg and Reeves (2003) illustrate a second stemmed point, from HiOu-34, a site that lies not far from HiOu-56, to south and west. This point exhibits the same general shape as the HiOv-16 item, including one lateral-edge notch. Unfortunately, though, the base of the HiOu-34 point is damaged. Reeves, Blakey, and Lobb (this volume) assign the stemmed points from HiOu-56 and HiOu-34 to pre–Agassiz flood times, that is, prior to about 9,600 BP.

The HiOv-16 point shares some characteristics with stemmed points from the Birch Mountains, although the Birch Mountains examples also lack concave bases and lateral-edge notches. Ives (1993 and chapter 8 in this volume) suggests that these points may be related to undated occupations of the Pointed Mountain Complex in the Northwest Territories (Millar 1968; Morrison 1987) and later Middle Prehistoric or early Late Prehistoric occupations (post–2,900 years BP) identified in the Charlie Lake Cave sequence.

Other morphological fits for this point fragment are the Manitoba (Pettipas 1972, 2003; Roskowski and Blower 2009) and Lovell Constricted stemmed points (Husted 1969). These point types have the same general shape and concave base of the HiOv-16 artifact, and the Manitoba type in particular shares the notching on the lateral edge of the stem. These artifacts have been found in association with radiocarbon dates of 8,000 to 8,500 years BP in the northern Great Plains and in the foothills and front ranges of the Rocky Mountains. Stemmed points that bear a resemblance to the HiOv-16 projectile have also been found at HhOv-483 and HhOv-484, in the lower Muskeg River valley, 20 kilometres south of the Fort Hills. Roskowski and Blower (2009) have assigned these points to the Manitoba category.

Side-notched points. Side-notched points in northern Alberta have been assigned to various Middle and Late Prehistoric period constructs in the Canadian northwest. These include northern frameworks such as the Northern Archaic Tradition, the Shield Archaic Tradition, and the Pre-Dorset or Arctic Small Tool Tradition (all three Middle Prehistoric period), as well as the Taltheilei Tradition (Middle to Late Prehistoric period). Northern Alberta side-notched points have also been attributed to several Middle Prehistoric period Plains types, including the Mummy Cave Complex and the Oxbow Complex. The similarity in form and wide spatio-temporal distribution of these points makes it difficult to assign specimens to specific times within the Middle Prehistoric (approximately 7,000 to 1,500 BP).

One side-notched point base was recovered from HiOv-44 (figs. 7.5 and 7.6), a site located on the dry upper reaches of Susan Creek. The point has one damaged ear and exhibits irregular flaking. No grinding is apparent on the base or lateral edges. Woywitka and Younie (2008a) assigned this point base to the Late Taltheilei period, based on similarities with side-notched points recovered from Black Lake, Saskatchewan, that were associated with the Frank Channel Phase (ca. 650 to 450 BP) of the Late Taltheilei Tradition (Minni 1976). The notching and the straight base are also consistent with the Beaver River Complex that Reeves, Blakey, and Lobb describe in chapter 6. It is worth noting that this point also resembles some Pelican Lake forms observed on the northern Great Plains (ca. 3,100 to 1,600 BP).

A side-notched point was recovered from HiOv-59, located on the southern flanks of the Fort Hills upland, along the upper reaches of Stanley Creek (figs. 7.5 and 7.6; Woywitka 2007b). The point has one relatively shallow notch preserved at the terminus of a stemmed base. One of the ears is completely missing, and the other is damaged. This point resembles side-notched points identified in the Birch Mountains that Ives (1993) has tentatively linked to Middle Prehistoric sites in the Fisherman Lake area of the Northwest Territories (Millar 1968). The point base also bears a resemblance to certain Late Taltheilei forms (Gordon 1996) and to points that Reeves, Blakey, and Lobb assign to their regional correlate of the Taltheilei Tradition, the Chartier Complex (see plate 6.9 in this volume, especially point 2). Unfortunately, the damaged and missing ears make it extremely difficult to determine the best typological match for the HiOv-59 point.

A fragment of a side-notched point base was recovered from HiOv-61, located on the north side of Stanley Creek directly across from HiOv-59 (figs. 7.5 and 7.6; Woywitka 2007b). The point is broken below the shoulder and through the centre of the blade, again making it difficult to associate it with a particular typological group. On the basis of the side notch, Woywitka (2007b) assigned it in a general manner to the Middle Prehistoric period.

Three side-notched point fragments were recovered from HiOv-80 (figs. 7.5 and 7.6), a site located on the northern margin of the McClelland Lake Wetland Complex, near its western terminus. HiOv-80: 20 is a portion of a BRS projectile point: only the base is preserved. Because the shoulders are not present, it cannot be definitively identified as a side-notched point, although its general shape and size suggest that it may fit into the Middle Prehistoric traditions and complexes identified in the oil sands region, such as the Northern Archaic Tradition (Ives 1993) and the Beaver River Complex (Reeves, Blakey, and Lobb, chapter 6 in this volume).

The stem and base are present on HiOv-80: 163 (see fig 7.5). The point fragment is made from quartzite and has a convex base with rounded ears and shallow, round notches. The base is bifacially flaked, and no grinding is visible. The rounded ears and narrow notches set this point fragment apart from other side-notched examples recovered in the Lower Athabasca region, although they do compare well with several Beaver River Complex points illustrated in plate 6.7 of this volume (see points 3 and 9). The item bears a closer resemblance to points recovered from the Migod site at Grant Lake and other sites in the Barrenlands that have been assigned to the Shield Archaic Tradition (Wright 1975, cited in Gordon 1976). Minni (1976) also assigned shallowly side-notched, convex-based projectile points to the Late Taltheilei Frank Channel Phase (ca. 650 to 450 BP).

HiOv-80: 22 is a BRS point fragment that is broken through the neck and also damaged on the left lateral edge (see fig 7.5). The right lateral edge has a long, stem-like notch that grades into rounded basal edges. The base itself is flat to slightly convex. Woywitka et al. (2009) note a similarity between this specimen and the Anderson point type originally identified by MacNeish (1964) in the southwest Yukon. Hare et al. (2008) have assigned this style to the Taye Lake Phase of the southern Yukon, with an estimated date of about 1,500 BP. This specimen also shares characteristics with some Late Taltheilei (1,300 to 200 BP) points illustrated in Gordon (1996).

Corner-notched points. Two corner-notched points have been recovered from the Fort Hills Project area. A small quartzite point was identified at HiOv-83, a site located between two shallow sinkhole lakes at the western extremity of the McClelland Lake Wetland Complex (see figs. 7.5 and 7.6). This specimen was formed from a small flake, the bulb of which is readily apparent at the base on the ventral side. The right lateral edge is broken, but a shoulder suggesting corner notching is preserved on the opposite edge. Flaking is evenly spaced with random orientation. The diminutive size of this artifact sets it apart from the other points found in the Fort Hills Project area. Although these dimensions are consistent with Pre-Dorset (Arctic Small Tool) Tradition items recovered in the Barrenlands (Gordon 1996) and elsewhere in the Lower Athabasca region (see chapter 6 in this volume), the generic attributes of the HiOv-83 specimen preclude a definitive assignment to a typological group.

A single corner-notched point was recovered from HiOv-126, a site located on the west side of Keyhole Lake (see figs. 7.5 and 7.6 and table 7.2). HiOv-126: 210 is fashioned from black silicified siltstone and is missing its tip. The base is straight with bifacial flaking. No grinding is present on the base. The notches are deep, broad, and rounded. Basal edges are low and pointed, and the base is relatively wide compared to the total width of the point. Flaking is evenly spaced and random in orientation. Morphologically, this point is very similar to Pelican Lake types found on the northern Plains and dating to about 3,200 to 1,600 BP. It also compares well with one of the Beaver River Complex points illustrated by Reeves, Blakey, and Lobb in this volume: see plate 6.7: 6.

Microblade Technology

Lithic material related to microblade production was recovered from the Little Pond site (HiOv-89), located adjacent to a sinkhole lake on the northern periphery of the McClelland Lake Wetland Complex (fig 7.6). The assemblage includes wedge-shaped cores, burins, microblades, and ridge flakes (Woywitka and Younie 2008a; Younie 2008; Younie, Le Blanc, and Woywitka 2010). Younie (2008) and Younie, Le Blanc, and Woywitka (2010) have demonstrated relationships between the microblade reduction sequence at HiOv-89 and at Denali Complex sites of central Alaska and the Yukon (Clark 2001). Similar microblade technology was also observed at Bezya site (HhOv-73), situated in the Muskeg River valley approximately 20 kilometres south of Little Pond (Le Blanc and Ives 1986). BRS is not a significant component of the lithic assemblage at either HiOv-89 or the Bezya site: the majority of artifacts recovered from HiOv-89 were made on silicified mudstone (Younie 2008), while chert was the dominant raw material at Bezya (Le Blanc and Ives 1986). Other Denali Complex artifacts have been recovered from HhOv-449 (a microcore) and HhOv-468 (a microcore preform) near the Quarry of the Ancestors (Wickham and Graham 2009; Wickham 2010).

The Denali Complex spans a lengthy time period (ca. 11,000 to 2,000 BP; Clark 2001) and cannot be used as a temporally diagnostic construct on its own (see chapter 11 in this volume). Although a radiocarbon date of 3,990 ± 170 ^{14}C yr BP was obtained from a charcoal sample at the Bezya site, the association between the artifacts and charcoal sample is equivocal (Le Blanc and Ives 1986). The age of the Denali Complex material in the Lower Athabasca thus remains uncertain.

RADIOCARBON DATES

Four radiocarbon dates have been returned from sites in the Fort Hills area (fig 7.7 and table 7.3). The dates from HiOu-8, HiOv-46, and HiOv-70 were obtained from bulk samples of calcined bone, while the material from HiOv-126 was a bulk sample of burned bone. The samples were collected from fairly discrete concentrations at each site, none of which extended beyond 10 centimetres in depth or 50 by 50 centimetres horizontally. The calcined nature of the bones indicates that they were heated to temperatures higher than those produced by a forest fire and is most probably indicative of intentional burning in a campfire. Although most of the bones were too fragmentary to be classified beyond the Family level (Mammal), hare, beaver, and large ungulate bones were identified. These animals are known to have been procured in historic times, and all of these species are hunted or trapped by local residents to this day.

Reliability

The reliability of dates obtained from properly treated cremated bone samples has been well demonstrated (Lanting, Aertis-Bijma, and van der Plicht 2001; Cherkinsky 2009; van Strydonck, Boudin, and de Mulder 2009). Laboratory procedures described by van Strydonck, Boudin, and de Mulder (2009) and Cherkinsky (2009) remove secondary and external carbon, greatly reducing contamination effects. These procedures were followed by the lab that processed the Fort Hills samples (Beta Analytic). However, several taphonomic factors complicate the interpretation of the dates. Bitumen inclusions were not prevalent in the sediment at any of the sites, but it is possible that contamination from older carbon contained in these elements could have affected the samples, yielding old dates for very recent material. In addition, despite the relative spatial

Table 7.3   AMS dates from sites in the Fort Hills

Site	Lab reference	Material	Measured radiocarbon age	$^{13}C/^{12}C$ ratio	Conventional radiocarbon age	Calibrated 2 sigma range
HiOu-8	Beta-258074	Calcined bone carbonate	140 ± 40 ^{14}C yr BP	-25.6 o/oo	130 ± 40 ^{14}C yr BP	260 cal yr BP 220 cal yr BP 140 cal yr BP 30 cal yr BP 0 cal yr BP
HiOv-46	Beta-258073	Cremated bone carbonate	2,240 ± 40 ^{14}C yr BP	-23.1 o/oo	2,270 ± 40 ^{14}C yr BP	2,350 to 2,290 cal yr BP 2,270 to 2,160 cal yr BP
HiOv-70	Beta-258075	Cremated bone carbonate	1,680 ± 40 ^{14}C yr BP	-23.0 o/oo	1,710 ± 40 ^{14}C yr BP	1,710 to 1,530 cal yr BP
HiOv-126	Beta-258076	Burned bone organics	104.1 ± 0.6 pMC	-24.3 o/oo	103.9 ± 0.6 pMC	Modern

NOTE: All sites were excavated under Archaeological Research Permit 08-163.

constriction of the samples collected, there is no way to preclude mixing or contamination of the samples given the massive structure and high hydrological conductivity of the sand matrix prevalent at all of the sites. The lack of stratified sediments or cultural features also prevents definitive associations between the artifacts and bone material. These caveats must be considered when interpreting the following dates.

HiOv-46. A raw radiocarbon age of 2,240 ± 40 ^{14}C yr BP, which yielded a date range of 2,350 to 2,160 cal yr BP, was obtained from a concentration of calcined bone found on the edge of a lithic scatter at HiOv-46 (Block A; Woywitka et al. 2009). Most of these specimens were unidentifiable mammal bone, with the exception of a single large ungulate carpal, which was not calcined. A few pieces of fire-broken rock were also observed at the site, although not in association with the calcined bone. The lithic scatter consisted primarily of BRS debitage, with smaller proportions of quartzite, chert, and siltstone also present. The assemblage included evidence of core reduction, early-stage bifacial reduction, and late-stage tool refinement, with the latter two stages more prevalent than the first. Two end scrapers, two bifaces, and one retouched flake were also recovered. The recovery of calcined bone and the nature of the lithic assemblage make it apparent that animal processing and related flint knapping activities took place at the site. The radiocarbon date potentially places the occupation of the site in the Late Prehistoric period.

HiOv-70. A sample of calcined bone recovered from HiOv-70 yielded a raw radiocarbon date of 1,680 ± 40 ^{14}C yr BP, or 1,710 to 1,530 cal yr BP. The sample

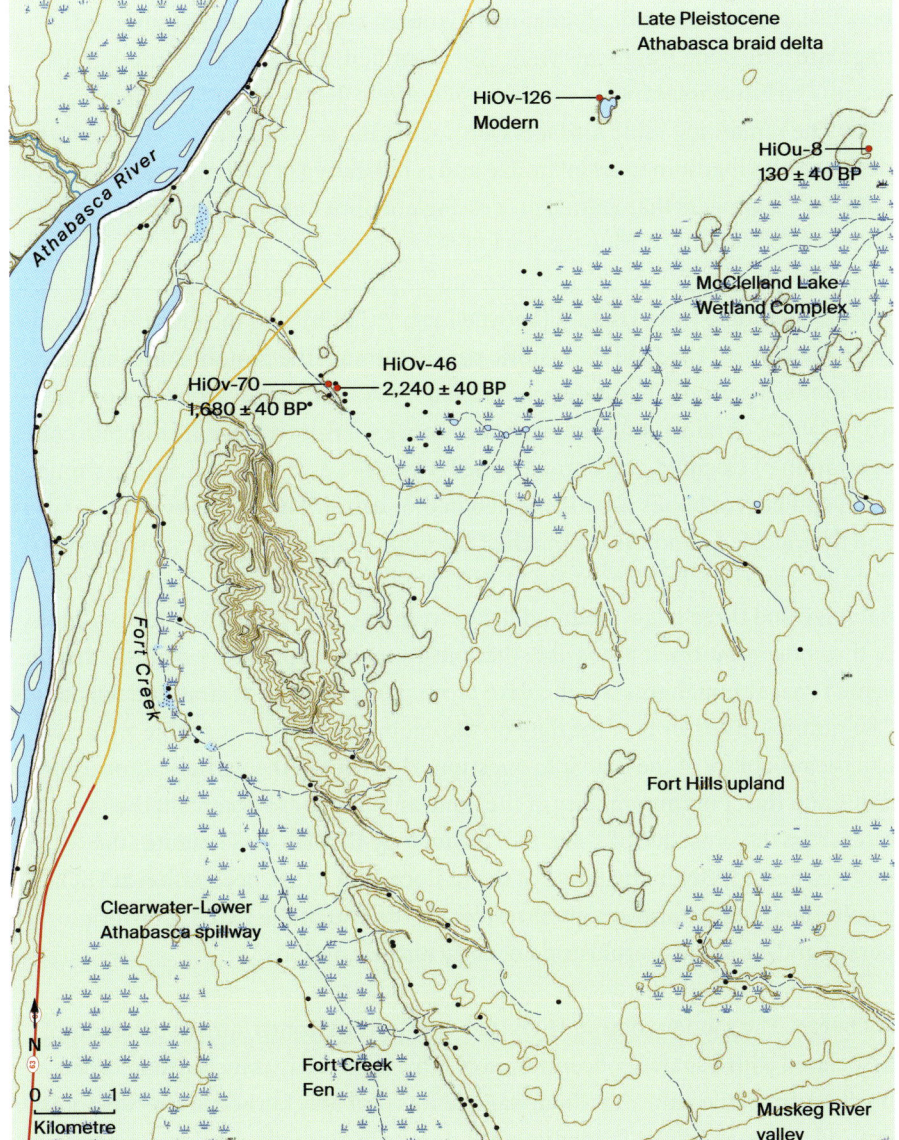

Figure 7.7. Sites that yielded radiocarbon dates

was recovered from a concentration of fire-broken rock, calcined bone, and lithic artifacts that may be representative of a disturbed hearth (Block B; Woywitka et al. 2009). Fire-broken rock was the most common element in the concentration (n = 793), followed by faunal material (n = 331) and lithic debitage (n = 18). The bone is largely unidentifiable calcined mammal bone, although two burned ungulate long bone fragments were also identified, along with one tooth fragment. The lithic assemblage consisted of entirely debitage, half of which was BRS. The remaining lithic artifacts were fashioned from quartzite. Most of this debitage was medium to small in size and is indicative of tool maintenance. As at HiOv-46, it appears that animal processing and flint knapping took place at HiOv-70, potentially during the Late Prehistoric period.

HiOu-8. The date from HiOu-8 (between 170 ^{14}C yr BP and the present) extends from the close of the Late Prehistoric period through the protohistoric, historic, and modern periods. The date was derived from a raw radiocarbon age of 130 ± 40 ^{14}C yr BP obtained on calcined bone recovered from a spatially discrete concentration of calcined and burned bone (Woywitka et al. 2009). Large mammal, hare, and beaver bones were identified in the scatter. This diversity of bone types and the calcined condition of much of the bone indicate that animal remains were processed by humans at HiOu-8. A small amount (n = 12) of quartzite lithic debitage and three edge-modified flakes were also recovered from the periphery of the concentration. It is possible that the lithics and bone were deposited at the same time, although this is difficult to prove given the lack of stratified deposits. The multiple intercepts of the date on the calibration curve (see table 7.3) also call the reliability of the date into question. However, if the earlier radiocarbon dates and the contemporaneity of the bone and lithics are accepted, then we can infer from the presence of stone tool technology that HiOu-8 predates the modern period. Given the range of the radiocarbon dates, occupation at HiOu-8 would thus have occurred sometime toward the end of the Late Prehistoric period and/or in the protohistoric or early historic periods.

HiOv-126. The modern date at HiOv-126 was derived from a burned bone sample associated with a sparse concentration of lithic material that included a corner-notched projectile point. The bone sample was probably burned in a recent forest fire and was worked into the sediment at the site by trampling and tree root growth.

Protein Residue

Two obsidian scrapers and a projectile point base recovered from HiOv-16 were submitted for protein residue analysis. Human protein was present on both scrapers (see Yost 2009). Given the extremely sharp edges of these artifacts, it is not surprising that someone may have nicked themselves at some point. The sediment control sample submitted with the scrapers yielded a negative reaction to human antiserum, indicating that contamination from the supporting matrix at the site was unlikely. However, human protein can be "transferred to artifacts easily through oils, sweat and blood" (Yost 2009, 3). The positive reaction from the tools themselves was weak, and it is very possible that the residue originated from the collector or analyst rather than the artifact manufacturer (Yost 2009).

The projectile point base yielded a positive reaction to turkey antiserum, indicating that "a member of the Phasianidae (wild turkey, grouse, pheasant, partridge, ptarmigan) or Anatidae (goose, duck) family was killed or processed using this tool" (Yost 2009, 3). The sediment control yielded negative reactions for these antisera, suggesting that soil contamination is unlikely. Ptarmigan are very common in the area, and ducks were observed nesting near shore on the sinkhole lake directly adjacent to HiOv-16 during the summer of 2008. Loons were also resident at the lake that year. It is possible, then, that the stemmed projectile point was used as a spear head to acquire migratory birds from the lake or ptarmigan from the surrounding area.

Several artifacts from other sites were submitted for protein residue analysis (Woywitka et al. 2009). These results are not reported here because no sediment controls were submitted with those samples, making it difficult to rule out soil contamination as the source of the protein residue.

DISCUSSION

Chronology and Palaeoenvironment

In view of the limited number of radiometric dates and the absence of stratified sites, it is difficult to make definitive statements about chronology in the oil sands region. The typological classification proposed by Saxberg and Reeves (2003) and by Reeves, Blakey, and Lobb in chapter 6 of this volume provides a framework for describing variations in projectile point morphology. However, the very specific chronological framework presented in these analyses is based entirely on typological comparisons and has yet to be confirmed by radiometric dates or firm stratigraphic relationships. Typological comparisons presented below should therefore be viewed as postulations that await further testing in future sedimentological, archaeological, and palaeoenvironmental research.

Terminal Pleistocene and Early Holocene (ca. 9,600 to 7,500 BP). Limiting dates for occupation of the Fort Hills Project area can be inferred from the deglacial chronology of the region. Fisher et al. (2009) and Fisher and Lowell (chapter 2 in this volume) have put forward limiting ages of 9,850 to 9,660 ^{14}C yr BP for the

Fort Hills and Firebag moraines, suggesting that nearly the entire Fort Hills lease would have been covered by ice or inundated by proglacial lakes until this time. Only the top of the Fort Hills upland may have been ice free. Murton et al. (2010) present an earlier timeline for deglaciation of the area, proposing dates older than 10,000 ^{14}C yr BP. In either scenario, shortly after the retreat of glacial ice, the catastrophic Agassiz flood passed through the area. A very high-energy environment would have prevailed at this time, with high water flows and the rapid deposition of thick deltaic deposits north of the Fort Hills. Although people may have been present in the general area, the environment directly adjacent to the Late Pleistocene Athabasca braid delta would have been extremely dynamic, and artifacts or other residue of human presence would have been washed away or deeply buried. It is therefore expected that most evidence of repeated, consistent use of the area postdates the abandonment of the Clearwater–Lower Athabasca spillway.

Following the recession of floodwaters, the Athabasca lowland landscape was characterized by open, birch-dominated forests and a warmer and drier climate than we have today (Bouchet and Beaudoin, chapter 4 in this volume). The Athabasca River would have been restricted to its current valley (Rhine and Smith 1988). It is unknown how deeply incised the channel would have been at this time, but it was probably higher than its present levels (Rhine and Smith 1988). Wetlands like the McClelland Lake Wetland Complex, Fort Creek Fen, and those on the Fort Hills upland would have been in their earliest stages of formation near the end of this period, likely with margins much farther "receded" than is observed at present. To date, there is no geological or palaeoenvironmental evidence of remnant lakes from the recession of Glacial Lake McConnell or the Agassiz flood waters in the area. Given the open vegetation and dry climate, pockets of active dunes may have persisted in the Late Pleistocene Athabasca braid delta area, although this, too, remains speculative.

Evidence of human occupation in the Fort Hills Project area during this period consists of the stemmed Manitoba projectile point recovered from HiOv-16 and the lanceolate Northern Plano point recovered from HiOv-92. The graver spur evident on one of the obsidian scrapers from HiOv-16 also has an affinity with scrapers dating to this period that were recovered in association with Cody Complex material at the Horner site in Wyoming (Frison 1987). Given that there are no hydrological or other subsistence-related features in the vicinity of HiOv-16, we can assume that the sinkhole lake next to the site drew people to the location. If the Manitoba point date range is accepted, this suggests that the lake and a somewhat stable vegetation cover were established at least 8,000 years ago. Stabilized dunes near the Peace River have been dated to the terminal

Pleistocene or early Holocene (Wolfe et al. 2007), and a similar age has been forwarded for dune fields in northern Saskatchewan (David 1981). The location of the dunes on the Late Pleistocene Athabasca braid delta suggests that they may also date to this time period, although Holocene-age dune fields also occur in northern Alberta (Halsey, Catto, and Rutter 1990), and active dune fields are present north of the areas near Richardson Lake and Lake Athabasca.

Middle Holocene (7,500 to 2,000 BP). In the Lower Athabasca area, the middle Holocene period is characterized by general cooling and an increase in precipitation (Bouchet and Beaudoin, chapter 4 in this volume). Coniferous trees, including pine, become more prevalent in forests, and wetlands expand throughout this period, reaching their present extents by roughly 3,000 to 4,000 years ago (Halsey, Catto, and Rutter 1990). Essentially modern boreal forest vegetation patterns are established during this time (Vance 1986).

The variety of notched points recovered from the Fort Hills Project excavations is representative of a well-established human presence in the Fort Hills area during the middle and late Holocene. The flexible and somewhat generalized side-notched morphology that persists in the boreal region throughout much of the middle Holocene often prevents the assignment of items to specific complexes. This leaves the chronological resolution of most occupations quite coarse, spanning over 5,000 years. However, the potential Shield Archaic match from HiOv-80 may represent an occupation from the early part of this period.

The McClelland Lake Wetland Complex would have reached its current extent near the end of this period. Assuming that it was the wealth of resources afforded by this large wetland that attracted people to the area, we can infer that most of the sites situated along the current margins of the McClelland Lake Wetland Complex represent middle to late Holocene occupations. The side-notched points recovered from HiOv-80 lend some support to this idea. Although people probably used the wetland before this period, those sites may be buried underneath thick peat accumulations.

Late Holocene (2,000 BP to present). During the late Holocene Period, the vegetation and climate of the Lower Athabasca region would have been more or less the same as they are now, although forests may have been more open, given the lack of forest fire prevention. Both radiometric dating evidence and projectile point typology matches indicate that people used the Fort Hills area extensively in the late Holocene. The three non-modern AMS dates obtained from the Fort Hills area fall within the past 2,350 years, and although no projectile points were recovered at sites that yielded radiocarbon ages, many of the typological matches from the Fort Hills Project area (Middle Prehistoric; Middle and Late Taltheilei, Taye Lake Phase) fit well with this time frame. The skew to more recent radiocarbon dates could be a function of preservation, with older organic material less likely to survive in the area because of the acidic forest soils. Again, with regard to these dates, the confounding factors of sediment mixing and bitumen contamination must be kept in mind.

Subsistence and Land Use

Overall, the site density in the Fort Hills is considerably lower, and the debitage less abundant, of smaller size, and more representative of later reduction stages, than at sites in the area of the Muskeg River, the Quarry of the Ancestors, and the Cree Burn Lake complex. The overall proportion of expedient and formed tools in the excavated Fort Hills Project assemblages is lower than that observed at the Quarry of the Ancestors (0.3% vs. 0.8%; see Saxberg and Robertson, chapter 10 in this volume). However, formed tools comprise 51.5% the tools in the Fort Hills Project sample, compared to 41.5% at the Quarry of the Ancestors. Marked differences also occur in the proportion of what Saxberg and Robertson term "manufacturing" tools (cores, hammerstones, tried cobbles, anvils). These items account for less than 1% of the Fort Hills Project assemblages, while, as Saxberg and Robertson note, they represent 36.1% of the Quarry of the Ancestors assemblages.

As is clear from the above, the lithic assemblages recovered from Fort Hills Project excavations indicate that very little primary lithic raw material procurement occurred at these sites. Instead, tool manufacture and maintenance were the dominant flint-knapping activities. These types of debitage assemblages, combined with the presence of projectile points and calcined bone at several sites, are consistent with short-term occupations related to hunting and gathering subsistence activities. This is expected, considering that most sites in the Fort Hills area are far removed from central stone procurement areas or other locales that would be more attractive for long-term habitation, such as major lakes or rivers.

The paucity of organic remains precludes specific conclusions regarding the types of resources harvested. Currently, among other animals, moose, beaver, deer, marten, bear, waterfowl, and caribou inhabit the area, but bison and potentially other large ungulates would have been more common in the prehistoric period. Although plant resources would have varied with changes in the vegetation over time (see Bouchet and Beaudoin, chapter 4 in this volume), berries, bark, wood, roots, and a variety of other items would have been plentiful in the area throughout most of the Holocene.

Despite poor organic preservation, protein residue analysis has provided some insight into specific species used by ancient inhabitants of the Fort Hills area. Assuming it can be trusted, the positive reaction to turkey antiserum provided by the stemmed point at HiOv-16 may provide information regarding the seasonality of the occupation of the site. Duck and goose are not present in the area during winter, and waterfowl are most frequently hunted in the late summer and autumn. Given the suspected age of the point, it is possible that the stemmed point was left at the site one autumn about 8,000 years ago.

Site location patterns help us to identify the types of environments people used in the past and are a good indicator of how people moved through the landscape. In general, site distribution in the Fort Hills is consistent with broad boreal forest patterns, with sites typically located on raised landforms near hydrological resources such as rivers, streams, and lakes or other areas of diverse habitat, notably wetlands. The margins of the McClelland Lake Wetland Complex, Fort Creek Fen, various sinkhole lakes, and tributaries of the Athabasca River all contain multiple sites (see fig 7.2), while site density in the interior of the Late Pleistocene Athabasca braid delta and Fort Hills upland is quite low. In the interior of the delta, the dry landscape and the extreme homogeneity of the jack pine forest cover account for the lack of sites, while the close proximity of the Fort Hills upland to more productive areas such as Fort Creek Fen and the Muskeg River valley probably explains why only limited use was made of that area.

Although the ages of some sites along the margins of the Fort Hills have been tied to receding shorelines of the Clearwater–Lower Athabasca spillway (Saxberg and Reeves 2003; Somer 2005), no published shoreline recession sequences are available, save for the elevation ranges noted in Saxberg and Reeves (2003). Some potential remnant beach ridges do occur along the western and southern flanks of the Fort Hills (fig 7.8), but these features are heavily dissected and obscured by channels that drain the Fort Hills upland. The fact that sites in this area tend to occur near the edges of these deeply incised channels (fig 7.8) raises the possibility that the sites are related to game travel corridors leading from the uplands to Fort Creek Fen and the Muskeg River valley along the drainage margins. These areas would have been attractive locations throughout the post-flood Holocene, and without chronological control on the sites, there is no reason to assume an exclusive relationship between the location of these sites and water bodies related to the recession of the Agassiz flood waters.

The cluster of sites along Susan Creek suggests that this channel was a main travel route between the Athabasca River valley and the neighbouring uplands. McClelland Lake, Fort Creek Fen, and the Fort Hills upland are all accessible from the nexus of the upper portion of Susan Creek and the southern tip of the

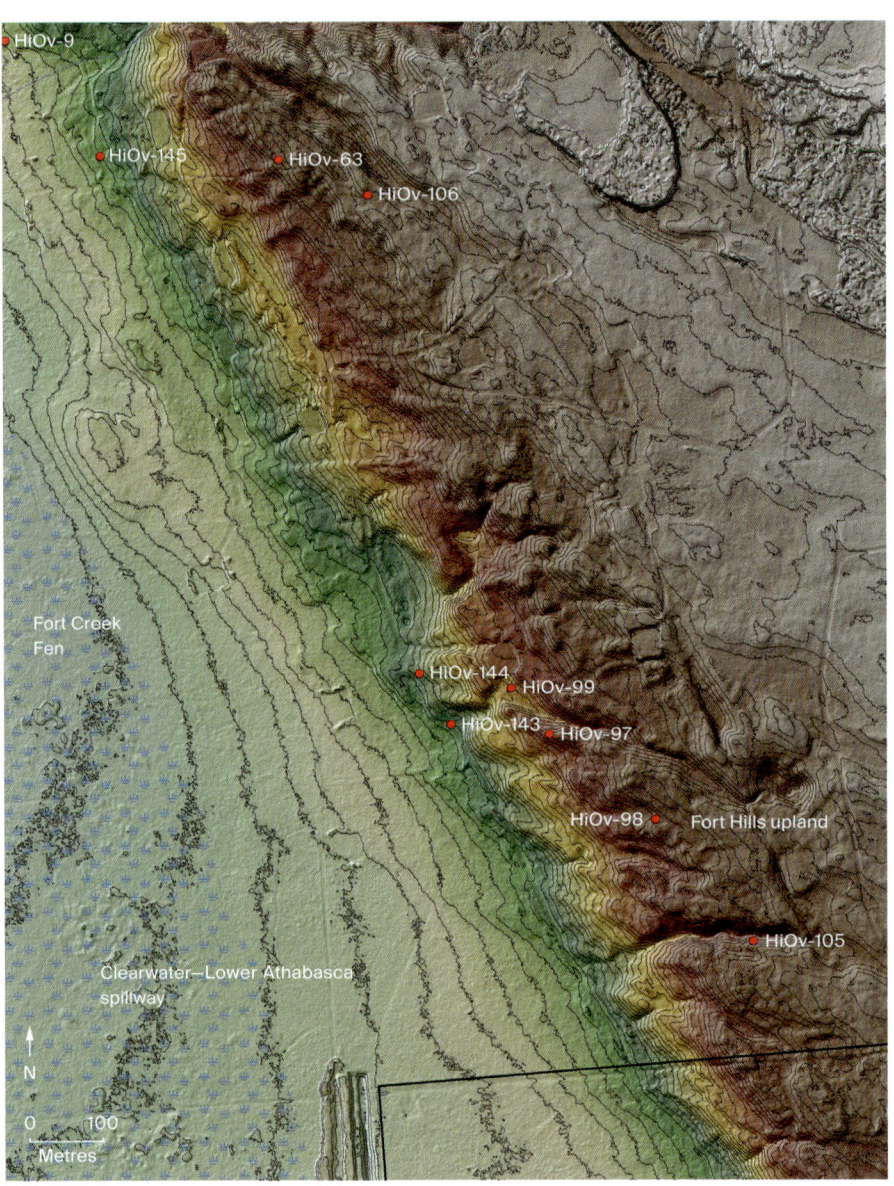

Figure 7.8. Fort Hills escarpment, with contour intervals of one metre

McClelland Lake Wetland Complex. This area could have served as an important hinterland "intersection," linking the Athabasca and McClelland Lake drainages to the Muskeg River drainage via Fort Creek Fen. The route would have become passable sometime after the stabilization of the Late Pleistocene Athabasca braid delta sand dunes and the formation of Susan Creek. Pollen records from the area suggest that some form of landscape stability had been achieved 7,500 years ago, and evidence indicates that the McClelland Lake Wetland Complex began accumulating peat approximately 7,000 to 8,000 years ago (True North Energy 2001). The archaeologically diagnostic materials—the Late Taltheilei point from HiOv-44, and the side-notched points from HiOv-80, which appear to be Middle Prehistoric—and the radiometric dates from HiOv-46 and HiOv-70 (see table 7.3) suggest dominance of middle to late Holocene occupations. This roughly coincides with the firm establishment of the McClelland Lake Wetland Complex. However, the travel route could have been used throughout most of the post–Agassiz flood Holocene period to access other focal areas.

Sinkhole lakes, such as Keyhole Lake, were an important resource for past inhabitants of the Fort Hills area. These features are interesting because they can be far removed from other focal areas such as the McClelland Lake Wetland Complex, the Fort Creek Fen, and the Athabasca River. There are no fish in these lakes, so it is likely that waterfowl, moose, berries, and perhaps other plant resources were the main attractions there. The presence of calcined bone and fire-broken rock at campsites such as HiOv-16, HiOv-89, HiOv-104 and HiOv-123 indicates that food was prepared and suggests that stays at these sites extended beyond quick tool-sharpening episodes. Assemblages from sinkhole lake sites also tend to exhibit diverse lithic raw materials. Although the reason for this is elusive, it could be that people conducted activities at these sites that required tools made of sturdy quartzites and cherts and/or that people stayed long enough that sharpening of these more durable tools became necessary.

It is also possible that different raw materials were deposited at the sites during separate occupation episodes, which raises the possibility that these sites were well known and that knowledge of their location was passed on through time. The fact that the obsidian scrapers recovered from HiOv-16 still appear to be functional may also provide some support for this suggestion. The abandonment of such high-quality material at this early stage of reduction suggests that these items had either been lost or else had been cached at the site. The latter possibility would provide further evidence that sinkhole lakes were target destinations in a seasonal round, rather than convenient stopovers on trips to other foraging areas.

REGIONAL ARCHAEOLOGICAL CONTEXT

Lower Athabasca Perspective

Much of the study area considered in this chapter is located within or north of the Fort Hills, on the northern periphery of the Quarry of the Ancestors–Cree

Burn Lake "heartland." With the possible exception of HiOv-97, HiOv-98, and one area of HiOv-49, sites in the Fort Hills Project area are representative of relatively short-term occupations related to resource gathering, rather than to raw material procurement and primary reduction. The predominance of BRS at most of these sites is indicative of strong ties to the Quarry of the Ancestors area. As Saxberg and Robertson demonstrate in chapter 10, the creation of bifacial preforms and other initial tool manufacture were key activities at the quarry. Given that these activities do not, for the most part, appear to characterize sites in the Fort Hills Project area, it is likely that many of the BRS tools recovered from the Fort Hills sites began as preforms or as finished tools at or near the Quarry of the Ancestors–Cree Burn Lake area. As resource-gathering forays into the Fort Hills Project region progressed, these tools and preforms would require maintenance or repurposing, and some would eventually get discarded in areas at some distance from the quarry. Saxberg and Robertson note that the low proportion of formed tools at the Quarry of the Ancestors itself may indicate that people were confident in the supply at the quarry and perceived little risk in abandoning used tools at satellite sites. The higher proportion of formed tools recovered from the Fort Hills Project assemblages supports this proposition. However, it is also possible that the tendency of primary reduction activities to produce large quantities of debitage masks the role of formed tools at the Quarry of the Ancestors. Further and more detailed analysis of the assemblages from the separate activity areas at the Quarry of the Ancestors will help to clarify this matter.

A consideration of raw material types recovered in the Fort Hills Project excavations provides some insight into prehistoric trade and/or travel in the Lower Athabasca region. The occurrence of Northern quartzite and salt and pepper quartzite suggests that people either visited or traded with people on the west side of the Athabasca River and the Birch Mountains, where these material types are more common in archaeological assemblages (Ives, chapter 8 in this volume). In addition, although BRS is the predominant lithic material at sites in the Fort Hills, these sites appear to display a greater diversity in raw material selection than those in the Quarry of the Ancestors area. Of the excavated Fort Hills Project sites that produced more than a hundred artifacts (see table 7.1), 30% yielded a lithic assemblage that contained over 5% non-BRS materials, and four of these sites had over 20% non-BRS materials. Two sites that contain significant proportions of non-BRS material, HiOv-16 and HiOv-93, have very spatially discrete BRS-dominated and non-BRS-dominated activity areas. As noted above, the relatively good condition of the obsidian scrapers recovered at HiOv-16 suggests that the site may have been occupied repeatedly, in which case it is possible that each of these activity areas represents a separate occupation of the site. Then again, it may also be that activities each of which required the use of different raw materials were conducted contemporaneously during a single occupation. Unfortunately, the lack of stratified sediments at both sites makes it difficult to ascertain the reason for the spatial separation of raw material types.

The diversity of raw material types may also be explained by the comparative remoteness of the area from BRS sources. If a group that set out from the Quarry of the Ancestors area were waylaid during one of the forays into the Fort Hills by weather, a journey to the Birch Mountains, or a productive stay at a sinkhole lake, the BRS with which they had provisioned themselves might have begun to run out, necessitating an increased use of other materials. This would result in the mixed BRS–non-BRS assemblages observed at sites like HiOv-64, HiOv-80, HiOv-104, and HiOv-126. In the following chapter, Ives describes a similar pattern in relation to travel between the Athabasca lowlands and the Birch Mountains. However, the drop in BRS proportions is more dramatic in the Birch Mountains, perhaps because other raw materials were readily available in the stream and river valleys that have incised the pre-glacial gravels of these mountains (Darryl Bereziuk, pers. comm., 2010; Ives, chapter 8 in this volume). Such raw material sources are much less common in the Fort Hills area, so a heavier reliance on BRS is to be expected. However, given that there are currently no comprehensive raw material statistics from the Quarry of the Ancestors or the Birch Mountains with which to compare the Fort Hills findings, this apparent increase in the variability of raw material has yet to be confirmed. Assuming it exists, its causes must likewise remain a matter of speculation.

Available chronological information, although tenuous, indicates that people inhabited the Fort Hills region from soon after the recession of glacial ice and floodwaters right up to the present. The datable occupations encountered during the Fort Hills Project excavations are mostly middle to late Holocene in age, a trait shared with those in other parts of the Lower Athabasca region. A strong early to middle Holocene presence in the region has been inferred from the abundant notched points recovered from excavations throughout the region. Often these points are assigned to the earlier portions of this period, about 9,000 to 8,000 years ago (Saxberg and Reeves 2003; Reeves, Blakey, and Lobb, chapter 6 in this volume; Saxberg and Robertson, chapter 10 in this volume). This florescence has been tied to arid and open conditions that increased accessibility to the Quarry of the Ancestors during the early and middle Holocene (see chapter 10 in this volume), to the enhanced biodiversity of the immediately postglacial landscape (Clarke and Ronaghan 2000), and to the use of receding or remnant water bodies from the Agassiz flood (Saxberg and Reeves 2003; Reeves, Blakey, and Lobb, chapter 6 in this volume).

Although all of these factors probably played some role in attracting people to the immediately postglacial Lower Athabasca region, some interpretations of the prehistoric record in the area may place an overemphasis on this early Holocene period (see, for example, Shortt, Saxberg, and Reeves 1998; Saxberg and Reeves 2003). Recent resource management studies (such as Roskowski and Blower 2009; Woywitka et al. 2009) and research projects suggest a well-established human presence in the region throughout most of the post–Aggasiz flood Holocene, including the late Holocene. Limited as they are, the diagnostic artifacts and radiocarbon dates recovered from the Fort Hills excavations appear to indicate that human presence in the area was relatively consistent throughout the middle and late Holocene. As noted above, conclusions regarding the timing of occupation in the region must remain cursory until diagnostic items are recovered from stratified contexts and their antiquity is corroborated by radiometric dating.

Subarctic Perspective

There are several indications that the prehistoric inhabitants of the Fort Hills area travelled to neighbouring regions or had contact with groups from these regions. The source of the obsidian used in the three scrapers recovered from HiOv-16 was traced to Mount Edziza in northwestern British Columbia, nearly 1,500 kilometres west of the Lower Athabasca region (Woywitka et al. 2009; Hughes 2009). This highly workable raw material was extensively quarried at various locations around Mount Edziza (Fladmark 1985), and artifacts fashioned from this volcanic glass have been recovered from archaeological sites in Haida Gwaii and on the British Columbia mainland, in Alaska and the Yukon, and elsewhere in Alberta (Carlson 1994). Obsidian has been used as an item of trade has been occurring for at least 9,500 years (Carlson 1994). As noted above, the fact that the scrapers are still in functional condition suggests that they could have been cached at the site, a possible indication that this material was considered valuable to its possessors. Whether this value was tied to the superior functionality of the stone or its use as a trade commodity is a question worthy of future study.

The microblade technology present at the Little Pond site is also indicative of some form of contact with groups to the northwest. Younie et al. (2010) and Younie, Le Blanc, and Woywitka (chapter 11 in this volume) suggest that, in all likelihood, the presence of Denali-style reduction sequences at Little Pond, Bezya, HhOv-449, and HhOv-468 is a product of cultural diffusion. These same authors also recognize the possibility that groups from the northwest penetrated the Lower Athabasca region on southerly territorial rounds. In this scenario, the dominance of non-BRS materials at Little Pond and Bezya may reflect occupation by people who were unfamiliar with BRS or who were denied access to the quarry by resident groups. Because the Denali Complex first appears in the terminal Pleistocene in Alaska, it is also possible that Little Pond and Bezya represent early colonizers from the north who had yet to find or realize the potential of the BRS sources in the Lower Athabasca. However, given the persistence of the Denali Complex throughout most of the Holocene, as well as the relatively late date obtained from Bezya (Le Blanc and Ives 1986)—tenuous as it is—and the general lack of reliable chronologies for the Lower Athabasca region, this hypothesis seems unlikely.

Influence from the Northwest Territories and Nunavut area is evinced by the Taltheilei Tradition point recovered from HiOv-44 and other potential Taltheilei matches at HiOv-59 and HiOv-80. This tradition was defined through extensive survey and excavation in the Barrenlands by Noble (1971) and Gordon (1975, 1996). The tradition is thought to represent the adoption, over the past 2,000 years or so, of a lifestyle based on following the seasonal migrations of caribou herds (Gordon 1975, 1996) and is considered ancestral to historic and present-day Athapaskan populations in north-central Canada (Gordon 1996). Several points recovered from the Lower Athabasca region have been matched to the Taltheilei Tradition (see, for example, Tischer 2004; Woywitka and Younie 2008a; Reeves, Blakey, and Lobb, chapter 6 in this volume). In the past, the winter range of Barrenlands caribou may have extended as far south as the Lower Athabasca, a circumstance that could explain the occurrence of Taltheilei points in the region. Alternatively, these sites may be representative of the cultural diffusion of point types into the Lower Athabasca. Projectile points assigned to the Taltheilei Tradition show extreme variability, with nearly every common point type present (lanceolate, stemmed, side-notched, corner-notched; see Gordon 1996). Some of these point types are similar to forms assigned to earlier time periods, another confounding trait that adds to the already unsteady chronological context of the Lower Athabasca.

The black siltstone Pelican Lake–like point recovered from HiOv-126 may represent some contact with or the presence of northern Plains groups in the Fort Hills Project area. Other Plains-style points have been observed in the Lower Athabasca region, including the Cody Complex materials cited by Saxberg and Reeves (2003) and by Reeves, Blakey, and Lobb (chapter 6 in this volume) and the Oxbow material from the Birch Mountains (see Ives, chapter 8 in this volume). Although it might be expected that earlier Plains styles would appear in the region shortly after deglaciation, as part of the initial postglacial occupation of the land, the presence of this middle-to-late-period Plains point style is harder to account for. Were Plains groups active in the area around 3,000 years ago? Given the essentially modern vegetation cover that would have been in place at the time, it seems unlikely that a band of Plains hunters would have penetrated so deeply into the boreal forest. Given its non-BRS source material, it is possible the item was traded into the area. It is of course equally possible that the point represents a convergence of form or a diffusion of style over disparate areas.

CONCLUSION

The Fort Hills Project area has probably been inhabited since the retreat of the last glacial ice and recession of the Agassiz floodwaters. Sites in the area are less abundant and less productive than sites in the Muskeg River valley and represent seasonal resource-gathering forays from the Athabasca and Muskeg river valleys to sinkhole lakes, Fort Creek Fen, McClelland Lake, the McClelland Lake Wetland Complex, and the Fort Hills upland. The largely BRS-dominated archaeological record shows strong ties to the Quarry of the Ancestors or other local sources. However, an apparent increase in raw material variability indicates that people visited or were in contact with people on the west side of the Athabasca and the Birch Mountains. Evidence of contact with groups from or visitation of areas outside the Lower Athabasca region itself includes the presence of Mount Edziza obsidian, a Denali Complex microblade assemblage, and a Taltheilei projectile point. These artifacts suggest that the inhabitants of the Lower Athabasca were most strongly tied to northern groups in present-day Yukon, Northwest Territories, Nunavut, and British Columbia, although a potential Pelican Lake point recovered in the area hints at some northern Plains influence. Chronological information is limited by the lack of precision of lithic typological analysis and a lack of demonstrably unmixed archaeological deposits. However, the available data suggest that human habitation in the area was relatively consistent, with the middle to late Holocene best represented in the archaeological record.

ACKNOWLEDGEMENTS

I would like to thank Wendy Unfreed of the Archaeological Survey of Alberta for introducing me to the Fort Hills Project and providing me with great advice, feedback, and insight—a practice she continues to this day. To the field assistants who moved all that sand and rock, I extend my deepest gratitude. Karen

Zbeetnoff, Barb Neal, Leah Mann, Mike Ross, and Angela Younie, in particular, bore much responsibility and made the completion of these projects a joy. Jennifer Tischer, Laura Roskowski, and Alison Landals, of FMA Heritage Inc., completed much of the permit reporting for the 2008 work and provided several important insights into the nature of the archaeological record. FMA Heritage also created the original artifact photographs. Trevor Peck, of the Archaeological Survey, and Jack Ives, of the University of Alberta, both lent their expertise on projectile point typology. Fort Hills Energy Ltd. funded much of the work, and Sheila Chernys, Ron Voogel, and Tom Wiebe provided excellent support throughout our field programs.

NOTES

1. Owing to the unique nature of microblade assemblages, these calculations do not include the Little Pond site (HiOv-89).
2. Reeves, Blakey, and Lobb (this volume) do not, however, associate this point with the Hell Gap style.

REFERENCES

Blaikie-Birkigt, Kurtis A.
2006   *Historical Resource Impact Assessment of Synenco Energy Inc. 10-18-97-10-W4M Remote Sump and Access Road: Final Report (ASA Permit 05-609)*. Copy on file, Archaeological Survey, Historic Resources Management Branch, Alberta Culture, Edmonton.

Carlson, Roy L.
1994   Trade and Exchange in Prehistoric British Columbia. In *Prehistoric Exchange Systems in North America*, edited by Timothy G. Baugh and Jonathon E. Ericson, pp. 307–361. Plenum Press, New York.

Cherkinsky, Alexander
2009   Can We Get Good Radiocarbon Age from "Bad Bone"? Determining the Reliability of Radiocarbon Age from Bioapatite. *Radiocarbon* 51(2): 553–568.

Clark, Donald W.
2001   Microblade-Culture Systematics in the Far Interior Northwest. *Arctic Anthropology* 38(2): 64–80.

Clarke, Grant M., and Brian M. Ronaghan
2000   *Historical Resources Impact Mitigation, Muskeg River Mine Project: Final Report (ASA Permit 99-073)*. Copy on file, Archaeological Survey, Historic Resources Management Branch, Alberta Culture, Edmonton.

David, Peter P.
1981   Stabilized Dune Ridges in Northern Saskatchewan. *Canadian Journal of Earth Sciences* 18: 286–310.

Donahue, Paul F.

1976   Alberta North, Project 75-8. In *Archaeology in Alberta 1975,* edited by J. Michael Quigg and W. J. Byrne, pp. 42–50. Archaeological Survey of Alberta Occasional Paper no. 1. Historic Resources Management Branch, Alberta Culture, Edmonton.

Fisher, Timothy G., Nickolas Waterson, Thomas V. Lowell, and Irka Hajdas

2009   Deglaciation Ages and Meltwater Routing in the Fort McMurray Region, Northeastern Alberta and Northwestern Saskatchewan, Canada. *Quaternary Science Reviews* 28: 1608–1624.

Fladmark, Knut R.

1985   *Glass and Ice: The Archaeology of Mt. Edziza.* Archaeology Press, Department of Archaeology Publication no. 14. Simon Fraser University, Burnaby, British Columbia.

Frison, George C.

1987   The Tool Assemblage, Unfinished Bifaces, and Stone Flaking Material Sources for the Horner Site. In *The Horner Site: The Type Site of the Cody Cultural Complex,* edited by George C. Frison and Lawrence C. Todd, pp. 233–278. Academic Press, Orlando.

Gordon, Bryan H. C. A.

1975   *Of Men and Herds in Barrenland Prehistory.* Mercury Series no. 28. Archaeological Survey of Canada, National Museum of Man, Ottawa.

1996   *People of Sunlight, People of Starlight: Barrenland Archaeology in the Northwest Territories of Canada.* Mercury Series no. 154. Archaeological Survey of Canada, Canadian Museum of Civilization, Gatineau, Québec.

Gorham, Les

1997   *Historical Resources Impact Mitigation, Solve-Ex Corporation Oil Sands Co-Production Experimental Project and Highway 63 Extension Site, HiOv-118 (ASA Permit 96-088).* Copy on file, Archaeological Survey, Historic Resources Management Branch, Alberta Culture, Edmonton.

Graham, James W., Shawn Morton, and Robin Woywitka

2008   *Fort Hills Oil Sands Project, Post-impact Assessment of Sites HiOv-49 and HiOv-52: Final Report (ASA Permit 08-030).* Copy on file, Archaeological Survey, Historic Resources Management Branch, Alberta Culture, Edmonton.

Gryba, Eugene M.

1980   *Highway Archaeological Salvage Projects in Alberta: Final Report (ASA Permit 79-066).* Copy on file, Archaeological Survey, Historic Resources Management Branch, Alberta Culture, Edmonton.

Gryba, Eugene M., Wendy J. Unfreed, and A. Kate Peach

2001   *Supplemental Historical Resources Impact Assessment, True North Energy L.P. Fort Hills Oil Sands Project (Eastern Half of Lease 52, Twp 97, Rges 10–11, W4M): Final Report (ASA Permit 01-228).* Copy on file, Archaeological Survey, Historic Resources Management Branch, Alberta Culture, Edmonton.

Gryba, Eugene M., and Wendy J. Unfreed

2002   *Historical Resources Impact Mitigation and Interim Avoidance Measures, True North Energy L.P. Fort Hills Oil Sands Project (Selected Sites in Leases 5 and 52, Twps 96–97, Rge 10, W4M): Final Report (ASA Permit 01-315).* Copy on file, Archaeological Survey, Historic Resources Management Branch, Alberta Culture, Edmonton.

Halsey, Linda A., Norman R. Catto, and Nathaniel W. Rutter

1990   Sedimentology and Development of Parabolic Dunes, Grande Prairie Dune Field, Alberta. *Canadian Journal of Earth Sciences* 27: 1762–1772.

Hare, P. Gregory, Thomas J. Hammer, and Ruth M. Gotthardt

2008   The Yukon Projectile Point Data Base. In *Projectile Point Sequences in Northwestern North America,* edited by Roy L. Carlson and Martin P. R. Magne, pp. 321–332. Department of

Archaeology Publication no. 35. Archaeology Press, Simon Fraser University, Burnaby, British Columbia.

Hughes, Richard

2009   Geochemical Research Laboratory Letter Report 2008-119. In Robin J. Woywitka, Angela M. Younie, Jennifer C. Tischer, and Laura Roskowski, *Historical Resources Impact Mitigation Fort Hills Energy Corporation, Fort Hills Oil Sands Project, 2008 Mitigation Studies: Final Report (ASA Permit 08-163),* Appendix III. Copy on file, Archaeological Survey, Historic Resources Management Branch, Alberta Culture, Edmonton.

Husted, Wilfred M.

1969   *Bighorn Canyon Archaeology.* Publications in Salvage Archaeology no. 12. River Basin Surveys. Smithsonian Institution, Lincoln, Nebraska.

Ives, John W.

1993   The Ten Thousand Years Before the Fur Trade in Northeastern Alberta. In *The Uncovered Past: Roots of Northern Alberta Societies,* edited by Patricia A. McCormack and R. Geoffrey Ironside, pp. 5–31. Circumpolar Research Series no. 3. Canadian Circumpolar Institute, University of Alberta, Edmonton.

Kjorlien, Yvonne P., and Robin J. Woywitka

2008   *Historical Resources Impact Assessment, ATCO Electric Ltd., ATCO Electric Fort Hills 240 kV Transmission Line, 9L32 Extension: Final Report (ASA Permit 08-219).* Copy on file, Archaeological Survey, Historic Resources Management Branch, Alberta Culture, Edmonton.

Lanting, J. N., A. T. Aerts-Bijma, and J. van der Plicht

2001   Dating of Cremated Bones. *Radiocarbon* 43(2A): 249–254.

Le Blanc, Raymond J., and John W. Ives

1986   The Bezya Site: A Wedge-Shaped Core Assemblage from Northeastern Alberta. *Canadian Journal of Archaeology* 10: 59–98.

MacNeish, Richard S.

1964   *Investigations in Southwest Yukon: Archaeological Excavations, Comparisons, and Speculations.* Papers of the Robert S. Peabody Foundation for Archaeology, vol. 6, no 2. Phillips Academy, Andover, Massachusetts.

McCullough, Edward J.

1980   *Historical Resources Inventory and Assessment, NOVA-PetroCanada Oil Sands Joint Venture, Core-Hole Drilling Program, B.S.L. Nos. 52, 20, 78, 88, 89, and 5 (ASA Permit 80-133).* 5 vols. Copy on file, Archaeological Survey, Historic Resources Management Branch, Alberta Culture, Edmonton.

McPherson, R. A., and C. P. Kathol

1977   *Surficial Geology of Potential Mining Areas in the Athabasca Oil Sands Region.* Open File Report 1977-04. Alberta Research Council, Edmonton.

Millar, James F. V.

1968   Archaeology of Fisherman Lake, Western District of Mackenzie, N.W.T. PhD dissertation, Department of Archaeology, University of Calgary.

Minni, Sheila J.

1976   *The Prehistoric Occupations of Black Lake, Northern Saskatchewan.* Mercury Series no. 53. Archaeological Survey of Canada, National Museum of Man, Ottawa.

Morrison, David A.

1987   The Middle Prehistoric Period and the Archaic Concept in the Middle Mackenzie Valley. *Canadian Journal of Archaeology* 11: 49–74.

Murphy, Brent

2009   *Historical Resources Impact Assessment for Enbridge Pipelines (Athabasca) Inc., Fort Hills Delivery System: Final Report (ASA Permit 08-070).* Copy on file, Archaeological Survey, Historic Resources Management Branch, Alberta Culture, Edmonton.

Murton, Julian B., Mark D. Bateman, Scott R. Dallimore, James T. Teller, and Zhirong Yang

    2010   Identification of Younger Dryas Outburst Flood Path from Lake Agassiz to the Arctic Ocean. *Nature* 464 (1 April): 740–743.

Noble, William C.

    1971   Archaeological Surveys and Sequences in Central District of Mackenzie, N.W.T. *Arctic Anthropology* 8(1): 102–135.

Peck, Trevor R.

    2011   *Light from Ancient Campfires: Archaeological Evidence for Native Lifeways on the Northern Plains.* Athabasca University Press, Edmonton.

Pettipas, Leo

    1972   A New Projectile Point Type from Manitoba. *Saskatchewan Archaeology Newsletter* 38: 1–5.

    2003   A Long Way from Utah? *Manitoba Archaeological Newsletter,* ser. 2, 15(3): 1–4.

Rhine, Janet L., and Derald G. Smith

    1988   The Late Pleistocene Athabasca Braid Delta of Northeastern Alberta, Canada: A Paraglacial Drainage System Affected by Aeolian Sand Supply. In *Fan Deltas: Sedimentology and Tectonic Settings,* edited by W. Nemec and R. J. Steel, pp. 158–169. Blackie and Son, Glasgow and London.

Ronaghan, Brian M.

    1981   *Final Report: Historical Resources Impact Assessment, Fort McMurray Energy Corridor, Fort Hills Townsite and Airstrip (ASA Permit 80-091).* Copy on file, Archaeological Survey, Historic Resources Management Branch, Alberta Culture, Edmonton.

Roskowski, Laura, and Morgan Blower

    2009   *Historical Resources Impact Assessment and Mitigation, HhOv-483 and HhOv-484, ATCO Electric Limited, Distribution Powerline Easement, Application No. 060077: Final Report (ASA Permit 06-515).* Copy on file, Archaeological Survey, Historic Resources Management Branch, Alberta Culture, Edmonton.

Saxberg, Nancy, and Brian O. K. Reeves

    2003   The First Two Thousand Years of Oil Sands History: Ancient Hunters at the Northwest Outlet of Glacial Lake Agassiz. In *Archaeology in Alberta: A View from the New Millennium,* edited by Jack W. Brink and John F. Dormaar, pp. 290–322. Archaeological Society of Alberta, Medicine Hat.

Shortt, Mack W., Nancy Saxberg, and Brian O. K. Reeves

    1998   *Aurora Mine North, East Pit Opening, Plant Site, Tailings and Related Workings, HRIA and Mitigation Studies: Final Report (ASA Permit 97-116).* Copy on file, Archaeological Survey, Historic Resources Management Branch, Alberta Culture, Edmonton.

Smith, Derald G.

    1994   Glacial Lake McConnell: Paleogeogrpahy, Age, Duration, and Associated River Deltas, Mackenzie River Basin, Western Canada. *Quaternary Science Reviews* 13: 829–843.

Smith, Derald G., and Timothy G. Fisher

    1993   Glacial Lake Agassiz: The Northwestern Outlet and Paleoflood. *Geology* 21(1): 9–12.

Somer, Brad

    2005   *Syncrude Aurora Mine North, Historical Resources Impact Assessment and Mitigation Studies, 2004 Field Studies: Final Report (ASA Permit 04-192).* 2 vols. Copy on file, Archaeological Survey, Historic Resources Management Branch, Alberta Culture, Edmonton.

Tischer, Jennifer C.

    2003   *Adaptive Management Program (2002), Final Report: True North Energy L.P. Fort Hills Oil Sands Project (ASA Permit 02-143).* Copy on file, Archaeological Survey, Historic Resources Management Branch, Alberta Culture, Edmonton.

2004 *Historical Resources Studies, Final Report: Albian Sands Energy Inc. Muskeg River Expansion (ASA Permit 04-249)*. 2 vols. Copy on file, Archaeological Survey, Historic Resources Management Branch, Alberta Culture, Edmonton.

True North Energy

2001 Applications for Approval of the Fort Hills Oil Sands Project. Alberta Environment, Edmonton.

Unfreed, Wendy J., Gloria J. Fedirchuk, and Eugene M. Gryba

2001 *Historical Resources Impact Assessment, True North Energy L.P. Fort Hills Oil Sands Project: Final Report (ASA Permit 00-130)*. 2 vols. Copy on file, Archaeological Survey, Historic Resources Management Branch, Alberta Culture, Edmonton.

Vance, Robert E.

1986 Pollen Stratigraphy of Eaglenest Lake, Northeastern Alberta. *Canadian Journal of Earth Sciences* 23: 11–20.

van Strydonck, Mark, Mathieu Boudin, and Guy de Mulder

2009 ^{14}C Dating of Cremated Bones: The Issue of Sample Contamination. *Radiocarbon* 51(2): 553–568.

Wickham, Michelle D.

2010 A Discussion of Wedge-Shaped Microblade Cores from Two Sites in Northern Alberta. Paper presented at the 35th annual meeting of the Archaeological Society of Alberta, Calgary, Alberta, 30 April–2 May.

Wickham, Michelle D., and Taylor Graham

2009 *Historical Resources Impact Mitigation of the TransCanada Pipelines Ltd. Fort McKay Mainline Expansion: Final Report and Post-construction Audit (ASA Permits 06-376 and 07-266)*. Copy on file, Archaeological Survey, Historic Resources Management Branch, Alberta Culture, Edmonton.

Wolfe, Stephen A., Roger C. Paulen, I. R. Smith, and Michel Lamothe

2007 *Age and Paleoenvironmental Significance of Late Wisconsinan Dune Fields in the Mount Watt and Fontas River Map Areas, Northern Alberta and British Columbia*. Geological Survey of Canada Current Research 2007-B4. Natural Resources Canada, Ottawa.

Wood, William J.

1979 *Archaeological Mitigation Project, Proposed Fort Hills Townsite and Airport (ASA Permit 79-118)*. Copy on file, Archaeological Survey, Historic Resources Management Branch, Alberta Culture, Edmonton.

Woywitka, Robin J.

2005 *Historical Resources Impact Assessment, Fort Hills Energy Corporation, Fort Hills Oil Sands Project, No Net Loss Lake, Fort Hills LOC, Historic Site Photo-documentation: Final Report (ASA Permit 05-455)*. Copy on file, Archaeological Survey, Historic Resources Management Branch, Alberta Culture, Edmonton.

2007a *Historical Resources Impact Assessment, Fort Hills Energy Corporation, Fort Hills Oil Sands Project, Lease 437/438: Final Report (ASA Permit 06-270)*. Copy on file, Archaeological Survey, Historic Resources Management Branch, Alberta Culture, Edmonton.

2007b *Historical Resources Impact Mitigation, Fort Hills Energy Corporation, Fort Hills Oil Sands Project, Out of Pit Tailings Area, HiOv-59, HiOv-61, and HiOv-64: Final Report (ASA Permit 06-548)*. Copy on file, Archaeological Survey, Historic Resources Management Branch, Alberta Culture, Edmonton.

2008 *Historical Resources Impact Assessment, Fort Hills Energy Corporation, Fort Hills Oil Sands Project, Gap Areas, Water Intake Facility: Final Report (ASA Permit 07-234)*. Copy on file, Archaeological Survey, Historic Resources Management Branch, Alberta Culture, Edmonton.

Woywitka, Robin J., and Angela M. Younie

2007 *Historical Resources Impact Assessment, Fort Hills Energy Corporation, Fort Hills Oil Sands Project, Out of Pit Tailings Area, Amended Plant Site, Water Intake Facility: Final Report (ASA Permit 06-271).* Copy on file, Archaeological Survey, Historic Resources Management Branch, Alberta Culture, Edmonton.

2008a *Historical Resources Impact Mitigation, Fort Hills Energy Corporation, Fort Hills Oil Sands Project, HiOv-44, HiOv-47, HiOv-49, HiOv-50, HiOv-52, HiOv-87, HiOv-89, HiOv-104, HiOv-115, and HiOv-124: Final Report (ASA Permit 05-328).* Copy on file, Archaeological Survey, Historic Resources Management Branch, Alberta Culture, Edmonton.

2008b *Historical Resources Impact Mitigation, Fort Hills Energy Corporation, Fort Hills Oil Sands Project, Stage I Mitigation, HiOv-97 and HiOv-98, Stage II Mitigation, HiOv-61 and HiOv-64: Final Report (ASA Permit 07-235).* Copy on file, Archaeological Survey, Historic Resources Management Branch, Alberta Culture, Edmonton.

2010 *Historical Resources Impact Mitigation, Fort Hills Energy Corporation, Fort Hills Oil Sands Project, HiOv-7: Final Report (ASA Permit 07-224).* Copy on file, Archaeological Survey, Historic Resources Management Branch, Alberta Culture, Edmonton.

Woywitka, Robin J., Jennifer C. Tischer, Laura Roskowski, and Angela M. Younie

2009 *Historical Resources Impact Mitigation, Fort Hills Energy Corporation, Fort Hills Oil Sands Project, 2008 Mitigation Studies: Final Report (ASA Permit 08-163).* Copy on file, Archaeological Survey, Historic Resources Management Branch, Alberta Culture, Edmonton.

Wright, James V.

1975 *The Prehistory of Lake Athabasca: An Initial Statement.* Mercury Series no. 29. Archaeological Survey of Canada, National Museum of Man, Ottawa.

Yost, Chad

2009 Protein Residue Analysis of Lithic Artifacts from Sites HiOv-80, HiOv-16, HiOv-83, HiOv-126 and HiOv-92 for the Fort Hills 2008 Mitigation (FMA Project 115708.MI08), Alberta Canada. In Robin J. Woywitka, Angela M. Younie, Jennifer C. Tischer, and Laura Roskowski, *Historical Resources Impact Mitigation, Fort Hills Energy Corporation, Fort Hills Oil Sands Project, 2008 Mitigation Studies: Final Report (ASA Permit 08-163),* Appendix III. Copy on file, Archaeological Survey, Historic Resources Management Branch, Alberta Culture, Edmonton.

Younie, Angela M.

2008 Prehistoric Microblade Technology in the Oilsands Region of Northeastern Alberta: A Technological Analysis of Microblade Production at Archaeological Site HiOv-89. MA thesis, Department of Anthropology, University of Alberta, Edmonton.

Younie, Angela M., Raymond J. Le Blanc, and Robin J. Woywitka

2010 Little Pond: A Microblade and Burin Site in Northeastern Alberta. *Arctic Anthropology* 47(1): 71–92.

8 The Early Human History of the Birch Mountains Uplands

JOHN W. IVES

Wedensday 5th Northerly wind with Cloudy weather all day & Cold about 10 o Clock there arrived two Achibawayans from Lack de Brochet with two trains of meat, at 4 O Clock Savoyard Arrived from St Germain with 15 fathom of Bark which he Raised at the Mountain, St Germain Sent word that he was not Shour whether he could find more or not.

Journal entry for 5 April 1786, "The English River Book," Hudson's Bay Company Archives F.2/1

The Birch Mountains made their appearance in Euro-Canadian fur trade literature in 1786, with these first remarks in the "English River Book" (Duckworth 1990). This journal, probably authored by Cuthbert Grant, recounts events as Peter Pond engaged in spring trade with Dene Sułine (Chipewyan), Dane-zaa (Beaver), and Cree bands at his post on the lower reaches of the Athabasca River. The "Mountain" of the passage above undoubtedly refers to today's Birch Mountains, which were frequently mentioned in the early fur trade literature as the "Bark" Mountains. That Paul St. Germain, also known as "Buffalo Head," sent birchbark for canoes from the Birch Mountains should not surprise us. St. Germain was the North West Company's principal guide in the Athabasca country from the outset of the trade that Pond inaugurated in 1778, after crossing Methye Portage and entering the Clearwater and Athabasca drainages. St. Germain would soon have become aware of the seasonal rhythm of human activities in the area, including the raw material needed for building or refurbishing canoes.

In contemplating the archaeology of the oil sands region, it is critical that we recognize this larger-scale use of the landscape, so that we do not view the Lower Athabasca River in geographic isolation. The Birch Mountains typically fell within the orbit of a far larger pattern of land use that covered both the oil sands region and adjacent uplands. Records of James Porter's 1799–1800 sojourn at Lac Claire, as well as Fort Chipewyan journal entries of the 1820s and 1830s, reveal that Chipewyan and Cree parties regularly made winter hunts in the Birch Mountains (see "Fort Chipewyan Post Journals," HBCA B.39/a/18–31, and, regarding Porter's activities, "McKenzie, James, Journal, 1799–1800"; see also Wallace 1929; Mathewson 1974). In more recent times, the traditional land use patterns of the Fort McKay First Nation include intensive use of the Birch Mountains (see, for example, Tanner, Gates, and Ganter 2001).

In fact, subarctic hunter-gatherers have characteristically employed low population densities, subsistence economies covering large tracts of land, and a geographically extensive web of kin relationships in dealing with the vicissitudes of the boreal forest ecosystem (see, for example, Ives 1990, 1993, 1998; Meyer 1984). Any comprehensive account of oil sands region archaeology requires that we understand the early human history of nearby geographic features like the Birch Mountains.

HISTORY OF ARCHAEOLOGICAL RESEARCH

Comparatively little archaeological research has taken place in the Birch Mountains. In 1975, Cort Sims (n.d. [1975]) undertook preliminary survey work in the Namur and Gardiner lakes areas, identifying the large and rich site, HjPd-1, at the Gardiner Lake Narrows. Donahue (1976) also led an archaeological survey in 1975 that extended from Eaglenest Lake at the north end of the central Birch Mountains depression, through Clear Lake (a long, narrow lake that lies immediately southeast of Eaglenest Lake), southwest to Sand Lake and Big Island Lake, and then south to portions of North Gardiner Lake: forty-nine sites were identified. In 1976, I excavated 96 square metres and two lengthy transects at the richest of the sites identified in Donahue's survey, the Eaglenest Portage site (HkPa-4) (Ives 1977a, 1977b, 1981a, 1985), while Sims (1980) excavated 253 square metres at the Gardiner Lake Narrows site.

Between 1980 and 1982, I undertook annual programs of survey and excavation at the north end of the central Birch Mountains depression, focusing primarily on Eaglenest and Clear lakes and adjacent regions (see Ives 1981b, 1982a). This fieldwork involved pedestrian and canoe efforts to test terrain features around and away from the perimeters of bodies of water. There was also excavation of a variety of sites, both large (such as HkPb-1, the Satsi site) and small, of variable functions. Apart from the occasional archaeological resource management impact assessment, little more happened in the Birch Mountains in the 1980s and 1990s.

In 2004, Brian Ronaghan and I took part in an Alberta Parks and Protected Areas Biophysical Inventory of Birch Mountains Wildland Provincial Park, revisiting previously discovered sites and locating a series of new sites in the Big Island, Gardiner, Namur, and Legend lake areas. At the time of writing, a number of archaeological research projects connected with oil sands development were underway on or near the eastern edge of the Birch Mountains (see, for example, Gryba and Tischer 2005). This work promises to fill a considerable gap in our knowledge of the relationship between the Birch Mountain uplands and adjoining Athabasca River lowlands.

LANDFORMS

The Birch Mountains are remnants of the Alberta plateau and are underlain by poorly consolidated Cretaceous shales and sandstones (Bayrock 1961). They rise some 525 metres above surrounding lowlands, reaching a maximum height of 850 metres above sea level. Van Waas (1974) described two major physiographic regions, the Birch Mountains uplands plains and the central Birch Mountains depression. The latter contains a chain of lakes, around which the vast majority of archaeological sites are concentrated (fig 8.1).

The Birch Mountains were covered with ice during the most recent Laurentide advance, although they may have been exposed for some time as a nunatak during deglaciation. The predominant landforms are undulating to rolling moraine. There are significant areas of glacial fluting; a series of ridges tangential to the fluting may result from glacial drift overlying bedrock. When these ridges are near lakes, they can have dense concentrations of archaeological materials. Luvisolic, brunisolic, gleysolic, and organic soil orders predominate. The Birch Mountains fall within what Rowe (1972) called the Mixedwood section of the Boreal Forest region.

Like other Tertiary remnants, the Birch Mountains have a radial drainage pattern that was significant in providing means of access for human activities. Porter established his 1799–1800 Lac Claire post near the mouth of the Birch

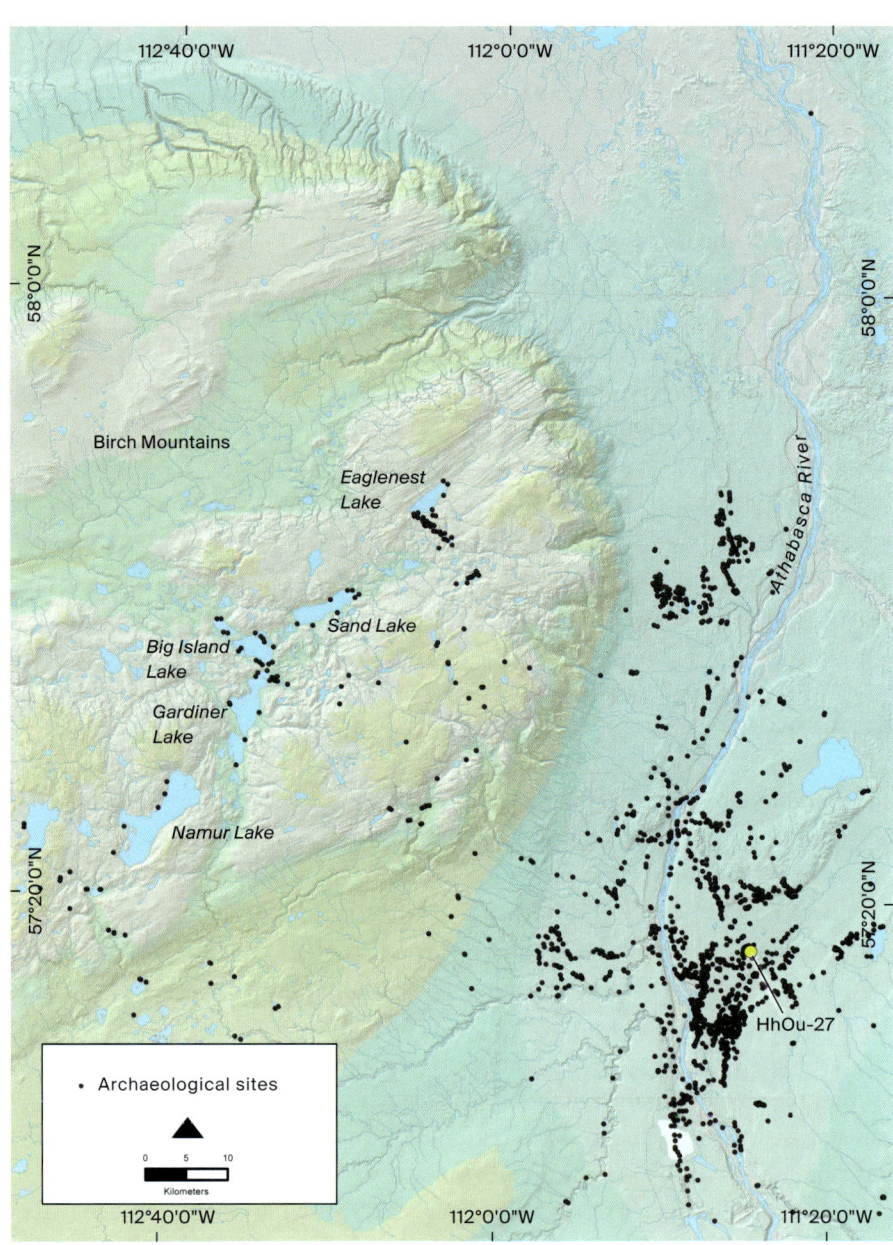

Figure 8.1. Site distributions in the Birch Mountains and Lower Athabasca valley. Note the concentration of sites around the lakes in the central Birch Mountains depression, as well as the location of HhOu-27, in the oil sands region.

River, which was clearly an important travel route for the Birch Mountains toward the north and northeast. Such use continued into the twentieth century: Fort Chipewyan elder Snowbird Marten (pers. comm., 1982) advised me that he used the Birch River for access to the Birch Mountains. The Mikkwa, or Little Red, River drainage would have provided access from the Peace River to the northwest, while the Ells River was a critical travel route into the Birch Mountains from the Fort McKay region of the Lower Athabasca River.

ANCIENT ENVIRONMENTS

Two lakes in the Birch Mountains were cored for pollen analysis in conjunction with the Birch Mountains Archaeological Study of the early 1980s (fig 8.2; Ives 1981b, 1982a). In March 1981, a 4.25-metre sediment core was extracted from the 30-metre-deep east basin of Clear Lake, at the north end of the central Birch Mountains depression. A radiocarbon sample from near the base of the core yielded a date of 2,800 ± 150 ^{14}C yr BP (GX-8361) (Ives 1982a). It is likely that sediments at this location are quite deep, a factor that prevented recovery of a longer core. Although no further analysis of this core proceeded, it is worth observing that substantial lengths of the core appeared to have rhythmitic deposits, with alternating light and dark sediment bands. Further work at Clear Lake could result in exceedingly fine-grained data regarding vegetation, climatic conditions, and fire history during the latter parts of the Holocene period.

Vance (1986) reported the results of a 7.65-metre sediment core extracted from nearby Eaglenest Lake in April 1982. Efforts to radiocarbon-date these sediments were complicated by the presence of traces of bitumen in the sediments, which necessitated specialized extraction by the Oils Sands Research Department of the Alberta Research Council (Vance 1986). This processing resulted in a date from near the base of the core of 10,740 ± 150 ^{14}C yr BP (Beta 8287). It would therefore appear that deglaciation of the Birch Mountains took place roughly 11,000 radiocarbon years ago (or 13,000 calendar years ago).

As was the case with results discussed by MacDonald (1987) for the Mackenzie River basin, Vance (1986) reported that the earliest pollen zone at Eaglenest, EL 1 (ca. 11,000 to 11,800 radiocarbon years ago), was characterized by low rates of pollen influx, together with indications of sparser vegetation on open mineral soil. Pre-Quaternary pollen and spores were thus common, with fairly rapid early accumulations of inorganic sediments. The initial re-vegetation of the Birch Mountains resulted from colonization by primarily non-arboreal taxa, particularly sage (*Artemisia*), grasses (Gramineae),

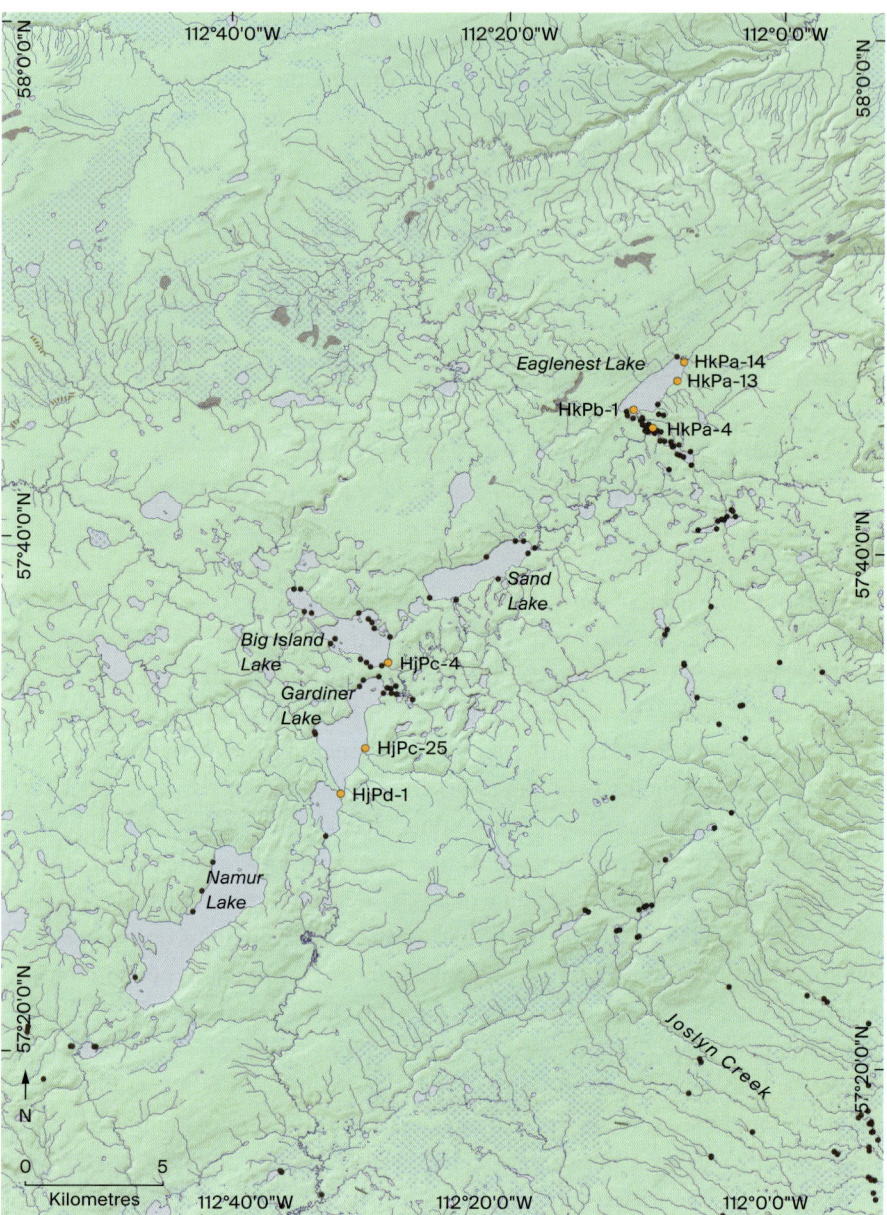

Figure 8.2. Sites in the central Birch Mountains depression. Sites discussed in the text are marked with yellow circles.

sedges (Cyperaceae), and diverse herbs. The only arboreal species present were poplar (*Populus*) and willow (*Salix*), and these occurred at low frequency (5% to 21%: Vance 1986). Vance concluded that pollen zone EL1 may have represented an open parkland or grassland growing under dry, windy conditions (resulting from southeasterly katabatic flows deflected by Laurentide ice to the north and east), as growing seasons warmed rapidly after glacial minima. The pollen profile—produced by unusual assemblages of plants species rapidly adjusting their ranges to postglacial conditions—has no analogue in existing vegetation communities. Better-drained, exposed locations in the Birch Mountains featured mainly sages and grasses; poorly drained and sheltered areas had poplar and willow; sedges were common around bodies of water; and fluctuating margins of ponds and lakes were inhabited by species of Chenopodiaceae and *Plantago*.

Pollen zone EL2 (11,000 to 7,500 ^{14}C yr BP) was marked by the distinct rise in spruce (*Picea*) and birch (*Betula*) pollen. EL2 was subdivided into two subzones: EL2a (11,000 to 9,750 ^{14}C yr BP) and EL2b (9,750 to ^{14}C yr 7,500 BP). The distinguishing factor was the presence of alder (*Alnus*) in frequencies greater than 10% in EL2b, suggesting that alder was by then locally present in the Birch Mountains. No direct analogy for EL2 exists in modern vegetation either. Vance considered it to reflect an open spruce woodland, in which herbaceous components were minor but were more consistently represented than in the succeeding zone, EL3. It is not a simple matter to determine which spruce and birch species were involved in what amounted to a rapid, time-transgressive colonization of the Birch Mountains by these species from refugia to the south (see, for example, Ives 1977c; Vance 1986, 17–18). On balance, the evidence favours white spruce (*Picea glauca*) and tree and shrub birches. Vance (1986) suggested that the open spruce forest of EL2 may have resembled the spruce woodland of Labrador, although, in the case of the Birch Mountains, the openness was probably controlled by aridity rather than by permafrost.

In many ways, the ice age world lay dying in the larger oil sands region during zones EL1 and EL2. At that time, the Birch Mountains would have had a rather peninsular quality, extending northeastward into various versions of Glacial Lake Peace and Glacial Lake McConnell (see the reconstruction by Lowell et al. 2005). To the east, Laurentide ice apparently stood at the Stony Mountain Moraine south of Fort McMurray at 10,030 ^{14}C yr BP, at the Cree Lake Moraine in northwestern Saskatchewan at 9,595 ^{14}C yr BP, and at the Firebag Moraine at 9,665 ^{14}C yr BP (Lowell et al. 2005). Vance (1986) thought it likely that westerly flows continued to be deflected to the southeast and that dry, windy, and warm conditions persisted throughout the early stage of EL2 (EL2a).

The Birch Mountains were not directly affected by the momentous glaciolacustrine events in this larger region. One or more massive outburst events from the northwestern arm of Glacial Lake Agassiz took place in the terminal Pleistocene or earliest Holocene times, carving the Clearwater River channel and creating deltaic deposits in the larger Fort McKay region (Smith and Fisher 1993; Fisher 2007; Fisher and Lowell, chapter 2 in this volume; Woywitka, Froese, and Wolfe, chapter 3 in this volume). As we shall see, this is a critical interval from the perspective of the initial human settlement of the Birch Mountains.

Pollen zone EL3 is denoted by the abrupt rise of pine (*Pinus*) pollen to above 20% at 7,500 ^{14}C yr BP, as well as a total increase in pollen influx. Higher pine pollen frequencies suggest that lodgepole and/or jackpine had reached the Birch Mountains, possibly as the consequence of a greater frequency of fires when the Arctic air mass shifted northward, stimulating a higher frequency of lightning strikes in the southern boreal forest (see MacDonald and Cwynar 1985; Vance 1986, 18). From this point on, essentially modern vegetation was established in the Birch Mountains.

CULTURAL HISTORY OF THE BIRCH MOUNTAINS

For some time, sites in the Birch Mountains provided the largest sample of projectile points in the oil sands region, although the recent flurry of archaeological resource management activities in the Lower Athabasca valley has shifted this balance. Archaeologists are frequently inclined to fashion regional chronologies in difficult circumstances through the use of available radiometric dates, along with inferences from comparative diagnostics (the latter usually for projectile points), and there could be considerable temptation to do so with the Birch Mountains record. In the case of northern Alberta, the "difficult circumstances" involve the near absence of well-stratified localities that also yield archaeological materials. Unhappily, high-fidelity archaeological deposits, like those of Peace Point in Wood Buffalo National Park, have turned out to be exceedingly rare. Given that a number of archaeologists have been on the lookout for settings in which stratified sites might occur, and given the scale and diversity of archaeological impact assessments in the oils sands region, one inevitably reaches the conclusion that the desired circumstances are rare in absolute terms.

In the relative absence of stratified sites, and with little knowledge of the actual temporal variability in diagnostic forms, it becomes unwise for archaeologists to attempt to define phases or complexes. This is particularly so when at least superficial resemblances exist among specimens from radically different time periods. For instance, some Early Prehistoric stemmed points resemble Taltheilei materials, as do some early Middle Prehistoric notched points.[1] No one has provided a comprehensive, objective basis for morphometric discrimination among these points, although comparisons are frequently made to different parts of the time spectrum in adjacent regions. The substantial intra-assemblage variability that we are increasingly able to discern for the Middle Prehistoric period on the northern Plains (a region where the archaeological record is infinitely better known than in northern Alberta) is especially instructive with regard to some of the interpretive pitfalls that can ensue in relation to side- and corner-notched points as well.

For these reasons, I will, in the following remarks, strive to avoid "over-reading" the Birch Mountains projectile point assemblage. I will stress those circumstances in which we have greater control over the stratigraphic and radiometric circumstances or in which precise typological comparisons can be made. This, of course, drastically reduces the volume of archaeological materials available for comparison, as we need to insist on a bona fide association between artifacts and a setting of a specific age.

Early Prehistoric Period

Because of their elevation and latitude, the Birch Mountains would have been the first substantial land mass available for occupation in northeastern Alberta. It is legitimate to ask whether initial occupation took place shortly after the Birch Mountains emerged from beneath glacial ice between 12,000 and 11,000 radiocarbon years ago. This interval is contemporaneous with the occurrence of fluted points in areas to the south, in Alberta and elsewhere. No fluted points have been reported from the Birch Mountains. It is important to bear in mind, however, that the majority of research conducted to date in the Birch Mountains has consisted of archaeological survey and excavation—there is, of course, no agriculture, nor has there been concerted development activity akin to the oil sands projects in the Lower Athabasca valley. This means that land surface disturbances and modern human activities, both of which might lead to the discovery of fluted points, are rare for the study area. At a continental scale, fluted points occur in moderate density in Alberta, but both agriculture and development activity have been significant factors in revealing the presence of Clovis, Folsom, and other fluted points in other regions of the province (Ives 2006; Ives et al. 2013).

It therefore remains conceivable that evidence of fluted-point makers could yet come to light in the Birch Mountains. Lanceolate points with basal thinning do occur in the Athabasca lowland to the east.[2] They may represent terminal parts of the fluted-point era, around 10,000 ^{14}C yr BP, or these points might represent vestiges of basal-thinning techniques being applied in the middle or late reaches of the Early Prehistoric (or Palaeoindian) period. In this scenario, traces of the fluting technique may have persisted in northeastern Alberta, near retreating ice fronts, at a time when fluting had disappeared across the rest of the Americas.

To a significant degree, archaeologists working in northeastern Alberta are habituated to making comparisons with the northern Plains world when, in fact, there are good reasons for looking to the North as well. Guthrie (2006) proposed an intriguing model for the human colonization of Alaska and the Yukon in a transitional period extending from roughly 13,000 to 12,000 radiocarbon years ago, as the xeric "Mammoth Steppe" gave way to an environment that featured more grasses and edible woody plants, particularly willow. By this time, mammoth and horse had begun to vanish, while first bison and then elk proliferated (although Haile et al. [2009] provide DNA evidence for a longer persistence of mammoth and horse). Both latter prey species are important in early archaeological sites in Alaska, such as Broken Mammoth, along with waterfowl, fish, and small game (see Yesner 2001, for example). By about 11,500 ^{14}C yr BP, vegetation in Alaska and the Yukon came to be dominated by mesic and hydric taiga or tundra, in which the dominant plant species are highly defended by toxins against herbivory. At this stage, moose, as browsers, became significant, as their digestive systems are better adapted to such conditions.

These ecological changes in Alaska and the Yukon would clearly precede the environmental reconstruction provided for the Birch Mountains in the previous section. By the same token, however, human populations in Alaska were established sometime between 12,000 and 10,000 radiocarbon years ago. They would be equally capable of taking part in the early settlement of northern Alberta, where similar open environments became available later in time, as deglaciation proceeded (Ives et al. 2013). In fact, there is ancient DNA evidence that northern and southern clade bison intermingled in the Peace River country prior to 11,000 radiocarbon years ago, implying that bison from Alaska and the Yukon were traversing the Ice-Free Corridor (Heintzman et al. 2016). MacDonald and McLeod (1996) argued that, between 14,000 and 11,000 years ago, the entire corridor region, from the Mackenzie Mountains to Montana, supported an open shrub, herb, and *Populus* vegetation in a relatively warm climate. It served as a "biogeographical" corridor facilitating the movement of grazing animals (which would have included mammoths, bison, horses, camels, elk, and sheep) until an early version of boreal forest, with incipient peatlands, began to form, closing off movement of these species.

The best evidence for early human presence in the Birch Mountains, from the Eaglenest Portage site (HkPa-4), suggests that this may be precisely what happened (Ives 1985, 1993). The strategy adopted for that site's excavation allowed for the spatial segregation of artifacts in an otherwise poorly stratified setting. The Block B excavation at Eaglenest Portage was in part triggered by the discovery of a distinctive, oblanceolate biface in an exploratory transect (fig 8.3: a). The expanded excavation unit resulted in the definition of a cluster of artifacts that could not only be identified by spatial analytical statistics but that also involved a raw material identifier.

The artifacts in the cluster, including the small biface, are of the vitreous grey quartzite common in northern Alberta assemblages. In this case, however, a less common globular impurity was present in the majority of artifacts, providing an additional line of evidence that they were related to each other. Although from shallow deposits in absolute terms, these were also among the more deeply buried artifacts at HkPa-4. The contents of the Block B cluster, which were illustrated in Ives (1993, plate 1), included a hammerstone made of a coarser quartzite, an exhausted discoidal core, the oblanceolate biface mentioned above, an edge-modified flake, and a variety of other flakes and fragments, all with the same crystalline impurity as the core and biface. It would appear that the core remnant was reduced at this location, although other activities, as reflected by the biface and edge-modified flake, also took place.

The oblanceolate biface is similar to two other bifaces from the Gardiner Lake Narrows site, HjPd-1 (fig 8.3: b and c), and all three specimens bear a decided resemblance to Component II (Denali Complex) bifaces from Dry Creek, Alaska (Powers and Hoffecker 1989). Outline shape and flaking patterns are close matches: the bases of both Alberta and Alaska specimens sometimes remain unflaked, featuring an unaltered ellipsoidal facet or flake scar. The Eaglenest Portage specimen (fig 8.3: a) is noticeably thick and resembles the Dry Creek, Mesa, and Sluiceway points from Alaska in this regard. All of these specimens may be knives or projectile points, but, in either case, they have clearly been resharpened. Dry Creek Component II materials have been dated to 10,690 ^{14}C yr BP (Powers and Hoffecker 1989). The deeper stratigraphic position of this Block B artifact cluster at Eaglenest Portage is consistent with that of early assemblages in similar topographic settings (see, for example, Rawluk et al. 2011). Consequently, it would seem that the Block B tool kit was deposited when the Birch Mountains were covered in the open spruce woodland forest described for EL 2a (11,000 to 9,750 ^{14}C yr BP).

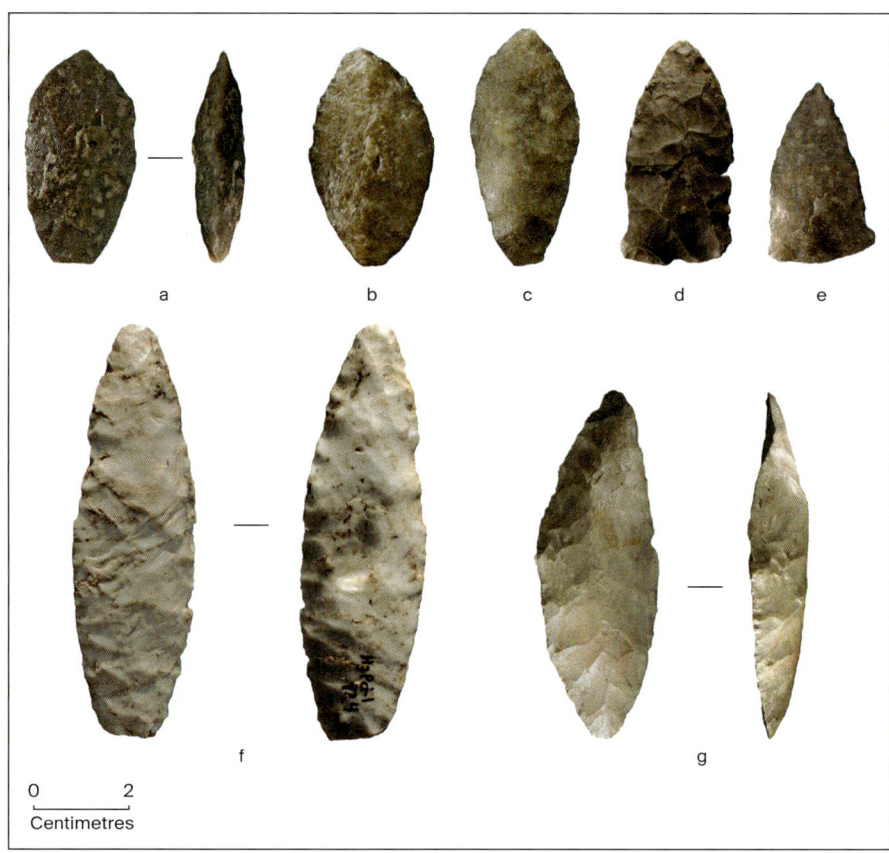

Figure 8.3. Evidence for an early human presence in the Birch Mountains. Oblanceolate points or knives from the Eaglenest Portage site, HkPa-4 (a), and the Gardiner Lake Narrows site, HjPd-1 (b and c). Note both the thickness and the heavy grinding in the longitudinal view of the HkPa-4 specimen. In (d), flared-base points from the Gardiner Lake Narrows site, and, in (e), from the Eaglenest Portage site. In (f), a large lanceolate point made of Tertiary Hills clinker from the Gardiner Lake Narrows site, with parallel, oblique flaking that in some instances is outrepassé. In (g), a beautifully made Beaver River Sandstone limace from the Gardiner Lake Narrows site, the ventral surface of which is unflaked.

a   b   c   d   e

f   g

0   2
Centimetres

The Component II materials at Dry Creek are associated with microblade technology, and roughly coeval microcores are known from Charlie Lake Cave in British Columbia and the Vermilion Lakes site near Banff (Driver et al. 1996; Fedje et al. 1995). No definitive instances of microblade technology have yet been identified in the Birch Mountains, but it would not be surprising if microblades and microcores from this specific time range one day came to light (see, for instance, Wilson, Visser, and Magne 2011).

Two specimens from the Gardiner Lake Narrows and Eaglenest Portage sites (fig 8.3: d and e) have flaring bases like points of the Fort Creek Fen Complex described by Saxberg and Reeves (2003; see also chapter 6 in this volume), thought to precede the Cody Complex in the oil sands region. In the absence of radiometric dating and well-defined strata, however, it is difficult to know whether these points genuinely are earlier than Cody Complex manifestations. Large lanceolate points from the retreating ice patches of southwestern

Yukon can be notably more recent than the Early Prehistoric or Palaeoindian periods. A lanceolate point from KaVn-2, for example, a site in the west-central Yukon, has a slightly flaring base like those found in the Fort Creek Fen Complex but comes from contexts that could be as little as 2,000 years old (Heffner 2002). Similarly, large lanceolate points with flared bases were associated with a radiocarbon date of 1040 ± 55 ^{14}C yr BP at the Karpinsky site in northwestern Alberta (Bryan and Conaty 1975). Thus, while flared-base points conceivably are quite early, further stratigraphic and radiometric confirmation of this will be required.

Another specimen, from Gardiner Lake Narrows, probably reflects an early time line. Sims (1980) recovered a beautiful lanceolate point (fig 8.3: f) that is of interest in two respects. In terms of typological affinity, this specimen is similar to Agate Basin materials found on the Plains, although its broadest portion occurs somewhat nearer its base than is the case for typical Agate Basin points (see, for example, the Agate Basin points illustrated in Frison and Stanford 1982). The point has parallel, oblique flaking. It would also fit within a "Northern Plano" frame of reference, but it is worth noting that it also bears a resemblance to points from the Spein Mountain, Mesa, and Healy Lake sites in Alaska (Ives 2006; Ackerman 2001; Holmes 2001; Kunz, Bever, and Adkins 2003). Andrews, MacKay, and Andrew (2009, 23) found a remarkably similar point or knife in their ice patch research in the Mackenzie Mountains, though it occurred on a mineral surface and could not be directly associated with organic artifacts. As such, it is difficult to fix an age for the Gardiner Lake Narrows point: it could be as old as 10,000 radiocarbon years or a few thousand years younger.

Of equal fascination concerning this specimen is the raw material it is made of: Tertiary Hills clinker (formerly "Tertiary Hills welded tuff" or "Tertiary Hills tuffaceous clinker") (see Cinq-Mars 1973; Ives and Hardie 1983; Le Blanc 2004, 15). The source for this material, the Tertiary Hills (outliers of the Mackenzie Mountains) in the Northwest Territories, lies more than 1,000 kilometres to the north of the Birch Mountains. As we have just seen with respect to glacial chronology and the outburst event from the northwestern arm of Glacial Lake Agassiz, access to points north of the Birch Mountains may not have been a simple matter. Yet, either through trade or a wide-ranging use of the landscape, this raw material found its way far, far to the south. A flake of Tertiary Hills clinker was also recovered from the Smuland Creek site, to the west of the Birch Mountains, in the Grande Prairie region, in deposits that contained a large fluted point (Bereziuk 2001; Ives 2006). Whether the fluted point and the clinker flake are truly temporally associated in this thinly stratified site is open to question, but there is a significant possibility that they are.

Sims (1980) recovered an additional, distinctive artifact of potentially early age from the Gardiner Lake Narrows site. This is a carinated, bipointed, unifacially flaked object made of Beaver River Sandstone: in every respect it fits the definition of a limace, a kind of artifact often associated with the Palaeoindian era (fig 8.3: g).

Intriguingly, one of the first projectile points that Sims recovered in the Athabasca lowlands was a Scottsbluff point (see Ives 1993, figure 2: d). Resource management work in recent years has, of course, resulted in the documentation of a distinct Cody Complex presence in the oil sands region (see, for example, Saxberg and Reeves 2003; Reeves, Blakey, and Lobb, chapter 6 in this volume). While far more archaeological survey and excavation has gone on in the Lower Athabasca valley, excavations in the Birch Mountains have yielded a fairly rich projectile point assemblage. Nevertheless, no specimens clearly attributable to the Cody Complex have been recovered on the Birch Mountains uplands themselves. It may be that this is simply a sampling error, but it is also possible that there is no corresponding evidence of a Cody Complex presence in the Birch Mountains.

Middle Prehistoric Period

In an earlier publication, I illustrated points that are very likely to date from the Middle Prehistoric period in the Birch Mountains, indicating that comparisons in a number of directions might be made and that it was difficult to be conclusive (see Ives 1993, figure 6). These points come from thinly stratified sites, in circumstances prone to forms of disturbance such as tree throws. Unfortunately, given the continued dearth of both stratified sites and radiocarbon dates in the larger oil sands region, we are little further ahead today.

Figure 8.4 (a and b) illustrates two of the large, side-notched to corner-removed points from the Gardiner Lake Narrows site. These probably date to the Middle Prehistoric period, and perhaps even its earliest expression. Looking to the south, there are reasonably good analogues for these kinds of specimens in the Boss Hill, Hawkwood, and Everblue Springs assemblages, for instance (Doll 1982; Van Dyke and Stewart 1985; Vivian 1998). Yet, if we expand our sample of notched and stemmed points to the greater range of items illustrated in Ives (1993, 13, figure 6), we find suitable comparisons in a number of different directions and time periods. The Pointed Mountain Complex materials from the Fisherman Lake area of the Northwest Territories (Millar 1968; Morrison 1987, figure 2), have decided similarities to the Gardiner Lake Narrows points, although the Fisherman Lake area poses equally difficult problems with regard to stratigraphy and radiocarbon dating. Certain Shield Archaic projectiles are also

Figure 8.4. Projectile points from the Gardiner Lake Narrows site (HjPd-1) and the Eaglenest Portage site (HkPa-4): large side-notched points (a and b), moderate-sized side-notched points (c and d), and stemmed projectile points (e to i) that illustrate the variability in styles

similar (Wright 1972). Large, notched points, dating in the range of 6,000 to 4,000 ^{14}C yr BP, likewise appear in British Columbia, in the Charlie Lake Cave sequence (Handly 1993; Driver et al. 1996). Given the possible connection with northern Alberta by way of the Peace River, the occurrence of such points at Charlie Lake Cave is especially significant.

The two moderate-sized side-notched points shown in figure 8.4 (c and d) are similar to materials that Saxberg and Reeves (2003) illustrated for their Early Beaver River Complex (undated, although estimated to 7,750 to 7,000 ^{14}C yr BP),

Figure 8.5. Arctic Small Tool Tradition artifacts from the Gardiner Lake Narrows site (HjPd-1, left) and the Eaglenest Portage site (HkPa-4, right). The HjPd-1 specimen is so thin that it is translucent, reflecting the work of a highly skilled artisan.

0   1

Centimetres

but they are also difficult to discriminate from large, Late Taltheilei notched points (Gordon 1996, 59; see also Gordon 1977a, 1977b). Further efforts toward such comparisons will only be warranted, in my estimation, when diagnostic materials from radiometrically dated, stratified sites are available or effective spatial analytical and refitting techniques have been applied.

The mid-sized stemmed points in figure 8.4 (e–i) are distinctive, occurring also in the Pointed Mountain materials (Millar 1968; Morrison 1987). Regrettably, though, they have yet to be found in securely dated contexts. Stemmed points do appear in the Charlie Lake Cave record sometime after 2,900 ^{14}C yr BP, and it may be that they become typical toward the end of the Middle Prehistoric and the beginning of the Late Prehistoric periods.

Some Birch Mountains materials are, however, distinctive enough to allow more precise comparisons. The Oxbow Complex (4,500 to 4,100 ^{14}C yr BP) does appear in northeastern Alberta, with good examples from Wabasca Lake and sites in the Lower Athabasca valley, as well as one point from Sand Lake in the Birch Mountains. There are also indications of Arctic Small Tool Tradition or Pre-Dorset affinities at both the Satsi (HkPb-1) and Gardiner Lake Narrows (HjPd-1) sites (see Ives 1993, 11–12). Given the findings of Wright (1975) and Gordon (1996) for Lake Athabasca and the Barrenlands, these artifacts should range in age from 3,500 to 2,650 ^{14}C yr BP. The Gardiner Lake Narrows specimen is strikingly beautiful, with fine symmetry and thinning (3 millimetres at its thickest) such that it has a translucent quality when backlit (fig 8.5).

Radiocarbon Dates from the Birch Mountains

In a region in which radiocarbon dates are exceedingly rare, it is worth noting that several dates do document various forms of human activity in the later Middle Prehistoric period, even if temporally diagnostic materials are generally lacking (table 8.1). Donahue (1976) described a cultural feature at site HjPc-4 on Big Island Lake. The feature consisted of a small pit, bearing artifacts; charcoal from the pit yielded a date of 3,610 ± 120 ^{14}C yr BP (Donahue 1976). One excavation unit at the Satsi site (HkPb-1) revealed an unusual pit feature along a ridge running away from the back of the site (Ives 1986). Along with the ash and charcoal fragments concentrated in the basin of the feature were a number of charred green spruce cones. The configuration of the feature and the use of a fuel certain to provide a smoky environment are both consistent with that of a smudge pit. Although this might have been purely for relief from insects, smudges are often connected with the hide preparation process for making clothing. In these cases, during the final stages of the tanning process, a conical arrangement of poles is placed over a small fire, and a hide is wrapped about the poles. Intriguingly, there was an unusual concentration of end scrapers—a tool often associated with hide preparation—scattered near the feature. Charcoal from the feature yielded a radiocarbon date of 2,795 ± 85 ^{14}C yr BP.

HkPa-13, on the northeastern shore of Eaglenest Lake, is a remnant site that appears to have been left after a larger beach had almost completely eroded. At the time of its discovery, artifacts were encountered on a mineral surface beneath 20 to 30 centimetres of muskeg (fig 8.6). Organics from the base of the peat rendered a date of 2,030 ± 105 ^{14}C yr BP, indicating that the artifacts had been deposited at some earlier time in the Middle Prehistoric period. Predominantly grey quartzite debitage was present, although a split chert pebble

Table 8.1   Radiocarbon dates from Birch Mountains sites

Site	Lab number	Date (^{14}C yr BP)	Calibrated 2 sigma range	Sample materials
HjPc-25	BGS 2571	596 ± 40	AD 1294 to 1414	Wood charcoal
HkPa-14	S-2177	470 ± 155	AD 1219 to out of range	Wood charcoal
HkPa-14	S-2175	660 ± 70	AD 1227 to 1418	Wood charcoal
HkPa-14	S-2174	1280 ± 95	AD 603 to 970	Wood charcoal
HkPa-14	S-2176	1335 ± 155	AD 404 to 1014	Wood charcoal
HkPa-14	GX-8811	1940 ± 130	351 BC to AD 384	Wood charcoal
HkPa-14	GX-8812	1965 ± 135	358 BC to AD 340	Wood charcoal
HkPa-4	DIC-720	1030 ± 110	AD 723 to 1223	Wood charcoal
HkPa-13	S-1973	2030 ± 105	359 BC to AD 212	Basal peat deposits
HkPb-1	GX-9126	2795 ± 85	1193 to 805 BC	Charred spruce cones and twigs
HjPc-4	RL-533	3610 ± 120	2336 to 1658 BC	Wood charcoal

SOURCE: Dates previously reported by Donahue (1976) and Ives (1977a, 1977b, 1981a, 1981b, 1982a, 1986), as well as previously unreported dates (for HjPc-25 and HkPa-14). All calibrations were made with OxCal 4.2, IntCal13 (see Reimer et al. 2013).

and flakes were also recovered. The hues of the chert and quartzite suggest that some heat-treating may have been applied. There are essentially three loci of activity, but, as figure 8.7 shows, refits exist across these artifact concentrations, suggesting that they are contemporary.

This is, by the way, an excavation strategy that could be applied to greater effect in the oil sands region—namely, deliberately seeking downslope artifact concentrations that run off terrain features and under established muskeg. Basal peat deposits can then yield a latest possible date for the deposition of the artifacts. The use of basal peats for dating is not an ideal circumstance, but the results from this strategy are superior to the typical outcome of excavations, where there is no prospect for dating whatsoever.

Looking forward to the Late Prehistoric period, the Pelican Beach site on Eaglenest Lake (HkPa-14) had complex, layered beach sands that yielded six radiocarbon dates spanning the past 2,000 years, ranging from 1,965 ± 135 to 470 ± 155 ^{14}C yr BP (see table 8.1). The artifact assemblage from this site consists almost entirely of grey quartzite debitage, however, so that it was an even smaller-scale version of the Wentzel Lake site (IfPo-1) in the Caribou Mountains, which yielded very few diagnostics (Conaty 1977). The occupied area of HkPa-14 is relatively limited and is adjacent to a small, unnamed stream that flows into the north end of Eaglenest Lake. The narrow breadth of this stream may have made it amenable to netting or otherwise trapping spawning fish. In any case, this location saw sustained use throughout the Late Prehistoric period.

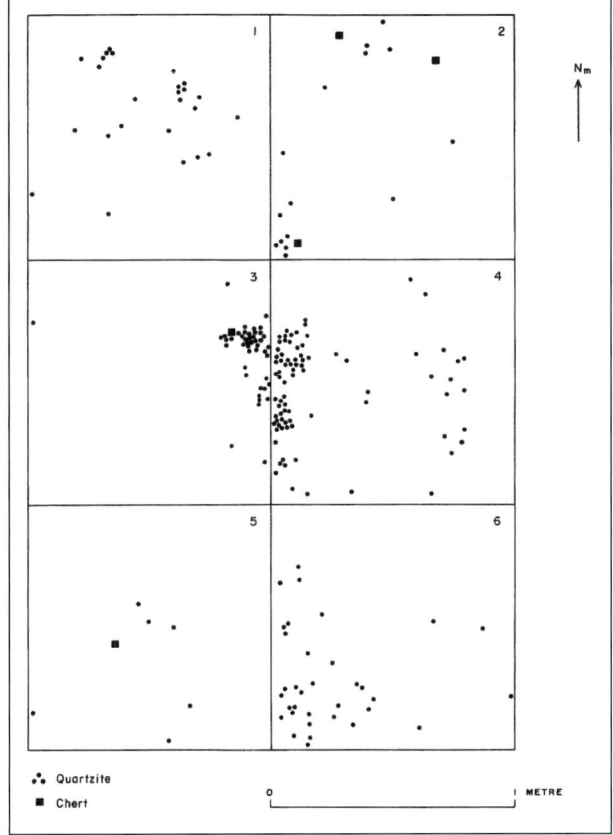

**Figure 8.6. The distribution of
lithic materials at HkPa-13, on the
northeastern shore of Eaglenest
Lake**

In 2004, Brian Ronaghan and I excavated a 1-by-2-metre unit in the stratified
beach deposits of HjPc-25, a site located on the east side of North Gardiner Lake.
A charcoal sample from an organic layer that also produced a quartzite core fragment yielded a date of 596 ± 40 ^{14}C yr BP (BGS 2571, with a 2 sigma range of AD
1296 to 1421).

Late Prehistoric Period

A slightly more refined approach to projectile point typology is possible for the
late period. Many smaller side- and corner-notched points that probably belong
to the Late Prehistoric period could be confused with those from earlier periods,
but there are a number of regional instances where a greater degree of stratigraphic and radiometric control is possible. If we confine our reasoning about

Figure 8.7. Patterns of refitting for artifacts at HkPa-13, with cross-mends extending across different clusters of artifacts

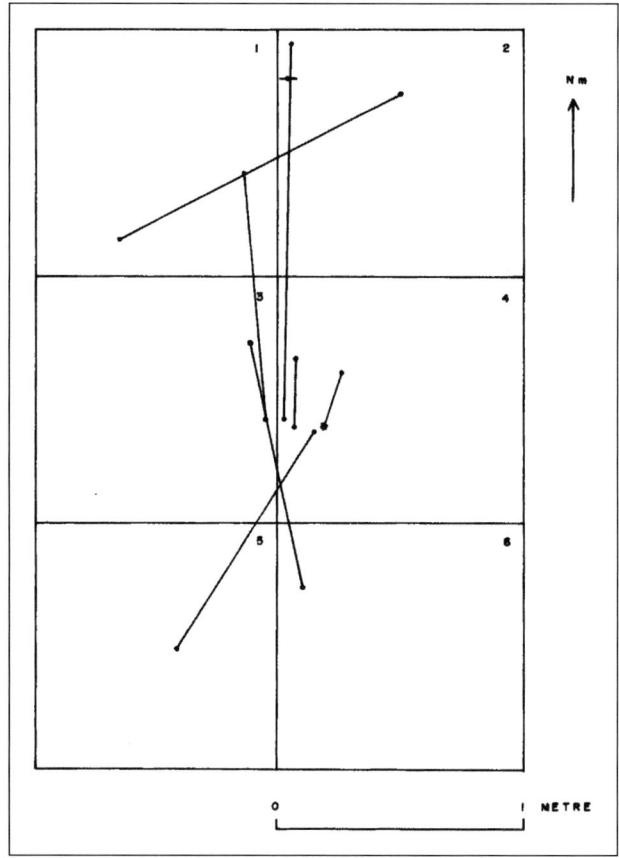

point typology to *just* these instances, essentially as illustrated in figure 8.8, a somewhat clearer picture emerges.

Records from the Charlie Lake Cave (Handly 1993; Driver et al. 1996) and Farrell Creek (Spurling 1980) sites, both in British Columbia, show that stemmed or very broadly notched points were being made on the Upper Peace River between roughly 3,000 to 2,500 ^{14}C yr BP (see fig 8.8). By roughly 1,600 to 1,500 ^{14}C yr BP, broadly side-notched points with relatively long blades were being manufactured. Sometime after 1,500 to 1,400 ^{14}C yr BP, small side-notched points occur in the upper strata of these sites. In some cases, these bear quite a striking resemblance to Early Cayley Series or Prairie Side-Notched points from the northern Plains and parkland regions of Alberta and Saskatchewan (Peck and Ives 2001).

Moving to the Birch Mountains themselves, I have illustrated four points from the Eaglenest Portage site, HkPa-4 (Ives 1977a), in figure 8.8. The top two specimens were recovered from a small patch of palaeosol surface in Block C of

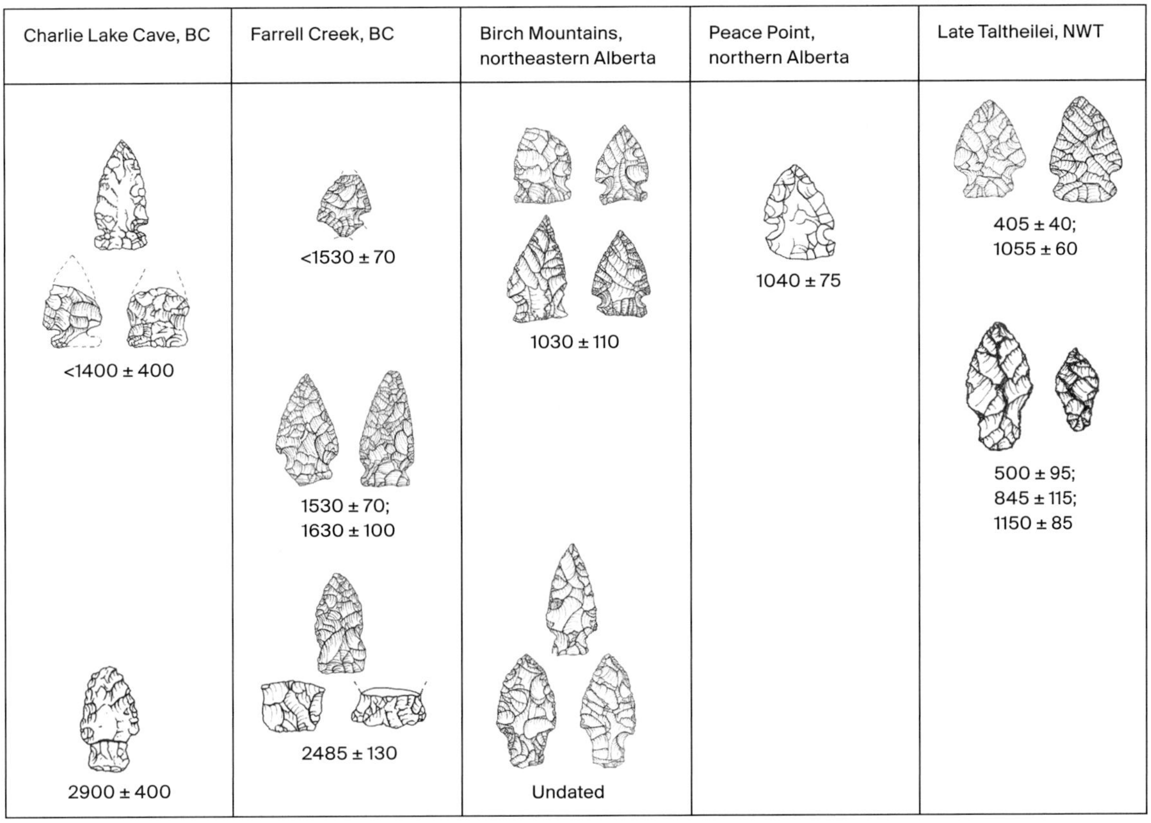

Charlie Lake Cave, BC	Farrell Creek, BC	Birch Mountains, northeastern Alberta	Peace Point, northern Alberta	Late Taltheilei, NWT
<1400 ± 400 2900 ± 400	<1530 ± 70 1530 ± 70; 1630 ± 100 2485 ± 130	1030 ± 110 Undated	1040 ± 75	405 ± 40; 1055 ± 60 500 ± 95; 845 ± 115; 1150 ± 85

the HkPa-4 excavation: charcoal from that surface yielded a radiocarbon date of 1030 ± 110 (DIC-720), providing an approximate age for the points. The two points immediately below came from Block B of the same site, where a restricted cluster of artifacts connected with the rehafting of small side-notched points could be spatially segregated. Beneath these four side-notched points are three examples of moderate-sized corner-notched and stemmed points from the Eaglenest Portage and Gardiner Lake Narrows sites (Sims 1980). These cannot be reliably dated but, on the basis of similar specimens found in the Upper Peace River region, might fall in the range of 3,000 to 1,500 ^{14}C yr BP.

Note, however, that similar stemmed points have been found in stratified, radiocarbon-dated Late Taltheilei sites in the Northwest Territories (the column to the far right in figure 8.8) and that these stemmed points date to the last 1,100 years of the late period (Gordon 1977a, 1977b, 1996). Fairly large side- to corner-notched Late Taltheilei points also occur in this time range, and a number of Birch Mountains specimens definitely resemble these notched forms. While it is

Figure 8.8. Small side-notched and stemmed points, primarily from secure stratigraphic contexts, accompanied by radiocarbon dates, in northeastern British Columbia, northern Alberta, and the Northwest Territories

tempting to conclude that many of these specimens are Late Taltheilei forms, it is nevertheless possible to confuse them with specimens from earlier time periods. It seems wise to resist this temptation until more secure data is forthcoming. To round out the discussion of figure 8.8, Stevenson's careful excavation of the Peace Point site (Stevenson 1985, 1986) yielded just one projectile point diagnostic, found in an upper layer of the site. This is the specimen illustrated, with a date of 1,040 ± 75 ^{14}C yr BP. It is almost certain that further excavations at Peace Point could illuminate projectile point variability in the region over the past 2,500 years.

Several observations concerning these Late Prehistoric points are pertinent. First, in these more controlled comparisons, it is evident that smaller side-notched points occur over the past 1,500 to 1,200 years in northern Alberta. In a number of cases, these points do resemble Early Cayley Series or Prairie Side-Notched forms; MacNeish (1964) Morrison (1984), and Spurling (1980) are among those who have applied Plains rubrics to these points. These applications are not particularly helpful, however, and are better taken as evidence of the truly broad, polyethnic geographic scope that major varieties of points tend to have in North America (see, for example, Ives 2003; Peck and Ives 2001, 185). It is also very much worth bearing in mind that a considerable degree of variability in point form exists beyond these side-notched forms and that stemmed and corner-notched forms are equally present throughout this time range. Southern Yukon projectile points that postdate the east lobe White River eruption, and so are less than 1,200 years of age, show a tremendous range of variability, with side-notched, corner-notched, and stemmed points all occurring (Sheila Greer, pers. comm., 2003; Hare, Hammer, and Gotthardt 2008). Morrison (1984) reported a similar finding for single-component Late Prehistoric sites in the Mackenzie basin, including sites in which the points lay above the east lobe White River Ash fall.

As to function, some of these smaller points are certain to be arrow tips, but this cannot be said of all of them. Other late specimens are perceptibly too large to be arrow tips, featuring broad neck and shoulder widths. In fact, when Pyszczyk (2003) applied the discriminant function analysis outlined by Shott (1997) to a sample of Alberta projectile points to determine which might be darts and which might be arrows, the four side-notched points at the top of the Birch Mountains column in figure 8.8 were each classed as dart *not* arrow tips. There is thus a distinct possibility that, in the Birch Mountains and the remainder of the boreal forest of northern Alberta, the bow and arrow did not completely supplant the spear thrower even well into the Late Prehistoric period.

Overcast weather. Settled with the Indians, who take their departure, highly satisfied with their reception they now mean to Pitch off towards the Bark Mountains and will not be here until Spring.

Alexander Stewart, journal entry for Monday, 4 December 1826, "Fort Chipewyan Post Journals" (Hudson's Bay Company Archives B 39/a/25)

There are few direct indications of seasonality for Birch Mountains occupations. The smudge pit mentioned above, radiocarbon dated to $2{,}795 \pm 85$ ^{14}C yr BP, used green spruce cones as a fuel and therefore reflects a midsummer use of the Satsi site (HkPb-1) at that time. The same site also had a hearth of unknown, but probably quite recent prehistoric, age (given its near-surface location, just under the LFH organic horizon) that yielded fragmentary and calcined bone, including fish, large ungulate (possibly bison or moose), small mammal (hare-sized), bird (grouse and smaller bird–sized), and caribou (*Rangifer tarandus*) (Tim Schowalter, pers. comm., 1981). Some of the larger mammal bone is foetal, and some of the bird bone had medullary tissue, suggesting spring occupation, as well as a diverse subsistence base.

In the absence of other faunal remains, it is difficult to infer patterns in the seasonal use of the Birch Mountains upland, although an examination of the seasonal structure of animal resources provides some clues. The Birch Mountains have in modern times been comparatively poor in game resources during the winter months (see the summary in Ives 1993). Wood bison and moose both regularly left the upland for the winter, leaving behind rather low-density woodland caribou populations and small game animals. Limited human occupation during the winter months might be predicted.

Moreover, fish resources would become available with the breakup of ice in the spring. Sites like Eaglenest Portage or the Pelican Beach site on Eaglenest Lake would be ideally suited for exploitation of spawning northern pike, walleye, and sucker populations. In fact, Eaglenest Lake has two of the three largest and richest sites in the Birch Mountains (namely, Eaglenest Portage and Satsi), even though virtually the entirety of Eaglenest Lake (save one tiny basin of three metres depth) is no more than one to two metres in depth and therefore freezes to its bottom each winter. It is very likely that the pike, suckers, and whitefish present in Eaglenest Lake during the summer come from the deep and relatively large basin at the east end of Clear Lake. These fish would have to move through the narrows and rapids between the two lakes, where they would be particularly susceptible to weirs and other fishing strategies. Once such spawns or runs had occurred, any human communities relying upon them could probably shift to returning big game populations, which had overwintered in lower-altitude drainages radiating away from the Birch Mountains (in the case of moose) or in the Peace-Athabasca Delta region (in the case of wood bison). The massive fall whitefish run in the Athabasca River might have been a trigger that encouraged departure from the Birch Mountains toward lowland settings, as might rutting season in the Peace-Athabasca Delta.

Alternative strategies must surely have been available, however. As Sims (1980) noted, North Gardiner Lake is a rich source of fish, and there has clearly been intensive occupation of the Gardiner Lake Narrows site through time. The channel at the narrows connecting North and South Gardiner Lake does not necessarily freeze during winter, at which time whitefish will congregate there. Consequently, the Gardiner Lake Narrows site may have centred on an unusually productive locale that could have offered a winter supply of fish.

The fur trade literature reveals that both Cree and Dene parties used the Birch Mountains for winter hunting and trapping. In 1799–1800, James Porter engaged in trade with groups that had been in the Birch Mountains, a pattern that continued into the 1820s. Mathewson (1974, 35) thought that, by then, at least some Cree parties had shifted away from remaining near the Athabasca River throughout the year. She reported a pattern dating to the 1820s in which Cree hunters overwintered in the Birch Mountains and summered in the Lower Athabasca valley to the east. By the 1830s, there was evidence that game animals had become scarce in the Birch Mountains and that Cree summering grounds had shifted north toward the Peace and Slave rivers and Lake Claire (Mathewson 1974, 35).

Consequently, it is possible that, at more or less any point in prehistory, a variety of seasonal uses were made of the Birch Mountains. Significant parts of the upland may, for example, have been relatively devoid of people during winter, save for one or a few bands exploiting areas connected with the richer fish lakes (most particularly the Gardiner lakes). Spring and summer may have attracted more bands onto the uplands, and the combination of spawning fish and returning moose and bison could conceivably have allowed temporary aggregates as large as regional bands or marriage isolates (a few hundred people). It is likely that use of the Lower Athabasca valley extended throughout all seasons. Consequently, one would expect traffic to have existed between the oil sands region and the Birch Mountains. It is also quite likely that the Birch Mountains may have brought people from different regions into proximity with each other, as the radial drainage pattern connected the upland with the Peace-Athabasca Delta to the north and east (especially via the Birch River), the Peace River to the north and west (via the Mikkwa, or Little Red, River), and the Wabasca drainage to the south and west.

SITE TYPES AND DISTRIBUTIONS IN THE BIRCH MOUNTAINS

Of necessity, most of the earlier survey work in the Birch Mountains (Sims n.d. [1975], 1980; Donahue 1976) was conducted in high-potential localities near lakes and watercourses. It was in this fashion that a variety of sites were found, including major base camps like Eaglenest Portage (HkPa-4), Satsi (HkPb-1), and Gardiner Lake Narrows (HjPd-1). Field crews involved in the Birch Mountains Archaeological Project that I conducted during the early 1980s were small, and they divided their time between survey activities and excavations at promising or representative archaeological sites. These survey activities involved more intensive examination of lakeshore perimeters. We also conducted testing along cutlines running away from Eaglenest or Clear lakes and made visits to small bodies of water off these larger lakes. Preliminary results for this work were presented in Ives (1981a). It is clear that a considerable diversity of sites can be detected in this way, ranging from major residential base camps to lookouts and other small sites that probably had more specialized functions.

For both the Birch and Caribou mountains, I have undertaken some examination of site distributions and densities against key variables, such as annual fish productivity in lakes. Few associations were immediately evident. For example, high fish productivity could not be reliably linked to high site densities or the presence of large and complex base camps, although there was a tendency for the highly productive lakes to provide settings in which sites were more likely to be discovered (that is, in which there was a low ratio of test stops to sites discovered). Given our present state of knowledge, however, such data can easily be interpreted too liberally. The Birch Mountains research surveys were undertaken without helicopter support or the intensive field activities typical of the development-sponsored oil sands region in the Lower Athabasca valley. Additional intensive surveys should therefore be conducted before significant conclusions of this sort are reached.

There has been one notable exception to these situations in which survey intensity was light relative to the high degree of vegetative cover. In 2004, Brian Ronaghan and I took part in an Alberta Parks biophysical inventory of the newly created Birch Mountains Wildland Provincial Park. Sims (n.d. [1975]) had made a pedestrian survey of the Namur Lake area in 1975 but was unable to locate any sites at that time, despite Namur's status as a trophy lake with considerable fish productivity. (Granted, Sims's survey probably included little, if any, subsurface testing.) Prior to the 2004 survey, however, an intense forest fire had burned to the west shore of Namur Lake. Only limited recovery of vegetation had occurred since the fire, and mineral surfaces were still readily visible. Under these circumstances, in less than three days of survey work, we were able to identify nineteen find spots, which were recorded as five sites: HiPd-1 to HiPd-3, HiPe-1, and HiPe-2 (fig 8.9: a and b). In all likelihood, more extensive pedestrian survey at the time would have yielded dozens of find spots and numerous sites.

The Namur find spots were small, normally featuring a few items of grey quartzite debitage. In one case (fig 8.9: c), a knapper had broken open a grey quartzite cobble, only to discover that it had both holes and frequent granular impurities. After some flaking the intended core was abandoned. The Namur sites occur in settings that resemble those in the Athabasca lowlands to the east. They were found on a series of parallel ridges running along the lake's western shore, with some sites as much as half a kilometre inland, at greater elevations. One might think that good candidates for Early Prehistoric period settlement would exist along ancient beach ridges, but this was not the case. Like the western shore of Eaglenest Lake, the west side of Namur Lake has extensive glacial fluting, so that sites actually occurred on till from morainal deposits, rather than on beach sands. Two of the sites yielded projectile points (fig 8.9: d and e). They are decidedly small and were undoubtedly used for tipping Late Prehistoric arrows.

The Namur sites show that using landforms to infer the age of sites lacking diagnostics and radiometric dates is a strategy prone to grave error. This is equally true of the Lower Athabasca valley, where we should assume that site distributions are *composite* patterns that arose over the course of prehistory, unless there is convincing evidence to the contrary. In fact, debitage-to-tool ratios in the Athabasca lowland are phenomenally high (owing to the huge quantities of Beaver River Sandstone freely available: see below), so that the great majority of sites have no objective indication of age whatsoever. Early Prehistoric period or Palaeoindian sites clearly have a significant presence in the Athabasca lowlands, as the very first field surveys demonstrated: among the relatively few diagnostics that Sims and Losey recovered were an Agate Basin or Lusk-like point, a Scottsbluff point, and a stemmed point referred to as a Hell Gap specimen. Yet, at the time, these finds did not cause researchers to overlook evidence that other time periods were represented. Any tendency to refer to the occupation of the oil sands region as *essentially* "Palaeoindian" is unwarranted and unwise, as is evident from the Birch Mountains ethnohistoric and archaeological data.

Alberta's Lower Athabasca Basin

Figure 8.9. The Namur Lake area: the distribution of find spots on the west shore of the lake (a); typical exposed mineral surface after the forest fire in the area (b); grey quartzite fragments and flakes from raw material probably discarded because of its flaws and impurities (c); and two small arrow tips in situ, on the mineral surface at sites on the northwest shore of the lake (d and e)

Another exciting recent development has been the discovery of sites situated along the eastern edge of the Birch Mountains (see, for example, Gryba and Tischer 2005, 2009). Although this work remained in progress at the time of writing, thirty-two small sites had thus far been reported, some in the head-waters of the Tar River and some extending onto the Birch Mountains upland itself. These latter sites occur on terrain features that appear to include eskers, other prominences, and stream terraces. Some of these sites probably served as way stations for people in transit between the Lower Athabasca valley and the central Birch Mountains depression, although we should remember the intense seasonal concentrations of moose that may have occurred in river and stream valleys emanating from the Birch Mountains.

A TALE OF TWO CACHES

If the archaeological record from the Birch Mountains is not especially well suited to resolving issues of chronology or subsistence, it is informative about the organization of lithic technology, particularly when we make comparisons with the Lower Athabasca valley. Two unusual concentrations of artifacts, from opposite ends of the time spectrum in the larger oils sands region, inform us about the intersection of lithic technology and economic strategies.

One of the most unusual finds in the Birch Mountains was a dense cluster of artifacts identified as "D2" at the Eaglenest Portage site, HkPa-4 (Ives 1977a, 1985). Discovered near the end of the 1976 field season, D2 had yielded more than 300 artifacts before excavation was suspended (fig 8.10). I returned to complete this excavation in 1980, hoping to find an associated diagnostic item. Although no diagnostics came to light, more than 400 artifacts were recovered in total—398 larger objects, as well as flake vials for a number of much smaller items. There has been no means to date this cluster of artifacts, save to observe that the uppermost part of the concentration occurred at the organic-mineral soil interface: many were visible as soon as LFH materials were removed. Although artifacts were also recovered at some depth below, we can presume that this large grouping of artifacts comes from the terminal prehistoric period. It is difficult to imagine that the many artifacts involved, which would have been attractive for further use, could have been exposed in this stratigraphic position for any great period of time without being significantly dispersed, or without at least showing signs of trampling.

The D2 artifacts were tightly packed in an area that seldom exceeded 50 by 50 centimetres. They were remarkable in several ways. First, raw material percentages are skewed relative to other upland sites. Black chert, primarily from split pebbles, occurs at twice the frequency (6% versus 3%) in D2 as it does in the entire Eaglenest Portage assemblage. Beaver River Sandstone values are even more noteworthy (fig 8.11). "Background" values for that raw material generally run between 4% and 5% for larger, more complex sites on the Birch Mountains uplands, while the material is seldom found at all at smaller "satellite" sites in the area (Ives and Fenton 1985; Fenton and Ives 1990). In D2, however, Beaver River Sandstone (BRS) comprises 21.6% of the artifacts in the cluster. Apart from tiny flakes of uncertain origin (see below), the artifacts are

Figure 8.10. The D2 cluster of artifacts at the Eaglenest Portage site (HkPa-4), as initially exposed at the mineral soil surface in 1976. Artifacts continued to be densely packed as excavation proceeded.

almost exclusively tools or larger flakes. In fact, the debitage-to-tool ratio for the D2 cluster is on the order of 5:1, whereas the ratio for the entire site is a more typical 22:1 (Ives 1985).

Of the entire cluster (adding the previously unreported 1980 finds to those from 1976), 78 artifacts were formed tools or core fragments, of which 51 were used or retouched flakes, 7 were split pebbles, 9 were end scrapers, 4 were large unifaces, 3 were large bifaces, 2 were core fragments, and 2 were side scrapers. These proportions are actually quite consistent with those for the entire site. In the spectrum of possibilities, D2 represents a lithic technology that is relatively undifferentiated when compared with the proliferation of formal tool categories one might see in an Arctic Small Tool Tradition, Upper Palaeolithic, or Palaeoindian assemblage, for example. In fact, the tightest statistical associations for tools at Eaglenest Portage were among used and retouched flakes as well as end scrapers, suggesting that most needs were being met by artifacts with relatively amorphous edges, as opposed to tools that can be assigned to formal classes, such as blades, burins, and other such specific tool morphologies (Ives 1985).

The unusual nature of the D2 cluster allows us to tease some additional meaning from the archaeological record. Stevenson (1986) suggested that the D2 cluster was a storage area at which artifacts were assembled prior to leaving the

Figure 8.11. Large quartzite and Beaver River Sandstone flakes, fragments, end scrapers, and split black chert pebbles from the D2 cluster, with examples of mottled Peace Point chert in the lower right-hand corner

0 3
Centimetres

site. That idea is plausible, but it fails to explain the unusually close proximity of the artifacts and certain details of their condition. Virtually every piece in the cluster is marked by unusual wear and abrasion, affecting not merely artifact edges but flake arrises as well (Ives 1987). I can think of no other satisfactory explanation for this phenomenon than that these 400-odd artifacts were at one point placed in a pliable container and thus had extensive opportunity to rub against each other.[3] In all likelihood, the D2 cluster represents the contents of a leather, babiche, or basketry container that was used to transport the artifacts over some distance. This container was then left, or its contents were spilled out, in the sharply delimited area occupied by the D2 cluster.

In short, we appear to be glimpsing something about customary modes of transporting stone tools in the Late Prehistoric period. It is unlikely that the persons who carried the stone tools came directly from the oil sands region, where close to 100% of any assemblage will be BRS. By the same token, it is likely that these individuals had been in the Lower Athabasca valley at some point not too long before coming to the central Birch Mountain depression. This would account for a percentage of BRS that still far outstrips the normal proportion (by a factor of roughly 5 to 1) in the Birch Mountains. Consequently, one might infer that the cluster reflects movement between the lowlands and the uplands but, rather than in a direct fashion, in one that would allow BRS to begin "falling out" of the stone tool inventory (see Ives 1993).[4] The sites along the eastern edge of the Birch Mountains (mentioned in the previous section) are very likely the locus at which this shift in raw material took place. It could easily be the case that the sites there, with high proportions of BRS, were created as ancient people travelled into the Birch Mountains, while sites having little or no BRS were created when people travelled out of the Birch Mountains.

Since it was probably no more than a day's journey between the uplands and the lowlands, one can ask what intervening activities might have detained people. In this regard, we should take notice of the tendency for moose to be concentrated during winter not in the uplands themselves but in the headwaters and middle courses of the streams draining radially from Birch Mountains (Hauge and Keith 1980; see also Ives 1993, figure 8).

The D2 cluster from HkPa-4 stands in notable contrast to another cache-like circumstance at HhOu-27, discovered in 1982, on the former Alsands lease in the Lower Athabasca valley (see fig 8.1; Ives 1982b). This spatially tight concentration of artifacts consisted of a relatively small number of large BRS flakes (fig 8.12) that were revealed by clearing activities in the area of the lease (Ives 1982b). Although a number of these flakes had been broken by a vehicle, refitted pieces and more complete specimens show that flake sizes are unusually large and so must have come from an extremely large core indeed (fig 8.13). The mean length of reasonably complete, large flakes is 98.70 millimetres (with a range of 17.80 to 130.06 mm), the mean width is 50.02 millimetres (with a range of 32.97 to 86.30 mm), and the mean thickness is 15.44 millimetres (with a range of 8.73 to 26.70 mm), while the mean weight is 62.80 grams (ranging from 25.50 to 186.20 gm,

Figure 8.12. The concentration of very large Beaver River Sandstone flakes and fragments at HhOu-27, as originally exposed by clearing activities in 1982

with most specimens not whole). The reasonably complete flakes total 1367.3 grams; the complete assemblage, including fragmentary materials, would exceed 2 kilograms.

Although some of these flakes have a linear orientation with a single or double arris, they are not blades prepared from large conical or wedge-shaped cores as we see in Upper Palaeolithic or Clovis assemblages (see, for example, Collins 2004; Le Blanc and Wright 1990). Instead, they have distinctly lipped and faceted striking platform areas clearly struck from a bifacial core (fig 8.13: b). The mean platform angle is 110 degrees (with a range of 72.00° to 113.14°), the

a

b

0   5
Centimetres

mean platform width is 28.68 millimetres (with a range of 12.56 to 65.40 mm), and the mean platform depth is 10.24 millimetres (with a range of 3.71 to 26.70 mm). Generally speaking, the HhOu-27 flakes are large, emanating broadly from rather small striking platforms. The superior skill level of the craftsperson is quite evident in a number of flakes that remain large but are both exceedingly thin for their size and almost perfectly flat, exhibiting virtually no curvature in longitudinal plane.

The use of large, ovate or discoidal flake cores to produce flakes of this type is often regarded as typical of Clovis technology, and it is likely that the HhOu-27 flakes reflect an instance of Early Prehistoric or Palaeoindian period activity in the Lower Athabasca oil sands region (see, for example, Collins 2004; Bradley, Collins, and Hemmings 2010). Some projectile points from the Lower Athabasca valley resemble Goshen specimens, while others exhibit basal thinning or fluting. Those projectile points can be expected to come from an occupation toward the end of the fluted-point era, predating Cody Complex expressions. I note also that the Duckett site in the Cold Lake region to the south has yielded several BRS artifacts, among them a small, basally thinned point—again suggesting that the use of Lower Athabasca valley sources of BRS was underway by late fluted-point times (Ives 2006; McCullough 1981). These large flakes might therefore come from the terminal portion of the fluted-point era. It is equally possible, however, that they could date to the middle or later reaches of the Palaeoindian interval, perhaps even being associated with Cody Complex aspects of nearby assemblages. In size and morphology, the HhOu-27 flakes are similar to large flakes from MacHaffie, a Cody Complex site in Montana (Knudson 1983).

The HhOu-27 flakes reflect a significantly different approach to the organization of lithic technology relative to the D2 cache from Eaglenest Portage. Extensive preparation of BRS raw material took place in this general vicinity (see Roskowski and Netzel 2009, for example). In the case of HhOu-27, these artifacts represent a specific segment of the raw material reduction strategy, in which flakes were carefully fabricated. It is not clear whether they were simply prepared, gathered, and deposited or whether they may have been transported briefly together and then deposited, although there is little indication of "bag wear" from lengthy transport. In either event, the HhOu-27 artifacts were also tightly clustered in an area of roughly one square metre. Given the absence of abrasion characteristics, it is more likely that these large flakes were deposited near their locus of manufacture, which is not far from viable sources of Beaver River Sandstone. No other tools accompany the flakes.

In an influential article, Kelly and Todd (1988) argued that Early Prehistoric period or Palaeoindian societies were "high technology foragers," with specific strategies of mobility and technological organization not found in later periods. As reflections of such an orientation, they highlighted the remarkable "sameness" in the assemblage of tools from Palaeoindian sites across North America (particularly fluted points themselves), the selection and distant transport of very high-quality raw materials, and the extensive curation of tools, as well as settlement and bone assemblage indications of high mobility. Biface technology factored into their analysis, in that the large bifaces typical of early caches were thought to allow for economical transport of raw stone material. The bifaces themselves, and the large flakes resulting from their reduction, were, moreover, suitable for various forms of subsequent manufacture—without predisposing the craftsperson as to which of a variety of forms would be selected. The HhOu-27 flakes do fit this model.

Bamforth (2002), however, made a critical examination of the premises underlying the high technology forager theory and found that they did not reflect many of the details of lithic technology revealed in early period sites. In many cases, it appears that Palaeoindian populations relied heavily on locally available raw stone materials. (This is certainly the case for fluted points in Alberta: see Ives 2006.) In addition, the life of tools may not have been extended as much as was previously thought, through reworking and curation, and evidence for the widespread use of large biface technology proved to be weak. The tight cluster of large flakes of BRS found at HhOu-27 occurs well within the Lower Athabasca River zone in which this tool stone material could be readily acquired (Fenton and Ives 1990; see also Bamforth 2009; Ellis 2011).

It may be that other circumstances came into play with respect to large biface technology in the early period. Baker and Kunz (2003) contrasted lithic technologies for Folsom, on the central Plains, with the Mesa site, located on Alaska's north slope. Folsom craftspersons frequently circulated through large areas in which raw stone materials were rare, necessitating conservation and economization in the fabrication and use of stone tools. In contrast, Mesa craftspersons were never more than 4 kilometres from abundant, high-quality lithic sources. Considerably less caution in tool fabrication and maintenance was thus required.

This latter situation better resembles that for BRS in the Lower Athabasca valley. BRS outcroppings would generally have been available along the Lower Athasbasca River and its tributaries in the target zone described by Fenton and Ives (1982, 1984, 1990), and BRS occurs at or very near the surface over large areas away from these watercourses. It is not clear whether prehistoric artisans sorted through large quantities of raw BRS looking for facies-level variation in raw material quality. Regardless of prospects for that strategy, however, Gryba (chapter 9 in this volume) describes a straightforward heat-treatment process that transforms typical BRS into a highly tractable raw material, and of which many BRS artifacts exhibit evidence, particularly in the form of the rosy hues that accompany heating. Any combination of facies variation, the impact of natural fires on near-surface BRS outcrops, and deliberate human heat-treating would mean that workable BRS was not a scarce commodity in the immediate oil sands region. Consequently, the high density of archaeological sites in the oil sands region may reflect not only intensive human use and highly effective archaeological search strategies but, equally, the fact that, when fabricating tools, prehistoric craftspersons did not have to be cautious about their use of stone: in the Athasbasca lowlands, abundant raw material was never far away. The scouring effect of outburst events from the northwestern arm of Glacial Lake Agassiz undoubtedly also meant that the visibility of raw material sources was at its apex in the Early Prehistoric period. The HhOu-27 cache of large, faceted flakes may have occurred in part simply because very large pieces of raw material were available and could be worked in this fashion.

Specifically with regard to the HhOu-27 cache and BRS as a raw material, I note that the Duckett site is not the only evidence for significant early, southward transport of Beaver River Sandstone. There are three easily recognized Cody Complex points and another probable fluted point made of BRS in the Thickwood Hills of Saskatchewan (Carlson 1993). Given the distances involved (300 to 500 kilometres in a straight line), it is quite possible that—as Ellis (2011) persuasively argued for the Great Lakes and Northeastern Palaeoindian record (and see also Speth et al. 2010)—trade or some social mechanism other than residential mobility was involved, such as visits of small parties directly to quarries. At the same time, this raw material distribution could represent the seasonal orbit of terminal Early Prehistoric peoples along the western edge of Glacial Lake Agassiz. If ice stood just beyond the moraines that Fisher (2007) reports to have existed between 10,000 and 9,000 radiocarbon years ago in the larger oil sands region (see also Fisher and Lowell, chapter 2 in this volume), winter conditions there might have been challenging. The oil sands region could have been occupied between spring and fall, while the Cold Lake and Thickwood Hills regions might have provided a more hospitable winter range. Whatever the situation may have been, it seems quite feasible that the technological organization evident at HhOu-27 was in play.

The D2 and HhOu-27 artifacts represent two very different approaches to lithic technology, at opposite extremes of the time spectrum. In the D2 cache, smaller artifacts, already committed to a few generalized tool forms, were being moved across the landscape. In the HhOu-27 assemblage, large but otherwise economical flakes had been prepared in such a way that the craftsperson would have a wide range of manufacturing alternatives. Yet these two quite exceptional sets of tools from the Birch Mountains and the oil sands region do have one feature in common. They clearly come as close as the present archaeological record allows to the "systemic" context for ancient tool kits to which Odess and Rasic (2007) refer. That is, in both cases the artifacts were simply set down by the people who had been carrying or making them, with negligible effects from any number of subsequent factors, ranging from discard practices to site formation processes, that routinely characterize archaeological contexts.

CONCLUSIONS

The Birch Mountains archaeological data provide a necessary counterpart to information from the Lower Athabasca valley, in which escalating oil sands development is occasioning a great many archaeological resource management studies. This is not simply because the Birch Mountains record is an integral component of the prehistory for the larger region. I would argue that, by investigating landscape use at this regional scale, we can engage in archaeological research that is oriented toward the resolution of fundamentally interesting questions that are perhaps not surfacing as often as they might, despite the extraordinary rate of impact assessment and mitigation currently underway. In attempting to cope with an archaeological record that is vanishing too rapidly, we seem to have done rather little thinking about why, for example, Beaver River Sandstone is so common in the oil sands region, but not elsewhere, or why its patterns of distribution might actually be different through time.

Tackling this issue of wider scale of landscape use has the beneficial effect of causing us to ask precisely where, how, and why the use of Beaver River Sandstone diminishes as we move away from the geological source area. The sites just now being discovered at or near the eastern edge of the Birch Mountains generally confirm the predictions of Fenton and Ives (1990; see also Ives 1993) about the dissipation of Beaver River Sandstone as people moved out of the Lower Athabasca valley, taking somewhat indirect routes but travelling toward adjacent uplands. It should prove possible to imagine more penetrating questions. For example, do some of these geographically intermediate sites contain artifacts composed primarily of Beaver River Sandstone, but relatively few tools, or do some contain materials more typical of the uplands, such as grey quartzite and pebble cherts, with very low debitage to tool ratios (Darryl Bereziuk, pers. comm., 2008)? Embedded in these questions are viable propositions about the modes of technological organization that First Nations ancestors might reasonably have adopted in entering and leaving the Birch Mountains upland—propositions that can be tested with the kinds of data being generated by current archaeological resource management studies.

I must also observe that while the archaeological record of northeastern Alberta is extremely difficult to work with—restricted, as it usually is, to lithics and frequently lacking in chronological and stratigraphic control—the Birch Mountains studies of previous decades continue to demonstrate that means exist to temper these difficulties. The measures I have in mind do, however, require carefully controlled excavations with piece plotting, spatial analysis (which can now be so greatly enhanced through intrasite applications of Geographic

Information System technology), attention to finer-scale variability in lithic raw materials, and systematic approaches to conjoinability or refitting studies. As the region's remarkable archaeological record continues to be consumed, it seems to me imperative that we *routinely* employ these methods, so that we do not simply assume that artifacts found together belong together in time (when, frequently, they do not) or that site distributions are anything other than the composite patterns one would expect for a settlement history extending backward in time to a receding, terminal Pleistocene world.

ACKNOWLEDGEMENTS

I thank Brian Ronaghan for his invitation to take part in this volume and for his editorial insights. Eric Damkjar, Michael Semenchuk, Robin Woywitka, Wendy Johnson, and Jessica Unger ably assisted with maps and diagrams; Jack Brink and Bob Dawe arranged access to Royal Alberta Museum collections; Joan Damkjar and Martina Purdon supplied much-needed site and report information. Ruthann Knudson generously provided detailed illustrative materials from the MacHaffie site in Montana. I enjoyed opportunities to discuss facets of this chapter with Eugene Gryba, Laura Roskowski, and Darryl Bereziuk. Brian Ronaghan and I both appreciated the invitation to take part in the Birch Mountains Wildland Provincial Park Biophysical Inventory, and particularly the support of Ted Johnson, Wayne Nordstrom, Joyce Gould, Drajs Vujnovic, and Dale Crossland. Ken Tingley and David Leonard provided masterful assistance with fur trade documents, for which I am grateful. Pamela Holway made most welcome editorial improvements to the manuscript.

NOTES

1. Gordon's (1996) sample of Taltheilei materials from the Northwest Territories might be applied in a comparative study with Cody Complex materials, for example, but the number of Taltheilei specimens from defined stratigraphic circumstances is rather small (in contrast to the larger proportion of surface-collected materials). The resemblance of some oil sands points to Taltheilei styles is also discussed in chapter 5 of this volume.

2. See, for example, the discussion of Cree Burn Lake projectile points in chapter 6 of this volume.

3. If this wear were a phenomenon of frost heaving and movement within the mineral soil, then all Eaglenest Portage artifacts should have comprehensive abrasion, as the soil matrix is sandy. This is also not the result of larger artifacts lying close together and moving through natural processes: that would not produce sufficiently comprehensive wear, and, in any case, there was at least some sand matrix between many of the artifacts.

4   There were also seven small flakes of Peace Point chert in the D2 cluster, which may be indicative of travel toward or contact with individuals who had been in the Peace River basin.

REFERENCES

Archival Sources

"The English River Book." Hudson's Bay Company Archives F.2/1. Provincial Archives of Manitoba, Winnipeg.

"Fort Chipewyan Post Journals." Hudson's Bay Company Archives B.39/a/18–31. Provincial Archives of Manitoba, Winnipeg.

"McKenzie, James. Journal, 1799–1800." Electronic transcription. McGill Fur Trade Project no. MFTP 0016. Masson collection no. MASS 2358. McGill University Libraries, Montréal.

Secondary Sources

Ackerman, Robert E.
    2001   Spein Mountain: A Mesa Complex Site in Southwestern Alaska. *Arctic Anthropology* 38(2): 81–97.

Andrews, Thomas D., Glen MacKay, and Leon Andrew
    2009   *Hunters of the Alpine Ice: The NWT Ice Patch Study*. Prince of Wales Northern Heritage Centre, Department of Education, Culture and Employment, Government of the Northwest Territories, Yellowknife.

Baker, Tony, and Michael Kunz
    2003   Contrasting the Lithic Technologies of Mesa and Folsom. Paper presented at the 68th annual meeting of the Society for American Archaeology, Milwaukee, Wisconsin, 9–13 April.

Bamforth, Douglas B.
    2002   High-Tech Foragers? Folsom and Later Paleoindian Technology on the Great Plains. *Journal of World Prehistory* 16(1): 55–98.
    2009   Projectile Points, People and Plains Paleoindian Perambulations. *Journal of Anthropological Archaeology* 28 (2009): 142–157.

Bayrock, L. A.
    1961   *Surficial Geology*. In J. D. Lindsay, S. Pawluk, and W. Odynsky, *Exploratory Soil Survey of Alberta Map Sheets 84-P, 84-I, and 84-H*, Appendix, pp. 49–51. Preliminary Soil Survey Report 62-1. Alberta Research Council, Edmonton.

Bereziuk, Darryl A.
    2001   The Smuland Creek Site and Implications for Palaeoindian Site Prospection in the Peace Region of Northwestern Alberta. In *Sovremeniye Problemiy Evraziyskovo Paleolitovedeniya* (Modern Problems in the Eurasian Palaeolithic), edited by A. P. Derevianko and G. I. Medvedev, pp. 382–402. Institute of Archaeology and Ethnography, Novosibirsk, Russia.

Bradley, Bruce A., Michael B. Collins, and Andrew Hemmings
    2010   *Clovis Technology*. Archaeologial Series no. 17. International Monographs in Prehistory, Ann Arbor, Michigan.

Bryan, Alan L., and Gerald Conaty
    1975   A Prehistoric Athapaskan Campsite in Northwestern Alberta. *Western Canadian Journal of Anthropology* 5(3–4): 64–91.

Carlson, Muriel

    1993   Collections Can Speak for Themselves: A Regional Profile of Paleo-Indian Lithics from West-Central Saskatchewan, as Viewed from Local Collections. Saskatchewan Heritage Branch, Type B Permit 89-10. Copy on file, Heritage Conservation Branch, Government of Saskatchewan, Regina.

Cinq-Mars, Jacques

    1973   An Archaeologically Important Raw Material from the Tertiary Hills, Western District of Mackenzie, Northwest Territories: A Preliminary Statement. In *Preliminary Archaeological Study, Mackenzie Corridor,* Appendix E. Environmental-Social Committee, Northern Pipelines, Task Force on Northern Oil Development, Report no. 73-10. Information Canada, Ottawa.

Collins, Michael B.

    2004   The Nature of Clovis Blades and Blade Cores. In *Entering America: Northeast Asia and Beringia Before the Last Glacial Maximum,* edited by David B. Madsen, pp. 159–183. University of Utah Press, Salt Lake City.

Conaty, Gerald T.

    1977   Excavation of the Wentzel Lake Site, Project 76-11. In *Archaeology in Alberta, 1976,* compiled by J. Michael Quigg, pp. 31–36. Archaeological Survey of Alberta Occasional Paper no. 4. Historic Resources Management Branch, Alberta Culture, Edmonton.

Doll, Maurice F. V.

    1982   *The Boss Hill Site (FdPe-4) Locality 2: Pre-Archaic Manifestations in the Parkland of Central Alberta, Canada.* Provincial Museum of Alberta, Human History Occasional Paper no. 2. Historic Resources Management Branch, Alberta Culture, Edmonton.

Donahue, Paul F.

    1976   *Archaeological Research in Northern Alberta, 1975.* Archaeological Survey of Alberta Occasional Paper no. 2. Historic Resources Management Branch, Alberta Culture, Edmonton.

Driver, Jonathan C., Martin Handly, Knut R. Fladmark, D. Erle Nelson, Gregg M. Sullivan, and Randall Preston

    1996   Stratigraphy, Radiocarbon Dating, and Culture History of Charlie Lake Cave, British Columbia. *Arctic* 49(3): 265–277.

Duckworth, Harry W. (editor)

    1990   *The English River Book: A North West Company Journal and Account Book of 1786.* McGill-Queen's University Press, Montréal and Kingston.

Ellis, Christopher J.

    2011   Measuring Paleoindian Range Mobility and Land-Use in the Great Lakes/Northeast. *Journal of Anthropological Archaeology* 30: 385–401.

Fedje, Daryl, James M. White, Michael C. Wilson, D. Erle Nelson, John S. Vogel, and John R. Southon

    1995   Adaptations and Environments in the Canadian Rockies During the Latest Pleistocene and Early Holocene. *American Antiquity* 60(1): 81–108.

Fenton, Mark M., and John W. Ives

    1982   Preliminary Observations on the Geological Origins of Beaver River Sandstone. In *Archaeology in Alberta 1981,* compiled by Jack Brink, pp. 166–189. Archaeological Survey of Alberta Occasional Paper no. 19. Historic Resources Management Branch, Alberta Culture, Edmonton.

    1984   The Stratigraphic Position of Beaver River Sandstone. In *Archaeology in Alberta 1983,* compiled by David Burley, pp. 128–136. Archaeological Survey of Alberta Occasional Paper no. 23. Historic Resources Management Branch, Alberta Culture, Edmonton.

    1990   Geoarchaeological Studies of the Beaver River Sandstone, Northeastern Alberta. In *Archaeological Geology of North America*, edited by Norman P. Lasca and Jack Donahue, pp. 123–135. Centennial Special Volume 4. Geological Society of America, Boulder, Colorado.

Fisher, Timothy G.

2007 Abandonment Chronology of Glacial Lake Agassiz's Northwestern Outlet. *Palaeogeography, Palaeoclimatology, Palaeoecology* 246: 31–44.

Frison, George C., and Dennis J. Stanford

1982 *The Agate Basin Site: A Record of Paleoindian Occupation of the Northwestern High Plains.* Academic Press, New York.

Gordon, Bryan H. C. A.

1977a Chipewyan Prehistory. In *Problems in the Prehistory of the North American Subarctic: The Athapaskan Question*, edited by James W. Helmer, Stanley Van Dyke, and François J. Kense, pp. 72–76. Archaeological Association, Department of Archaeology, University of Calgary, Calgary.

1977b Temporal, Archaeological and Pedological Separation of the Barrenland Arctic Small Tool and Taltheilei Traditions. In *Problems in the Prehistory of the North American Subarctic: The Athapaskan Question*, edited by James W. Helmer, Stanley Van Dyke, and François J. Kense, pp. 77–82. Chacmool Archaeological Association, Department of Archaeology, University of Calgary, Calgary.

1996 *People of Sunlight, People of Starlight: Barrenland Archaeology in the Northwest Territories of Canada.* Mercury Series no. 154. Archaeological Survey of Canada, Canadian Museum of Civilization, Gatineau, Québec.

Gryba, Eugene M., and Jennifer C. Tischer

2005 *Historical Resources Impact Assessment, Deer Creek Energy Limited Joslyn North Mine Project: Final Report (ASA Permit 05-094).* Copy on file, Archaeological Survey, Historic Resources Management Branch, Alberta Culture, Edmonton.

2009 *Historical Resources Impact Assessment, UTS Energy Limited / Teck Cominco Limited, Equinox Project: Final Report (ASA Permit 08-265).* Copy on file, Archaeological Survey, Historic Resources Management Branch, Alberta Culture, Edmonton.

Guthrie, R. Dale

2006 New Carbon Dates Link Climatic Change with Human Colonization and Pleistocene Extinctions. *Nature* 441 (11 May): 207–209.

Haile, James, Duane G. Froese, Ross D. E. MacPhee, Richard G. Roberts, Lee J. Arnold, Alberto V. Reyes, Morten Rasmussen, Rasmus Nielsen, Barry W. Brook, Simon Robinson, Martina Demuro, M. Thomas P. Gilbert, Kasper Munch, Jeremy J. Austin, Alan Cooper, Ian Barnes, Per Möller, and Eske Willerslev

2009 Ancient DNA Reveals Late Survival of Mammoth and Horse in Interior Alaska. *Proceedings of the National Academy of Sciences* 106(52): 22352–22357.

Handly, Martin J.

1993 Lithic Assemblage Variability at Charlie Lake Cave (HbRf-39): A Stratified Rock Shelter in Northeastern British Columbia. MA thesis, Department of Anthropology, Trent University.

Hare, P. Gregory, Thomas J. Hammer, and Ruth M. Gotthardt

2008 The Yukon Projectile Point Data Base. In *Projectile Point Sequences in Northwestern North America*, edited by Roy L. Carlson and Martin P. R. Magne, pp. 321–332. Department of Archaeology Publication no. 35. Archaeology Press, Simon Fraser University, Burnaby, British Columbia.

Hauge, T. M., and L. B. Keith

1980 *Dynamics of Moose Populations in the AOSERP Study Area in Northeastern Alberta.* Report prepared for the Alberta Oil Sands Environmental Research Program, Project LS 21.1.1, Alberta Energy, Edmonton, and Environment Canada, Ottawa.

Heffner, Ty

2002   *KaVn-2: An Eastern Beringian Tradition Archaeological Site in West-Central Yukon Territory, Canada.* Occasional Papers in Archaeology no. 10. Hud Hudän Series, Heritage Branch, Government of Yukon, Whitehorse.

Heintzman, Peter D., Duane G. Froese, John W. Ives, André E. R. Soares, Grant D. Zazula, Brandon Letts, Thomas D. Andrews, Jonathan C. Driver, Elizabeth Hall, P. Gregory Hare, Christopher N. Jass, Glen MacKay, John R. Southon, Mathias Stiller, Robin Woywitka, Marc A. Suchard, and Beth Shapiro

2016   Bison Phylogeography Constrains Dispersal and Viability of the Ice Free Corridor in Western Canada. *Proceedings of the National Academy of Sciences of the United States of America* 113(29): 8057–8063.

Holmes, Charles E.

2001   Tanana River Valley Archaeology circa 14,000 to 9000 B.P. *Arctic Anthropology* 38(2): 154–170.

Ives, John W.

1977a   The Excavation of HkPa4, Birch Mountains, Alberta. In *Archaeology in Alberta, 1976,* compiled by J. Michael Quigg, pp. 37–44. Archaeological Survey of Alberta Occasional Paper no. 4. Historic Resources Management Branch, Alberta Culture, Edmonton.

1977b   A Spatial Analysis of Artifact Distribution on a Boreal Forest Archaeological Site. MA thesis, Department of Anthropology, University of Alberta.

1977c   Pollen Separation of Three North American Birches. *Arctic and Alpine Research* 9: 73–80.

1981a   The Prehistory of Northern Alberta. In *Alberta Archaeology: Retrospect and Prospect,* edited by Terry A. Moore, pp. 39–58. Archaeological Society of Alberta, Lethbridge.

1981b   Birch Mountain Archaeological Study, 1980, Permit 8080. In *Archaeology in Alberta, 1980,* compiled by Jack Brink, pp. 127–138. Archaeological Survey of Alberta Occasional Paper no. 17. Historic Resources Management Branch, Alberta Culture, Edmonton.

1982a   Birch Mountain Archaeological Study, 1981. In *Archaeology in Alberta, 1981,* compiled by Jack Brink, pp. 61–70. Archaeological Survey of Alberta Occasional Paper no. 19. Historic Resources Management Branch, Alberta Culture, Edmonton.

1982b   Evaluating the Effectiveness of Site Discovery Techniques in Boreal Forest Environments. In *Directions in Alberta: A Question of Goals*, edited by Peter D. Francis and Eric C. Poplin, pp. 95–114. Chacmool Archaeological Association, Department of Archaeology, University of Calgary, Calgary.

1985   *A Spatial Analysis of Artifact Distribution on a Boreal Forest Archaeological Site.* Archaeological Survey of Alberta Manuscript Series no. 5. Historic Resources Management Branch, Alberta Culture, Edmonton.

1986   A First Millennium B.C. Smudge Pit from Eaglenest Lake, Birch Mountains. In *Archaeology in Alberta, 1985,* edited by John W. Ives, pp. 201–205. Archaeological Survey of Alberta Occasional Paper no. 29. Historic Resources Management Branch, Alberta Culture, Edmonton.

1987   Review of *Window on the Past: Archaeological Assessment of the Peace Point Site, Wood Buffalo National Park, Alberta*, by Marc G. Stevenson. *Prairie Forum* 12(2): 309–311.

1990   *A Theory of Northern Athapaskan Prehistory*. Westview Press, Boulder, Colorado, and University of Calgary Press, Calgary.

1993   The Ten Thousand Years Before the Fur Trade in Northeastern Alberta. In *The Uncovered Past: Roots of Northern Alberta Societies*, edited by Patricia A. McCormack and R. Geoffrey Ironside, pp. 5–31. Circumpolar Research Series no. 3. Canadian Circumpolar Institute, University of Alberta, Edmonton.

1998   Developmental Processes in the Pre-contact History of Athapaskan, Algonquian and Numic Kin Systems. In *Transformations of Kinship,* edited by Maurice Godelier, Thomas

R. Trautmann, and Franklin Tjon Sie Fat, pp. 94–139. Smithsonian Institution Press, Washington, D.C.

2003   Alberta, Athapaskans and Apachean Origins. In *Archaeology in Alberta: A View from the New Millennium*, edited by Jack W. Brink and John F. Dormaar, pp. 256–289. Archaeological Society of Alberta, Medicine Hat.

2006   13,001 Years Ago—Human Beginnings in Alberta. In *Alberta Formed, Alberta Transformed*, edited by Michael Payne, Don Wetherell, and Cathy Cavanaugh, vol. 1, pp. 1–34. Calgary and Edmonton: University of Calgary Press and University of Alberta Press.

Ives, John W., and Mark M. Fenton

1985   *Progress Report for the Beaver River Sandstone Geological Source Study (ASA Permit 83-054)*. Copy on file, Archaeological Survey, Historic Resources Management Branch, Alberta Culture, Edmonton.

Ives, John W., and Karie Hardie

1983   Occurrences of Tertiary Hills Welded Tuff in Northern Alberta. In *Archaeology in Alberta, 1982*, compiled by David Burley, pp. 171–176. Archaeological Survey of Alberta Occasional Paper no. 21. Historic Resources Management Branch, Alberta Culture, Edmonton.

Ives, John W., Duane G. Froese, Kisha Supernant, and Gabriel Yanicki

2013   Vectors, Vestiges and Valhallas—Rethinking the Corridor. In *Paleoamerican Odyssey*, edited by Kelly E. Graf, Caroline V. Ketron, and Michael R. Waters, pp. 149–169. Peopling of the Americas Publications, Center for the Study of the First Americans, Texas A & M University, College Station, Texas.

Kelly Robert L., and Lawrence C. Todd

1988   Coming into the Country: Early Paleoindian Hunting and Mobility. *American Antiquity* 53(2): 231–244.

Knudson, Ruthann

1983   *Organizational Variability in Late Paleo-Indian Assemblages*. Reports of Investigations no. 60. Laboratory of Anthropology, Washington State University, Pullman, Washington.

Kunz, Michael, Michael Bever, and Constance Adkins

2003   *The Mesa Site: Paleoindians Above the Arctic Circle*. BLM–Alaska Open File Report 86. Bureau of Land Management, U.S. Department of the Interior, Alaska State Office, Anchorage.

Le Blanc, Raymond J.

2004   *Archaeological Research in the Lesser Slave Lake Region: A Contribution to the Pre-Contact History of the Boreal Forest of Alberta*. Mercury Series no. 166. Archaeological Survey of Canada, Canadian Museum of Civilization, Gatineau, Québec.

Le Blanc, Raymond J., and Milton J. Wright

1990   Macroblade Technology in the Peace River Region of Northwestern Alberta. *Canadian Journal of Archaeology* 14: 1–2.

Lowell, Thomas V., Timothy G. Fisher, Gary C. Comer, Irka Haidas, Nicholas Waterson, Katherine Glover, Henry M. Loope, Joerg M. Schaefer, Vincent Rinterknecht, Wallace Broecker, George Denton, and James T. Teller

2005   Testing the Lake Agassiz Meltwater Trigger for the Younger Dryas. *Eos, Transactions, American Geophysical Union* 86(40): 365–373.

MacDonald, Glen M.

1987   Postglacial Vegetation History of the Mackenzie River Basin. *Quaternary Research* 28: 245–262.

MacDonald, Glen M., and Les C. Cwynar

1985   A Fossil Pollen Based Pollen Reconstruction of the Late Quaternary History of Lodgepole Pine (*Pinus contorta* ssp. *latifolia*) in the Western Interior of Canada. *Canadian Journal of Forest Research* 15: 1039–1044.

MacDonald, Glen M., and T. Katherine McLeod

1996   The Holocene Closing of the "Ice-Free" Corridor: A Biogeographical Perspective. *Quaternary International* 32: 87–95.

MacNeish, Richard S.

1964   *Investigations in Southwest Yukon: Archaeological Excavations, Comparisons, and Speculations.* Papers of the Robert S. Peabody Foundation for Archaeology, vol. 6, no. 2. Phillips Academy, Andover, Massachusetts.

Mathewson, Pamela Ann

1974   The Geographical Impact of Outsiders on the Community of Fort Chipewyan, Alberta. MA thesis, Department of Geography, University of Alberta.

McCullough, Edward J.

1981   *The Duckett Site (GdOo-16): An Evaluative Study (ASA Permit 80-155).* Copy on file, Archaeological Survey, Historic Resources Management Branch, Alberta Culture, Edmonton.

Meyer, David

1984   The Development of the Marriage Isolate Among the Pas Mountain Indians. *Western Canadian Anthropologist* 1: 2–10.

Millar, James F. V.

1968   Archaeology of Fisherman Lake, Western District of Mackenzie, N.W.T. PhD dissertation, Department of Archaeology, University of Calgary.

Morrison, David A.

1984   The Late Prehistoric Period in the Mackenzie Valley. *Arctic* 37(3): 195–209.

Morrison, David A.

1987   The Middle Prehistoric Period and the Archaic Concept in the Middle Mackenzie Valley. *Canadian Journal of Archaeology* 11: 49–74.

Odess, Daniel, and Jeffrey T. Rasic

2007   Toolkit Composition and Assemblage Variability: The Implications of Nogahabara I, Northern Alaska. *American Antiquity* 72(4): 691–717.

Peck, Trevor R., and John W. Ives

2001   Late Side-Notched Projectile Points in the Northwestern Plains. *Plains Anthropologist* 46(176): 163–193.

Powers, William R., and John F. Hoffecker

1989   Late Pleistocene Settlement in the Nenana Valley, Central Alaska. *American Antiquity* 54(2): 263–287.

Pyszczyk, Heinz W.

2003   Aboriginal Bows and Arrows and Other Weapons in Alberta: The Last 2,000 Years, or Longer? In *Archaeology in Alberta: A View from the New Millennium*, edited by Jack W. Brink and John F. Dormaar, pp. 46–71. Archaeological Society of Alberta, Medicine Hat.

Rawluk, Matt, Aileen Reilly, Peter Stewart, and Gabriel Yanicki

2011   Identification of a Palaeoindian Occupation in Compressed Stratigraphy: A Case Study from Ahai Mneh (FiPp-33). *Diversipede* 1(1): 1–15.

Reimer, Paula J., Edouard Bard, Alex Bayliss, J. Warren Beck, Paul G. Blackwell, Christopher Bronk Ramsey, Caitlin E. Buck, Hai Cheng, R. Lawrence Edwards, Michael Friedrich, Pieter M. Grootes, Thomas P. Guilderson, Haflidi Haflidason, Irka Hajdas, Christine Hatté, Timothy J. Heaton, Dirk L. Hoffmann, Alan G. Hogg, Konrad A. Hughen, K. Felix Kaiser, Bernd Kromer, Sturt W. Manning, Mu Niu, Ron W. Reimer, David A. Richards, E. Marian Scott, John R. Southon, Richard A. Staff, Christian S. M. Turney, and Johannes van der Plicht

2013   IntCal13 and Marine13 Radiocarbon Age Calibration Curves 0–50,000 Years Cal BP. *Radiocarbon* 55(4): 1869–1887.

Roskowski, Laura, and Morgan Netzel

    2011   *Historical Resources Impact Mitigation, Shell Canada Energy, Muskeg River Mine Expansion of RMS 10, Mitigation for Sites HhOv-87 and HhOv-200: Final Report (ASA Permit 09-168).* Copy on file, Archaeological Survey, Historic Resources Management Branch, Alberta Culture, Edmonton.

Rowe, J. S.

    1972   Forest Regions of Canada, Department of Environment, Canadian Forestry Service, Publication no. 1300.

Saxberg, Nancy, and Brian O. K. Reeves

    2003   The First Two Thousand Years of Oil Sands History: Ancient Hunters at the Northwest Outlet of Glacial Lake Agassiz. In *Archaeology in Alberta: A View from the New Millennium*, edited by Jack W. Brink and John F. Dormaar, pp. 290–322. Archaeological Society of Alberta, Medicine Hat.

Shott, Michael J.

    1997   Stones and Shaft Redux: The Metric Discrimination of Chipped-Stone Dart and Arrow Points. *American Antiquity* 62(1): 86–101.

Sims, Cort

    n.d. [1975]   An Archaeological Survey of the Namur Lake Area in Northeastern Alberta. Unpublished manuscript in possession of the author.

    1980   Models for Explaining Material Culture Form and Distribution at Two Archaeological Sites in Northern Alberta. Draft report on file (CRM 194), Archaeological Survey, Historic Resources Management Branch, Alberta Culture, Edmonton

Smith, Derald G., and Timothy G. Fisher

    1993   Glacial Lake Agassiz: The Northwestern Outlet and Paleoflood. *Geology* 21(1): 9–12.

Speth, John D., Khori Newlander, Andrew A. White, Ashley K. Lemke, and Lars E. Anderson

    2010   Early Paleoindian Big-Game Hunting in North America: Provisioning or Politics? *Quaternary International* 285: 111–139.

Spurling, Brian E.

    1980   *The Site C Heritage Resource Inventory and Assessment Final Report: Substantive Contributions.* Report prepared for B.C. Hydro and Power Authority. Copy on file, B.C. Archaeology Branch, Ministry of Forest, Lands and Natural Resource Operations, Government of British Columbia, Victoria.

Stevenson, Marc G.

    1985   The Formation of Artifact Assemblages at Workshop/Habitation Sites: Models from Peace Point in Northern Alberta. *American Antiquity* 50(1): 63–81.

    1986   *Window on the Past: Archaeological Assessment of the Peace Point Site, Wood Buffalo National Park, Alberta.* Studies in Archaeology, Architecture and History. National Historic Parks and Sites Branch, Parks Canada, Environment Canada, Ottawa.

Tanner, James N., C. Cormack Gates, and Bertha Ganter

    2001   Some Effects of Oil Sands Development on the Traditional Economy of Fort McKay. Fort McKay Industry Relations Corporation, Fort McMurray, Alberta.

Vance, Robert E.

    1986   Pollen Stratigraphy of Eaglenest Lake, Northeastern Alberta. *Canadian Journal of Earth Sciences* 23: 11–20.

Van Dyke, Stanley, and Sally Stewart

    1985   *Hawkwood Site (EgPm-179): A Multicomponent Prehistoric Campsite on Nose Hill.* Archaeological Survey of Alberta Manuscript Series no. 7. Historic Resources Management Branch, Alberta Culture, Edmonton.

Van Waas, C.

    1974   *Biophysical Analysis and Evaluation of Capability: Namur Lake Area.* Report prepared for the Land Use Assignment Branch, Alberta Lands and Forests, Edmonton.

Vivian, Brian C.

    1998   *Historic Resource Conservation Excavations, EgPn-230: Final Report (ASA Permit 97-083)*. Copy on file, Archaeological Survey, Historic Resources Management Branch, Alberta Culture, Edmonton.

Wallace, J. N.

    1929   *The Wintering Partners on Peace River*. Ottawa: Thorburn and Abbott.

Wilson, Michael C., John Visser, and Martin P. R. Magne

    2011   Microblade Cores from the Northwestern Plains at High River, Alberta, Canada. *Plains Anthropologist* 56(217): 23-36.

Wright, James V.

    1972   *The Shield Archaic*. Publications in Archaeology no. 3. National Museum of Man, Ottawa.

    1975   *The Prehistory of Lake Athasbasca: An Initial Statement*. Mercury Series no. 29. Archaeological Survey of Canada, National Museum of Man, Ottawa.

Yesner, David R.

    2001   Human Dispersal into Interior Alaska: Antecedent Conditions, Mode of Colonization, and Adaptations. *Quaternary Science Reviews* 20: 315-327.

3 Lithic Resource Use

9 **Beaver River Sandstone** | Characteristics and Use, with Results of Heat Treatment Experiments

EUGENE M. GRYBA

Beaver River Sandstone (BRS) was the primary stone type utilized by knappers throughout the precontact period in the Fort McKay region. It occurs in bedrock within the Lower Cretaceous McMurray Formation, particularly along the east side of the Athabasca River valley to the south of Fort McKay, and as isolated pieces in gravel deposits at least as far north as Bitumount. In addition, in view of the occurrence of BRS at sites in the area, a source of the stone may exist along the Clearwater River upstream from Fort McMurray, although such a source has yet to be identified. In the Fort McKay area, BRS is generally found in progressively lower frequencies as the distance of archaeological sites from the stone's core areas of occurrence increases to the point where BRS is largely replaced by other locally obtainable lithics.

Since it was first reported in 1973, BRS has been the object of a number of studies that have sought to accurately define its origin, age, and physical characteristics, as well as its mineral composition. Initially named Beaver Creek Quartzite, the stone is now generally known as Beaver River Sandstone or as Beaver River Silicified Sandstone. More recently, the term "Muskeg Valley Microquartzite" has been proposed to distinguish a locally occurring, very fine-textured BRS from the more common coarse-textured variety.

In addition, experimental studies have been carried out to determine how BRS responds to heat treatment. As these studies demonstrate, heating BRS to around 400°C to 450°C tremendously improves its workability and also produces certain distinctive features that are evident on many archaeological specimens.

333

Heating causes BRS to recrystallize, making it much easier to work by percussion and pressure methods than the raw lithic material. Upon being heated, the stone often develops a thin, rusty red rind on the original cortex and along existing fracture planes. Unlike raw BRS, the heat-treated material also has a smoother, lustrous fracture surface. Both the red rind and the smooth, glassy fracture surface are attributes visible on many archaeological specimens of BRS recovered from the Fort McKay area. These traits are excellent evidence that deliberate heat treatment had been practiced by knappers throughout the precontact period in this part of the province. The practice of heat treatment may also explain why no high-quality, lustrous BRS has been found in bedrock situations but is nonetheless prevalent in archaeological sites.

BRS appears to be a lithic material markedly distinct from the various varieties of quartzite of Alberta Rocky Mountain provenance. It may have an origin similar to that of silcrete. Some of the possible Cretaceous-age quartzite found in Quaternary deposits in southwestern Manitoba approach BRS in terms of colour and texture, as well as by displaying isolated quartz crystals and plant impressions. In texture and colour, BRS overlaps with the Cretaceous-age Dakota Sandstone found in northern and central Colorado.

HISTORY OF RESEARCH ON BEAVER RIVER SANDSTONE

In his detailed mapping of the Athabasca oil sands, Carrigy (1966) presented the first geological account of the bedrock material that has come to be known as Beaver River Sandstone. In 1973, during an archaeological assessment conducted in connection with Syncrude's Lease 17, a precontact quarry site (HgOv-29) was discovered on the lower part of Beaver Creek, to the north of the Lease 17 area (Syncrude Canada Ltd. 1973, 87). A subsequent report described the site as located "on the north side of an abandoned channel of Beaver Creek," in SE 1-94-11-W4M, around 1.6 kilometres west of the confluence of Beaver Creek with the Athabasca River, and indicated that it was situated on "an outcrop of quartzite bedrock which was overlain by a few feet of sand and gravelly sediments" (Syncrude Canada Ltd. 1974, 15, and see figures 1 and 3). As the report noted, in the earlier archaeological survey report, the stone was "mistakenly identified as limestone and chert": the dull grey, fossil-bearing variety was considered to be limestone, while the highly siliceous variety was termed chert. Following an examination of thin sections and chemical tests, however, the material was deemed to be "clearly a quartzite" (Syncrude Canada Ltd. 1974, 47; cf. Syncrude Canada Ltd. 1973, 79–80). The 1974 report applied the name "Beaver Creek

Quartzite" to distinguish the stone found at the quarry from other quartzites that occurred throughout the area and that derived from nonlocal bedrock formations.

By 1980, the results of academic and development-related archaeological research on both sides of the Athabasca River about 20 kilometres upstream and downstream from Fort McKay had demonstrated that lithics similar to "Beaver Creek Quartzite" constituted the dominant material used by precontact knappers in close to three hundred sites (Fenton and Ives 1982, 166). In 1981, Mark Fenton, of the Alberta Geological Survey, and John Ives, of the Archaeological Survey of Alberta, visited the Beaver Creek site with the aim of clarifying the origin and age of the stone. The following year, they revisited the quarry and also broadened their search to portions of the Athabasca, Muskeg, MacKay, and Firebag rivers, with the objective of delimiting the natural bedrock occurrences of BRS (Ives and Fenton 1983, 78). Fenton and Ives renamed the stone "Beaver River Sandstone" because they considered the material to be a sedimentary rather than a metamorphic rock and because "Beaver River" is the official name of the stream that the Syncrude reports had called "Beaver Creek" (Fenton and Ives 1982, 175).

Since the initial identification of BRS, the Fort McKay district has witnessed considerable archaeological activity, relating mainly to development of the oil sands. During this period, the bedrock formation containing BRS has been narrowed to the lower part of the McMurray Formation. It has also become evident that the material is more widespread than previously suspected and that it is quite variable in quality. There have also been further modifications to the name applied to the rock. By 2003, the term "Beaver River Silicified Sandstone" had been adopted by some archaeologists (see, for example, Saxberg and Reeves 2003, 292). More recently, the term "Muskeg Valley Microquartzite" has been used to define a very fine-grained facies of BRS found at the Quarry of the Ancestors (De Paoli 2005, 5–6). The accumulated data now enable us to make accurate statements as to the age, physical variability, natural occurrence, patterns of exploitation, spatial dispersal by precontact peoples, and so on, of what was obviously a locally significant lithic material for precontact knappers.

PHYSICAL AND CHEMICAL CHARACTERISTICS

Some caution must be exercised when one is interpreting the physical characteristics of BRS as previously described by various researchers, as often they did not state whether they were referring to raw BRS obtained directly from bedrock outcrops or to BRS that had been recovered from archaeological sites. Experiments I have undertaken reveal that significant physical differences exist between raw and heat-treated BRS, and, as an examination of existing collections demonstrates, it is the latter that is most apt to be represented at archaeological sites.

One of the first reports to refer to BRS noted that it is "sometimes glassy, breaks with fairly sharp edges, and is easily flaked and shaped into stone tools. There are unconformities in the stone, however, which complicate the process. Molds of fossil organisms are also common. Some of the quartzite is quite granular and [it] is therefore difficult to control its fracture (although frost action has broken much of the residual material at the site)" (Syncrude Canada Ltd. 1974, 47). Describing the BRS that occurred in a metre-thick layer (their Unit 2) at the Beaver River Quarry, Fenton and Ives commented:

> The rock is a bimodal, silica cemented, quartz sandstone. The texture is usually medium to fine grained sand floating in a matrix of very fine sand and silt. Locally isolated pebbles or small lenses of very coarse to medium grained sandstone are present within the unit. . . . Microscopic examination of hand specimens shows the rock to be composed of quartz and a few fine black grains. Thin sections and x-ray diffraction analysis show the rock to be composed of quartz. (1982, 172)

With regard to colour, Fenton and Ives described the stone as "generally light grey (10 YR 7/1) on a fresh surface" but noted that it "ranges through 10 YR 5/1 to 5 YR 7/2–3. A colour banding consisting of 1 to 5 millimetre streaks of light grey and grey (10 YR 5/1) is present in places, especially where the matrix exceeds 90%" (1982, 172). They observed "no reddish iron staining" on the sandstone obtained from Unit 2 but found this trait on one artifact from the Beaver River Quarry as well as on some artifacts collected from sites in the Birch Mountains and in the area of the Shell Alsands lease (1982, 174).

A slightly better quality for stone tool manufacture than that observed at the Beaver River Quarry was discovered by Fenton and Ives at two outcrops at the Cree Burn Lake site (Ives and Fenton 1983, 82, 85). As they noted, however, they failed to find any of the very fine-grained, high-quality material from which artifacts in the oil sands area were manufactured. It is likely that they were comparing raw BRS, from bedrock, with BRS from archaeological sites that had probably been heat-treated.

Tsang carried out extensive field and laboratory investigations on BRS. He reported that "in outcrop, the BRS is a light to dark grey, silicified sandstone of variable thickness. Where it is thick (>1m), outcrop is characterized by massive, angular to rounded boulders that are usually vertically jointed. If it is poorly exposed or is thin, BRS outcrop is found as a flat lying, continuous rock floor" (1998, 16). He further described BRS as

> a fine to coarse grained, microcrystalline quartz cemented quartz sandstone. It is characterized by poor sorting, with both well rounded and angular grains. Though sedimentary structures such as cross-bedding have been observed in BRS, they are not common. Organic material is present in low amounts in BRS, but the content varies from sample to sample. The variation in light and dark colouring is due to a heterogeneous distribution of the bitumen, giving the rock a coarsely mottled appearance. (1998, 17)

Tsang's X-ray diffraction data showed that over 99% of the coarse, fine, and microcrystalline grains consisted of quartz and that the matrix was composed of microcrystalline quartz. He also detected trace amounts of feldspars and anatase (TiO_2) in BRS and noted that titanium oxides and bitumen constituted its opaque material (1998, 34).

Commenting on the fracture property of BRS, Tsang (1998, 37) observed that earlier-forming microcrystalline quartz grains could be found subsequently encased in thin, pore-filling, anhedral quartz cement, which resulted in BRS being well indurated. He observed that the cementation was so strong that framework grains often fractured during rock chip preparation, rather than separating along the grain boundaries. Although this rock fracture pattern is one of the defining characteristics of quartzite (Pearl 1962, 93), Fenton and Ives (1982, 175) regarded the material as sandstone because of its sedimentary rather than metamorphic source.

Artifacts recovered from the Quarry of the Ancestors were originally considered to be made of silicified limestone on the basis of the fine texture of the material and also because of the relatively close proximity of the site to a limestone outcrop. Through petrographic examination, however, De Paoli determined that the artifacts were manufactured from a very fine-grained facies of BRS, one that he classified as a "microquartz-cemented orthoquartzitic siltstone" (2005, 5–6). He observed that the artifacts made from this stone

> have a light tan to light grey colour with a "speckled" appearance due to scattered larger (0.5–1 mm) quartz grains. . . . Matrix was not discernible to the naked eye or hand lens, but an examination of thin sections showed the material to be composed of 95–99% of

very fine grained (10–100 μm) anhedral to subhedral quartz with scattered large (0.5–1.0 mm) euhedral to subhedral grains. (2005, 6)

De Paoli also noted that both the matrix and the large grains of BRS showed the same ragged edges and pitted surfaces that Fenton and Ives (1990, 132) and Tsang (1998) had observed.

AGE AND STRATIGRAPHIC POSITION

The age of the bedrock formation that is the primary source of BRS has been an object of speculation since the study of this important lithic material began. As Fenton and Ives noted in 1990, rather than making a systematic effort to identify the source of BRS, archaeologists had suggested "an extraordinarily broad range of geological ages and origins" for the material, often on the basis of "little or no supporting data" (1990, 123). By the early 1980s, a wide range of possible formations had indeed been proposed as the most likely source of BRS. These included the Devonian Waterways Formation, the Cretaceous "pre-McMurray" or lower McMurray Formation, the uppermost McMurray Formation, and the Cretaceous basal Wabiskaw member of the Clearwater Formation, as well as Quaternary glacial and fluvial sources (Fenton and Ives 1990, 127).

Carrigy had assigned a "thin bed of quartz-cemented sandstone that outcrops extensively between the Athabasca and Muskeg Rivers east of Fort McKay in Township 94 NS Range 10" (1966, 6) to a questionable "pre-McMurray" age, viewing it as part of the basal Cretaceous sandstones overlying an eroded Devonian surface (1966, 9). Similarly, according to the 1974 Syncrude study of the Beaver River Quarry, the BRS at that site "derived from a three- to five-foot bed of thin material which occurs on top of Devonian Age deposits at the site." The report further noted that the quartzite was of unknown age and origin but speculated that it was probably a "pre-McMurray formation, Cretaceous Age deposit" and perhaps an Early Cretaceous one (Syncrude Canada Ltd. 1974, 47).

Fenton and Ives (1982, 172) identified the metre-thick layer at the Beaver River Quarry that they called Unit 2 as the source of the stone from which the artifacts at the site had been manufactured. Unit 2, which formed the land surface at the site, was underlain by a 6-metre-thick sequence of fine-grained, laminated bituminous sand (Unit 1a) and a light grey, slightly sandy silty clay (Unit 1b). In some parts of the quarry, the break between Units 1 and 2 was sharp, while, in another locality, a gradation from siliceous sandstone to bituminous sand was observed over a vertical distance of around 30 centimetres (Fenton and Ives 1990, 128).

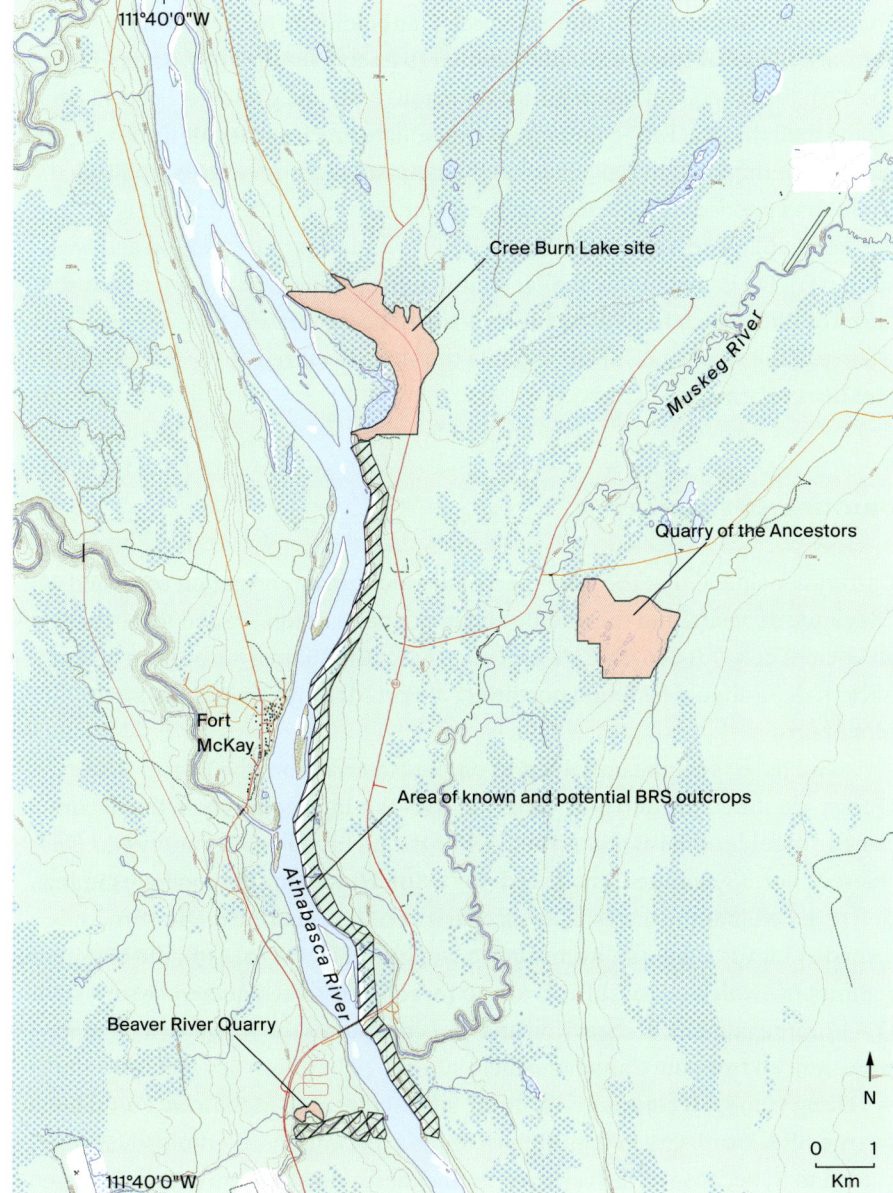

Figure 9.1. Known and potential sources of Beaver River Sandstone in the Athabasca oil sands area near Fort McKay

According to Fenton and Ives, the BRS that Carrigy had mapped on the east side of the Athabasca River was from the same sedimentary deposit as that found at the Beaver River Quarry (1982, 179). At the Cree Burn Lake site (fig 9.1), they discovered an in situ unit of BRS that was both overlain and underlain by bituminous sands and that they interpreted as belonging to the McMurray Formation, probably to its middle member (Ives and Fenton 1983, 82 and figures 5 and 6). In subsequent investigations, Fenton, Ives, and geologist Peter Flach established that the BRS-bearing unit was situated near the top of the lower member of the McMurray Formation (Fenton and Ives 1990, 130–131). Flach (1984) had previously reported that the lower member of the McMurray Formation filled depressions on an eroded Devonian surface (Fenton and Ives 1990, 131). De Paoli noted a similar stratigraphic position for the finer-textured facies of BRS, which he called Muskeg Valley Microquartzite, that was used at the Quarry of the Ancestors, observing that the stone occurred "near the base of the McMurray Formation" (2005, 5).

ORIGIN AND FORMATION

Carrigy concluded that the lower part of the McMurray Formation consisted of "fluviatile deposits, and the middle and upper parts of foreset and topset beds of an ancient delta" (1966, 26). He speculated that the submarine portion of that ancient delta should be found northwest of Fort McKay, beneath the Birch Mountains.

As we have seen, the initial report on the material from the Beaver River Quarry identified the dull grey variety of BRS as a limestone, given that it contained "moulds of fossil organisms," while the highly siliceous variety was considered chert (Syncrude Canada Ltd. 1973, 80). However, subsequent examination revealed that the material was a quartzite of non-marine origin (Syncrude Canada Ltd. 1974, 47–48). According to Douglas et al. (1970, 459), the McMurray sediments—which consist of 100 to 200 feet of well-sorted quartzose sandstone that is impregnated with heavy oil and that constitutes the Athabasca oil sands— were derived from the Canadian Shield.

Tsang (1998, 15) considered only the silica-cemented McMurray Formation to be true BRS. He observed that BRS occurs as a discontinuous outcrop and concluded that, because of the variation in thickness and elevation over a small geographic area, the mircrocrystalline quartz was not detrital in origin (1998, 150). He also drew attention to the presence of vertical cracks in the sandstone deposit and suggested that the BRS was the result of a mineralization process similar to the one that Fedikow et al. (1996) had described for the creation of silica concretions at the Mafeking Quarry in southwestern Manitoba (see Tsang 1998, 21–22 and plate 2.5). The mineralization at the Mafeking Quarry is regarded to be the result of a dissolution/replacement process that was probably caused by upwardly moving, silica-rich fluids travelling along vertical fracture zones in the limestone bedrock and may have had a hydrothermal origin, one that led to the formation of nodules of chert (Grasby, Gryba, and Bezys 2002, 277). BRS differs from the chert found at the Mafeking Quarry in that it occurs not as nodules formed in randomly distributed solution chimneys but as localized masses of silica-cemented sandstone. Its process of formation is similar to that of silcrete (see, for example, Summerfield 1983, 61).

With specific reference to the finer-textured material found at the Quarry of the Ancestors, De Paoli (2005, 5) commented that the Muskeg Valley Microquartzite (MVMq) originated as "a detrital sediment, not a limestone" because of its annealed texture and relict quartz outgrowths, which are visible in larger grains, the absence of crinoid and brachiopod fossils such as occur in nearby limestone outcrops, and the stratigraphic position of the material near the base of the McMurray Formation. He did not speculate on how the silica cementation was formed.

PRIMARY GEOLOGICAL OCCURRENCES

Bedrock outcrops of Beaver River Sandstone occur along the Beaver River and on the east side of the Athabasca River several kilometres southeast of Fort McKay, between Highway 63 and the Muskeg River (see fig 9.1). The most extensive outcrops are reported from an area located roughly 1 to 2 kilometres east of the Athabasca River and extending around 4 to 8.5 kilometres north of the mouth of Muskeg River (Tsang 1998, figure 2). The Cree Burn Lake quarry site, HhOv-16, extends for over 2.5 kilometres along an abandoned oxbow on the east side of the Athabasca River. Reeves (1996, 63) observed that the BRS that occurred at HhOv-55—located 2 kilometres south of the Cree Burn Lake complex and about 8 kilometres north of the Beaver River Quarry (HgOv-29) but on the east side of the Athabasca River—was similar to that found at the quarry, which is located on the west side of the Athabasca River. He also suggested that any outcrops at the Cree Burn Lake site that had once been exposed for precontact knappers would have since been obscured by slumping. Two occurrences of BRS in situ within bituminous sands of the McMurray Formation had apparently been reported earlier at HhOv-55 (Ives and Fenton 1983, 82 and figures 5 and 6).

In addition, an extensive area of BRS outcrops was recently discovered at the Quarry of the Ancestors, which is located several kilometres east of the Muskeg River and around 4 to 8 kilometres east of the concentration of coarse-textured material identified by Tsang. The quarry complex includes sites HhOv-305 and

HhOv-319 and takes in parts of sections 27, 28, 29, 32, 33, and 34 of township 94, range 10, west of the 4th meridian (Saxberg 2007, figure 3). Because the rock here is of somewhat finer quality, De Paoli labelled it "Muskeg Valley Microquartzite" in order to distinguish it from the coarser variety of BRS mapped by Tsang and also to identify its geographic bedrock occurrence (2005, 5–6).

Two possible bedrock sources of BRS have been reported further afield from Fort McKay, although neither has been confirmed. One consists of a thin layer (less than 30 centimetres thick) of silica-cemented sand in the McMurray Formation encountered in a core drilled in the Birch Mountains. Tsang (1998, 19) mentioned the existence of this lithic material, citing as his source a personal communication from "S. Sabag" (presumably Shahé Sabag), but provided no further information. The other is the potential presence of an outcrop of BRS along the Clearwater River, to the east of Fort McMurray (Ives and Fenton 1983, figure 7). Exposures of the McMurray Formation occur along the Clearwater River for a considerable distance upstream from Fort McMurray (see Ives and Fenton 1983, figure 7)—an area that includes two sites, HdOs-1 and HeOs-1, at which Paul Donahue (1976, 49–51) noted a relatively high proportion of BRS among the artifacts recovered. While the existence of a local source of the material offers one possible explanation for this pattern, no such source has thus far been identified.

SECONDARY GEOLOGICAL OCCURRENCES

BRS was found to be absent from glacial tills in the oil sands area (Ives and Fenton 1983, 88). Cobbles and even large boulders of fine- and coarse-textured BRS are, however, found in the gravel that was deposited into Glacial Lake McConnell during a catastrophic outflow from Glacial Lake Agassiz that took place around 9,800 to 9,600 BP (see chapter 2 in this volume). This gravel pavement extends along both sides of the Athabasca River from east of Fort McMurray to just north of Bitumount (Smith and Fisher 1993, figure 1). It contains a wide assortment of lithic types that had washed down the Clearwater and Athabasca rivers or had been eroded from local Quaternary and bedrock formations. In the southwestern part of the Fort Hills Oil Sands Lease, the coarse gravel occurs immediately beneath a thin carpet of moss and could easily have been exposed for precontact knappers in tree falls or following forest fires. North of Bitumount, the gravel grades into sandy delta deposits where far fewer large rocks are present.

During a survey of the Fort Hills Oil Sands Lease completed by Fedirchuk McCullough and Associates Ltd., I collected cobbles and plates of mainly the finer-textured BRS from the gravel exposed at the abandoned Solvex plant located not far southeast of Bitumount, roughly 20 kilometres northeast of Fort McKay (Gryba 2001b). I also noticed cobbles and even larger boulders of fine- and coarse-textured BRS along the east bank of the Athabasca River near Bitumount. Some of that material may have eroded from local gravel deposits, or it may have been tumbled or ice-rafted downstream from upstream sources in more recent times.

DISTRIBUTION AT ARCHAEOLOGICAL SITES

As the occurrence of thousands of pieces of debitage suggest, the greatest use of BRS occurred at sites located at or near the natural outcrops of this lithic material (see, for example, Reardon 1976; Ronaghan 1981, plate I-23; Reeves 1996; Saxberg and Reeves 2004). Diagnostic artifacts recovered from sites in the general area indicate that the use of BRS was not restricted to any specific cultural group; rather, BRS was the lithic material most often selected by knappers throughout the precontact period (see, for instance, Syncrude Canada Ltd. 1974; Ronaghan 1981; Reeves 1996; Reeves and Saxberg 1998; Tischer 2006). Microblades and blade cores of BRS were recovered from the Cree Burn Lake site (Reeves and Saxberg 1998) and from the Quarry of the Ancestors (Saxberg 2007, plates B.9, B.10, B.12, B.24, B.26, B.29, and B.32).

As was noted over two decades ago, artifacts manufactured of BRS frequently accounted for upwards of 90% of the assemblage at sites in the Fort McKay region, within roughly a 30-kilometre radius of the known source area at the Beaver River Quarry and downstream towards the Firebag River (Fenton and Ives 1990, 123; Ives 1993, 19). This observation has been borne out by the findings of more recent surveys or excavations at sites HiOv-49 and HiOv-52, located along the west side of Fort Creek near Bitumount (Unfreed, Fedirchuk, and Gryba 2001, tables 14, 16, 18, and 20; Woywitka and Younie 2008), at sites HiOv-59, HiOv-61, and HiOv-64, situated near the headwaters of Stanley Creek (Woywitka 2007), and at sites found at the confluence of the Ells and Athabasca rivers (Gryba and Tischer 2005, 41, 46–47).

As one moves away from the bedrock and secondary sources of BRS, archaeological sites reflect a marked drop-off in the use of this material and a corresponding increase in the use of locally available stone. For instance, sites located along the Calumet River 2.5 to 3 kilometres upstream from its confluence with the Athabasca River, northwest of Bitumount, contained from 88% to 96% salt and pepper quartzite but little or no BRS (Tischer 2006, 26, 42). I have noticed a similarly dramatic decrease in the frequency of BRS at sites found along Joslyn

Creek and the northern tributaries of Ells River in the area roughly midway between the Athabasca River and the Birch Mountains. Further afield, artifacts of BRS have been discovered in proportions as low as 2% to 5% in sites in the Birch Mountains (Donahue 1976, table 10; Fenton and Ives 1990, 124). At these sites, BRS is represented by artifacts that reflect terminal phases of tool manufacture and maintenance (Ives 1993, 19). This interpretation is supported by the discovery of tiny pressure flakes of BRS at three sites located in the northeastern part of the Birch Mountains near the headwaters of Joslyn Creek (Gryba and Tischer 2008).

BRS is also a major lithic type in sites located southwestward to the middle reaches of the MacKay River and along the Athabasca River south as far as Fort McMurray (Fenton and Ives 1990, 123). East of Fort McMurray, BRS artifacts were encountered at sites HdOs-1 and HeOs-1, both located within the Clearwater River valley, with HdOs-1 lying around 6 kilometres to the east of the confluence of the Clearwater with the Christina River, and HeOs-1 lying a similar distance to the west (Donahue 1976, 49–51 and figure 12). At some sites along the Clearwater River, BRS accounted for 40% to 50% of the lithic assemblage (Fenton and Ives 1990, 124).

At sites even further away from Fort McKay, only sporadic finds of BRS artifacts have been reported. Among these occurrences is a reworked, basally thinned point of grey BRS found at the Duckett site at the northeast edge of Ethel Lake, about 5 kilometres west of Cold Lake (Gryba 1988, A4–A6), an Eden point from Alberta's Barrhead district (Fenton and Ives 1990, figure 4), and possibly several Cody Complex points from North Battleford, Saskatchewan (John Ives, pers. comm., 2007). Citing various studies, Fenton and Ives note that artifacts made of BRS have been reported from sites near Lac La Loche, in northwestern Saskatchewan, and at Wentzel Lake, in the Caribou Mountains north of the Peace River, at site IkOv-8 on the Slave River, and in the Wabasca River and Wabasca Lake areas of north-central Alberta (Fenton and Ives 1990, 124; for additional discussion of BRS dispersal, see Ives, chapter 8 in this volume). In southwestern Alberta, just southeast of Calgary, a bifacially worked piece of material that appears identical to heat-treated BRS was discovered in the upper level of site EfPl-254 (Gryba 2007, 24).

HEAT TREATMENT OF BEAVER RIVER SANDSTONE

Sims (1974, 51) reported that he did not recognize any evidence of heat treatment on the BRS artifacts recovered from the Beaver River Quarry, although he detected its presence on a few items he had collected the previous year from sites he discovered some 8 kilometres upstream from the quarry. Unfortunately, Sims did not mention what specific attributes he had used to determine that the material had been thermally altered. In subsequent experiments, I was able to demonstrate that BRS is indeed amenable to heat treatment and that the procedure produces distinct attributes, ones that can be recognized on many archaeological specimens (Gryba 2001b; see also Gryba 2002).

Heat-treating various silica-rich lithics in order to improve their workability was a very deliberate strategy that was practiced widely in time and space by prehistoric knappers. It is particularly evident among those cultures in which pressure flaking played an important role in the manufacture of stone tools. Archaeological evidence shows that the practice of heat treatment dates from possibly 164,000 BP in southern Africa (Brown et al. 2009, 859), from around 110,000 BP, during the Middle Palaeolithic period, in the Near East (Copeland 1998, 76), and from the Late Palaeolithic period, at around 30,000 to 27,000 BP, in eastern Europe (Bradley, Anikovich, and Giria 1995, 996). In southwestern Europe, evidence of heat treatment is visible on Solutrean artifacts dating from 18,000 to 19,000 BP (Bordes 1968, 159), and the practice was known to the Late Palaeolithic microblade and other lithic industries of eastern Siberia (Flenniken 1987, 121; Kononenko, Kononenko, and Kajiwara 1998). In North America, heat treatment has been recognized on artifacts dating back to as early as the Clovis culture (ca. 11,500 to 10,800 BP; see, for example, Bonnichsen 1977, 192; Hall 1995, 9 and 19; Gryba 2001a, 259) and among many subsequent cultural traditions.

Knappers applied heat treatment to improve the workability not only of chert and similar siliceous materials but even of types of stone that some archaeologists might consider relatively easy to work. There is, for instance, ethnographic evidence that a number of different Aboriginal groups in the area extending west from what is now Montana through the Great Basin and into northern California heat-treated obsidian (Hester 1972). In western Canada, artifacts of brightly coloured Swan River Chert constitute some of the best local evidence we have for intentional heat treatment by precontact knappers. Many types of lithics from which flaked stone tools were made do not display such obvious signs of heat treatment, however, either because they lack those trace elements that would cause a marked colour change to occur when the rock is heated to a high enough temperature or because the colour or lustre that resulted from intentional heating has become obscured by weathering or patination. In addition, some types of siliceous rock can be effectively heat-treated at a temperature low enough that no colour change occurs. In these cases, experimentation with the lithic materials in question often provides the best guide to the precontact application of this practice.

Heat treatment is a relatively simple procedure. It can be accomplished virtually overnight and requires no great expenditure of labour aside from collecting the stones and fuel and setting up some sort of shelter within which the heating is carried out. But both the rise and the fall of temperature must occur gradually so that the rock does not expand too quickly when heated or contract too rapidly during cooling. Sudden temperature changes can cause the stone to shatter and thus render it unsuitable for flaking. In addition, too little heat may not bring about the desired changes, while too much heat will destroy the crystal structure to the point where the flaking quality of the stone will be spoiled beyond recovery. Prehistoric knappers may have roasted stones in a pit, or inside a small shelter, or perhaps even under a smouldering open fire or smudge, where ashes would have kept the stones insulated from any abrupt temperature changes, particularly during cooling. What types of stone responded positively to heating and how much fuel was required to achieve successful heat treatment were critical pieces of knowledge that were no doubt acquired through a lengthy process of trial and error and then passed on to succeeding generations of knappers as traditional information about percussion and pressure flaking.

For the heat treatment experiments, I chiefly used samples of fine-textured BRS I had collected from secondary geological sources during fieldwork completed by Fedirchuk McCullough and Associates Ltd. Most of the stone was collected from the gravel exposed at the abandoned Solvex plant near Bitumount, which lies about a kilometre to the east of the Athabasca River (Gryba 2001b). Other samples of BRS used in the heat treatment experiments were obtained from bedrock sources at the Beaver River Quarry, as well as along the road leading to the Quarry of the Ancestors, east of Muskeg River.

The samples of BRS were heated in a small electric kiln (fig 9.2), a method that allowed me to minimize any abrupt temperatures changes during heating and cooling. Despite this reliance on modern technology, I am confident that the heat treatment results closely match those that precontact knappers would have obtained using wood fires. The pieces of BRS selected for the experiments varied from 3 to 20 centimetres in length. In roasting BRS, I followed a method quite similar to one that I had found appropriate for determining the optimum flaking quality of different varieties of chert. The first few batches of rock were heated in a number of successive trials designed to determine the temperature range that was needed for achieving optimum flaking quality. The upper reading was initially set at around 300°C. For each succeeding trial, the temperature was increased by approximately 15°C to 20°C until the desired changes to the stones were attained. During each trial, the temperature of the kiln was allowed to rise gradually over some eight to ten hours. Shortly after the target temperature was

Figure 9.2. Kiln used in heat treatment experiments

reached, the kiln was turned off, and the BRS was allowed to slowly cool down to room temperature, a step that took approximately nine to ten hours. Following each heating, the BRS was tested for its workability.

During heating, the darker, more porous variety of BRS, even at a relatively low temperature, gave off a nauseating odor of burning petroleum. While an improvement in the workability was first noticed after BRS had been heated to about 360°C, optimum flaking quality was attained at 400°C to 425°C, temperatures well within the range at which certain grades of silcrete acquire their maximum workability (Domanski and Webb 2007, table 5). The 400°C to 425°C range also approximates the temperature level at which I successfully heat-treated some medium grades of Swan River Chert (Grasby, Gryba, and Bezys 2002, 279), as well as Cat Head Chert and mud shale.[1] In other words, the temperature that precontact knappers would have required to alter BRS to its optimum workability fell squarely within the range of temperatures that they would have required to successfully heat-treat other types of stone.

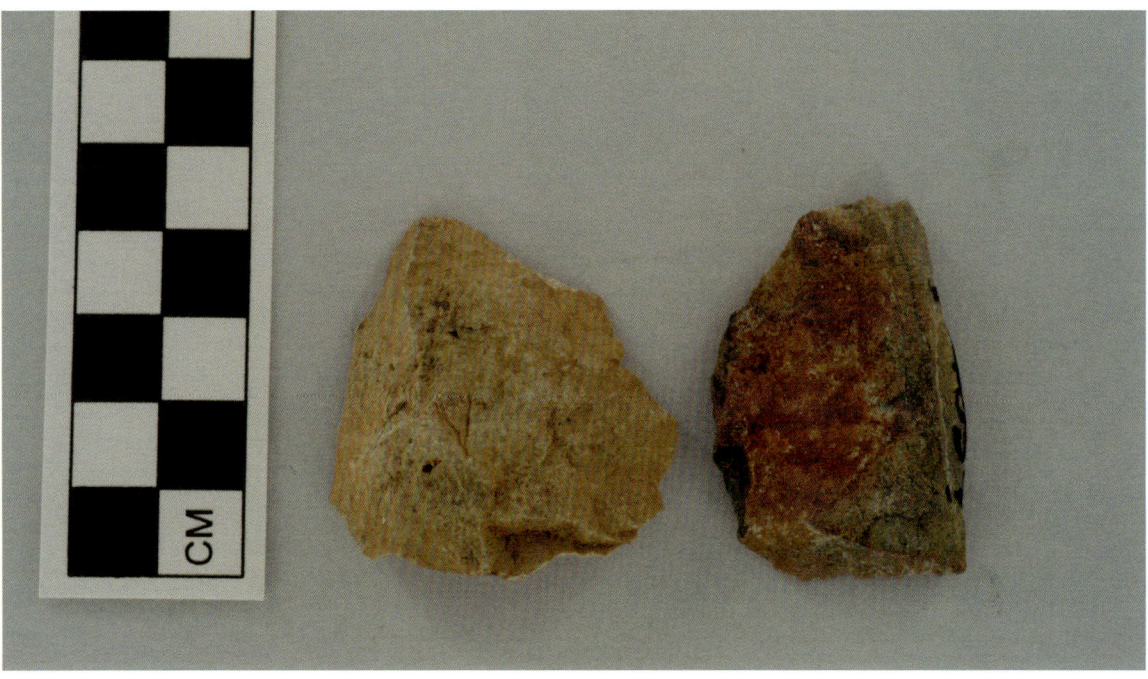

Figure 9.3. Fragments of raw BRS (left) and heat-treated BRS (right). Note the red tint on the cortex of the heat-treated fragment.

In its raw state, the finer-textured BRS is normally drab grey to yellowish in colour. In contrast, some of the coarser varieties are so impregnated with hydrocarbons that they are almost black. When heated to about 350°C, the cortex of some of the BRS samples I treated started to acquire a reddish colour. At around 400°C, the temperature was high enough to cause a permanent red "rind" to develop on the cortex as well as along open fracture planes on some of the pieces (fig 9.3). It is likely that this thin red layer, which usually measured less than a millimetre in thickness, was caused by the oxidation of iron that had accumulated on or just beneath the surface of the rock. A red rind was not noticed on any of the raw pieces of BRS. However, a rusty yellow colour occurs on the cortex of some natural pieces (fig 9.3), suggesting that the raw stone came from an environment relatively rich in iron, a fact that had been earlier observed (see, for example, Carrigy 1966, 7, 10, 11, 15 and 17; Tsang 1998, 15). Only in a few instances did the interior of the stone exhibit any reddening as a result of heating; rather, it usually remained close to its natural drab buckskin or grey tone.

When heated to around 375°C to 400°C, the BRS recrystallized. At this temperature, the stone acquired a fine glassy texture and lustrous appearance (fig 9.4) and could easily be fractured by percussion and pressure methods. By comparison, the untreated material had proved very difficult to work, even by percussion

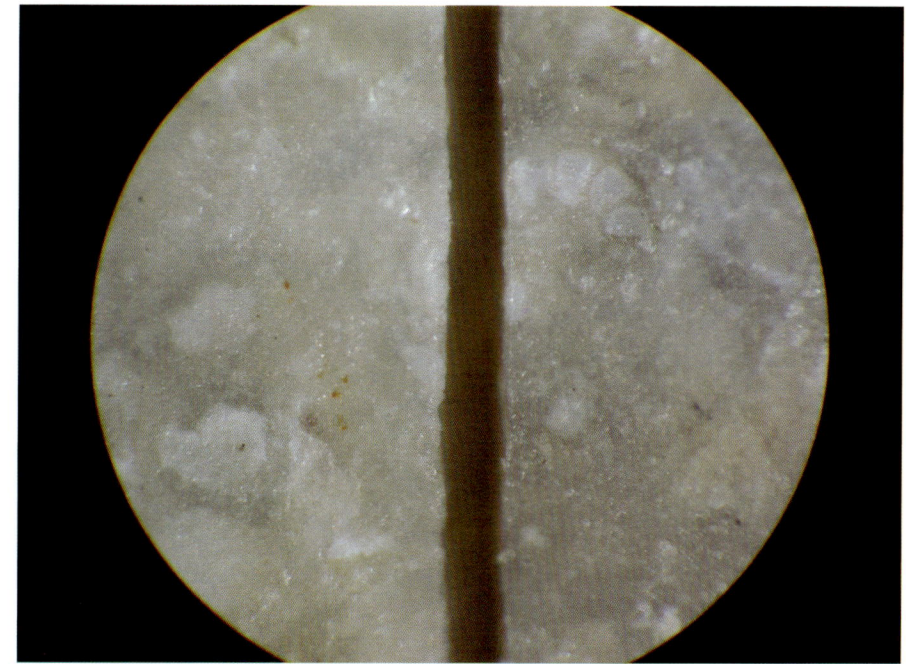

Figure 9.4. Refittable fragments of the same BRS specimen: the raw portion is on the left and heat-treated portion on the right. Note the somewhat improved lustre and reduced granularity in the heat-treated specimen.

flaking methods. Even some of the coarse-grained BRS, when heated to the optimum level, became relatively easy to work. The fracture surface on heated fine-textured BRS was found to be very smooth, while on raw samples it was fairly rough. On some of the finest-quality BRS, I was easily able, by simple hand pressure, to press off 4-centimetre-long flakes and microblades that ranged from 5 to 6 centimetres in length. Figures 9.5 and 9.6 clearly illustrate the degree of precision that can be achieved by pressure-flaking heat-treated BRS. It should be noted, however, that the pattern of pressure-flake scars seen on the lanceolate "preforms or bifaces" is actually more akin to that seen on some Solutrean bifaces than on anything found in North America, aside perhaps from some Clovis points.

Attributes such as the reddened cortex and the smooth fracture surface of experimentally heat-treated samples were also observed on many archaeological specimens of BRS collected from the Fort Hills and Fort McKay districts, as well as on Cody Complex artifacts recovered by Lifeways Canada Ltd. from site HhOu-37. A reddish tint had been reported earlier (Gryba 1980, 30; Fenton and Ives 1982, 172), which we can now interpret as evidence of intentional heat treatment. In addition, when I made a visual comparison of such archaeological artifacts with experimentally heat-treated samples, it appeared that BRS was not susceptible to weathering or patination but tended to retain a fairly fresh appearance.

Figure 9.5. Lanceolate points and preforms created (by the author) from artificially heat-treated BRS

In summary, the experimental results, when compared with archaeological artifacts, indicate beyond any reasonable doubt that precontact knappers in northeastern Alberta throughout the Prehistoric period took full advantage of heat treatment to make BRS more workable for percussion and pressure flaking. Key indicators of heat treatment are a smooth fracture surface, a lustrous glassy appearance, and a red rind on the natural cortex and along existing fracture planes. Given this evidence, it is reasonable to assume that knappers also used heat treatment to improve the workability of chert and other locally available

rocks used in the production of flaked stone artifacts. This assumption is, in fact, supported by the recovery of a microblade core of heat-treated Swan River Chert within half a kilometre east of the Quarry of the Ancestors during mitigation excavations at site HhOv-449 (Wickham and Graham 2009, 360–361 and figures 3 and 144). As I have suggested, the practice of heat treatment by precontact knappers may explain why Ives and Fenton were unable to find a bedrock source of high-quality BRS despite its predominance among artifacts found in the oil sands area (see Ives and Fenton 1983, 85).

SIMILARITIES TO OTHER LITHIC MATERIALS ON THE NORTHERN PLAINS

A wide assortment of quartzites derived from different geological eras occurs throughout the Rocky Mountains and foothills of western Alberta (Douglas et al. 1970). In addition, Tertiary- age gravels that contain quartzite and other rocks of western provenance cap many upland features throughout Alberta and south-western Saskatchewan, including the Swan Hills, Hand Hills, Vermilion Hills, Cypress Hills, and Wood Mountain (Douglas et al. 1970, 481; Westgate 1966, 12), and were even transported by pre-glacial streams as far east as Duck Mountain in southwestern Manitoba (Nielsen 1988, 4). In addition, Precambrian Shield quartzite is incorporated in the glacial till throughout the northeastern part of the northern Plains. These quartzites, however, differ markedly from BRS in terms of colour and texture as well as their responsiveness to heat treatment. While white, light brown or buckskin, and banded purple varieties prevail, red, green, dull greenish-grey, dark grey, and black quartzites can also be found.

A fine- to coarse-textured quartzite occurs fairly commonly in secondary geological contexts in Manitoba's Swan Valley. In Glacial Lake Agassiz beach gravels, it is found as cobbles with a very smooth cortex. The most likely source for this material is the Lower Cretaceous Swan River Formation, a 60- to 75-metre thick sandstone unit that underlies most of southwestern Manitoba and is composed of "loose to poorly cemented, fine-grained to coarse-grained quartz sand with minor amounts of shale and some lignite beds" (Nielsen 1984, 11). The lower part of the formation was deposited in a continental environment, while the upper unit was laid down in a marginal marine setting. Bannatyne describes the Swan River Formation as consisting of "grey and dark grey kaolinitic shale, silica sand, and scattered lignite fragments, iron sulphide concretions, and siderite nodules" (1971, 249). This quartzite may have a geological dispersal similar to that suggested for Swan River Chert, extending across the southern part of the Prairie provinces in Canada and into adjacent Plains regions in the United States (see, for example, Grasby, Gryba, and Bezys 2002, figure 1).

Raw specimens of this quartzite are light tan to buff in colour and are thus very reminiscent of BRS. In addition, the stone is similar to BRS in that it occurs in fairly homogenous pieces and may contain impressions of plant stems. Like BRS, it appears to be a silcrete and responds very favourably to heat treatment as well. Individual heat-treated pieces sometimes display scattered quartz grains of varying size, a distinguishing trait common to most, but not all, samples of BRS. The fact that many pieces of the quartzite I examined did not exhibit this characteristic may simply be due to a field selection process that focused mainly on cobbles of the fine-textured material. When subjected to heat treatment, the quartzite variously turns dull yellow, light grey, light tan, tan with a hint of orange or red, red, or maroon and breaks easily, with a smooth, lustrous fracture. Light grey and orangish tan examples are identical in colour and texture to some of the heat-treated BRS from the Fort Hills area that I produced. Heat-treated examples of this quartzite are sometimes observed in archaeological collections from the Swan River and Souris districts of southwestern Manitoba (see, for example, Gryba 1976, figure 5: a, b and c).

In my experience, however, the material most similar to BRS in both texture and colour is the Dakota sandstone from the Gunnison area of south-central Colorado. Indeed, the similarity between coarser-textured BRS and the two chunks of Dakota sandstone that I examined is so striking that I would not be able to distinguish between the two rock types. Like BRS, Dakota sandstone displays scattered grains of clear quartz, small and large. It also responds favourably to heat treatment. This locally available quartzite, which "occurs in a wide variety of grain size, colour and texture," was the primary lithic material used by Folsom knappers at the Mountaineer Folsom site near Gunnison (Stiger 2006, 326).

Dakota sandstone is Lower Cretaceous in age and represents delta-front, channel fill, and delta-plain sediments deposited in a marine environment (Young 1960, 172). Near Gunnison, Dakota sandstone reflects sediments that were originally deposited in a near-shore marine environment (Bartleson 1989, 1147). Dakota sandstone from Garfield County, located immediately northwest of Gunnison County, has been described as a "predominantly medium- to thick-bedded, medium- to fine-grained, light grey (7.5 YR 7/1) to light-brownish-gray (10 YR 6/2–3) well-sorted quartz sandstone, pale orange weathered in part," while silica cement makes the stone highly resistant to erosion (Perry et al. 2003, 9). Raw Dakota sandstone thus falls within the colour range that Fenton and Ives (1982, 172) assigned to BRS. LaBelle also noted an overlap in colour and texture between Dakota sandstone from the Boulder area of north-central Colorado and the heat-treated samples of BRS that I provided to him (Jason LaBelle, pers. comm., 2009).

SUMMARY AND CONCLUSIONS

Beaver River Sandstone is a very distinctive lithic material that occurs in the Lower Cretaceous McMurray Formation in northeastern Alberta. It may have formed as a silcrete. Near Fort McKay, BRS is present as bedrock outcrops, but it is also found in secondary sources on both sides of the Athabasca River. When heat treated, the stone is easily worked, and in the area around Fort McKay, it was the dominant lithic type used by precontact knappers. As archaeological research has shown, however, there was a marked decline in the use of BRS within 10 to 20 kilometres of its natural occurrence. Beyond these limits, BRS is largely replaced by locally available quartzites and occurs mainly as isolated implements or as debitage produced during tool sharpening or rejuvenation.

Lithic material somewhat reminiscent of BRS occurs as smooth cobbles in Quaternary deposits in the Swan Valley of southwestern Manitoba, where it may have been derived from a local Lower Cretaceous formation. Small quantities of heat-treated samples of this quartzite have been recognized in archaeological sites in southwestern Manitoba. Dakota sandstone of Cretaceous age from south-central and northern Colorado overlaps more closely with BRS in colour and texture and also responds positively to heat treatment. In Alberta's oil sands area, archaeological research has yielded numerous artifacts fashioned from what appears to be high-quality BRS, and yet a natural source of this material has yet to be located. It thus seems reasonable to view these artifacts as evidence that precontact knappers practiced heat treatment of the stone.

ACKNOWLEDGEMENTS

I am grateful to Mark Stiger, of the Department of Anthropology at Western State College, Gunnison, Colorado, for providing me with samples of Dakota sandstone, and to Jason LaBelle, of Colorado State University, Fort Collins, who compared samples of heat-treated BRS with the Dakota sandstone found in archaeological sites in north-central Colorado.

NOTE

1 Cat Head Chert comes from the Cat Head member of the Ordovician-age Red River Formation that outcrops on the west side of Lake Winnipeg near the community of Grand Rapids. This mostly white chert is quite common in the glacial drift and Lake Agassiz beach gravels in Manitoba's Swan Valley (see Bakken 1995). Mud shale occurs in concretions in the Lower Carboniferous Mount Head Formation along the Rocky Mountain Front Range in southwest Alberta, as well as in gravel deposits around Calgary. It heat-treats to a beautiful jet black colour and is rather tough and thus difficult to flake, clearly displaying the flake scars. Because of this characteristic, and its toughness, I like to use it for display items and for show-and-tell purposes.

REFERENCES

Bakken, Kent

1995   Lithic Raw Material Resources in Minnesota. Paper presented at the 60th annual meeting of the Society for American Archaeology, Minneapolis, Minnesota, 3–7 May. http://www.tc.umn.edu/~bakk0029/MinnLith/.

Bannatyne, Barry B.

    1971   Industrial Minerals of the Sedimentary Area of Southern Manitoba. In *Geoscience Studies in Manitoba,* edited by A. C. Turnock, pp. 243–251. Geological Association of Canada Special Paper no. 9. Business and Economic Service Limited, Toronto.

Bartleson, Bruce

    1989   Dakota Sandstone and Associated Rocks Adjacent to San Juan Sag near Gunnison, Colorado. Abstract in *American Association of Petroleum Geologists Bulletin* 73: 1147.

Bonnichsen, Robson

    1977   *Models for Deriving Cultural Information from Stone Tools.* Mercury Series no. 60. Archaeological Survey of Canada, National Museum of Man, Ottawa.

Bordes, François

    1968   *The Old Stone Age.* Translated by J. E. Anderson. New York and Toronto: McGraw-Hill.

Bradley, Bruce A., Michael Anikovich, and Evgenii Giria

    1995   Early Upper Palaeolithic in the Russian Plain: Streletskayan Flaked Stone Artifacts and Technology. *Antiquity* 69 (266): 989–998.

Brown, Kyle S., Curtis W. Marean, Andy I. R. Herries, Zenobia Jacobs, Chantal Tribolo, David Braun, David L. Roberts, Michael C. Meyer, and Jocelyn Bernatchez

    2009   Fire as an Engineering Tool of Early Modern Humans. *Science* 325 (14 August): 859–62.

Carrigy, M. A.

    1966   *Lithology of the Athabasca Oil Sands.* Alberta Geological Survey Bulletin no. 18. Research Council of Alberta, Edmonton.

Copeland, Lorraine

    1998   The Middle Paleolithic Flint Industry of Ras el-Kelb. In *The Mousterian Site of Ras el-Kelb, Lebanon*, edited by Lorraine Copeland and Norah Moloney, pp. 73–101. British Archaeological Reports International Series no. 706. Archaeopress, Oxford.

De Paoli, Glen R.

    2005   Petrographic Examination of the Muskeg Valley Microquartzite (MVMq). Appendix A in Nancy Saxberg (2007), *Birch Mountain Resources Ltd., Muskeg Valley Quarry, Historical Resources Mitigation, 2004 Field Studies: Final Report (ASA Permit 05-118),* vol. 2. Copy on file, Archaeological Survey, Historic Resources Management Branch, Alberta Culture, Edmonton.

Domanski, Marian, and John Webb

    2007   A Review of Heat Treatment Research. *Lithic Technology* 32(2): 153–194.

Donahue, Paul F.

    1976   *Archaeological Research in Northern Alberta, 1975.* Archaeological Survey of Alberta Occasional Paper no. 2. Historic Resources Management Branch, Alberta Culture, Edmonton.

Douglas, R. J. W., H. Gabrielse, J. O. Wheeler, D. F. Scott, and H. R. Belyea

    1970   Geology of Western Canada. In *Geology and Economic Minerals of Canada,* ed. R. J. W. Douglas, pp. 366–488. Geological Survey of Canada, Economic Geology Report No. 1. Department of Energy, Mines and Resources, Ottawa.

Fedikow, M. A. F., R. K. Bezys, J. D. Bamburak, and H. J. Abercrombie

    1996   Prairie-Type Microdisseminated Au Mineralization—a New Deposit Type in Manitoba's Phanerozoic Rocks (NTS 63C/14). In *Manitoba Energy and Mines, Minerals Division, Report of Activities, 1996,* pp. 108–121. Manitoba Energy and Mines, Winnipeg.

Fenton, Mark M., and John W. Ives

    1982   Preliminary Observations on the Geological Origins of Beaver River Sandstone. In *Archaeology in Alberta, 1981*, compiled by Jack Brink, pp. 166–189. Archaeological Survey of Alberta Occasional Paper no. 19. Historic Resources Management Branch, Alberta Culture, Edmonton.

1990   Geoarchaeological Studies of the Beaver River Sandstone, Northeastern Alberta. In *Archaeological Geology of North America*, edited by Norman P. Lasca and Jack Donahue, pp. 123–135. Centennial Special Volume 4. Geological Society of America, Boulder, Colorado.

Flach, Peter D.

1984   *Oil Sands Geology—Athabasca Deposit North.* Bulletin no. 46. Geological Survey Department, Alberta Research Council, Edmonton.

Flenniken, J. Jeffrey

1987   The Paleolithic Dyuktai Pressure Blade Technique of Siberia. *Arctic Anthropology* 24(2): 117–132.

Grasby, Stephen E., Eugene M. Gryba, and Ruth K. Bezys

2002   A Bedrock Source of Swan River Chert. *Plains Anthropologist* 47(182): 275–281.

Gryba, Eugene M.

1976   The Hill Sites: Two Surface Sites in the Swan Valley, Manitoba. *Na'pao* 6(1–2): 23–40.

1980   *Highway Archaeological Salvage Projects in Alberta: Final Report (ASA Permit 79-066).* Copy on file, Archaeological Survey, Historic Resources Management Branch, Alberta Culture, Edmonton.

1988   *An Inventory of Fluted Point Occurrences in Alberta.* 2 vols. Report prepared for the Alberta Historical Resources Foundation. Copy on file, Royal Alberta Museum, Edmonton.

2001a   Evidence of the Fluted Point Tradition in Western Canada. In *On Being First: Cultural Innovation and Environmental Consequences of First Peopling,* edited by Jason Gillespie, Susan Tupakka, and Christy de Mille, pp. 251–284. Proceedings of the 31st Annual Chacmool Conference. Chacmool Archaeological Association, Department of Archaeology, University of Calgary, Calgary.

2001b   Lithic Resources Available to or Used by Precontact Knappers at the True North Energy L.P. Fort Hills Oil Sands Leases near Fort MacKay, in Northeastern Alberta. Appendix II in Wendy J. Unfreed, Gloria J. Fedirchuk, and Eugene Gryba, *Historical Resources Impact Assessment, True North Energy L.P. Fort Hills Oil Sands Project: Final Report (ASA Permit 00-130),* vol. 2. Copy on file, Archaeological Survey, Historic Resources Management Branch, Alberta Culture, Edmonton.

2002   Lithic Types Available to Prehistoric Knappers at the True North Energy Limited Partnership Fort Hills Oil Sands Lease near Ft. McKay, in Northeastern Alberta. *Alberta Archaeological Review* 36 (Spring): 21–27.

2007   *Excavations at Sites EfPl-254 and EfPl-255, Heritage Point Phase III Subdivision Development in SW 4-22-29-W5M, near Calgary: Final Report (ASA Permit 06-193).* Copy on file, Archaeological Survey, Historic Resources Management Branch, Alberta Culture, Edmonton.

Gryba, Eugene M., and Jennifer C. Tischer

2005   *Historical Resources Impact Assessment, Deer Creek Energy Limited Joslyn North Mine Project: Final Report (ASA Permit 05-094).* Copy on file, Archaeological Survey, Historic Resources Management Branch, Alberta Culture, Edmonton.

2008   *Historical Resources Studies, Value Creation Inc. Terre de Grace Project: Final Report (ASA Permit 07-254).* Copy on file, Archaeological Survey, Historic Resources Management Branch, Alberta Culture, Edmonton.

Hall, Don Alan

1995   Clovis Tools Plentiful in Tennessee. *Mammoth Trumpet* 10(1): 9 and 19.

Hester, Thomas Roy

1972   Ethnographic Evidence for the Thermal Alteration of Siliceous Stone. *Tebewa* 15(2): 63–65.

Ives, John W.

1993   The Ten Thousand Years Before the Fur Trade in Northeastern Alberta. In *The Uncovered Past: Roots of Northern Alberta Societies,* edited by Patricia A. McCormack and R. Geoffrey

Ironside, pp. 5–31. Circumpolar Research Series no. 3. Canadian Circumpolar Institute, University of Alberta, Edmonton.

Ives, John W., and Mark M. Fenton

1983　Continued Research on Geological Sources of Beaver River Sandstone. In *Archaeology in Alberta, 1982,* compiled by David Burley, pp. 78–88. Archaeological Survey of Alberta Occasional Paper no. 21. Historic Resources Management Branch, Alberta Culture, Edmonton.

Kononenko, Alekesy V., Nina A. Kononenko, and Hiroshi Kajiwara

1998　Implications of Heat Treatment Experiments on Lithic Materials from the Zerkalnaya River Basin in the Russian Far East. *Proceedings of the Society for California Archaeology* 11: 19–25.

Nielsen, Erik

1984　The Kettle Formations at Swan Lake. In *Lasting Impressions: Historical Sketches of the Swan River Valley,* edited by Gwen Palmer and Edward Dobbyn, pp. 1–12. Friesen Printers, Altona, Manitoba.

1988　*Surficial Geology of the Swan River Area.* Geological Report GR80-7. Geological Services, Manitoba Energy and Mines, Winnipeg.

Pearl, Richard M.

1962　*Geology: An Introduction to Principles of Physical and Historical Geology.* 2nd ed. Barnes and Noble, New York.

Perry, William J., Ralph R. Shroba, Robert B. Scott, and Florian Maldonado

2003　*Geologic Map of the Horse Mountain Quadrangle, Garfield County, Colorado.* Pamphlet to Accompany Miscellaneous Field Studies Map MF-2415. U.S. Geological Survey, U.S. Department of the Interior.

Reardon, Gerard V.

1976　A Cognitive Approach to Lithic Analysis. MA thesis, Department of Archaeology, University of Calgary, Calgary.

Reeves, Brian O. K.

1996　*Aurora Mine Project, Historical Resources Baseline Study.* Copy on file, Archaeological Survey, Historic Resources Management Branch, Alberta Culture, Edmonton.

Reeves, Brian O. K., and Nancy Saxberg

1998　*Aurora Mine Project Historical Resources Management Studies: Current Interpretive Status and Implications for Understanding the Early Postglacial Native Culture History and Occupation of the Athabasca Lowlands—a Mid-1998 Program Update.* Copy on file, Lifeways of Canada Ltd., Calgary.

Ronaghan, Brian M.

1981　*Final Report: Historical Resources Impact Assessment of Selected Portions of the Alsands Lease 13 (ASA Permit 80-091).* Copy on file, Archaeological Survey, Historic Resources Management Branch, Alberta Culture, Edmonton.

Saxberg, Nancy

2007　*Birch Mountain Resources Ltd. Muskeg Valley Quarry, Historical Resources Mitigation, 2004 Field Studies: Final Report (ASA Permit 05-118).* 2 vols. Copy on file, Archaeological Survey, Historic Resources Management Branch, Alberta Culture, Edmonton.

Saxberg, Nancy, and Brian O. K. Reeves

2003　The First Two Thousand Years of Oil Sands History: Ancient Hunters at the Northwest Outlet of Glacial Lake Agassiz. In *Archaeology in Alberta: A View from the New Millennium,* edited by Jack W. Brink and John F. Dormaar, pp. 290–322. Archaeological Society of Alberta, Medicine Hat.

2004　*Birch Mountain Resources Ltd. Muskeg Valley Quarry, Historical Resources Impact Assessment, 2003 Field Studies: Final Report (ASA Permit 03-249).* Copy on file, Archaeological Survey, Historic Resources Management Branch, Alberta Culture, Edmonton.

Sims, Cort

    1974   *Syncrude Lease 22—Beaver Creek Quarry Site (HgOv-29).* Report prepared for Syncrude Canada Ltd. Copy on file, Syncrude Canada Ltd., Edmonton.

Smith, Derald G., and Timothy G. Fisher

    1993   Glacial Lake Agassiz: The Northwestern Outlet and Paleoflood. *Geology* 21(1): 9–12.

Stiger, Mark

    2006   A Folsom Structure in the Colorado Mountains. *American Antiquity* 71(2): 321–351.

Summerfield, M. A.

    1983   Silcrete. In *Chemical Sediments and Geomorphology,* edited by Andrew S. Goudie and Kenneth Pye, pp. 59–92. Academic Press, San Diego.

Syncrude Canada Ltd.

    1973   *Syncrude Lease No. 17: An Archaeological Survey.* Environmental Research Monograph 1973-4. Syncrude Canada Ltd., Edmonton.

    1974   *The Beaver Creek Site: A Prehistoric Stone Quarry on Syncrude Lease No. 22.* Environmental Research Monograph 1974-2. Syncrude Canada Ltd., Edmonton.

Tischer, Jennifer C.

    2006   *Historical Resources Studies, Final Report: Canadian Natural Resources Limited Horizon Oil Sands Project, Historical Resources Mitigation of HiOw-39, HiOw-42, HiOw-43 and Historical Resources Impact Assessment of Tar River Drainage Diversion (Revised) (ASA Permit 05-225).* Copy on file, Archaeological Survey, Historic Resources Management Branch, Alberta Culture, Edmonton.

Tsang, Brian W. B.

    1998   The Origin of the Enigmatic Beaver River Sandstone. MSc thesis, Department of Geology and Geophysics, University of Calgary, Calgary.

Unfreed, Wendy J., Gloria J. Fedirchuk, and Eugene M. Gryba

    2001   *Historical Resources Impact Assessment, True North Energy L.P. Fort Hills Oil Sands Project: Final Report (ASA Permit 00-130).* 2 vols. Copy on file, Archaeological Survey, Historic Resources Management Branch, Alberta Culture, Edmonton.

Westgate, J. A.

    1966   *Surficial Geology of the Foremost–Cypress Hills Area, Alberta.* Alberta Geological Survey Bulletin no. 22. Research Council of Alberta, Edmonton.

Wickham, Michelle D., and Taylor Graham

    2009   *Historical Resources Impact Mitigation of the TransCanada Pipelines Ltd. Fort McKay Mainline Expansion: Final Report and Post-construction Audit (ASA Permits 06-376 and 07-266).* Copy on file, Archaeological Survey, Historic Resources Management Branch, Alberta Culture, Edmonton.

Woywitka, Robin J.

    2007   *Historical Resources Impact Mitigation, Fort Hills Energy Corporation, Fort Hills Oil Sands Project, Out of Pit Tailings Area, HiOv-59, HiOv-61, and HiOv-64: Final Report (ASA Permit 06-548).* Copy on file, Archaeological Survey, Historic Resources Management Branch, Alberta Culture, Edmonton.

Woywitka, Robin J., and Angela M. Younie

    2008   *Historical Resources Impact Mitigation, Fort Hills Energy Corporation, Fort Hills Oil Sands Project, HiOv-44, HiOv-47, HiOv-49, HiOv-50, HiOv-52, HiOv-87, HiOv-89, HiOv-104, HiOv-115, and HiOv-124: Final Report (ASA Permit 05-328).* Copy on file, Archaeological Survey, Historic Resources Management Branch, Alberta Culture, Edmonton.

Young, Robert G.

    1960   Dakota Group of Colorado Plateau. *American Association of Petroleum Geology Bulletin* 44: 156–194.

10 The Organization of Lithic Technology at the Quarry of the Ancestors

NANCY SAXBERG AND ELIZABETH C. ROBERTSON

The Quarry of the Ancestors, a dense complex of precontact archaeological sites in the oil sands region north of Fort McMurray, centres on two significant outcrops of Beaver River Sandstone (BRS), a lithic material that is ubiquitous in assemblages from sites in the Lower Athabasca region. The ancient quarry was discovered in 2003 during an historic resources impact assessment required in connection with the proposed development of a modern limestone quarry. Early in the application review process, the developer, Birch Mountain Resources, agreed to exclude some of the lands surrounding the outcrops from the area slated for development. This prompt action on the part of the developer ensured the preservation of a crucial archaeological resource, one that is now under the protection of the Province of Alberta.

The vast majority of the materials recovered from archaeological investigations at the Quarry of the Ancestors are lithic artifacts. In this chapter, we explore the cultural inferences that can be drawn from these artifacts and from other evidence found at the quarry. In addition to describing the quarry and reviewing the initial archaeological studies conducted there, we propose an interpretation of the lithic technological organization characteristic of the precontact peoples who occupied the area. For the purposes of this chapter, the organization of lithic technology is understood to encompass the production, use, maintenance, reuse, and discard of stone tools, as well as interpretations of how that technology was embedded in, and is reflective of, ancient land use patterns—that is, all facets of "the manner in which humans organize themselves with regard to lithic technology" (Andrefsky 2008, 4).

The Quarry of the Ancestors is thought to have been in use from the earliest postglacial period to the mid-Holocene period of extensive peatland initiation in the western boreal forest. Traditionally, Alberta archaeologists have used the Plains-based descriptors Early, Middle, and Late period as precontact chronological categories, terminology that has been applied to boreal forest archaeology in Alberta and, to some extent, in Saskatchewan. In this chapter, we instead use the terms Palaeoindian (ca. 11,000 to 7,750 BP, or 13,000 to 9,000 cal yr BP), Archaic (ca. 7,750 to 2,650 BP, or 9,000 to 2,000 cal yr BP), and Late Precontact (ca. 2,650 to 300 BP, or 2,000 to 200 cal yr BP).[1] This terminology is consistent with that employed in discussions of archaeological evidence from boreal forest environments elsewhere in northwestern North America (see, for example, Clark 1981; Wright 1995).

LOCATION AND DESCRIPTION

The name "Quarry of the Ancestors" refers to a bounded area of 203.7 hectares that contains two known outcrops of BRS, with associated workshop deposits and a series of precontact artifact scatters. Located approximately 50 kilometres north of Fort McMurray, the site lies 3.5 kilometres to the east of the Muskeg River, one of the tributaries of the Athabasca River (fig 10.1).

As indicated above, the Quarry of the Ancestors was discovered in 2003 during an historical resources impact assessment carried out for Birch Mountain Resources. Initial shovel tests in the area proposed for development revealed the presence of a dense scatter of lithic material. For ease of management, the area was divided into forty-four separate archaeological sites (Saxberg and Reeves 2004). Further assessment and mitigative excavations were conducted in 2004 at twenty-four of the forty-four original sites, covering a total of 346.5 square metres (Saxberg 2007) (fig 10.2). Most of these mitigative excavations were located within 200 metres of the western outcrop of BRS, and most of them sampled only the dense lithic scatters spread over the landscape. Nonetheless, these excavations recovered a total of 337,959 lithic artifacts.

In 2005, an additional 336.5 square metres were excavated at sites again within 200 metres of the western outcrop (de Mille and Reeves 2009). These excavations recovered approximately 390,000 additional artifacts. Further excavations were conducted in 2007 at some of the same sites examined in previous years and at sites identified to the north of the western outcrop. Although Nancy Saxberg served as the field director for all four rounds of excavations, she was involved in

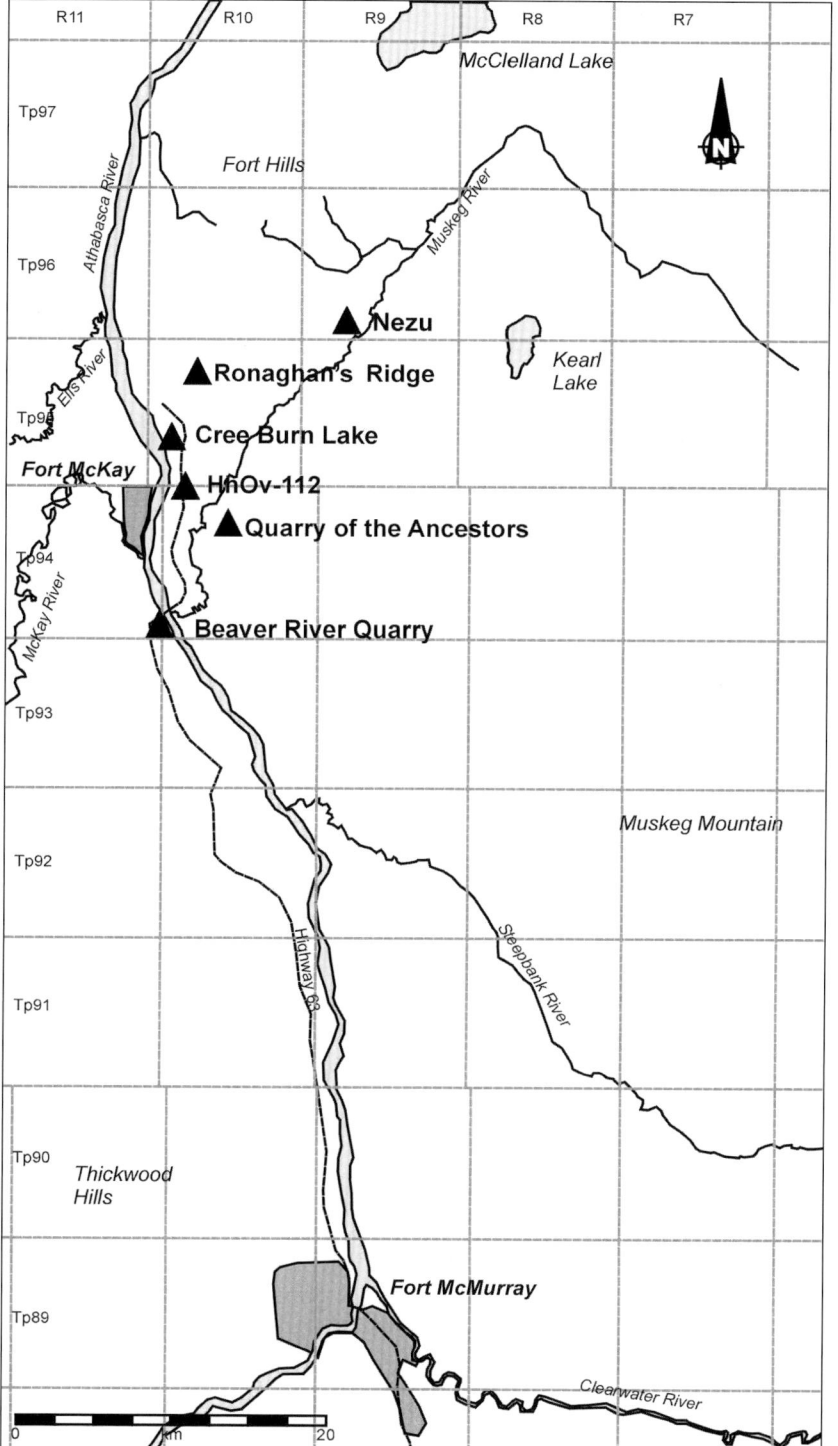

Figure 10.1. Location of the Quarry of the Ancestors and other major archaeological sites in northeastern Alberta

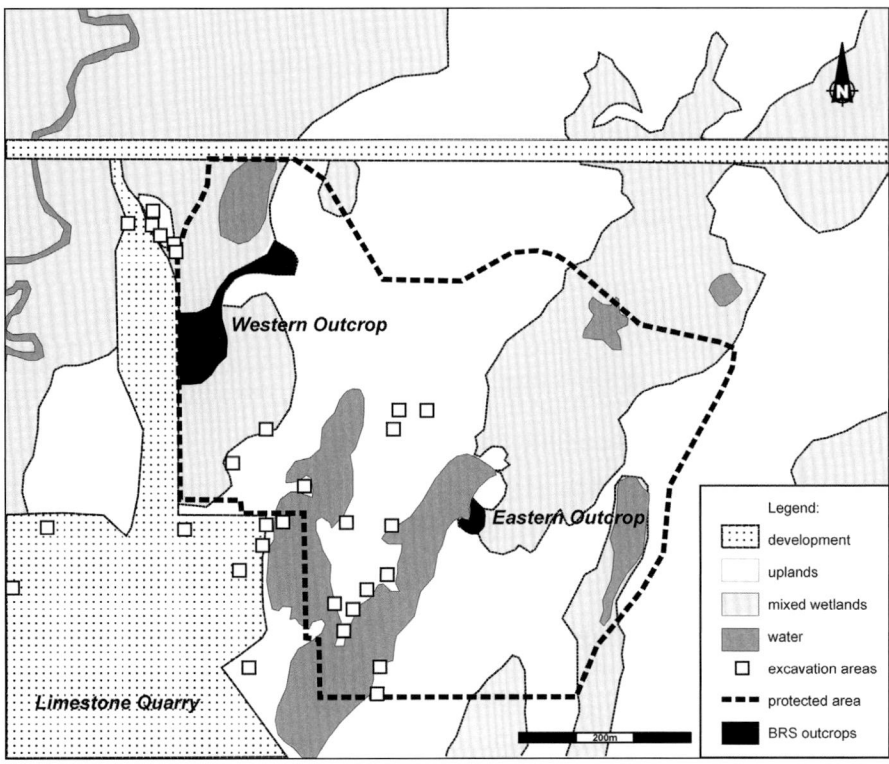

Figure 10.2. The Quarry of the Ancestors, showing the location of excavation areas and the two outcrops of Beaver River Sandstone

the reporting only on the 2003 and 2004 seasons. The discussion in this chapter is therefore based primarily on the results of the first two rounds of excavation.

VEGETATION AND TERRAIN

The Quarry of the Ancestors is situated in the boreal forest environment of the Athabasca River lowlands. The type of vegetation that grows in the area differs in accordance with changes in the drainage of various landforms that can vary only slightly in elevation. Wetlands are common and include bogs, fens, and marshes interspersed with areas of open water. Some of the wetlands surround smaller, well-drained landforms, particularly in the northeastern portion of the site. The Quarry of the Ancestors is located east of the Muskeg River in moderately undulating terrain, and the drainage in the study area is to the west and south, into the river.

Spruce bogs, dominated by black spruce and tamarack with a mossy ground cover, occur in low-lying areas with poor drainage and saturated organic soils.

Upon excavation, however, it was found that sand often underlay the black organic deposits in these locations (Saxberg 2007; Jennifer Tischer, pers. comm., 2006). The open fens support sedges, with willows along the edges, and often contain standing water. Sediments at the edges of these wetlands generally consist of silty sand. In the uplands, soils are sandy or rocky, and forest composition varies. Most dry landforms support either an open jack pine or aspen forest. The underbrush in the jack pine forests is sparse, and the ground cover is reindeer lichen. In the aspen forests, however, prickly rose, alders, and various other shrubs occur.

The two rock outcrops are low-lying and have minimal soil cover. They are well vegetated and are not immediately evident as anything other than typical, poorly drained boreal forest lands. The larger western outcrop is exposed only as the result of disturbances along an old exploration road. The smaller eastern outcrop, in the northeastern portion of the site, is located in a depression on the low side of a beaver dam and has a stream draining through it from the east.

GEOLOGY AND GEOMORPHOLOGY

Beaver River Sandstone is a silica-cemented quartz sandstone that occurs within the oil-sands-bearing Cretaceous McMurray Formation (Fenton and Ives 1990; see also Abercrombie and Feng 1997, 255 and fig. 4c; and Gryba, chapter 9 in this volume). The texture of BRS can range from macro- to micro- to cryptocrystalline, although the macrocrystalline material, commonly known as "coarsegrained" BRS, does not occur within the Quarry of the Ancestors itself. Samples of BRS from the western outcrop appear to be mostly microcrystalline, with inconsistent quality, unpredictable flaws, and linear cleavage, such that the stone fractures into tabular chunks (fig 10. 3). Samples from the eastern outcrop have a finer grain and are considered cryptocrystalline. These samples also contain flaws, but the matrix is much more consistent.

The bedrock geology, consisting of the Cretaceous McMurray Formation overlying the limestone and shale of the Devonian Waterways Formation, is generally capped by Quaternary deposits, including glacial, glaciolacustrine, glaciofluvial, and aeolian sediments. The Quaternary geomorphology of the region is crucial to understanding the distribution of surface sediments. Research has shown that the topography of the Lower Athabasca valley was fundamentally altered around the end of the Pleistocene era during a catastrophic deluge caused by the drainage of Glacial Lake Agassiz through a northwestern outlet located at the headwaters of what is now the Clearwater River (Fisher 1993; Smith and Fisher 1993; Fisher and Smith 1994; Fisher and Souch 1998). Smith and Fisher (1993) initially proposed a

Figure 10.3. BRS in situ at the western outcrop

date of 9,990 ^{14}C yr BP for this massive drainage of the lake, although subsequent geomorphological studies (Fisher et al. 2009) pointed to a slightly earlier date, in the range of 9,850 to 9,660 ^{14}C yr BP. Fisher and Lowell (chapter 2 in this volume) refine this date still further, arguing that the Lake Agassiz flood could not have occurred prior to about 9,800 ^{14}C yr BP. In chapter 2 of this volume, Fisher and Lowell further refine this chronology, arguing that the flood event could not have occurred prior to about 9,000 ^{14}C yr BP. Other studies suggest, however, that this massive flood may have taken place considerably earlier, resulting in a climate-altering burst of freshwater into the Arctic Ocean that may have marked the onset, around 11,000 ^{14}C yr BP, of the cold, dry period in the late Pleistocene known as the Younger Dryas (Tarasov and Peltier 2004, 2005, 2006; Teller et al. 2005; see also Dyke 2004, 413–416). More recently, Murton et al. (2010) have argued that two high-energy flood episodes occurred, the first shortly before Younger Dryas conditions set in (that is, sometime before about 11,000 ^{14}C yr BP) and the second around 9,900 ^{14}C yr BP.

The area that is now the Quarry of the Ancestors would have been in the flood path. Initial geoarchaeological interpretations of early occupations in the oil sands region appeared to support a model according to which flood waters receded fairly slowly, creating transgressive shorelines that were occupied by successive cultural groups. Sedimentary evidence from one of the excavation areas at the Quarry of the Ancestors suggests, however, that flood waters receded relatively quickly and that dry conditions followed. As flood waters receded, remnant ponds were probably created, including Lake Nezu, which would have lain northeast of the quarry along the current channel of the

Muskeg River (Saxberg and Reeves 2003). The presence of such ponds would explain the existence of inland archaeological sites that are now located on well-drained knolls above wetlands, often at some distance from permanent sources of water. The relative availability of water sources closer to the time of the flood may also explain why most of the evidence for use of the quarry outcrops appears to be restricted to the Early Archaic period.

CHRONOLOGY AND PALAEOENVIRONMENT

Given the lack of organic remains at the Quarry of the Ancestors, no radiocarbon dates have yet been obtained. There are, however, lines of evidence, both within the quarry complex and in the surrounding region, that delimit the period of occupation of the site in general terms. Indicators of possible Palaeoindian use of the quarry include a basally thinned projectile point formed of BRS, which was found near at the Duckett site, in the vicinity of Cold Lake (Fedirchuk and McCullough 1992), as well as a Palaeoindian projectile point, also made of BRS, found within the quarry area itself (Saxberg and Reeves 2004). A Palaeoindian presence is also suggested by the use of BRS for distinctive, pressure-flaked Cody Complex artifacts at the Nezu site, located to the northeast of the quarry in the Muskeg River valley (Shortt, Saxberg, and Reeves 1998; Bourges 1998) and by projectile points from Ronaghan's Ridge (Saxberg 1998), another relatively dense site to the north of the quarry (see fig 10.1). Use of the quarry site probably continued until the middle Holocene, when it was constrained by increasing peatland initiation (Kuhry 1997; Halsey, Vitt, and Bauer 1998; Gorham et al. 2007).

The Palaeoindian point found within the Quarry of the Ancestors (fig 10.4) was recovered from a shovel test in 2003 (Saxberg and Reeves 2004) and was later submitted for protein residue analysis. This point tested positive for proboscidean protein, indicating the presence of mammoth (*Mammuthus sp.*) or mastodon (*Mammut americanum*) (Parr 2005). The only previously known evidence for the presence of these creatures in the region are three pelvic bones from either a mammoth or a mastodon that were found in 1976 in the gravels of a mine in the Fort McMurray area (see Burns and Young, chapter 1 in this volume; see also Harington 2003, 13). At present, the latest date for a proboscidean in Alberta is 10,240 ± 325 ^{14}C yr BP (ca. 11,980 cal yr BP), from considerably further south, at the James River Bridge, near Sundre (Burns 1996).

The contemporaneity of proboscideans and humans in the Athabasca lowlands has significant implications with respect to the antiquity of human occupations and the dating of the drainage of Lake Agassiz through the northwestern

0   0,5
Centimetre

Figure 10.4. Palaeoindian projectile point, recovered in 2003, that tested positive for proboscidean protein

outlet. In spite of a few late dates for proboscidean remains in North America (especially for mastodons), it is generally accepted that the extinction of proboscideans occurred before 11,600 calendar years ago (Agenbroad 2005, 85). The extirpation of proboscideans in Alaska and the Yukon occurred somewhat earlier, and a study by Guthrie (2006) appears to correlate significant late Pleistocene environmental change with the disappearance of megafauna in that area.

One of the localities excavated during the 2004 season at the Quarry of the Ancestors yielded particularly important data with respect to the timing of pre-contact occupation and palaeoenvironmental conditions. A stratified profile in excavations adjacent to one of the wetlands displayed a series of sand and clay layers, none except the uppermost containing organic material (figs. 10.5 and 10.6). This suggests that, as these layers were being deposited, no surfaces were stable long enough to develop an organic soil or that organic evidence was destroyed by chemical leaching. The lowest levels contain boulders and are banded, indicating high-energy fluvial deposition. This pattern of banded sand and clay continues to approximately 75 centimetres from the surface, well above the boulders. Above that, a layer of sandy clay gives way to massive sand, followed by a layer of pink-coloured clay, indicating a non-organic lacustrine environment. The top layer, immediately beneath the forest litter, was also massive sand.

The only evidence in this profile of a catastrophic deluge, of the sort associated with the flooding of Glacial Lake Agassiz, is the lowest layer, with its boulders mantled in clay. This clay layer shows evidence of desiccation cracks, indicating that the flood had receded and the land had dried before the majority of the sand was deposited (fig 10.7). Above that, another episode of fluvial activity is evident, likely moving in the direction of present drainage, from the northeast. The uppermost levels of massive sand probably represent a dry, windy environment, where low-lying interdune areas trapped moisture, creating the ephemeral clay layers between massive sand beds.

A sediment-filled ice wedge was observed in the sand layer below the uppermost layer of pink-coloured clay (see fig 10.6), indicating cold, dry, possibly periglacial conditions. Periglacial features dating to approximately 11,000 to 9,900 ^{14}C yr BP have been observed in sedimentary profiles in northwestern Saskatchewan (Fisher 1996), and evidence also indicates that cold, dry conditions existed before about 10,750 ^{14}C yr BP at Eaglenest Lake, in the Birch Mountains (Vance 1986), and at Kearl Lake prior to about 10,100 ^{14}C yr BP (Beirele 1996; Bouchet-Bert 2002). If we assume that the ice wedge present in the sand layer is of roughly the same age, then it presumably also reflects the cold, dry period known

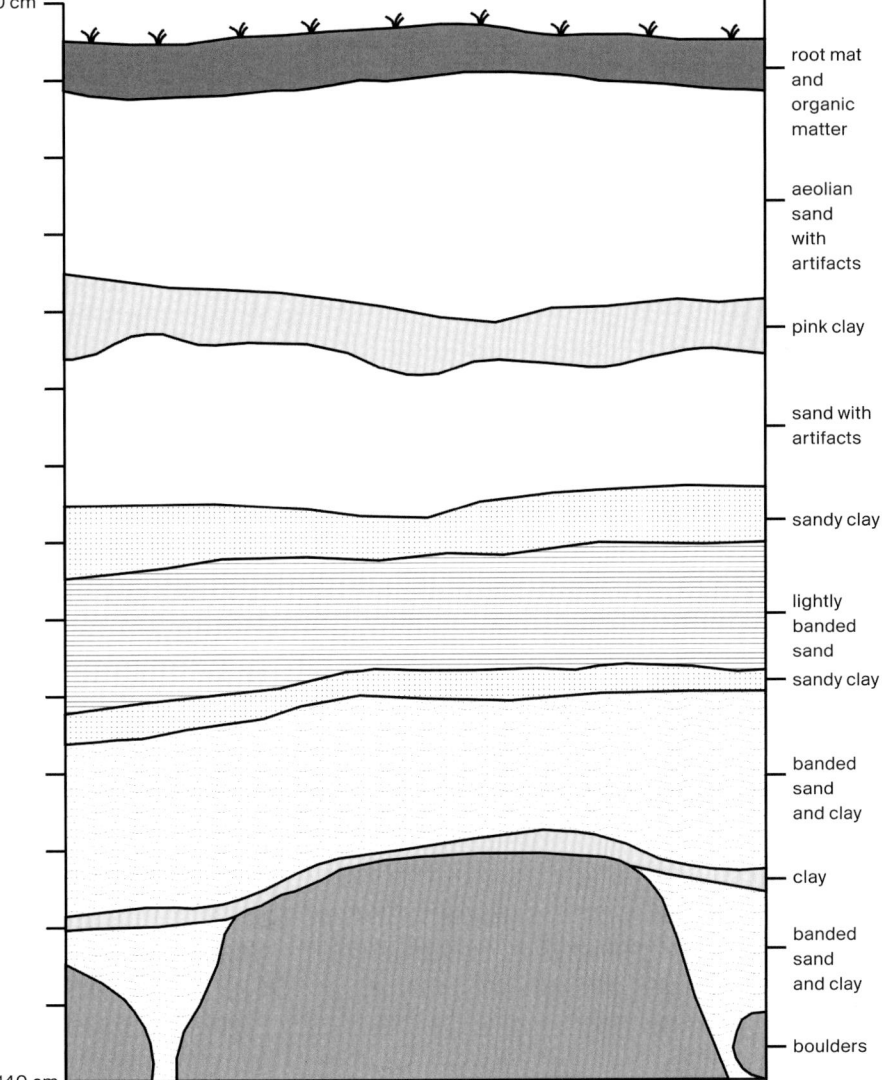

0 cm

root mat
and
organic
matter

aeolian
sand
with
artifacts

pink clay

sand with
artifacts

sandy clay

lightly
banded
sand

sandy clay

banded
sand
and clay

clay

banded
sand
and clay

boulders

140 cm

Figure 10.5. Profile of the stratified excavation area, showing the alternating layers of clay and sand and the two phases of human occupation

as the Younger Dryas, which extended from roughly 11,000 to 10,000 ^{14}C yr BP (ca. 12,900 to 11,500 cal yr BP). And if Younger Dryas conditions are visible in a profile *above* sedimentary evidence for high-velocity flooding, then this flood must have happened *before* the Younger Dryas conditions set in—that is, sometime prior to about 11,000 ^{14}C yr BP.

Figure 10.6. Photograph of the stratified area, showing the layer of sand containing the sediment-filled ice wedge

The uppermost clay layer, with its pink colour, indicates a gentler period of inundation of portions of the site, although with little organic material. This could represent the second meltwater event, which roughly coincided with the Hypsithermal (or Altithermal) interval—the early to mid-Holocene period of maximum postglacial warming. The clay is overlain by a thick layer of aeolian sand probably deposited during this warm, dry period, which Bouchet and Beaudoin (chapter 4 in this volume) date to roughly 9,820 to 7,580 ^{14}C yr BP (11,200 to 8,390 cal yr BP) in the Kearl Lake area—although, as they point out, regional variations existed in the duration of warmer conditions, as well as in the timing of their onset. The profile is capped by modern forest litter, but the sand continues under the organic wetland deposits to the east of the excavated area (Saxberg 2007).

No culturally or chronologically diagnostic artifacts were recovered from the excavations at this locality, although two occupations were observed in the sandy sediments, immediately above and below the layer of pink clay. If, as discussed above, the pink clay represents the final episode of deglaciation in the area and the sand below was deposited in the cold, dry conditions of the Younger Dryas, then these occupations represent adaptations to dry conditions in both cold and warm periods. Moreover, given that artifacts occur in the sand beneath the

Alberta's Lower Athabasca Basin

organic wetlands deposits to the east, both occupations must predate the formation of local wetlands.

The artifact scatter in the upper level of occupation was denser and more diverse than that in the lower, which implies an increased intensity of occupation in later times, during the Hypsithermal interval, when the climate had become warmer and drier. In all likelihood, this more intense period of use of the Quarry of the Ancestors in Early Archaic times persisted until significant peatland initiation created the more patchy boreal forest conditions in evidence today (Halsey, Vitt, and Bauer 1998; Kuhry 1997).

BEAVER RIVER SANDSTONE

When archaeological sites were first identified in the oil sands region, it quickly became evident that artifact assemblages were dominated by a single raw material type, now commonly called Beaver River Sandstone. In 1973, in the course of an assessment conducted for Syncrude, archaeologists discovered a quarry site, HgOv-29, on Beaver Creek, not far west from its confluence with the Athabasca River. This site, which came to be known as Beaver River Quarry, appeared to constitute a local source of the material so ubiquitous in artifact assemblages. One variety of the stone was initially deemed to be grey chert, while another was regarded as limestone (Syncrude Canada Ltd. 1973, 79), but the material was subsequently identified as a quartzite and assigned the name "Beaver Creek Quartzite" (Syncrude Canada Ltd. 1974, 47).

The stone was renamed "Beaver River Sandstone" by Fenton and Ives (1982), who carried out much of the subsequent research on the material. They examined various outcrops and determined that BRS was a silicified sandstone occurring as part of the McMurray Formation (Ives and Fenton 1983, 1985). This stratigraphic position was further refined to "near the top of the lower member of the McMurray Formation" (Fenton and Ives 1984, 130; see also Fenton and Ives 1990).

This characterization of the stone became accepted in the archaeological literature of the oil sands region, but the problem of determining where the raw material for tools actually came from remained. As it turned out, the outcrops of BRS identified at the Beaver River Quarry contained only relatively coarse-grained (macrocrystalline) material, whereas the vast majority of artifacts recovered from archaeological sites in the region were formed from finer-grained varieties of BRS (micro- and cryptocrystalline). In an analysis of the Beaver River Quarry material, Reardon (1976, 63–64) noted that the stone found in artifacts at other sites in the region was "far less coarse grained than 99.9% of the material found at the quarry; by its very nature it is a better quality material." He concluded that this finer material must have come from other, as yet undiscovered, outcrops on the east side of the Athabasca River.

The BRS problem gained attention as development in the oil sands surged in the late 1990s. By this time, a new factor had been added to the interpretation of archaeological sites in the region, namely, the Glacial Lake Agassiz flood, which accounted for an ex situ dispersal of raw material. Tsang (1998) located multiple outcrops of the macrocrystalline variety of BRS in a roughly north-south orientation west of the Muskeg River and east of the Athabasca, with a single outcrop also at the location of the Beaver River Quarry, west of the Athabasca River (Tsang 1998). He confirmed Fenton and Ives's determination with respect to the age of the BRS, but, because his work was oriented toward understanding how BRS was formed, he did not address the question of variation in material quality.

Another site, known only by its Borden number, HhOv-55, was discovered during a survey of the Athabasca River in 1976. It was revisited by Ives and Fenton in the early 1980s (1985, 22–23) and then again by Unfreed in 2000 (Unfreed and Fedirchuk 2001). An outcrop of BRS was visible in an erosional bank, underlain and overlain by bituminous McMurray Formation sands.

All the recorders of HhOv-55 agreed that the site was a quarry and workshop locale, a designation that was initially based on the observation of unmodified in situ BRS. Unfreed and Fedirchuk (2001, 78) subsequently identified the site as a quarry because of "the presence of extremely dense concentrations of debitage including cores, shatter, and flakes resulting from various stages of reduction, and the relatively small quantity of finished tools." On the basis of the localized hummocky terrain, they also argued that "materials were being removed from a number of local shallow holes found across the area, possibly from the large Beaver River Sandstone cobbles and boulders which lie shallowly buried under the surface of the area." The presence of BRS as "cobbles" does not suggest an in situ occurrence of the material but seems more likely to be evidence of a detrital deposit. The outcrop of BRS at the site was described by Ives and Fenton (1985, 23) as "not a source for high-quality Beaver River Sandstone."

During assessments connected with oil sands projects (see, for example, Gryba 2002; Saxberg, Somer, and Reeves 2003), both naturally occurring and minimally modified blocks of BRS were observed in surface contexts in the Fort Hills area. This may indicate that much of the BRS found in archaeological sites in the northern part of the oil sands area originated from float blocks that, during the flood, had broken away from an outcrop to the south. However, BRS does not occur in gravel deposits in the area (Ives and Fenton 1985, 44), which suggests that these deposits either originated from a different source or were dispersed by a different mechanism (possibly glacial activity) than the BRS float.

Several other archaeological sites in the region, such as the Cree Burn Lake site, Ronaghan's Ridge, and HhOv-112 (see fig 10.1), contain abundant waste material, which was previously taken as evidence that primary extraction of BRS took place at these sites. Excavations at HhOv-112 in particular, however, demonstrated that these very dense concentrations of debitage centred on single large nodules, indicating a float source of raw material. This site also contains evidence of "secondary quarrying" (Saxberg and Reeves 2003).

Away from the Quarry of the Ancestors, sites with the densest concentrations of BRS primarily occur to the north. This is arguably because the Lake Agassiz flood dislodged large nodules of BRS from the outcrops at the quarry and deposited them as it flowed north. The plethora of float material, coupled with the likelihood that the quarry itself was inaccessible until relatively late in the period of deglaciation, suggests that earlier occupations containing large amounts of BRS would occur to the north of the quarry. Relatively few archaeological sites containing BRS have been found to the south of the quarry. In fact, both the number and the density of sites decrease dramatically within 2 kilometres south of the outcrops (Saxberg and Reeves 2006).

The first outcrop identified in the Quarry of the Ancestors was regarded by project geologists as a silicified zone of the local outcrop of Devonian limestone (see Saxberg and Reeves 2004). This stone in this outcrop was different in quality from the coarse-grained varieties of BRS examined by Tsang (1998), and it was immediately evident to archaeologists that ancient peoples had made extensive use of the area for raw material extraction and tool manufacture. Further examination by geologists over the winter of 2004–2005 demonstrated that the original stratigraphic determination of the material was correct (that it is part of the lower member of the Cretaceous McMurray Formation, rather than Devonian) and that the material is not a silicified limestone or sandstone but a "microquartz-cemented orthoquartzitic siltstone" (De Paoli 2005, 6). A new name, "Muskeg Valley Microquartzite" (MVMq), was accordingly suggested. This designation was used in assessment and mitigation reports mainly to distinguish the finer-grained variety from the macrocrystalline variety, such as that found at the Beaver River Quarry. For the most part, however, the archaeological literature has retained the name Beaver River Sandstone, and, for the sake of consistency, we do the same here.

What is even more intriguing about the material as a tool stone, however, is the issue of heat treatment. Gryba (2002 and chapter 9 in this volume) has suggested that both the coarse-grained and finer-grained varieties of BRS were rendered highly workable through thermal alteration. His experiments indicate that the raw material, even that from the Quarry of the Ancestors, was not in a form that allowed it to be easily shaped until it had been subjected to controlled heating, at relatively low temperatures with prolonged heating and cooling periods. One of the authors of this chapter (Robertson) employed X-ray absorption near edge structure (XANES) spectra analysis to determine precisely what changes occur in macro-, micro-, and cryptocrystalline BRS when it is heat-treated. As the XANES analysis revealed, all varieties of BRS become more workable not because of chemical alteration or crystalline reorganization of silica but because of changes to elements present only in trace amounts, particularly titanium, although changes to these trace elements appear to be responsible only for greater ease of flaking (Robertson and Blyth 2009). The improvement of the material is most noticeable in the macro- and microcrystalline varieties, and the micro- and cryptocrystalline varieties develop a red rind when heated.

LITHIC TECHNOLOGY AT THE QUARRY OF THE ANCESTORS

The lithics recovered from archaeological investigations at the Quarry of the Ancestors most probably represent multiple occupations that cannot now be spatially and temporally separated. The lithic technology, which appears to have centred on opportunistic, informal flaking and the creation of utilizable flakes, is relatively similar across the various localities within the quarry, although the signature and patterning of both tools and debitage are highly variable. Of the 337,959 lithic artifacts recovered during the 2004 excavations, only 4,058 (1.2%) were tools. The tools were divided into three categories, formal, expedient, and manufacturing, which accounted for 26.5%, 37.4%, and 36.1% of the total number, respectively. The category of formal tools comprised all finely worked tools, as well as tools for which a specific use was evident and that had obviously been made in order to perform that function, even if extensive retouch was not present. This category included projectile points, scrapers, gravers, drills, wedges, and knives. Expedient tools were all minimally modified flake tools, including retouched and utilized flakes. Manufacturing tools consisted of cores, modified cobbles, anvils, and hammerstones.

During shovel testing the previous year, 25,146 artifacts were recovered, of which 305 (again, 1.2%) were classified as tools. A total of six projectile points or projectile point fragments were recovered, four of them formed from BRS, one from quartzite, and one from chert. The BRS points consist of the Palaeoindian point described above (see fig 10.4) and three point fragments that probably broke during manufacture. The two non-BRS points, which are complete, are similar to Archaic types discussed by Wright (1972). The quartzite point is a smaller, side-notched point that has been reworked, while the chert point has an expanding stem and is heavily reworked. The Palaeoindian point exhibits parallel-collateral pressure flaking and has a plano-convex to diamond-shaped cross-section. The base is thinned by a few pressure flakes on both sides, and the base and sides are ground. An exact match for this point has not been found in the literature. The flaking patterns are similar to those of Cody Complex materials, but it is also stylistically similar to many other lanceolate types from the Plains and the North.

In the 2004 excavations, eleven projectile points or point fragments were recovered, of which seven were formed from BRS and the rest from other materials. The complete points are predominantly small and corner-notched, with an expanding stem, a variety that is not specifically diagnostic but, in view of similarities to points described in Wright (1972), probably dates to the Archaic period. Only one of these points—a relatively large, side-notched BRS specimen—showed evidence of extensive reworking. The excavations in 2005 yielded another twenty-five points and point fragments in various stages of manufacture, the majority of them formed from BRS (Saxberg 2007). They are mostly side- and corner-notched varieties, similar to those found in the previous excavations.

Some of the Quarry of the Ancestors excavations yielded evidence of microblade technology. Among the finds was a primary ridge flake from a classic wedge-shaped microblade core, although no other evidence was recovered of highly structured core preparation. A microblade core was, however, recovered from excavations in 2006 at HhOv-449, a site just to the east of the designated quarry area, as was a probable core preform at HhOv-468, located not far west of the quarry (Wickham and Graham 2009; see also Younie, Le Blanc, and Woywitka, chapter 11 in this volume).

The debitage was catalogued according to the mass analysis method developed by Sullivan and Rozen (1985). Overall, the debitage signature appeared to indicate biface and tool production rather than core reduction, although there was considerable variation (Saxberg 2007). A high rate of flake and chunk utilization for resource processing was observed, as was evidence of edge modification indicative of cutting, scraping, and chopping. Few finished formal tools (only 0.3% of all artifacts) were found, and while there is some evidence of economical approaches, such as the use of bipolar reduction and microblade manufacture, such approaches were not the prevailing technology.

None of the mitigative excavations conducted during any of the field seasons at the Quarry of the Ancestors were located on the outcrops themselves, although some shovel testing of the outcrops was carried out in 2003 (Saxberg and Reeves 2004). The shovel testing revealed very thin soil development supporting relatively thick vegetation. Materials in the shovel tests consisted almost exclusively of tabular chunks of BRS, either unmodified or else exhibiting signs of testing, with some clusters of artifacts, including flakes and hammerstones, found in sandy areas immediately adjacent to the western outcrop (Saxberg and Reeves 2004). These observations, and those gleaned from the larger excavations, indicate that people were sorting and removing lithic packages from the outcrops to sandy uplands prior to further reduction. This spatial signature is arguably the result of the need to heat-treat the raw material, a process that is most effective when undertaken in sandy sediments (Mandeville and Flenniken 1974).

All of the sites investigated in the Quarry of the Ancestors lie within one kilometre of one of the two outcrops. Logically, archaeological sites close to an abundant lithic source should display evidence of relatively less economical approaches to tool manufacture, with used tools discarded in favour of newly constructed ones. One would predict large amounts of waste material, with little use of waste flakes as expedient tools, and one would expect to find all levels of tool manufacture represented in the debitage, including initial reduction, shaping, and finishing, with rejects at each stage. In addition, one would expect that all sites within a short walk of the quarry would have a similar signature with respect to the types of tools and debitage represented, since they presumably represent a similar point in the seasonal round and a similar stage in the lithic reduction process.

What was found is a great deal of dissimilarity in relative abundance of tool types, with manufacturing tools, such as cores, hammerstones, and modified cobbles, occurring in approximately inverse frequency to utilizable tools, that is, tools used for resource processing, such as most formal and expedient items not used in the manufacture of other lithic tools. Also, some localities show a very high rate of utilization of flakes, and some do not. Some have a great deal of manufacturing debris, while others do not, and some sites are dominated by high-quality, fine-grained material, whereas others contain no high-quality material.

The physical characteristics of BRS were significant with respect to lithic reduction strategies at the Quarry of the Ancestors. Many pieces of BRS exhibit both micro- and cryptocrystalline textures. The natural structure of the rock contains numerous horizontal and vertical fracture planes, which produce tabular chunks that are easily removed (see fig 10.3). The tabular chunks can be used as naturally prepared cores, while the inconsistency of the material necessitates an opportunistic and occasionally economical strategy to maximize what may be a very small pocket of fine-grained material. As a result, the debitage signature for what should be primary reduction mimics that of tool manufacture, with a concentration of smaller flakes and a minimal occurrence of detrital pieces with no flake characteristics. Naturally occurring small packages of raw material also facilitate transport.

The need for heat treatment may, however, have complicated the otherwise easy removal and transport of the tabular chunks of stone, in that a cultural process had to be applied to the material in order to make it more workable. Perhaps as a result of bioturbation, no features have been located in the region that can be definitively identified as heat-treatment pits. Yet the occasional regional discovery of finely pressure-flaked formal tools (such as the Palaeoindian point described above or the Cody Complex points from the Nezu site), as well as the reddening seen on certain flakes and chunks, indicate that heat treatment was being performed to some degree.

In the region surrounding the Quarry of the Ancestors, we know that BRS is both locally and abundantly available, that it is of variable quality, and that it is easily removed from its source. We also know that the workability of BRS improves with heat treatment. Archaeological sites in the region overwhelmingly contain BRS as a tool stone, and it is rare outside of the region. We also know that finely crafted, formal tools made of BRS are relatively scarce and that there is a high degree of expedient tool manufacture and use, at least within the vicinity of the quarry.

The organization of lithic technology is influenced by three major characteristics of raw materials: the size and shape of blanks, the relative availability of raw material, and the quality of the stone itself. These three factors will influence the proportion of expedient versus formal tools observed at a site. When raw material is abundant, there is little reason to spend time carefully crafting a tool; rather, the stone readily at hand will generally be used to fashion an assortment of multi-purpose tools, with the result that one expects to find a preponderance of expedient tools (Andrefsky 1998, 222).

Raw Material Size and Shape

Lithic technologies are determined in part by the size and shape of raw material blanks (Lothrop 1989; Kuhn 1992). The use of bipolar technology is thought to be an adaptation to small lithic packages and/or to a scarcity of raw materials. The BRS that occurs in the western outcrop at the Quarry of the Ancestors is a highly fractured material that exists mainly in relatively small, tabular chunks (see fig 10.3). Larger chunks with less linear fracture patterns were noted at the eastern outcrop, however, suggesting that the natural sizing of fractured BRS is highly variable. This sizing may be linked to quality, in that the smaller, more fractured tabular chunks were of poorer quality than the larger, more amorphous packages. In view of the overgrown nature of the outcrops, however, there was no opportunity to investigate this possibility in more detail.

The use of tabular chunks as ready-made biface blanks, thereby eliminating the need for an additional step in the reduction sequence, has been documented at other quarry locations (Petraglia, LaPorta, and Paddayya 1999). This process appears to be in evidence to some degree at the Quarry of the Ancestors, as various bifacially worked tabular chunks have been recovered from the excavations. One of the localities contained an exceptional number of wedges, some of which were made from exhausted bipolar cores. Bipolar technology is thus evident at the Quarry of the Ancestors, as is microblade technology, although the two do not necessarily co-occur at the excavation localities, however, and, as noted earlier, neither is the prevailing technological approach. This variation suggests a fluid, adaptable approach to lithic reduction, depending on the availability of high-quality materials.

Raw Material Availability

The Quarry of the Ancestors probably first became exposed after the Lake Agassiz flood as a result of the removal of overlying sediments. The ease of extraction of this raw material owing to internal fracture planes would have encouraged its exploitation. At the Quarry of the Ancestors, BRS was therefore both abundant and accessible—the two basic aspects of availability. Ethnographic studies in Australia (Gould 1980) suggest that availability was the primary factor governing the use of a particular lithic material in stone tool manufacture, especially if the raw material was situated near a water source where a habitation might be located.

The evidence from the Quarry of the Ancestors supports this conclusion, given that tools of all types were more commonly formed from BRS than from any other material. Differences in the excavated localities at the site, however, suggest that, in spite of this abundance, a premium may have been placed on access to higher-quality varieties of the stone. Most localities contained a mixture of relatively fine-grained and relatively coarse-grained BRS, although some contained only coarser material. In contrast, one locality featured very fine-grained materials, which suggests that careful sorting and probably thermal alteration had occurred. While expedient tools were made of all varieties of BRS, the locality with the finest material was also the locality with the highest percentage of utilized flakes, which demonstrates that quality may have taken precedence over abundance in flake tool manufacture.

Raw Material Quality

Andrefsky (1994) suggested that the quality of tool stone was a major factor in the structuring of tool stone quarry assemblages, in that, when raw materials were abundant but of inconsistent quality, they were approached in much the same manner as less abundant raw materials, with an emphasis on the conservation of higher-quality stone through the use of more economical and strategic technologies. Moreover, higher-quality materials would be reserved for tools that required finer crafting, with the result that formal tool types would be well represented at such sites. Similarly, Brantingham et al. (2000) examined formal versus informal tool types in northeast Asia and found that poor raw material quality dictated the formal tool technologies employed during the Middle and Upper Palaeolithic. At Skink Rockshelter, in West Virginia, the quality of the raw material was similarly found to be more important than local abundance when it came to selecting the lithic material used for flake tools (MacDonald 2008).

At the Quarry of the Ancestors, an abundant supply of high-quality raw material may have been the most important factor in determining lithic reduction strategies. The high degree of utilization of flakes and the lack of effort spent on the manufacture of formal tools, as well as the sheer number and variety of artifacts fashioned of BRS, suggest that high-quality material was both abundant and accessible, in spite of the occasional use of economical reduction strategies.

As we have seen, the effects of inconsistencies in quality could have been mitigated through heat treatment. Even though no evidence of heat treatment pits has been found at the Quarry of the Ancestors, recent experimental studies have suggested that the heat treatment of lithics would not necessarily have required elaborately prepared facilities, provided the knappers exercised control over the size of the lithic packages being heated (Mercieca and Hiscock 2008). This may have been particularly important in situations where the firing temperature was difficult to control, as in the case of windy conditions, or when limited fuel was available. Smaller packages can be heated in relatively short periods of time under surface hearths insulated by a layer of sand.

THE ORGANIZATION OF LITHIC TECHNOLOGY

The lithic reduction strategies employed at the Quarry of the Ancestors were structured by the three basic attributes of raw material described above, and those strategies both influenced and were influenced by cultural factors. Some of the key characteristics of the lithic technology visible at the Quarry of the Ancestors, which in turn inform cultural interpretations, are the lack of evidence for retooling, the high number of expedient tools, the presence of specialized technologies and tool types, and the apparent heat treatment of raw materials. On the basis of these characteristics, certain conclusions can be suggested regarding settlement patterns and adaptive flexibility.

Adapting to Abundance

The relationship of lithic technology to mobility patterns in ancient societies, as well as in living ones, has been extensively studied (see, for example, Binford

1978, 1980; Gould 1980; Parry and Kelly 1987; Bamforth 1991). In a classic study of hunter-gatherers, Binford (1980) drew a distinction between residential mobility, that is, the movement of an entire group from one residential base to another, and "logistical" mobility, in which smaller parties periodically set out on hunting and gathering expeditions and then return to the base camp. He accordingly identified two possible patterns of group mobility: the collector and the forager. Collectors are characterized by low residential mobility and high logistical mobility: they tend to remain in one location for extended periods, travelling out on temporary forays and then returning to the base camp. In contrast, foragers are characterized by high residential mobility and correspondingly low logistical mobility. Binford proposed a correlation between these two patterns of mobility and spatial and/or temporal variations in the availability of resources, arguing that the relative abundance or scarcity of resources will be reflected in the nature of archaeological assemblages.

Groups that move about a great deal are considered more likely to curate tools (Kelly and Todd 1988; Andrefsky 1991), and formal tools are more likely to be curated than expedient tools. We would therefore expect that sites created by groups of high residential mobility would exhibit significant evidence of the curation of tools, through reworking and sharpening. Sites at the Quarry of the Ancestors, however, yielded almost no tools that had been substantially reworked or heavily worn tools that had presumably been discarded, nor was there evidence of extensive sharpening, again suggesting that the occupants did not need to recycle their tools because they were never very far from their tool stone source. The most heavily reworked tools are three projectile points that were recovered from the excavations in 2003 and 2004. The other artifacts (well over 350,000 of them in those two years alone) show very little evidence of reworking.

Tool richness refers to the number of different activities that are represented by the tool assemblage, whether at individual sites and at sites within a region. Sites created by groups that are highly mobile typically exhibit less tool richness than sites created by groups who largely remain in one place (Shott 1986). Highly mobile groups tend to depend on multi-purpose tools, whereas the tools of less mobile groups are much more specialized. These specialized tools may include an assortment of scrapers and cutting tools and, in particular, a varied array of unifacial tools and core types. In contrast, bifaces serve multiple purposes, as they can function both as utilizable tools and as cores (Kelly 1988).

Kuhn (1995, 22) distinguished between the provisioning of places, which he associated with relatively sedentary groups, and the provisioning of individuals, which he viewed as characteristic of more mobile populations. When members of a group travelled, they carried with them a small number of tools, which they carefully maintained so that they would be ready for use whenever the need arose. In contrast, sites that served as base camps tend to show evidence of place-provisioning activities of the sort in which relatively low-mobility groups engaged prior to setting out on short-term logistical forays. Provisioned places—places that contained abundant resources or were otherwise important—were sites to which members of low-mobility groups repeatedly returned, and the stockpiling of lithics, particularly multi-purpose lithic packages, became a central focus in mobility patterns. For such groups, the key factor was not the versatility of tools but the versatility of the stockpile. Among more mobile groups, who did not necessarily return to the same place on a regular basis and could not be assured of an abundant supply of raw material, versatility in tool function became the salient factor.

The data from the Quarry of the Ancestors suggest that it was a provisioned place. The lithic stockpile, while originally a natural phenomenon, was extraordinarily diverse and capable of producing an unlimited array of tools for any purpose, as is evident from the variety of expedient tools recovered from the excavations. That the lithic assemblages from the Quarry of the Ancestors represent groups with low residential mobility but some degree of logistical mobility has implications throughout the region and finds additional support in archaeological data from areas beyond the quarry. Small sites located in the Fort Hills, approximately 30 kilometres north of the quarry, are frequently characterized by high percentages of formal tools made from BRS with some evidence of curation (Saxberg, Somer, and Reeves 2003, 2004; Somer 2005), suggesting the presence of groups from the quarry who had travelled out on logistical forays. BRS tools are likewise common at sites in northwestern Saskatchewan but frequently show signs of reworking and resharpening (Brad Somer, pers. comm., 2009). As Bamforth (1986) has argued, evidence of a combination of maintenance and recycling is a signal of raw material shortage.

In a study of site patterning in the Great Basin during the Early Archaic period, Duke and Young (2007, 135) observed a similar pattern of "low visibility assemblages comprised of curated tools" found in upland areas, with larger, more prominent scatters of flakes and expedient tools found at sites along lake margins in lowland areas, sites they interpreted as "a record of predominantly women's activities." Given that archaeological sites are primarily evidence of the discard or loss of tools, the presence of curated tools at smaller, upland sites raises an intriguing question. If curated tools were discarded in temporary camps associated with logistical mobility, does this not imply that groups returned to the base camp with no tools? If so, such behaviour would seem to indicate a confident attitude with regard to the supply of raw materials.

In another Great Basin study, Jones et al. (2003) argued that the geographical spread of a single local raw material, which they termed a "lithic conveyance zone," was roughly equivalent to a group's territorial range. The territorial range of people using BRS would thus appear to be limited to the Athabasca lowlands, with some forays into the Birch Mountains to the west (see Ives, chapter 8 in this volume) and, to the east, into northwestern Saskatchewan (Brad Somer, pers. comm., 2009). Although artifacts formed from BRS have been reported from more distant locations, they occur only rarely, while non-local materials are equally rare at the quarry, with 99.9% of the artifacts formed from BRS, as one would expect at a provisioned place.

Binford (1978, 1980) concluded that hunter-gatherers often embedded tool stone procurement in their seasonal round, such that the procurement of materials was part of their regular pattern of movement, rather than requiring a special trip. Such a pattern would have meant that tools fashioned of non-local materials found at other places on the seasonal round would tend to be discarded in favour of new ones once a group arrived at the primary lithic source (see, for example, Gramly 1980). In addition, if lithic procurement is embedded in the usual seasonal round, evidence of routine lithic processing activities would presumably be found at the lithic source. While the diversity of tools at the quarry might appear to support such a pattern of hunter-gatherer behaviour, the scarcity of tools made of non-local materials does not. Significantly, the meagre evidence for the retooling or discard of exhausted non-local tools is a universal characteristic at the Quarry of the Ancestors, which suggests that the quarry was not simply a stopping point on a seasonal round that included other sources of raw materials.

The evidence indicates that the exploitation of BRS was not a mere matter of logistical expediency but was culturally embedded in the lives of the region's occupants and that, as with other provisioned places, the quarry was a major focal point of settlement. The role of raw material sources in determining settlement patterns has also been noted for the Archaic period in the southeastern United States (Daniel 2001). The evidence of habitation, rather than just tool manufacture, at the Quarry of the Ancestors is similar to that documented at Cummins Quarry, a raw material extraction site in northwestern Ontario that dates to approximately the same period (Julig 1994, 215). In the Great Lakes region, bedrock raw materials may have been intensively exploited in the late Palaeoindian period because the use of a particular lithic material provided evidence of a shared identity, which in turn allowed groups to minimize risk (Ellis 1989, 2004; Carr 2005). This use of a specific lithic material as a marker of group identity resulted in a limited spread of that material and a demonstrated preference for it over all other tool types within a certain time period and region.

That the quarry was a permanent or semi-permanent base camp implies that a relatively stable food source was also available in the local area. It may be that the Archaic period inhabitants of the oil sands region were adapted to wetlands edges, as were those in the Great Basin (Duke and Young 2007), and knew how to exploit the abundant plant and animal resources. The scarcity of stone projectile points within the quarry area suggests that larger game may not have been the focus of subsistence. Fiedel (2007) has noted that birds, and particularly migratory waterfowl, were an abundant resource during the early occupation of North America, and reliance on birds as a food source is often underestimated in archaeological interpretation. Open water in northeastern Alberta would certainly have seasonally attracted large flocks of migratory birds, particularly during relatively dry climatic periods.

The warm, dry climate during the mid-Holocene is likely to have affected cultural patterning in northeastern Alberta and may have encouraged more stable settlement. On the Plains to the south, mid-Holocene Hypsithermal conditions have been considered a trigger for cultural changes (Reeves 1973; Robertson 2004; Walker 1992). As the climate warmed, the tendency of people to concentrate in areas of plentiful resources, such major river valleys, foothills, and parklands, may have brought about an influx of population into boreal forest regions, resulting in greater interaction among groups. The arrival of outsiders might have triggered an entrenchment of existing land use patterns, with the people already occupying the oil sands region seeking to establish their rights of access to the quarry and surrounding areas in the face of encroachment by other groups. Such encroachment may have provoked conflict and may have required a reinforcement of identity and territoriality. At the same time, an increased population in the vicinity may have presented new opportunities for trade, perhaps also enabling the quarry inhabitants to limit their mobility, given that they could trade for food supplies in times of scarcity (although warmer conditions could also have increased the productivity of the local area).

Limited mobility has other interesting implications within hunter-gatherer societies. Traditionally, semi-sedentary peoples are thought to have developed craft specialization and storage as a means of coping with periods of scarcity. No evidence of storage has been recovered at the Quarry of the Ancestors, but some degree of craft specialization may be indicated by a few examples of specialized lithic technologies, such as microblade production, and at least one cluster of distinctive tool types, found in the northwestern portion of the site, north of the

Figure 10.8. Notched engraving tools, evidence of artistic activity at the Quarry of the Ancestors

western outcrop. There, excavations in a small area of about 4 square metres recovered twelve (of a total of fifteen) small engraving tools that were unique to that part of the site. The area also contained four scrapers. The engravers were originally thought to be perforators but, upon closer examination, proved to have characteristics of engraving tools, including frequent sharply triangular cross-sections and finely retouched steep lateral-edge angles on one or both margins (fig 10.8). They were formed on prismatic blade flakes. A few of these tools displayed a steeply retouched blunted distal margin and were notched proximally, implying hafting or at least attachment to another material. These tools have not been previously reported in oil sands sites, unless they have been categorized as flake points or as perforating tools.

Although no decorative or apparently non-utilitarian objects have been recovered from oil sands sites (owing to the lack of organic preservation), these engraving tools imply the presence of artistic activities. The small size of the area suggests that it was created by a single individual, and the engraving tools were probably part of a specialized activity practiced by a particular craftsperson. In Palaeo-Eskimo sites, the occurrence of a plethora of specialized engraving tools is attributed to a well-developed bone-, antler-, and ivory-carving industry (see, for example, Maxwell 1984). One of the engraving tools at the Quarry of the Ancestors tested positive for deer proteins (Parr 2007), which would be consistent with use on antler, and one of the scrapers tested positive for cat proteins. This would be lynx, bobcat, or cougar, none of which is traditionally hunted for meat but whose products, such as fur, claws, and teeth, may have been used for decorative or ceremonial purposes.

The presence of microblade technology is even more intriguing. The occurrence of this specialized technology suggests a low producer-consumer ratio (*sensu* Goodale et al. 2008) and stands in opposition to the majority of other lithic evidence at the quarry. A possible explanation for the presence of microblades is discussed below.

Remembering Scarcity

Although microblade technology occurs in a myriad of contexts over long time periods in western North America, its presence at archaeological sites is frequently considered a cultural marker. In Alaska, particularly, archaeologists debate whether early precontact microblades were part of a diverse adaptive technology or instead indicate specific cultural systems (Bever 2001). In recent years, Alaskan archaeologists have tended toward the first interpretation, regarding microblades as part of adaptive systems rather than as cultural diagnostics (Yesner and Pearson 2002; Esdale 2008; Potter 2008; Rasic 2008).

Many microblade sites are located in the northern regions of North America, Europe, and Asia, and it has been suggested that microblades represent a technological adaptation to environments that have an extreme cold season (Elston and Brantingham 2002). Microblades were inserted into slotted tools made of organic materials, such as spears made of antler, and such composite tools are less brittle in cold conditions. Microblade technology was also highly portable, and microblade cores could be formed from relatively small pebbles, two characteristics that are especially suited to areas in which resources are highly dispersed and exposures of lithic source materials relatively rare. In spite of the economical and portable aspects of microblade technology, however, the tool kit required to create the organic portions of the composite tool was extensive. At a site where composite tools were made, the manufacturing tools may include various hammers and scrapers, engravers for slotting, wedges for splitting, anvils, and some kind of mastic (Elston and Brantingham 2002, 105).

There is no evidence at the Quarry of the Ancestors to suggest that the microblades date to a different period than the majority of the other lithics. At the single stratified excavation area, described above (see fig 10.5), microblades occur only in the upper occupation, not in the lower one, indicating that they probably belong to the Archaic period, like the rest of the lithic scatters associated with modern landforms. If microblade technology does not represent a chronologically or culturally separate occupation, however, then its anomalous presence must be explained in some other way.

In an intriguing evolutionary examination of lithic technologies at the Keatley Creek site in southern British Columbia, Prentiss and Clarke (2008) propose that microblades, which had no apparent function in a sedentary fishing village, were preserved as a means of cultural persistence, either as a vestigial technology or as a critical component in the relatively limited exploitation of terrestrial resources. Later occurrences of microblades in other parts of British Columbia have also been interpreted as vestigial (Fladmark 1985).

A similar argument can be made with respect to the microblades in the Quarry of the Ancestors. The strict, formulaic use of wedge-shaped microblade cores is evident in the discovery of a classic primary ridge flake and several classic microblades. Regardless of whether this technology was ancestral to the quarry inhabitants, someone at the quarry knew the process. It is possible that, much in the way that knowledge of the past is preserved through oral history, microblade technology and bipolar percussion were deliberately perpetuated in order to preserve knowledge of the techniques, in case that knowledge became necessary. A similar situation, in which stemmed microliths appeared to serve as a technological contingency plan, occurred in Papua New Guinea after a volcanic eruption devastated the region. As Torrence (2002, 187) noted, "although the social system that created stemmed tools no longer existed, the distinctive, traditional shape of the tool may have been preserved by the refugees from this disaster."

The persistence of anomalous lithic technologies as cultural traits suggests that the semi-sedentary occupants of the Quarry of the Ancestors were never very far from their mobile roots and that their commitment to reduced mobility and to the use of BRS as part of their social identity was somewhat tenuous. Moreover, the opportunistic and flexible character of lithic technological organization at the quarry implies that the precontact inhabitants were probably able to reorganize as mobile foragers if necessary.

Indeed, many sites in the oil sands region, including some very close to the quarry (Saxberg and Reeves 2006), reflect a much more ephemeral use of the land. These sites are very small, with limited tool and flake diversity. A few of them are even more suggestive of groups with high residential mobility, containing tools formed primarily from bipolarly split water-worn chert pebbles rather than from BRS cores. Low-visibility sites such as these may be the product of pioneering populations (of any time period) who, instead of travelling with all the lithic trappings of their culture and practicing raw material or tool caching (perhaps a Palaeoindian trait: see, for example, Kelly and Todd 1988), reverted to a practical, deeply ingrained lithic technology during times of unpredictable or unknown tool stone availability.

It is interesting to note that the well-known Archaic-period Gowen sites in Saskatchewan (Walker 1992) contain more bipolarly split pebble tools than any other kind of tool, although the assemblage also yielded several corner- and side-notched projectile points. In Alaska, both bipolar reduction and microblade technology are considered to have been highly flexible reduction strategies used in the late Pleistocene and early Holocene periods (Goebel 2008; Bever 2006). In a comprehensive study of radiocarbon dates and sites in Alaska, Bever (2006) demonstrated that whereas traditional Palaeoindian sites with large bifacially worked points were not occupied after the end of the Younger Dryas, sites containing microblades continued to occur well into the Archaic period. This suggests that microblades were part of a flexible pattern of adaptation that withstood environmental alteration with relative success.

DISCUSSION

While the artifacts found at the Quarry of the Ancestors are exclusively lithic and consist primarily of debitage, they allow us to draw certain conclusions regarding palaeoenvironmental conditions and precontact land use in northeastern Alberta. The excavation of a stratified locality within the quarry boundary indicated that an episode of catastrophic flooding took place well before a period of cold, dry, periglacial conditions, possibly representing the Younger Dryas, and that, after the flood, the waters receded fairly quickly and the muddy lands dried. This sequence of conditions supports the theory proposed by Murton et al. (2010) of an earlier Lake Agassiz flood, one that occurred prior to the onset of cold, dry Younger Dryas conditions and may in fact have triggered them.

Excavations at this site also reveal two periods of human occupation, the first during a cold, dry period and the second during a warm, dry period. If the earlier occupation is contemporaneous with the Younger Dryas and hence with a remnant proboscidean population, then it is probably more or less contemporary with the projectile point found elsewhere in the quarry that tested positive for proboscidean proteins. It appears, then, that the site was occupied in Early Palaeoindian times, even if the evidence of this occupation is now scant. It may be that additional evidence of Palaeoindian occupation is more deeply buried in sandy sediments or is currently submerged under wetlands, or it could simply be that this occupation had a much less visible archaeological signature, perhaps because the Palaeoindian population was more mobile than later Archaic peoples. The key environmental indicator for quarry use appears to be aridity, which suggests that the quarry was either submerged or heavily vegetated during moister climatic episodes.

The lithic data from the Quarry of the Ancestors also indicates that people used the quarry most intensively during the Early Archaic period, when consistently warm and dry Hypsithermal conditions probably facilitated exposure of the BRS outcrops. It also appears that, as was the case for Archaic cultures in the

Great Basin (Graf and Schmitt 2007) and the Great Lakes area (Julig 1994, 24), the people using the quarry were not as mobile as earlier populations had been. This is suggested in part by the very small proportion of formal tools within quarry assemblages, but it is even more evident in the lack of evidence for retooling at the quarry. A few used projectile points were abandoned, probably in favour of new ones, but this pattern of use is not common.

Nor is there evidence that hunter-gatherers used the quarry as a stop on their seasonal round during the Late Precontact period. Rather, similarities in the lithic data across the site suggest a temporally circumscribed adaptation to a specific set of conditions, perhaps both social and environmental. At the same time, variations in both the abundance and the spatial distribution of particular lithic tools suggest functional differentiation, which supports the conclusion that Archaic period peoples were semi-sedentary. The Quarry of the Ancestors assemblages also exhibit variety in tool types, as well as a preference for expedient tools. These characteristics are frequently associated with the use of lithic sources as camps and workshops, which, by inference, again points to limited mobility. Similar degrees of tool richness and expedient tool use have been documented at other lithic extraction sites in the boreal forest zone in Canada, such as Mount Edziza (Fladmark 1985), the Cummins Quarry (Julig 1994), and Vihtr'iishik, a siliceous argillite quarry on the Mackenzie River in the Northwest Territories (Pilon 1990; Pokotylo 1994).

The extravagant use of a single raw material at the Quarry of the Ancestors and the limited geographical spread of that material both imply some sense of possession—that those who occupied the lands around the quarry identified themselves with the rock, as did later Palaeoindian and Early Archaic peoples in other parts of North America (Ellis 1989, 2004; Carr 2005; Daniel 2001; Julig 1994). Although widespread similarity in tool types, particularly projectile points (see Wright 1972; Esdale 2008), during the Early Archaic period does not appear to support the idea of increasing social segregation and territorialism, the intensive use of specific raw material sources in parts of the boreal forest does. This entrenchment of identity at lithic extraction points arguably represents a strategy to reaffirm group membership and thereby share risk, as suggested by Ellis (1989) and Carr (2005), although it may also have been in part a response to external pressures in the form of an influx of groups from the Plains during the Hypsithermal interval.

While the adoption of a bedrock raw material as a form of social identifier seems to have been more commonly a Palaeoindian trait (Ellis 1989, 2004; Carr 2005), the almost exclusive use of a specific material within a relatively small zone continued into the Early Archaic period in some areas (Daniel 2001;

Hinshelwood 2004). The intensive use of specific bedrock raw material sources at the Quarry of the Ancestors, the Cummins Quarry (Julig 1994), and the highest-elevation primary sources at Mount Edziza (Fladmark 1984, 1985) appears in all cases to predate the Late Archaic period, suggesting that this intensification of use was dictated by similar historical conditions.

The persistence of microblade and bipolar technology at the Quarry of the Ancestors as a vestigial reminder of mobility implies that the intensified use of BRS was, in part, a collective response to past experiences of uncertainty with regard to the availability of lithic raw materials. The use of a lithic type as a marker of group identity was therefore probably not the deeply embedded trait that it may have been in the case of the Great Lakes Palaeoindians. It was more an adaptation to abundance that was, like reduced mobility, temporary. This interpretation also implies that microblades and bipolar technology were portable adaptive strategies rather than cultural indicators, which would explain their widespread occurrence in western North American prehistory.

The lithic assemblages from the Quarry of the Ancestors shed little clear light on the issue of where the first inhabitants of the region came from or where they went. The presence of microblades, whether as a vestigial technology or not, suggests a northwestern North American origin, although the opportunistic approach evident in the majority of the materials appears to be an in situ adaptation to an abundance of high-quality material. The quarry inhabitants may have had the most in common (including microblades) with a widespread Northern Archaic Tradition (Esdale 2008), but a more detailed analysis would be needed to confirm this. The Northern Archaic Tradition is not necessarily known to include an emphasis on bipolar pebble chert reduction in areas of limited raw material availability, but perhaps further study in regions poor in raw material will reveal otherwise. The fact that many known Northern Archaic sites in Alaska and the Yukon are located on bedrock ridges rather than on glacial deposits (Esdale 2008, 14) suggests that the search for high-quality raw material sources may have been part of the adaptive strategy of these peoples.

The lack of evidence for Late Archaic period occupations may reflect environmental changes that placed constraints on raw material availability and produced greater patchiness in the local habitat. Widespread climatic cooling and increasing moisture are also blamed for the similar lack of occupations dating to the Late Archaic period at the Mount Edziza obsidian source in British Columbia (Fladmark 1984, 1985) and the Cummins Quarry in northwestern Ontario (Hinshelwood 2004; Julig 1994), both of which were occupied during late Palaeoindian and Early Archaic times. Environmental change is a reasonable explanation for discontinued use of three quarries in three such disparate areas around the same time, as Archaic hunter-gatherers probably shifted their patterns of land use and reorganized themselves as mobile foragers.

CONCLUSIONS

Data from the Quarry of the Ancestors are invaluable in addressing aspects of adaptation and settlement in ancient northern Alberta. The site contains evidence of intensive use in the Early Archaic period, when warm and dry environmental conditions prevailed. The data suggest that aridity is the key environmental variable that enabled exploitation of the BRS outcrops. The physical characteristics of BRS in its natural state facilitated extraction and reduction at the outcrops, and easily transportable lithic packages were probably removed to sandy uplands to be heat-treated to improve the quality of the stone. Lithic reduction was for the most part opportunistic, flexible, and extravagant, with the goal of producing utilizable flakes, although evidence of highly structured, economical, and portable technologies also occurs within the quarry, if only infrequently. While a more thorough archaeological examination of the extraction areas may shed light on how rock was selected for further reduction and thus generate a more complete picture of activities at the site overall, the persistence of these technologies suggests the preservation of strategies to manage scarcity.

These interpretations contribute to the growing body of literature suggesting that, rather than functioning as a cultural diagnostic, lithic technology is often part of an adaptive strategy developed in response to a specific historical context. In the Quarry of the Ancestors, the historical context receives material expression in the form of a particular lithic strategy. But seemingly disparate adaptive systems can exist within what might otherwise be perceived as a single archaeological culture. Without large-scale archaeological investigations, however, this broader historical context will be obscured, or else it will be fragmented, with specific adaptive strategies interpreted as evidence of cultural, or chronological, or functional, variety.

The implications for the recognition of complete precontact cultural systems are significant. As Bever (2006) pointed out in a study of the earliest occupation of Alaska—a place where one should find evidence of the earliest occupations in the New World—the archaeological record gives the impression of people who were familiar with the landscape, rather than initial colonizers. If the true pioneers, in any landscape and at any time, travelled with the most spartan, portable, and flexible technology possible, no matter what their technological expression during times of abundance, perhaps we have been looking for the wrong thing and have been distracted by high-visibility sites in places where dependable lithic sources are abundant and, more importantly, known to exist.

The archaeological record of northern Alberta may ultimately inform the study of precontact peoples living with scarcity. There are very few bedrock lithic sources in northern Alberta, and the use of pebble sources, bipolar technology, and blade-like flakes, if not true microblades, is very common in both time and space in the boreal forest (see the discussions in Gruhn 1981; Stevenson 1986). Detailed research examining the geographical placement, the timing, and the technology of these sites may assist in the development of a model for adaptation in an environment where the luxury of stylistic variation and functional specificity in tool kits is not available.

The fortunate discovery of the Quarry of the Ancestors has provided a special opportunity to investigate a highly significant site within the context of historic resource management. Preservation of at least a portion of the quarry is underway, and future archaeological investigations may reveal important new insights into the precontact occupation of the region. New techniques in chronometric dating may eventually help us to develop a more precise understanding of patterns of occupation, while further excavation in the stratified locality could potentially reveal organic remains and/or yield any information about palaeoenvironmental conditions. Future scientific analyses—for example, of protein residues or of the effects of heat treatment on the structure of BRS—will eventually fill in more of the story as well. As our knowledge base expands, we will be better able to appreciate the role of the Quarry of the Ancestors as a central locus of cultural expression within a larger system of adaptation to abundance and scarcity in the boreal forest of North America.

ACKNOWLEDGEMENTS

Special thanks to Birch Mountain Resources, in particular Don Dabbs, who named the quarry and was its champion. Also, thanks to Glen de Paoli and Ken Foster at Birch Mountain for their assistance with the field projects. At Lifeways of Canada, thanks to Brian Reeves and to all the field crews, cataloguers, and lithic analysts who worked on the 2003, 2004, 2005, and 2007 seasons. For research conducted at the University of Saskatchewan, we are grateful for the financial support of the University of Saskatchewan's President's SSHRC Research Fund and for support in the form of beamtime from Canadian Light Source, Inc. Finally, thanks to AMEC Earth & Environmental for supporting this research.

NOTE

1 Calendar dates (cal yr BP) were generated from radiocarbon years using the online calibration tool OxCal 4.0 (http://c14.arch.ox.ac.uk/oxcal/), which employs the IntCal04 curve developed for the northern hemisphere (Reimer et al. 2004).

REFERENCES

Abercrombie, Hugh J., and Rui Feng
    1997   Geological Setting and Origin of Microdisseminated Au-Ag-Cu Minerals, Fort MacKay Region, Northeastern Alberta. In *Exploring for Minerals in Alberta: Geological Survey of Canada Geoscience Contributions, Canada-Alberta Agreement of Mineral Development (1992–1995)*, edited by R. W. Macqueen, pp. 247–277. Geological Survey of Canada, Ottawa.

Agenbroad, Larry D.
    2005   North American Proboscideans: Mammoths: The State of Knowledge, 2003. *Quaternary International* 126–128: 73–92.

Andrefsky, William, Jr.
    1991   Inferring Trends in Prehistoric Settlement Behavior from Lithic Production Technology in the Southern Plains. *North American Archaeologist* 12: 129–144.
    1994   Raw-Material Availability and the Organization of Technology. *American Antiquity* 59(1): 21–34.
    1998   *Lithics: Macroscopic Approaches to Analysis*. Cambridge University Press, Cambridge.
    2008   An Introduction to Stone Tool Life History and Technological Organization. In *Lithic Technology: Measures of Production, Use, and Curation*, edited by William Andrefsky, Jr., pp. 3–22. Cambridge University Press, Cambridge.

Bamforth, Douglas B.
    1986   Technological Efficiency and Tool Curation. *American Antiquity* 51(1): 38–50.
    1991   Technological Organization and Hunter-Gatherer Land Use: A California Example. *American Antiquity* 56(2): 216–234.

Beirele, Brandon D.
    1996   Holocene Environments of Kearl Lake, N.E. Alberta. Unpublished report prepared for Syncrude Canada Ltd. Copy on file, Syncrude Canada Ltd., Edmonton.

Bever, Michael R.
    2001   An Overview of Alaskan Late Pleistocene Archaeology: Historical Themes and Current Perspectives. *Journal of World Prehistory* 15(2): 125–191.
    2006   Too Little, Too Late? The Radiocarbon Chronology of Alaska and the Peopling of the New World. *American Antiquity* 71(4): 595–620.

Binford, Lewis R.
    1978   *Nunamiut Ethnoarchaeology*. Academic Press, New York.
    1980   Willow Smoke and Dogs' Tails: Hunter-Gatherer Settlement Systems and Archaeological Site Formation. *American Antiquity* 45(1): 4–20.

Bouchet-Bert, Luc
    2002   When Humans Entered the Northern Forests: An Archaeological and Paleoenvironmental Perspective. MA thesis, Department of Archaeology, University of Calgary.

Bourges, Claire

  1998 *HhOu-36 (Nezu Site) Historical Resources Impact Assessment, 1998 Field Studies: Interim Report (ASA Permit 98-040).* Copy on file, Archaeological Survey, Historic Resources Management Branch, Alberta Culture, Edmonton.

Brantingham, P. Jeffrey, John W. Olsen, Jason A. Rech, and Andrei I. Krivoshapkin

  2000 Raw Material Quality and Prepared Core Technologies in Northeast Asia. *Journal of Archaeological Science* 27: 255–271.

Burns, James A.

  1996 Vertebrate Paleontology and the Alleged Ice-Free Corridor: The Meat of the Matter. *Quaternary International* 32: 107–112.

Carr, Dillon H.

  2005 The Organization of Late Paleoindian Lithic Procurement Strategies in Western Wisconsin. *Midcontinental Journal of Archaeology* 30(1): 3–35.

Clark, Donald W.

  1981 Prehistory of the Western Subarctic. In *Handbook of North American Indians,* William C. Sturtevant, general editor, vol. 6, *Subarctic,* edited by June Helm, pp. 107–129. Smithsonian Institution Scholarly Press, Washington, D.C.

Daniel, I. Randolph, Jr.

  2001 Stone Raw Material Availability and Early Archaic Settlement in the Southeastern United States. *American Antiquity* 66(2): 237–265.

de Mille, Christy, and Brian O. K. Reeves

  2009 *Birch Mountain Resources Ltd., Muskeg Valley Quarry Historical Resources Mitigation, 2005 Field Studies: Final Report (ASA Permit 05-230).* 3 vols. Copy on file, Archaeological Survey, Historic Resources Management Branch, Alberta Culture, Edmonton.

De Paoli, Glen R.

  2005 Petrographic Examination of the Muskeg Valley Microquartzite (MVMq). Appendix A in Nancy Saxberg and Brian O. K. Reeves (2006), *Birch Mountain Resources Ltd., Hammerstone Project, Historical Resources Impact Assessment, 2004 Field Studies: Final Report (ASA Permit 04-235).* Copy on file, Archaeological Survey, Historic Resources Management Branch, Alberta Culture, Edmonton.

Dyke, Arthur S.

  2004 An Outline of North American Deglaciation with Emphasis on Central and Northern Canada. In *Quaternary Glaciations: Extent and Chronology, Part II: North America*, edited by Jürgen Ehlers and Philip L. Gibbard, pp. 373–424. Developments in Quaternary Science no. 2. Elsevier, Amsterdam.

Duke, Daron G., and D. Craig Young

  2007 Episodic Permanence in Paleoarchaic Basin Selection and Settlement. In *Paleoindian or Paleoarchaic? Great Basin Human Ecology at the Pleistocene/Holocene Transition*, edited by Kelly E. Graf and Dave N. Schmitt, pp. 123–138. University of Utah Press, Salt Lake City.

Ellis, Christopher J.

  1989 The Explanation of Northeastern Palaeoindian Lithic Procurement Patterns. In *Eastern Paleoindian Lithic Resource Use*, edited by Christopher J. Ellis and Jonathan C. Lothrop, pp. 139–164. Westview Press, Boulder, Colorado.

  2004 Hi-Lo: An Early Lithic Complex in Southern Ontario. In *The Late Palaeo-Indian Great Lakes: Geological and Archaeological Investigations of Late Pleistocene and Early Holocene Environments*, edited by Lawrence J. Jackson and Andrew Hinshelwood, pp. 57–83. Mercury Series no. 165. Archaeological Survey of Canada, Canadian Museum of Civilization, Gatineau, Québec.

Elston, Robert G., and P. Jeffrey Brantingham

2002   Microlithic Technology in Northern Asia: A Risk-Minimizing Strategy of the Late Paleolithic and Early Holocene. In *Thinking Small: Global Perspectives on Microlithization*, edited by Robert G. Elston and Steven L. Kuhn, pp. 103–116. Archeological Papers of the American Anthropological Association no. 12. American Anthropological Association, Arlington, Virginia.

Esdale, Julie A.

2008   A Current Synthesis of the Northern Archaic. *Arctic Anthropology* 45: 3–38.

Fedirchuk, Gloria J., and Edward J. McCullough

1992   *The Duckett Site: Ten Thousand Years of Prehistory on the Shores of Ethel Lake, Alberta.* Fedirchuk McCullough and Associates Ltd. Copy on file, Esso Resources Canada Limited, Calgary.

Fenton, Mark M., and John W. Ives

1982   Preliminary Observations on the Geological Origins of Beaver River Sandstone. In *Archaeology in Alberta, 1981,* compiled by Jack Brink, pp. 166–189. Archaeological Survey of Alberta Occasional Paper no. 19. Historic Resources Management Branch, Alberta Culture, Edmonton.

1984   The Stratigraphic Position of Beaver River Sandstone. In *Archaeology in Alberta 1983,* complied by David Burley, pp. 128–136. Archaeological Survey of Alberta Occasional Paper no. 23. Historic Resources Management Branch, Alberta Culture, Edmonton.

1990   Geoarchaeological Studies of the Beaver River Sandstone, Northeastern Alberta. In *Archaeological Geology of North America*, edited by Norman P. Lasca and Jack Donahue, pp. 123–135. Centennial Special Volume 4. Geological Society of America, Boulder, Colorado.

Fiedel, Stuart J.

2007   Quacks in the Ice: Waterfowl, Paleoindians, and the Discovery of America. In *Foragers of the Terminal Pleistocene in North America*, edited by Renee B. Walker and Boyce N. Driskell, pp. 1–14. University of Nebraska Press, Lincoln.

Fisher, Timothy G.

1993   Glacial Lake Agassiz: The Northwest Outlet and Paleoflood Spillway, N.W. Saskatchewan and N.E. Alberta. PhD dissertation, Department of Geography, University of Calgary, Calgary.

1996   Sand-Wedge and Ventifact Palaeoenvironmental Indicators in Northwest Saskatchewan, Canada, 11 ka to 9.9 ka BP. *Permafrost and Periglacial Processes* 7: 391–408.

Fisher, Timothy G., and Derald G. Smith

1994   Glacial Lake Agassiz: Its Northwest Maximum Extent and Outlet in Saskatchewan (Emerson Phase). *Quaternary Science Reviews* 13: 845–858.

Fisher, Timothy G., and Catherine Souch

1998   Northwest Outlet Channels of Lake Agassiz, Isostatic Tilting and a Migrating Continental Drainage Divide, Saskatchewan, Canada. *Geomorphology* 25: 57–73.

Fisher, Timothy G., Nickolas Waterson, Thomas V. Lowell, and Irka Hajdas

2009   Deglaciation Ages and Meltwater Routing in the Fort McMurray Region, Northeastern Alberta and Northwestern Saskatchewan, Canada. *Quaternary Science Reviews* 28: 1608–1624.

Fladmark, Knut R.

1984   Mountain of Glass: Archaeology of the Mount Edziza Obsidian Source, British Columbia, Canada. *World Archaeology* 16(2): 139–156.

1985   *Glass and Ice: The Archaeology of Mt. Edziza.* Archaeology Press, Department of Archaeology Publication no. 14. Simon Fraser University, Burnaby, British Columbia.

Goebel, Ted

 2008 What Is the Nenana Complex? Paper presented at the 73rd annual meeting of the Society for American Archaeology, Vancouver, British Columbia, 26–30 March.

Goodale, Nathan B., Ian Kuijt, Shane J. Macfarlan, Curtis Osterhoudt, and Bill Finlayson

 2008 Lithic Core Reduction Techniques: Modeling Expected Diversity. In *Lithic Technology: Measures of Production, Use, and Curation*, edited by William Andrefsky, Jr., pp. 317–336. Cambridge University Press, Cambridge.

Gorham, Eville, Clarence Lehman, Arthur Dyke, Johannes Janssens, and Lawrence Dyke

 2007 Temporal and Spatial Aspects of Peatland Initiation Following Deglaciation in North America. *Quaternary Science Reviews* 26: 300–311.

Gould, Richard A.

 1980 *Living Archaeology*. Cambridge University Press, Cambridge.

Graf, Kelly E., and Dave N. Schmitt (editors)

 2007 *Paleoindian or Paleoarchaic? Great Basin Human Ecology at the Pleistocene/Holocene Transition*. University of Utah Press, Salt Lake City.

Gramly, Richard Michael

 1980 Raw Materials Source Areas and "Curated" Tool Assemblages. *American Antiquity* 45(4): 823–833.

Gruhn, Ruth

 1981 *Archaeological Research at Calling Lake, Northern Alberta*. Mercury Series no. 99. Archaeological Survey of Canada, National Museum of Man, Ottawa.

Gryba, Eugene M.

 2002 Lithic Types Available to Prehistoric Knappers at the True North Energy Limited Partnership Fort Hills Oil Sands Lease near Ft. McKay, in Northeastern Alberta. *Alberta Archaeological Review* 36 (Spring): 21–27.

Guthrie, R. Dale

 2006 New Carbon Dates Link Climatic Change with Human Colonization and Pleistocene Extinctions. *Nature* 441 (11 May): 207–209.

Halsey, Linda A., Dale H. Vitt, and Ilka E. Bauer

 1998 Peatland Initiation During the Holocene in Continental Western Canada. *Climatic Change* 40: 315–342.

Harington, Charles R. (editor)

 2003 *Annotated Bibliography of Quaternary Vertebrates of Northern North America with Radiocarbon Dates*. University of Toronto Press, Toronto.

Hinshelwood, Andrew

 2004 Archaic Reoccupation of Late Palaeo-Indian Sites in Northwestern Ontario. In *The Late Palaeo-Indian Great Lakes: Geoarchaelogical and Archaeological Investigations of Late Pleistocene and Early Holocene Environments*, edited by Lawrence J. Jackson and Andrew Hinshelwood, pp. 225–274. Mercury Series no. 165. Archaeological Survey of Canada, Canadian Museum of Civilization, Gatineau, Québec.

Ives, John W., and Mark M. Fenton

 1983 Continued Research on Geological Sources of Beaver River Sandstone. In *Archaeology in Alberta, 1982,* compiled by David Burley, pp. 78–88. Archaeological Survey of Alberta Occasional Paper no. 21. Historic Resources Management Branch, Alberta Culture, Edmonton.

 1985 *Progress Report for the Beaver River Sandstone Geological Source Study (ASA Permit 83-054)*. Copy on file, Archaeological Survey, Historic Resources Management Branch, Alberta Culture, Edmonton.

Jones, George T., Charlotte Beck, Eric E. Jones, and Richard E. Hughes

    2003   Lithic Source Use and Paleoarchaic Foraging Territories in the Great Basin. *American Antiquity* 68(1): 5-38.

Julig, Patrick J.

    1994   *The Cummins Site Complex and Paleoindian Occupations in the Northwestern Lake Superior Region*. Ontario Archaeological Reports no. 2. Ontario Heritage Foundation, Toronto.

Kelly, Robert L.

    1988   The Three Sides of a Biface. *American Antiquity* 53(4): 717-734.

Kelly, Robert L., and Lawrence C. Todd

    1988   Coming into the Country: Early Paleoindian Hunting and Mobility. *American Antiquity* 53(2): 231-244.

Kuhn, Steven L.

    1992   Blank Form and Reduction as Determinants of Mousterian Scraper Morphology. *American Antiquity* 57(1): 115-128.

    1995   *Mousterian Lithic Technology: An Ecological Perspective*. Princeton University Press, Princeton.

Kuhry, Peter

    1997   The Paleoecology of a Treed Bog in Western Boreal Forest Canada: A Study Based on Microfossils, Macrofossils and Physico-Chemical Properties. *Review of Palaeobotany and Palynology* 96: 183-224.

Lothrop, Jonathan C.

    1989   The Organization of Paleoindian Lithic Technology at the Potts Site. In *Eastern Paleoindian Lithic Resource Use*, edited by Christopher J. Ellis and Jonathan C. Lothrop, pp. 99-138. Westview Press, Boulder, Colorado.

MacDonald, Douglas H.

    2008   The Role of Lithic Raw Material Availability and Quality in Determining Tool Kit Size, Tool Function, and Degree of Retouch: A Case Study from Skink Rockshelter (46NI445), West Virginia. In *Lithic Technology: Measures of Production, Use, and Curation*, edited by William Andrefsky, Jr., pp. 216-232. Cambridge University Press, Cambridge.

Mandeville, M. D., and J. Jeffrey Flenniken

    1974   A Comparison of the Flaking Qualities of Nehawka Chert Before and After Thermal Pretreatment. *Plains Anthropologist* 19: 146-148.

Maxwell, Moreau S.

    1984   Pre-Dorset and Dorset Prehistory of Canada. In *Handbook of North American Indians*, William C. Sturtevant, general editor, vol. 5, *Arctic*, edited by David Damas, pp. 359-368. Smithsonian Institution Scholarly Press, Washington, D.C.

Mercieca, Alison, and Peter Hiscock

    2008   Experimental Insights into Alternative Strategies of Lithic Heat Treatment. *Journal of Archaeological Science* 35: 2634-2639.

Murton, Julian B., Mark D. Bateman, Scott R. Dallimore, James T. Teller, and Zhirong Yang

    2010   Identification of Younger Dryas Outburst Flood Path from Lake Agassiz to the Arctic Ocean. *Nature* 464: 740-743.

Parr, R. E.

    2005   Protein Residue Analysis from site HhOv-323, Alberta. Appendix A in Nancy Saxberg, *Birch Mountain Resources Ltd. Muskeg Valley Quarry, Mitigative Excavations, 2004 Field Studies: Interim Report, ASA Permit 2005-118*. Copy on file, Archaeological Survey, Historic Resources Management Branch, Alberta Culture, Edmonton.

2007   Protein Residue Analysis of 97 Artifacts from Archaeological Sites in Yellowstone, the Eastern Slopes of Alberta, and Northeastern Alberta. Appendix C in Nancy Saxberg, *Birch Mountain Resources Ltd. Muskeg Valley Quarry, Historical Resources Mitigation, 2004 Field Studies: Final Report (ASA Permit 05-118),* vol. 2. Copy on file, Archaeological Survey, Historic Resources Management Branch, Alberta Culture, Edmonton.

Parry, William J., and Robert L. Kelly

1987   Expedient Core Technology and Sedentism. In *The Organization of Core Technology,* edited by Jay K. Johnson and Carol A. Morrow, pp. 285-304. Westview Press, Boulder, Colorado.

Petraglia, Michael, Philip LaPorta, and K. Paddayya

1999   The First Acheulian Quarry in India: Stone Tool Manufacture, Biface Morphology, and Behaviors. *Journal of Anthropological Research* 55: 39-70.

Pilon, Jean-Luc

1990   *Vihtr'iishik*: A Stone Quarry Reported by Alexander Mackenzie on the Lower Mackenzie River in 1789. *Arctic* 43(3): 251-261.

Pokotylo, David L.

1994   Archaeological Investigations at Vihtr'iitshik (MiTi-1), Lower Mackenzie Valley, 1992. In *Bridges Across Time: The NOGAP Archaeology Project*, edited by Jean-Luc Pilon, pp. 171-192. Canadian Archaeological Association Occasional Paper no. 2. Canadian Archaeological Association, Victoria, British Columbia.

Potter, Ben A.

2008   Understanding Assemblage Variability in the Alaskan Subarctic: Beyond Type. Paper presented at the 73rd annual meeting of the Society for American Archaeology, Vancouver, British Columbia, 26-30 March.

Prentiss, Anna Marie, and David S. Clarke

2008   Lithic Technological Organization in an Evolutionary Framework: Examples from North America's Pacific Northwest Region. In *Lithic Technology: Measures of Production, Use, and Curation,* edited by William Andrefsky, Jr., pp. 257-285. Cambridge University Press, Cambridge.

Rasic, Jeff

2008   A Model of Late Pleistocene Land Use and Technology from Northwest Alaska. Paper presented at the 73rd annual meeting of the Society for American Archaeology, Vancouver, British Columbia, 26-30 March.

Reardon, Gerard V.

1976   A Cognitive Approach to Lithic Analysis. MA thesis, Department of Archaeology, University of Calgary, Calgary.

Reeves, Brian O. K.

1973   The Concept of an Altithermal Cultural Hiatus in Northern Plains Prehistory. *American Anthropologist,* n. s., 75(5): 1221-1253.

Reimer, Paula J., Mike G. L. Baillie, Édouard Bard, Alex Bayliss, J. Warren Beck, Chanda J. H. Bertrand, Paul G. Blackwell, Caitlin E. Buck, George S. Burr, Kirsten B. Cutler, Paul E. Damon, R. Lawrence Edwards, Richard G. Fairbanks, Michael Friedrich, Thomas P. Guilderson, Alan G. Hogg, Konrad A. Hughen, Bernd Kromer, Gerry McCormac, Sturt Manning, Christopher Bronk Ramsey, Ron W. Reimer, Sabine Remmele, John R. Southon, Minze Stuiver, Sahra Talamo, F. W. Taylor, Johannes van der Plicht, and Constance E. Weyhenmeyer

2004   IntCal04 Terrestrial Radiocarbon Age Calibration, 0-26 cal kyr BP. *Radiocarbon* 46 (3): 1029-1058.

Robertson, Elizabeth C.

2004   Communal Hunting as a Social Model for the Paleoindian to Early Archaic Transition on the Plains. In *Archaeology on the Edge: New Perspectives from the Northern Plains*, edited by Brian Kooyman and Jane Kelley, pp. 211–230. University of Calgary Press, Calgary.

Robertson, Elizabeth C., and Robert Blyth

2009   XANES Investigation of the Effects of Heat Treatment on Archaeological Tool Stone. Unpublished paper, University of Saskatchewan, Saskatoon.

Saxberg, Nancy

1998   *Aurora Mine North Utility Corridor, Historical Resources Impact Assessment and Phase I Mitigative Excavations, HhOv-87 (Ronaghan's Ridge), 1998 Field Studies: Interim Report (ASA Permit 98-039).* Copy on file, Archaeological Survey, Historic Resources Management Branch, Alberta Culture, Edmonton.

2007   *Birch Mountain Resources Ltd. Muskeg Valley Quarry, Historical Resources Mitigation, 2004 Field Studies: Final Report (ASA Permit 05-118).* 2 vols. Copy on file, Archaeological Survey, Historic Resources Management Branch, Alberta Culture, Edmonton.

Saxberg, Nancy, and Brian O. K. Reeves

2003   The First Two Thousand Years of Oil Sands History: Ancient Hunters at the Northwest Outlet of Glacial Lake Agassiz. In *Archaeology in Alberta: A View from the New Millennium*, edited by Jack W. Brink and John F. Dormaar, pp. 290–322. Archaeological Society of Alberta, Medicine Hat.

2004   *Birch Mountain Resources Ltd. Muskeg Valley Quarry, Historical Resources Impact Assessment, 2003 Field Studies: Final Report (ASA Permit 03-249).* Copy on file, Archaeological Survey, Historic Resources Management Branch, Alberta Culture, Edmonton.

2006   *Birch Mountain Resources Ltd. Hammerstone Quarry, Historical Resources Impact Assessment, 2004 Field Studies: Final Report (ASA Permit 04-235).* Copy on file, Archaeological Survey, Historic Resources Management Branch, Alberta Culture, Edmonton.

Saxberg, Nancy, Brad Somer, and Brian O. K. Reeves

2003   *Syncrude Canada Ltd. Aurora Mine North, Historical Resources Impact Assessment, 2002 Field Studies: Final Report (ASA Permit 02-140).* Copy on file, Archaeological Survey, Historic Resources Management Branch, Alberta Culture, Edmonton.

2004   *Syncrude Canada Ltd. Aurora Mine North, Historical Resources Impact Assessment and Mitigation Studies, 2003 Field Studies: Final Report (ASA Permit 03-279).* Copy on file, Archaeological Survey, Historic Resources Management Branch, Alberta Culture, Edmonton.

Shortt, Mack W., Nancy Saxberg, and Brian O. K. Reeves

1998   *Aurora Mine North, East Pit Opening, Plant Site, Tailings and Related Workings, HRIA and Mitigation Studies: Final Report (ASA Permit 97-116).* Copy on file, Archaeological Survey, Historic Resources Management Branch, Alberta Culture, Edmonton.

Shott, Michael J.

1986   On Tool-Class Use Lives and the Formation of Archaeological Assemblages. *American Antiquity* 54(1): 9–30.

Smith, Derald G., and Timothy G. Fisher

1993   Glacial Lake Agassiz: The Northwestern Outlet and Paleoflood. *Geology* 21(1): 9–12.

Somer, Brad

2005   *Syncrude Aurora Mine North, Historical Resources Impact Assessment and Mitigation Studies, 2004 Field Studies: Final Report (ASA Permit 04-192).* 2 vols. Copy on file, Archaeological Survey, Historic Resources Management Branch, Alberta Culture, Edmonton.

Stevenson, Marc G.

1986   *Window on the Past: Archaeological Assessment of the Peace Point Site, Wood Buffalo National Park, Alberta*. Studies in Archaeology, Architecture and History, National Historic Parks and Sites Branch, Parks Canada. Environment Canada, Ottawa.

Sullivan, Alan P., and Kenneth C. Rozen

1985   Debitage Analysis and Archaeological Interpretation. *American Antiquity* 50(4): 755–779.

Syncrude Canada Ltd.

1973   *Syncrude Lease No. 17: An Archaeological Survey.* Environmental Research Monograph 1973-4. Syncrude Canada, Ltd., Edmonton.

1974   *The Beaver Creek Site: A Prehistoric Stone Quarry on Syncrude Lease No. 22*. Environmental Research Monograph 1974-2. Syncrude Canada Ltd., Edmonton.

Tarasov, Lev, and W. Richard Peltier

2004   A Geophysically Constrained Large Ensemble Analysis of the Deglacial History of the North American Ice-Sheet Complex. *Quaternary Science Reviews* 23: 359–388.

2005   Arctic Freshwater Forcing of the Younger Dryas Cold Reversal. *Nature* 435 (2 June): 662–665.

2006   A Calibrated Deglacial Drainage Chronology for the North American Continent: Evidence of an Arctic Trigger for the Younger Dryas. *Quaternary Science Reviews* 25: 659–688.

Teller, James T., Matthew Boyd, Zhirong Yang, Phillip S. G. Kor, and Amir Mokhtari Fard

2005   Alternative Routing of Lake Agassiz Overflow During the Younger Dryas: New Dates, Paleotopography, and a Re-evaluation. *Quaternary Science Reviews* 24: 1890–1905.

Torrence, Robin

2002   Thinking Big About Small Tools. In *Thinking Small: Global Perspectives on Microlithization*, edited by Robert G. Elston and Steven L. Kuhn, pp. 179–189. Archeological Papers of the Americal Anthropological Association no. 12, American Anthropological Association, Arlington, Virginia.

Tsang, Brian W. B.

1998   The Origin of the Enigmatic Beaver River Sandstone. MSc thesis, Department of Geology and Geophysics, University of Calgary, Calgary.

Unfreed, Wendy J., and Gloria J. Fedirchuk

2001   *Historical Resources Impact Assessment, ATCO Pipelines Limited, Muskeg River Pipeline Project (Section 29-92-20-W4M to Section 23-95-10-W4M): Final Report (ASA Permit 00-064)*. Copy on file, Archaeological Survey, Historic Resources Management Branch, Alberta Culture, Edmonton.

Vance, Robert E.

1986   Pollen Stratigraphy of Eaglenest Lake, Northeastern Alberta. *Canadian Journal of Earth Sciences* 23: 11–20.

Walker, Ernest G.

1992   *The Gowen Sites: Cultural Responses to Climatic Warming on the Northern Plains (7500–5000 B.C.)*. Mercury Series no. 145. Archaeological Survey of Canada Canadian Museum of Civilization, Gatineau, Québec.

Wickham, Michelle D., and Taylor Graham

2009   *Historical Resources Impact Mitigation of the TransCanada Pipelines Ltd. Fort McKay Mainline Expansion: Final Report and Post-construction Audit (ASA Permits 06-376 and 07-266)*. Copy on file, Archaeological Survey, Historic Resources Management Branch, Alberta Culture, Edmonton.

Wright, James V.

    1972   *The Shield Archaic.* Publications in Archaeology no. 3. National Museum of Man, Ottawa.

    1995   *A History of the Native Peoples of Canada.* Vol. 1, *10,000 to 1,000 B.C.* Mercury Series no. 152. Archaeological Survey of Canada, Canadian Museum of Civilization, Gatineau, Québec.

Yesner, David R., and Georges A. Pearson

    2002   Microblades and Migrations: Ethnic and Economic Models in the Peopling of the Americas. In *Thinking Small: Global Perspectives on Microlithization*, edited by Robert G. Elston and Steven L. Kuhn, pp. 133–161. Archeological Papers of the American Anthropological Association no. 12. American Anthropological Association, Arlington, Virginia.

11 Microblade Technology in the Oil Sands Region | Distinctive Features and Possible Cultural Associations

ANGELA M. YOUNIE, RAYMOND J. LE BLANC,
AND ROBIN J. WOYWITKA

Recent archaeological work in the oil sands region has brought new attention to the presence of prehistoric microblade technology within Alberta. The extent and scope of this work, which has consisted mainly of contracted surveys and mitigative excavations, have been discussed throughout this volume and need not be reiterated here. In contrast to sites in Alaska and the Yukon, where microblade technology is a defining feature of prehistoric assemblages, microblade artifacts in Alberta are a relative rarity, found in possibly a hundred out of tens of thousands of known sites. Although microblades have been discovered in isolated finds across the province from as early as 1968 (Le Blanc and Ives 1986; Pyszczyk 1991; Sanger 1968a; Wilson, Visser, and Magne 2011; Younie, Le Blanc, and Woywitka 2010), the study of microblade technology remains something of a novelty, and our understanding of the cultural significance of these discoveries is still in its formative phase. Evidence for the existence of microblade technology in northern Alberta has, however, been recognized as important, and even the presence of tiny blade-like flakes within an archaeological assemblage is often cause for excitement.

In this chapter, we seek to propose a framework within which to incorporate microblade technology into the archaeological study of northern Alberta. This will include an examination of the wider role of microblades in northern systems of material culture and a discussion of methods of identification and analysis, but we will also explore the specific role of microblade technology in the cultural history of the oil sands region. These artifacts are significant not only for their rarity but also for the role they have played in the prehistory of arctic and sub-arctic populations in Asia and North America. Composite technologies that incorporate stone blades into tools made of organic materials are characteristic of northern cultural systems and lifeways. These technologies are also complex. They depend on specialized knowledge for lithic core reduction and require a diverse tool kit for the creation, hafting, and replacement of broken blades. Different methods of producing and using microblades have been shown to exhibit spatial and temporal patterning in Asia and North America, and, to a certain extent, they can be associated with specific archaeological cultures. Because of these characteristics, the analysis of microblades in Asian, Alaskan, and Beringian studies has focused strongly on the importance of identifying the technology that underlies microblade production—that is, the sequence of reduction of a microblade core—rather than simply describing the presence or absence of blade-like artifacts and the physical appearance of cores and blades. The following discussion of microblades in Alberta's oil sands region will adopt a similar analytical focus, stressing the importance of function and lithic reduction strategy as key factors in the study of microblade technology.

MICROBLADE TECHNOLOGY

What Is a Microblade?

Microblades are small, delicate artifacts produced through a systematic lithic reduction process and are intended for use as insets in composite tools (Wygal 2011). Narrowly defined, they are long, thin, parallel-sided flakes, less than 11 millimetres in width and at least twice as long as they are wide (Taylor 1962) that are produced by means of pressure flaking from specialized cores known as microblade cores or microcores (fig 11.1). In order to regulate the size and shape of these blades for use as insets, the proximal and distal ends of the microblade are often snapped off (Wyatt 1970), and one lateral edge may be straightened through light retouch or grinding. Microblades may be hafted into a slotted handle or, more commonly, inset into bone, antler, or ivory projectile points (Bleed 2001). The organic handles or points are shaped and slotted with the use of specialized tools such as burins and gravers, and the microblades are fixed into the haft with an adhesive such as resin (Flenniken 1987). Despite a complex production process requiring specialized knowledge, acquired skill, and a specific tool kit for the creation and insetting of the microblades (Bamforth and Bleed 1997), these artifacts are widespread in both time and space. Microblade technology is found in late Pleistocene assemblages throughout Beringia from

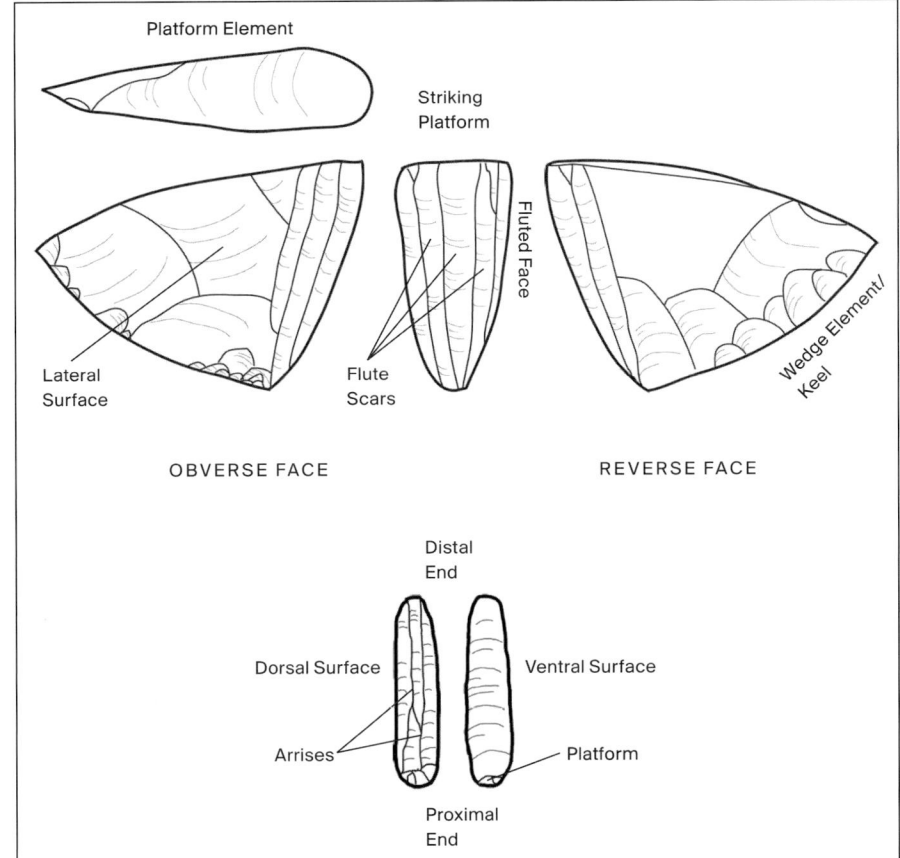

Figure 11.1. Basic features of microblade cores (top) and microblades (bottom)

Platform Element

Striking
Platform

Fluted Face

Lateral
Surface

Flute
Scars

Wedge Element/
Keel

OBVERSE FACE

REVERSE FACE

Distal
End

Dorsal Surface

Ventral Surface

Arrises

Platform

Proximal
End

northeastern Asia to the Canadian Arctic, as well as the Northwest Coast and British Columbia plateau, and it persisted in Alaska and the Yukon throughout most of the Holocene (Clark 2001; Clark and Gotthardt 1999; Goebel 2002; Goebel, Waters, and O'Rourke 2008; Graf and Bigelow 2011; Sanger 1970).

Composite lithic-organic technologies are often discussed as an adaptation to cold climates, and a number of possible advantages for prehistoric arctic and subarctic cultures have been proposed on the basis of both ethnographic and experimental archaeological studies (Kuhn and Elston 2002; Wygal 2011). Lithic points can shatter in the cold, either as the result of a missed shot or upon impact with targeted prey (Ellis 1997; Elston and Brantingham 2002), and are therefore less dependable for cold-weather hunting than organic points, which are less affected by the cold. Furthermore, organic points are lightweight and less likely to become damaged during transport, and they can be produced from raw materials that can be as easily found in the winter as in the summer. In contrast, lithic material sources are apt to be buried under the snow or frozen into the ground, and, because they become brittle in the cold, may also become difficult to work during the winter, as well as less reliable during use. Despite these advantages, organic points are not as lethal as lithic points (Elston and Brantingham 2002), a drawback that can be addressed by the incorporation of microblade insets. Not only do these blades provide a razor-sharp edge suitable for the penetration of thick hide and fur, but once the point is embedded in the prey, the thin stone pieces can snap off and migrate, increasing tissue damage and blood loss. Moreover, after use or breakage, new microblades can be swiftly reinserted into the organic point while it is still attached to the shaft. In this way, hunters were able to avoid the complicated and time-consuming re-hafting of an entire point tip during a hunting expedition, and they were no longer faced with the choice of whether to use a foreshaft to speed re-hafting at the expense of the strength of the finished spear or arrow (Elston and Brantingham 2002). Finally, compared to bifacial reduction, the production of small, standardized microblades has been shown to increase the amount of cutting edge available per quantity of raw material used, thereby conserving lithic raw material (Flenniken 1987). However, findings differ depending upon the criteria employed to define and quantify "usable" cutting edge (see Elston and Brantingham 2002).

Microblades Versus Blade-Like Flakes

Microblade technology first arose in the late Upper Palaeolithic period in Asia, during the late Pleistocene Epoch. Most evidence indicates that microblades appeared after the end of the last glacial maximum, approximately 18,000 years ago, and that they can be associated with the recolonization of Siberia and far eastern Asia during postglacial warming (Goebel 2002). They are most likely to have evolved from the larger blade and blade-core technology of the early and middle Upper Palaeolithic, a specialized lithic reduction technique allowing for the creation of long, regularly shaped flakes that could be used as blanks for scrapers, burins, points, and flake tools. Blade cores, or prismatic cores, could be identified by the presence of long, narrow flake scars, produced during blade removal. Often, blade cores might be reduced nearly to the point of exhaustion, producing blades of increasingly smaller size, the smallest of which are sometimes identified as bladelets (Goebel 1999). Although these small, regularly produced blades are similar in shape to microblades, and have been confused with microblades by researchers in both Asia and North America, there is a clear distinction not only in the mode of use (flake blank versus inset piece) but also in the technology of production (Goebel 1999). Although both are produced from standardized prismatic cores, microblade cores are more specialized in their reduction pattern, allowing for the repeated removal of delicate flakes of a regular width and thickness. It is not enough that microblades be long and narrow; they must also have straight edges, lack strong features such as heavy ripple marks and bulbs of percussion, and have thin, flat cross-sections, preferably trapezoidal rather than triangular.

Despite these differences, archaeological sites in Asia with bladelets and blade-like flakes found in their assemblages have been described by some researchers as containing microblades. Such descriptions are then used as evidence in the controversial archaeological debate surrounding the geographical and chronological origins of microblade technology in Siberia and the existence and nature of human occupation in the region during the harsh conditions of the last glacial maximum (see discussion in Goebel 2002). Such differences of interpretation have also occurred in Alaska, where both blade and microblade technology appear to have been brought across the Bering land bridge from Asia. Differing interpretations of the late Pleistocene cultural history of Alaska have been proposed on the basis of the presence or absence of microblades in early assemblages such as Healy Lake, Swan Point, and Broken Mammoth, which have in turn influenced theories about the timing and nature of the peopling of the Americas (Graf and Bigelow 2011; Hamilton and Goebel 1999; Holmes 2011; Yesner and Pearson 2002). Microblades, lanceolate projectile points, and cores of a specific shape are considered to be diagnostic of the Denali Complex, but in some cases diagnoses have been made on the presence of just a few microblades, bladelets, or even simply blade-like flakes. Blade-like flakes—flakes that are small and twice as long as they are wide but that lack other microblade characteristics—are often produced unintentionally, as the by-products of bipolar technology and occasionally through bifacial or other types of lithic reduction as well. For these reasons, it has become important to distinguish between these artifacts and microblades, not simply for technological and classificatory purposes but in order to develop an accurate picture of Beringian and subarctic prehistory.

Microblades may be distinguished from bladelets and blade-like flakes by physical characteristics such as those described above—straight edges, carefully prepared, pressure-flaked platforms, light ripple marks and low bulbs of percussion, and a tendency toward delicate features and trapezoidal cross-sections. The interpretation of a particular artifact as a microblade is further supported by the presence of microblade cores and other by-products of microblade production, as well as associated artifacts such as burins and organic tool technologies. In Asia, such artifacts might include specialized Dyuktai or Yubetsu cores, core preparation flakes such as ski-spalls, and platform rejuvenation flakes (Bleed 2001; Gómez Coutouly 2011). Comparable artifacts, representing very similar reduction processes, are found in North America (Keates, Kuzmin, and Shen 2007; Morlan 1970). These include the Dyuktai-like assemblage at the earliest levels of Swan Point in central Alaska (Holmes 2001, 2011; Holmes, VanderHoek, and Dilley 1996), and, later in time, Denali and other wedge-shaped microblade cores, ridge flakes, and core tablets (Graf and Bigelow 2011).

Methods of Microblade Production

The successful removal of microblades from a core depends on the existence of long, uninterrupted longitudinal ridges that channel the force of flaking pressure and facilitate the production of long, narrow blades. Widely spaced or irregular ridges disperse this force, resulting in the removal of wider flakes with a greater variation in shape and size. Microblade cores are thus carefully shaped to create long, parallel, closely spaced ridges on at least one surface, referred to as the *fluted face* of the core (Morlan 1970). While the presence of a fluted face is indicative of microblade production, these cores should also exhibit evidence of the deliberate shaping that allowed for the production of such a surface, as well as a platform exhibiting traces of the grinding and usually rejuvenation needed to maintain a suitable edge for pressure flaking. Two major approaches to core reduction are typically seen in the archaeological record, both of which lead to the creation of suitable platforms and fluted faces. The simplified classification of these approaches presented below is based on the typological systems identified and described by Kobayashi (1970), Morlan (1970), and Smith (1974).

The most common, and more formal, method of microblade core production is the shaping of a piece of raw material into a core preform with a relatively thin cross-section and at least one unifacial or bifacial edge or ridge that will eventually be shaped into the fluted face (fig 11.2). Starting materials may include cobbles, pebbles, flake blanks, or even small biface blanks. Most commonly, a platform is made by the burination of one end of the core to create a long, narrow, flat surface that forms an angle with the intended fluted face. Platform creation may also be accomplished through side-blow flaking or by snapping the blank and using the break as a platform. From this platform, the ridged edge may then be removed through the careful pressure-flaking of a single long, narrow flake—the primary ridge flake—leaving two ridges on the core at the edges of the primary flake scar to guide further pressure flaking. A few more secondary ridge flakes may then be removed until a number of parallel ridges are present and a fluted face has been formed, from which microblades can be detached. The ridges on

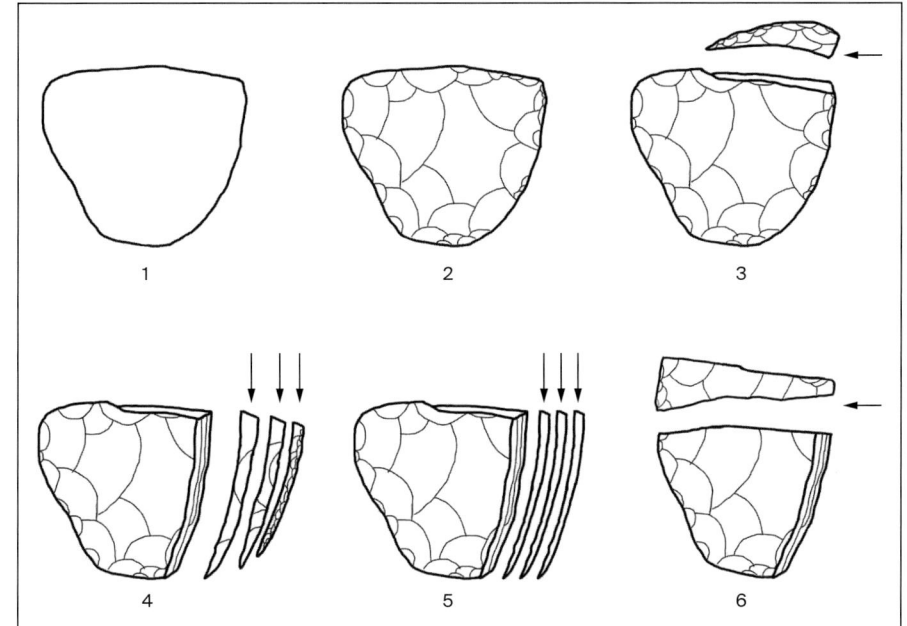

the core form the *arrises* of the dorsal surface of the microblade (see fig 11.1) and determine whether the cross-section of the artifact is triangular—a single arris—or trapezoidal—two or more arrises. Platforms may be rejuvenated by partial or complete removal of the platform, either by a single spall struck from the front or by multiple flakes struck from the side (see fig 11.2: 6). Complete spalls removed from the platform are known as *core tablets* or *platform tablets,* while flakes removed by side-blow flaking are smaller and more difficult to identify. Both unifacially and bifacially shaped cores tend to exhibit a wedge-shaped cross-section, as well as a ridge or *keel* representing remnant bifacial or unifacial retouch, extending from the back of the platform to the base of the fluted face.

Less formally, a microblade core may be created from a pebble blank, with platform production preceding the shaping of the core (fig 11.3). A platform may be created by the removal of a thick flake through hard-hammer percussion. With the resulting flat surface serving as a platform, unidirectional flaking may proceed from one edge or around the circumference of the entire platform until a sufficient number of closely spaced, parallel ridges are present for the removal of microblades. Rather than having a wedge-shaped outline, these cores are often either conical or cylindrical in form, when microblades are removed from the entire circumference of the core, or else tabular or "boat-shaped," when microblades are removed from a more limited surface, leading to a flattening of one

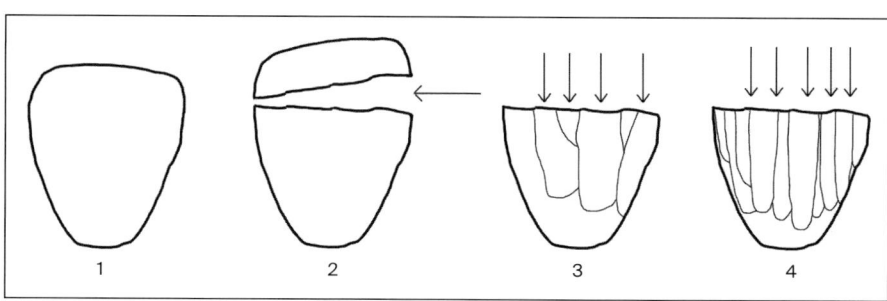

Figure 11.3. The informal method of microblade production: (1) cobble or pebble section; (2) creation of the platform; (3) flaking to shape the fluted face; (4) microblade removal

side of the core. They may also simply have an irregular shape. Microblades produced through these methods may be less standardized in size and shape than those produced from wedge-shaped cores; however, the platform-first core reduction sequence is faster and simpler and, because of this, is also less dependent on the availability of high-quality raw material (Elston and Brantingham 2002).

NORTH AMERICAN TRADITIONS OF MICROBLADE PRODUCTION

The Denali Complex and the Wedge-Shaped Microblade Core
The Denali Complex was defined early in the history of Alaskan archaeology by West (1967). Since then, other researchers have broadened the definition, with the result that the term has come, in its widest sense, to describe any Alaskan assemblage that contains microblades and the artifacts associated with their production, including microblade cores, ridge flakes, and core tablets. Such a definition is misleading, and occasionally misinterpreted, as microblades themselves appear in assemblages as early as 14,000 cal yr BP and persist until approximately 1,000 cal yr BP (Clark 2001; Holmes 1984, 2011; Potter 2008). In the strictest definition, however, the Denali Complex is confined to some of the earliest assemblages in Alaska, dating between 13,000 and 9,000 cal yr BP, and is associated with wedge-shaped microblade cores, lanceolate points, foliate bifaces, flat-topped end scrapers, and burins, especially the notched transverse burin (West 1967). Classic Denali cores have distinct keels and wedge-shaped cross-sections and are generally associated with core tablets and bifacial ridge flakes, although unifacial flaking and various platform-shaping methods are also seen in the complex (Clark 2001). While sharing similarities with the classic Denali Complex, Denali microblade sites from the later Holocene exhibit a wider variety of core-shaping methods and resulting core shapes and do not display the classic Denali Complex assemblage. These later sites are found across

Alberta's Lower Athabasca Basin

northwestern North America, in Alaska, Yukon, the Northwest Territories, and British Columbia (Clark and Gotthardt 1999). It is most likely this later variant of Denali technology that is found in northern Alberta.

The West Coast and Interior Plateau

The Northwest Coast Microblade Tradition is a diverse, loosely defined category encompassing various core types found on the Northwest Coast after the disappearance of the earliest, Denali-type cores. The term was proposed by Ackerman et al. (1985) to describe the early and middle Holocene microblade assemblages throughout the area, dating from 8,500 cal yr BP to approximately 4,000 cal yr BP and generally including cores fashioned from split pebble preforms and, less commonly, from thick, unshaped flakes (Ackerman 1996). Cores are mainly conical and cylindrical in form, but also reported are blocky, amorphous, and informally wedge-shaped forms, which are thicker and have wider fluted faces than their Denali Complex counterparts to the north. Platform preparation is rare, and platform rejuvenation even more so. Cortex is common on the lateral surfaces and even on the striking platform. Ackerman has proposed that the Northwest Coast method of microblade production was an adaptation designed to take advantage of the beach cobbles and pebbles readily available in the local area, whose coarser-grained texture made extensive shaping more difficult but required little preparation in order to roughen a platform (Ackerman 1980). Conservation of lithic material is obviously less crucial here, and methods of core preparation on the Northwest Coast indeed seem to reflect a more liberal, less systematic approach to lithic reduction.

The Plateau Tradition was defined by Sanger (1968b) to describe a method of production consistently seen throughout the interior plateau of British Columbia from about 7,500 to 2,000 cal yr BP. It is identified by the presence of informally wedge-shaped cores with little to no platform preparation and a single fluted face. The microblades tend to be triangular, with a general absence of ridge flakes, and the materials used are typically acquired from local sources (Sanger 1968b, 1970). It is possible that this tradition spread inward from the coast, retaining the lack of ridge flakes and platform preparation seen in the Northwest Coast Microblade Tradition but again adapting to reflect locally available materials.

Arctic Small Tool Tradition

The Arctic Small Tool Tradition is associated with the Palaeo-Eskimo groups who inhabited the high Arctic after the final recession of the far northern glaciers and prior to the arrival of the Thule ancestors of modern-day Inuit groups. Early Palaeo-Eskimo settlements appear in coastal sites across the Arctic by 4,500 cal yr BP, suggesting rapid migration and colonization (Odess 2005). Although ambiguous and short-lived in southern Alaska, these settlements appear to be initially associated with Bering Strait populations, who swiftly spread north along the coastline and whose presence in northwestern Alaska is visible as the Denbigh Flint Complex (Maxwell 1985). From there, Pre-Dorset cultures expanded across the Canadian Arctic, reaching Greenland as the Independence I and Saqaaq industries prior to 4,000 cal yr BP (Knuth 1967; Larsen and Meldgaard 1958). Around 3,000 cal yr BP, the early Palaeo-Eskimo cultures across the north transitioned into the Palaeo-Eskimo Dorset culture, either through migration or in situ development of traits shared through trade and political alliances. The Dorset culture, characterized by a combination of inland and coastal resource use, specialized harpoon technology, and a continued focus on microblade technology, was stable across the Arctic for nearly 3,000 years (Maxwell 1985).

Arctic Small Tool cores are formed by the creation of a platform through a single percussive blow and the subsequent removal of flakes from the platform to create conical or pyramidal fluted faces (Morlan 1970), and specialized non-triangular microblades are often found to have been extensively retouched and broken distally (Wyatt 1970). Some wedge-shaped core variants also exist but are similarly produced through the platform-first method (McGhee 1970). As with the Northwest Coast Tradition, platform grinding and rejuvenation is rare. Ridge flakes are also rare but have been found among artifacts of the Denbigh Flint Complex component at Onion Portage (Anderson 1970). Microblades of the Arctic Small Tool Tradition show modifications that indicate end-hafting and use for the production, carving, and curating of organic artifacts and artwork, rather than side-hafting or insetting into projectile points (McGhee 1970).

OIL SANDS MICROBLADE TECHNOLOGY

To date, evidence of microblade technology and possible microblade technology has been found at over fifty sites in the oil sands study area (fig 11.4). Of these reported finds, two have been published (Le Blanc and Ives 1986; Younie, Le Blanc, and Woywitka 2010), and two others presented at conferences (Reeves and de Mille 2010; Wickham 2010), while the rest have been described in unpublished impact mitigation archaeological reports. By far, the great majority of these reports are of a few isolated microblades or blade-like flakes within larger

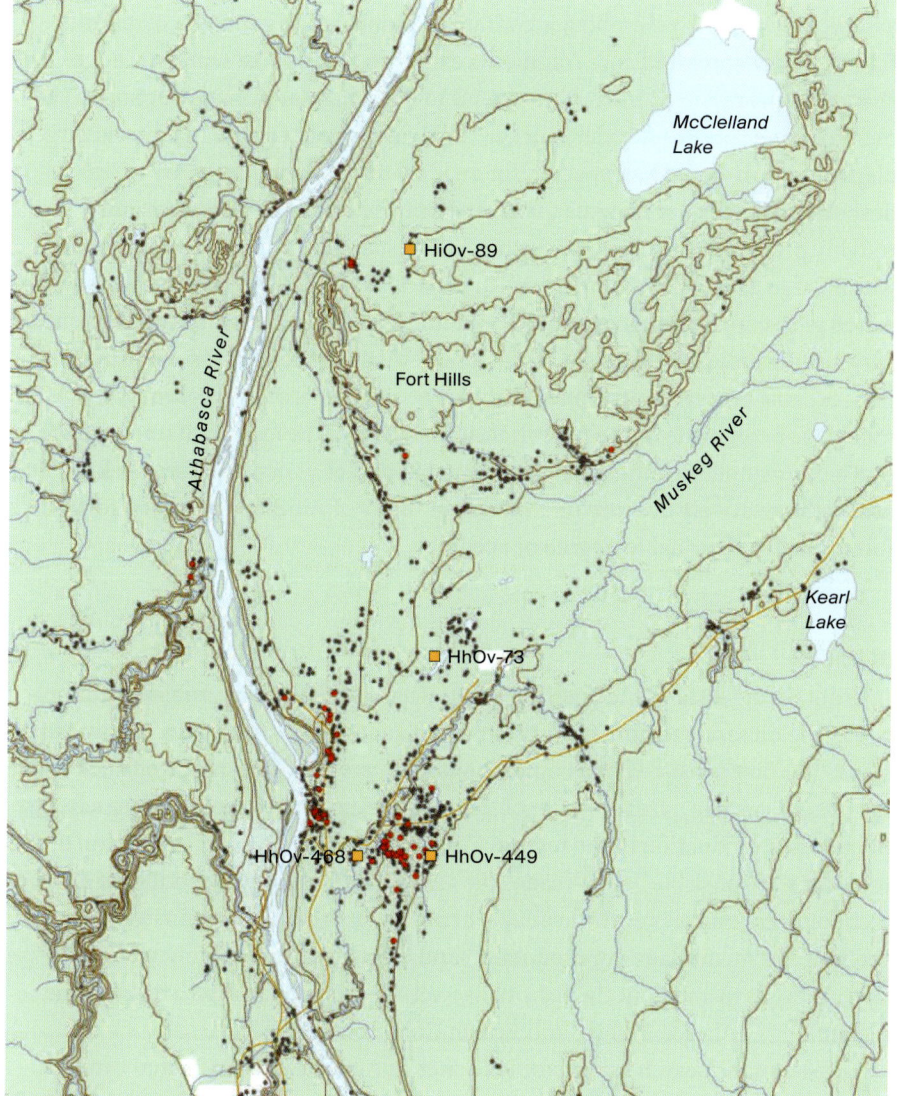

Figure 11.4. Oil sands archaeological sites containing possible or definitive evidence of microblade technology

McClelland Lake

HiOv-89

Fort Hills

Athabasca River

Muskeg River

Kearl Lake

HhOv-73

HhOv-468   HhOv-449

- ■ Verified microblade site
- ● Possible microblade site
- • Archaeological site

assemblages. A few assemblages contain microblades, microblade cores and fragments, or cores and core fragments exhibiting blade-like scars or wedge-like shapes. Many reported microblades tend to be triangular in cross-section, fragmentary, and robust—heavier, thicker, and with stronger ripple marks and larger platforms than would be expected for an ideal microblade. Cores are often described as blocky or irregular, with one or two adjacent blade scars and roughly prepared platforms. These more robust artifacts are most often found within large assemblages of Beaver River Sandstone (BRS) debitage that is associated with core and biface reduction. Methods of description and assessment vary among researchers, while associated dates are generally absent given the poor organic preservation typical to the dry, sandy soils of the region. While much has been made of the potential significance of these artifacts, we believe that a consistent, technologically informed approach is needed if we seek to assess the relatively obscure microblade industry of Alberta's northern subarctic forests with a reasonable degree of accuracy.

Methods

In order to understand the possible cultural associations of microblade technology in the oil sands region, we undertook an analysis that proceeded in two main steps. The first consisted of a comprehensive overview of the reported sites, with the goal of gaining a clear picture of the possible extent of specific approaches to microblade production in the region. This overview included an evaluation of the relevance of "possible" and "blade-like" artifacts. Second, we conducted a basic technological analysis of the available artifacts in an effort to reconstruct the sequence of microblade core reduction and evaluate the consistency of reduction methods throughout the region, as well as their similarities to microblade technology in Alaska, Yukon, and British Columbia.

A systematic search for microblade sites in Alberta was conducted through the Alberta Archaeological Site Inventory, a digital database of all registered archaeological sites in the province, maintained by Alberta Culture. The database contains information collected during all archaeological survey and excavation work that has occurred in the province under permit since the 1960s, which is reported to the Archaeological Survey of Alberta through standardized site inventory forms. The categories [Site] Description, Collection Remarks, and Further Remarks were queried for the terms *microblade, microcore, bladelet, ridge flake, micro,* and *blade.* This initial search yielded over one hundred sites, but, after restricting the search to the oil sands region and ruling out false-positive results relating to micro-debitage, knife blades, and unifacial blades, we isolated a final sample of fifty-six sites (see fig 11.4). The majority of these sites came up in the search because the recovery of "blade-like" flakes was reported in either the Description or Collection Remarks categories; however, a number of "micro-blade" and "microcore" items were also reported. Archaeological assessment reports were then reviewed for these fifty-six sites, and artifact descriptions, catalogues, and photographs were all assessed to help guide physical laboratory analysis. Although all reported microblades, cores, and blade-like artifacts were sought out for study, this was not always possible, given that much oil sands archaeological work is quite recent. Many studies are ongoing, and the collections had not yet been submitted to the Royal Alberta Museum and so could not be accessed at the time of study.

In all, 249 artifacts from nineteen assemblages were examined. Preliminary analysis included a basic identification of the artifact type, which involved sorting artifacts into microblade technology and non-microblade technology using the criteria outlined above, such as platform characteristics, outline, robusticity, and cross-section. A technological analysis of qualitative characteristics was subsequently conducted, which identified sequences of core production, including the method and extent of core shaping, platform production and maintenance, and fluted-face shaping and rejuvenation. These characteristics were then used to assess the likelihood that artifacts were in fact intentionally produced microblades, whether by formal or informal methods, and, if so, what tradition of microblade production, if any, they might be most closely associated with. The resulting artifact counts are presented below, in tables 11.1 and 11.2.

Microblade Assemblages

HhOv-73 (Bezya). The first microblades and cores to be discovered in the oil sands region were recovered from the Bezya site between 1982 and 1983 (Le Blanc and Ives 1986). The site assemblage included five chert microblade cores, three core tablets, twenty-seven ridge flakes, and 103 microblades, as well as a single notched transverse burin (table 11.1). With the exception of a core created from a flake blank, the microblade cores had been bifacially shaped from water-worn chert river cobbles. Many cores retained portions of cortex on their lateral faces. Platforms and fluted faces were shaped by the removal of bifacially shaped edges in the form of ridge flakes. Platform ridge flakes were removed through a longitudinal blow, similar to a burin blow, leaving a short, acute-angled platform. Following platform production, primary and then secondary ridge flakes were removed from the adjacent bifacial edge, again via a longitudinal burin blow, creating the fluted face, from which microblades were then removed. Platform

Table 11.1 Previously published and/or verified microblade artifacts

	Microblade	Microblade core
HhOv-73	103 (chert)	5 (chert); 3 tablets; 27 ridge flakes (chert)
HhOv-449	3 (chert)	1 (chert)
HhOv-468		1 preform (chert)
HiOv-89	39 (silicified mudstone)	23 (silicified mudstone); 1 tablet; 17 ridge flakes

grinding and rejuvenation are both evident; platform rejuvenation was similar to platform creation, entailing the removal of short spalls through a burin blow (Le Blanc and Ives 1986).

While this overall production sequence is very similar to that seen in the Denali Complex, the short platforms found at Bezya are unreported in the far northwest. Le Blanc and Ives (1986) propose that short spall removal, rather than removal of the entire platform element, for platform production and rejuvenation may serve to reduce the amount of material removed from the core, thereby preserving core height and extending the life of the core. A composite charcoal sample from the site provided a radiocarbon date of 3,990 ± 170 ^{14}C yr BP (Le Blanc and Ives 1986), or 4,235 to 4,810 cal yr BP (calibrated with CALIB 7.0.2 using the IntCal 13 calibration curve: Reimer et al. 2013), placing the site within the time range of both the Arctic Small Tool Tradition and later Denali industries found in the southern Yukon and western Northwest Territories. However, given the composite nature of the sample taken at the site and the lack of distinct stratigraphy, this date may not be representative of the age of the Bezya microblade assemblage itself (Le Blanc and Ives 1986).

HiOv-89 (Little Pond). The Little Pond site lies approximately 30 kilometres north of the main concentration of microblade sites in the oil sands region. Discovered in 2000 and excavated in 2005 (Woywitka and Younie 2008), this site was found to contain a relatively large assemblage of twenty-three wedge-shaped microblade cores and fragments, a platform tablet, seventeen ridge flakes, and thirty-nine microblades and fragments (see table 11.1), as well as sixteen burins and nine scrapers. The cores exhibited evidence of a distinct production pattern. The majority had been created from thick flakes, which were shaped by marginal unifacial flaking and occasional thinning to create unifacial ridges for shaping of the platform and fluted face. Platforms were most commonly prepared by the removal of a platform ridge flake via a burin blow, creating platforms similar to those seen at Bezya (Younie, Le Blanc, and Woywitka 2010). Platforms were also created by side-blow flaking. Following platform production, a series of ridge flakes were removed from an adjacent ridge to form the fluted face. Platform rejuvenation is evident in the form of partial and complete spall removal, as well as side-blow flaking, while some artifacts indicated that fluted face rejuvenation was also attempted. Most cores had thin fluted faces exhibiting two to three flute scars. In addition to the high number of ridge flakes, Little Pond provided some unique artifacts considered indicative of the Denali Complex (Younie, Le Blanc, and Woywitka 2010). Most notable is a platform tablet with evidence of bifacial flaking on the remnant keel and six remnant flute scars on the fluted face where it was removed from the core. The site also contained a large microblade core preform with a bifacially shaped keel and a flat, unused platform.

A notable characteristic of the artifacts at the Little Pond site was a high incidence of reuse and reworking of artifacts. Nearly all of the microblade cores exhibited use-wear on the edges of the fluted faces and platforms, while one had been recycled from a utilized biface and another was turned into a scraper after use as a core. Many burins in the assemblage exhibited multiple burination scars, while others also presented steep, scraper-like retouch and use-wear on or near the burin scars, indicating multiple types of usage for a single artifact (Younie, Le Blanc, and Woywitka 2010). All of these intensively reused and recycled artifacts, including all of the microblade cores, were created from silicified mudstone, a material not commonly found in archaeological sites in the area but known to be locally available in the form of river cobbles (Unfreed, Fedirchuk, and Gryba 2001).

Isolated finds. A single microblade core was recovered from HhOv-449, a site located just to the east of the Quarry of the Ancestors site complex. The specimen is thin and wedge-shaped and is composed of a pink, fine-grained, chert-like material, with three blade scars on the fluted face. A fourth scar extends onto one of the lateral surfaces of the core. There is a short, partially prepared platform (Wickham and Graham 2009; Wickham 2010).

A single microblade core preform was found at HhOv-468, about a kilometre to the west of the formal boundaries of the Quarry of the Ancestors. The artifact shows extensive, regular bifacial shaping along two margins, with the edge of one of the shaped margins partially removed through a single, burin-like flake detachment. The result is an apparent microblade core platform, with the adjacent, bifacially shaped fluted face awaiting ridge flake removal. If this artifact is, as is strongly suggested, truly a microblade core preform, it can be most closely compared to the Bezya cores, based on the shaping methods and platform

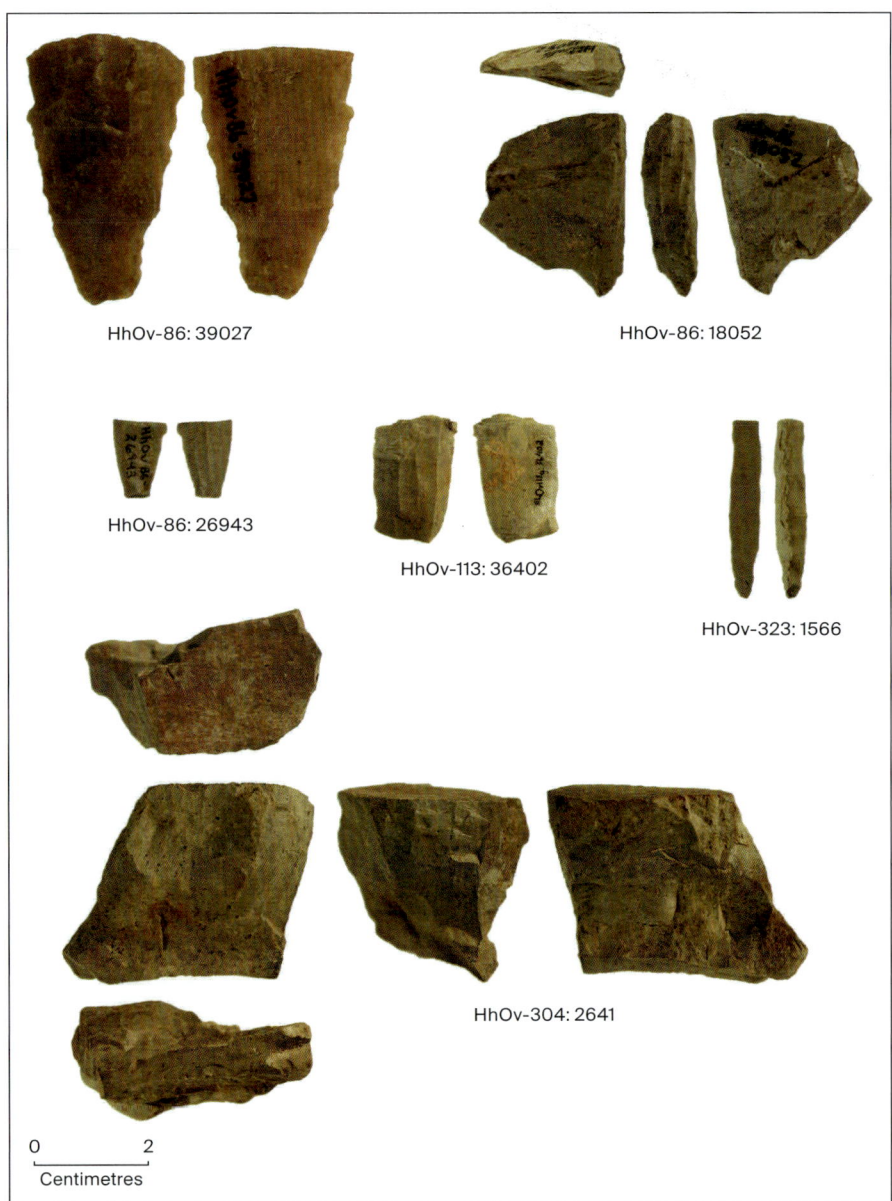

Figure 11.5. Possible microblade cores and core fragments

HhOv-86: 39027

HhOv-86: 18052

HhOv-86: 26943

HhOv-113: 36402

HhOv-323: 1566

HhOv-304: 2641

0   2
Centimetres

preparation technique (Wickham and Graham 2009; Wickham 2010). It is unknown why the preform might have been abandoned at this stage in reduction.

Another interesting isolated find is a microblade core recovered from HhOv-304 (Saxberg 2007), a site located within the Quarry of the Ancestors complex.

Although exhibiting little evidence of deliberate core shaping, and no evidence of platform preparation, this core has three small, short, neatly parallel blade-like scars on a lateral face (fig 11.5). It might be most accurately considered a "probable" microblade core. The blade scars are short and hinged, indicating that if this core was used in an attempt to create microblades, this attempt was unsuccessful, and the core appears to have been abandoned without further modification. It is composed of a piece of roughly cone-shaped BRS, with a protruding section of material next to the fluted lateral face. This protrusion may have served as a handle for gripping the core during microblade flaking, a technique commonly found in the wedge-shaped microblade tradition of Mesolithic and Neolithic Europe, which has been called the "handle-core" tradition (Vang Petersen 1984). Aside from the handle, however, this core—lacking evidence of the shaping of a fluted face, and with evidence of only a few unsuccessful attempts at blade removal—is closest in form to the Northwest Coast informal boat-shaped and conical cores.

A single primary ridge flake fragment was discovered at site HhOv-323 (fig 11.5), which is also located within the Quarry of the Ancestors complex (Saxberg 2007). This ridge flake is a small proximal fragment, with a tiny pressure-flaked platform and extensive delicate bifacial retouch along its central ridge. Only a small portion of the distal tip appears to be missing.

Blade-Like Artifacts

Quarry of the Ancestors. Of the fifty microblade cores and core fragments reported to have been discovered at various sites within the Quarry of the Ancestors area (HhOv-304, 305, 307, 311, 313, 319, 322, and 323) (Reeves and de Mille 2010; Saxberg 2007; Saxberg and Reeves 2004), twenty-three were available for the current study (table 11.2). Of these, the one from HhOv-304, described above, was found to be a probable microblade core, and the remaining twenty-two were found to be cores or core fragments exhibiting blade-like flake scars. The majority exhibited one to three wide but blade-like scars, while some exhibited no blade-like scars but did have unusual shapes, which may have prompted their initial interpretation as specialized cores. In most cases, these characteristics show clear evidence of having been created either by bipolar core reduction or by random factors of the reduction process leading to coincidentally overlapping flake scars that are too wide and too few to be considered flute scars (fig 11.6). A few core fragments are too small to allow any reliable assessment of their possible relationship microblade reduction. None show any direct evidence of true flute scars or core preparation.

Table 11.2 Other microblade sites and artifacts reported in the Alberta oil sands region

| | Microblade artifacts reported on site form or in permit report | | | | | Verified microblade artifacts |
	Artifacts analyzed in the current study					
	Microblade	Blade-like flake	Microblade core	Blade-like core	Total	
HgOv-45		3 BRS			3	
HgOv-106			1 ridge flake BRS		1	
HhOv-83		Several			Not specified	
HhOv-86	26 BRS		4 BRS; 1 chert		31	2
		26 BRS	1 BRS possible fluted-face flake; 1 BRS possible core	3 BRS	31	
HhOv-113	19 BRS		42 BRS		61	1
		19 BRS	1 BRS possible fluted-face flake	41 BRS	61	
HhOv-114	12 BRS; 2 blades				14	0
		14 BRS			14	
HhOv-117		3 BRS	1 possible BRS		4	
HhOv-122		1 BRS			1	
HhOv-123		3 BRS			3	0
		3 BRS			3	
HhOv-146					0	1
	1 BRS	1 BRS			2	
HhOv-159		1 BRS			1	
HhOv-160		2 BRS			2	
HhOv-165		2 BRS			2	
HhOv-166		1 BRS			1	
HhOv-167			1 blade-like BRS		1	
HhOv-256	9 BRS				9	
HhOv-260		1			1	
HhOv-267		2 blades			2	
HhOv-302	11 BRS		1 BRS		12	4
	4 BRS	7 BRS	fragment		11	
HhOv-304			1 BRS		1	1
	1 probable BRS		fragment		1	
HhOv-305	17 BRS		17 BRS		34	0
		9 BRS; 1 BRS ridge-like flake		11 BRS	21	
HhOv-307	13 BRS		2 BRS fragments		15	1
	1 chert	9 BRS		4 BRS	14	
HhOv-309			3 chert exhausted		3	
HhOv-311			1 BRS		1	0
				1 bipolar core	1	
HhOv-313	29 BRS		1 BRS		30	0
		29 BRS		1 BRS	30	
HhOv-319	12 BRS; 1 chert		2 possible		15	1
	1 chert	5 BRS		2 BRS	8	
HhOv-322	1		1 BRS fragment		2	0
		1 BRS, utilized		1 BRS	2	

| | Microblade artifacts reported on site form or in permit report Artifacts analyzed in the current study | | | | | Verified microblade artifacts |
	Microblade	Blade-like flake	Microblade core	Blade-like core	Total	
HhOv-323	9 BRS		8 BRS; 1 ridge flake		18	5
	3 BRS; 1 BRS blade	20 BRS	1 primary ridge flake	2 BRS	27	
HhOv-325			Possible		Not specified	
HhOv-330		1 BRS	Over half of cores have BLF (out of 36) (05-320)		Not specified	
HhOv-331					0	0
		1 BRS			1	
HhOv-332	24 BRS		14 BRS		38	1
	1 microburin	3 BRS			4	
HhOv-338	34 (05-320)				34	
HhOv-339	"Microblade technology and use are well represented" (05-320)				Not specified	
HhOv-340	"Prominence of microblade technology" (05-320)				Not specified	
HhOv-345	32 BRS microblade/ BLF		1 blade-like BRS		33	0
		2 BRS; 1 possible burin spall			3	
HhOv-348		5 BRS			5	
HhOv-371		Several BRS			Not specified	
HhOv-373		1 BRS			1	
HhOv-385		1 BRS			1	
HhOv-394	1 chalcedony	Several BRS			Not specified	
HhOv-399		1 BRS			1	
HhOv-424	1 BRS				1	
HhOv-464		Several BRS			Not specified	
HhOv-481	4 BRS				4	
HhOv-482	10 BRS; 5 BRS blades				15	
HhOw-16		5 BRS			5	
HhOw-36			1 BRS possible		1	
HiOu-68	1 BRS				1	
HiOv-57		1 chert			1	
HiOv-68		4 BRS			4	0
				4 BRS	4	
HiOv-70		11 BRS			11	0
				11 BRS	11	
Total	272	49	103		424	17
	12	166	5	66	249	

Figure 11.6. Bipolar and blade-like cores of BRS, which do not fit a strict definition of a microblade core

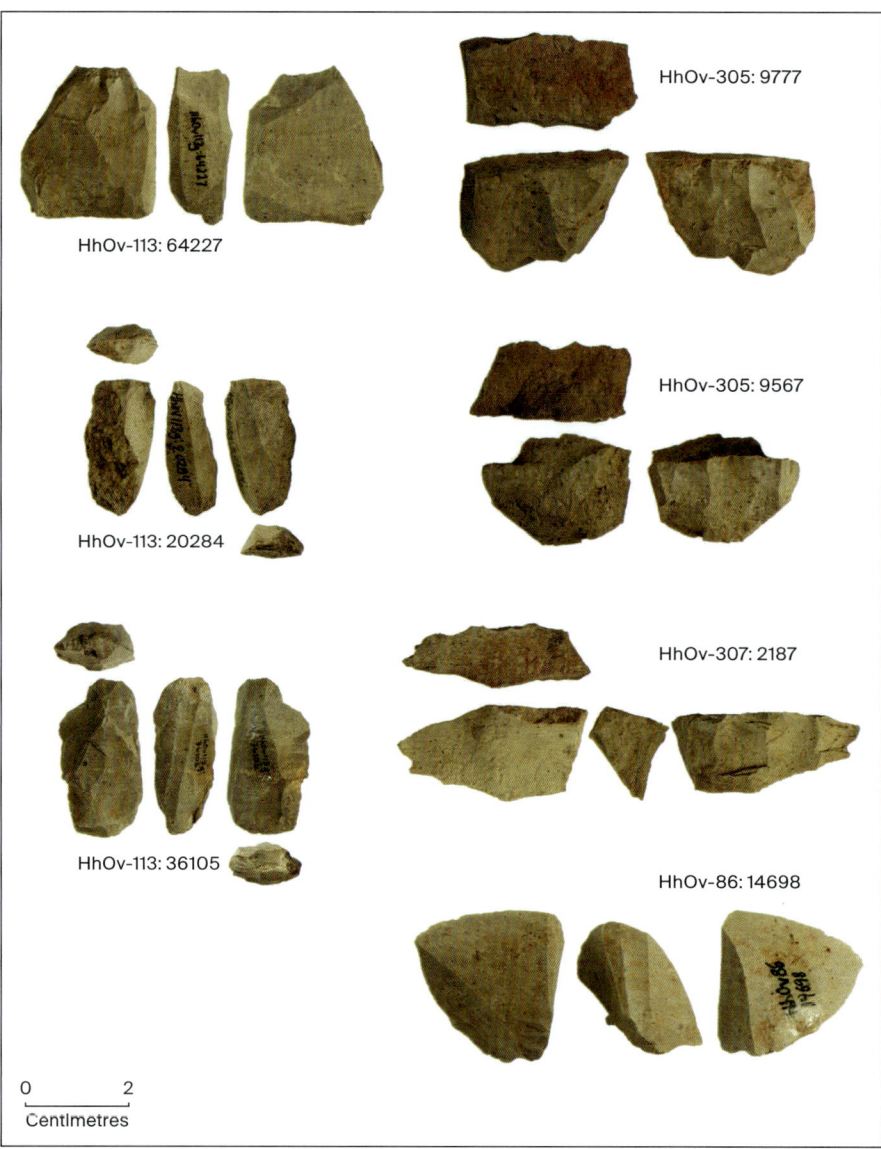

HhOv-113: 64227

HhOv-305: 9777

HhOv-113: 20284

HhOv-305: 9567

HhOv-307: 2187

HhOv-113: 36105

HhOv-86: 14698

0   2
Centimetres

A ridge-flake-like artifact was also found at HhOv-305 (fig 11.7). In contrast to the isolated ridge flake find at HhOv-323, this artifact is large, robust, and roughly bifacially shaped, with evidence of heavy percussive flake removal in the form of a large platform, ripple marks, and skewed curvature. While possibly a remarkably large platform ridge flake, this artifact is more likely to be a burin spall or other flake removed during the shaping or recycling of an irregular biface preform.

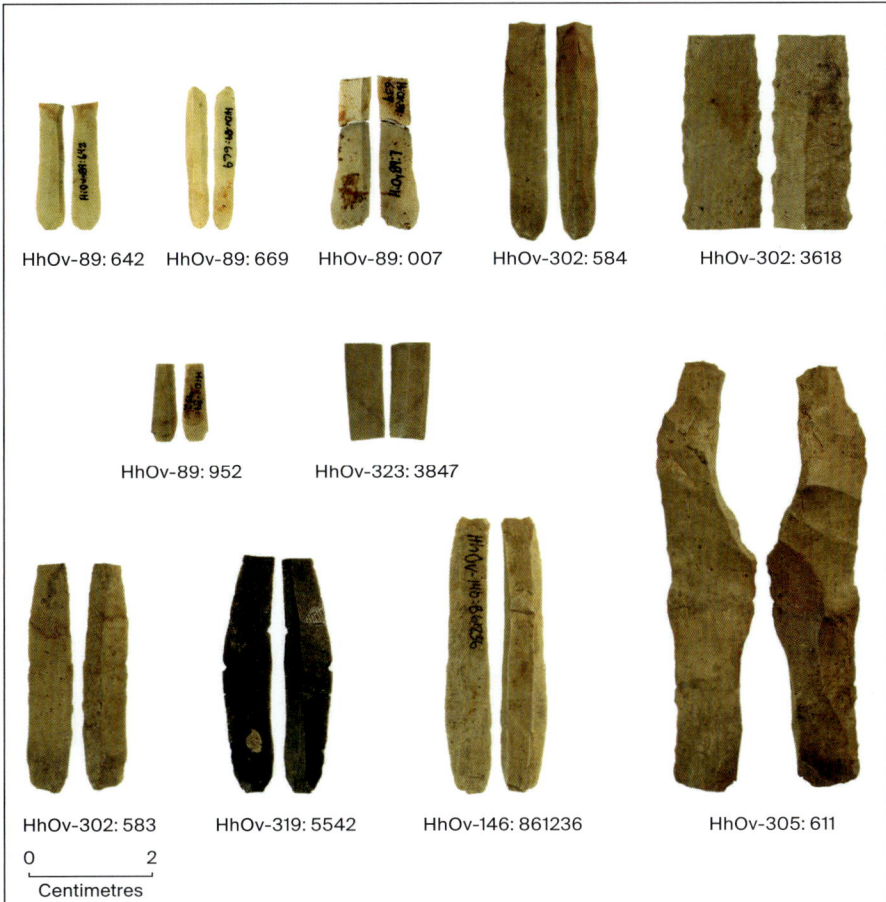

Figure 11.7. Oil sands microblades and, from HhOv-305, a possible ridge flake

HhOv-89: 642   HhOv-89: 669   HhOv-89: 007   HhOv-302: 584   HhOv-302: 3618

HhOv-89: 952   HhOv-323: 3847

HhOv-302: 583   HhOv-319: 5542   HhOv-146: 861236   HhOv-305: 611

0 ____ 2
Centimetres

Of the eighty microblades reported to have been found in the Quarry of the Ancestors that were available for study, six appeared to be true microblades (fig 11.7), while the other seventy-four appeared to be simply blade-like flakes, produced by coincidence through bipolar technology. These flakes generally were robust, with strong ripple marks and irregular or ovid outlines, and many showed evidence of bipolar percussion.

HhOv-86. Five microblade cores and twenty-six microblades were reported from site HhOv-86, located near the Cree Burn Lake site to the west of the Quarry of the Ancestors (Clarke and Ronaghan 2004). All the microblades are small and relatively parallel-sided, but they are also relatively short and thick, with hinge or step terminations indicative of strong percussive force used during flake removal.

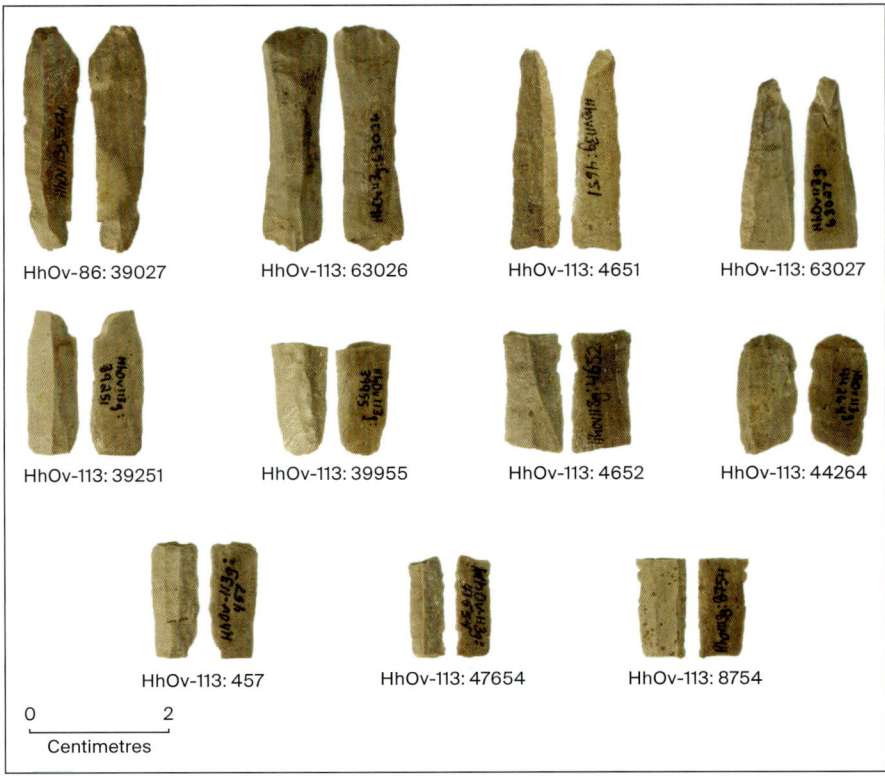

Figure 11.8. Blade-like flakes of BRS, none of which can strictly be defined as microblades

HhOv-86: 39027   HhOv-113: 63026   HhOv-113: 4651   HhOv-113: 63027

HhOv-113: 39251   HhOv-113: 39955   HhOv-113: 4652   HhOv-113: 44264

HhOv-113: 457   HhOv-113: 47654   HhOv-113: 8754

0   2
Centimetres

All may be best categorized as blade-like flakes. Three of the reported microblade cores are wedge-shaped and appear microcore-like in outline. However, two of these are wedge-shaped flakes with narrow fluted-face-like surfaces but show no evidence of shaping and no evidence of flake removal from the narrow faces, which appear to be simply oddly shaped dorsal flake scars (see fig 11.5). The third can be identified as a possible microblade core on a wedge-shaped flake: although the fluted face is irregular, and the flutes poorly defined, there is evidence of some platform preparation and of unidirectional flake removal. Although wedge-shaped, the core is expedient, with little evidence of shaping, and does not reflect the more systematic Denali technology seen at Little Pond and Bezya. There is also a utilized chert uniface with three narrow, short, parallel flake scars ending in step fractures. One of these scars appears to be a burination, and, while the flute-like scars are similar in appearance to those observed in Northwest Coast micro-blade technology, the core is not sufficiently shaped, nor have a sufficient number of flakes been removed to permit diagnosis. Finally, a small flake with four narrow, parallel dorsal flake scars was examined and found to be similar to a possible fluted-face rejuvenation flake from Little Pond. The flake scars are regular and closely spaced, converging slightly toward the platform as if the flake had been removed using the base of a core as the platform.

HhOv-113. From site HhOv-113, located a few kilometres west of the Quarry of the Ancestors complex, forty-eight reported microblade cores (Green et al. 2006) were analyzed in the current study, with photographs taken of those with the strongest resemblance to true microblade cores (see fig 11.6). No cores from the collection provided evidence of a prepared microblade technology. The majority exhibited bipolar crushing, which had been noted in the permit report as possible evidence of the use of a vise to stabilize the cores during microblade production. Despite the small size of the cores and the presence of one or two blade-like scars on each, their irregular shapes and the evidence of heavy bipolar force used to reduce the cores indicate that they are simply bipolar cores. The majority have collapsed bipolar ridges rather than prepared platforms, while the majority of blade-like scars are irregular and only superficially blade-like. In addition, nineteen microblades reported to have been found at HhOv-113 were studied, as well as fourteen from the nearby HhOv-114, none of which exhibited microblade characteristics. Many exhibited crushed platforms, thick triangular cross-sections, or heavy ripple marks indicating possible relationships to bipolar reduction (fig 11.8).

A PROPOSED OUTLINE OF OIL SANDS MICROBLADE PRODUCTION

At this time, it seems that a distinct and cohesive pattern of microblade production has begun to emerge in the oil sands region. While evidence of such a pattern appears to be localized within this portion of northeastern Alberta, this perception may be as much a reflection of the rate of excavation within the area of oil sands leases as it is of patterns of prehistoric cultural activity. Despite these limitations, many of the cores seen here are strongly suggestive of those seen in the Denali Complex to the northwest and may be indicative of a continuity in lithic cultural tradition.

On the whole, oil sands microblade cores seem to be best categorized as wedge-shaped, produced through a relatively uniform reduction sequence with varying degrees of formality. There are two main variants to the production system. The first and more formalized sequence, in which a pebble is bifacially reduced to create a core preform, is seen at Bezya and on the core preform from HhOv-468. The second pattern is seen at Little Pond and seems to be reflected as well in the core from HhOv-449 and the possible microblade cores from HhOv-86. Here, core preforms are more casually produced from wedge-shaped flakes, with bifacial or unifacial marginal flaking to shape the keel and occasional thinning to create a more suitable cross-section. Unifacial or bifacial flaking is also used to shape ridges for the creation of the fluted face and often also the platform. Occasionally, bifacial or unifacial tools are reused, with the shaped edges of the tool serving to provide an ideal starting point for the removal of ridge flakes. Little Pond illustrates the most formally produced cores within this second category, while those from HhOv-86 and HhOv-113 are produced more expediently. The presence of two main variants may, however, simply reflect the fact that only two assemblages are thus far known to fall within this category. Future discoveries may help to round out our understanding of this sequence.

The two variants share a number of reduction methods. Ridge flakes and platform ridge flakes appear to be common among oil sands microblade sites, indicating an association with Denali microblade technology. After shaping of the core preform, the platform is almost always created, and later rejuvenated, through a partial spall removal. Occasional side flaking or the opportunistic use of an existing ideal surface also occurs. Platform preparation is followed by formation of the fluted face and, finally, by microblade removal. The use of a burin-type blow seems to be the most typical method of platform rejuvenation. Fluted faces are thin, usually exhibiting between two and three flute scars, with the occasional core exhibiting up to four or five.

While an analysis of every reported microblade find in the oil sands region could not be conducted, the results of this sample indicate that the optimistic reporting of blade-like artifacts as microblades has led to an inflated view of the prevalence of microblade technology in the oil sands area. Of the 178 reported microblades and blade-like artifacts studied (not including previously published or presented assemblages at Bezya, Little Pond, HhOv-449, and HhOv-468), only twelve could be described with confidence as microblades; of seventy-one cores studied, only five could be described with confidence as microblade cores. This over-reporting stems partly from the need to ensure that possible microblade assemblages are not overlooked. However, in a number of ways, it has had the effect of skewing the data set available to those archaeologists seeking to analyze microblade assemblages within the region. First, a relatively uncritical review of the databases and cultural resources management reports might easily lead to the conclusion that microblade production is much more prevalent in the oil sands than it truly is. Second, as it is reported, this microblade technology appears quite varied and includes a high number of unprepared cores of various shapes, which might imply relationships to a wide range of traditions of microblade production elsewhere in North America. In fact, the cores being assessed are not microblade cores at all, and much of the apparent variation has resulted from the analysis of bipolar technology through the unrelated lens of microblade reduction. Finally, when blade-like artifacts are identified as microblades, the degree of use of BRS as a material in microblade production is greatly overrepresented. The majority of securely identifiable microblade artifacts are in fact composed of rarer materials such as chert and silicified mudstone.

Ideally, then, in order to prevent this skewing of data, all materials with blade-like scars, especially core fragments exhibiting no evidence that they were specifically intended for use as microblade cores, should be reported as "blade-like" items, not as microblades or microblade cores. Standard methods for distinguishing microblade from non-microblade technology should also be employed, so that the distribution of reported microblade technology in the region reflects the actual presence or absence of microblade technology, rather than the range of definitions held by different researchers. The most reliable methods of identifying true microblades require either that the researcher have some prior degree of familiarity with other microblade assemblages or else that a comparative analysis be carried out using a collection of microblades and cores. However, even without benefit of direct comparison, several key microblade features tend to be lacking in the majority of blade-like flakes reported at many oil sands sites: straight, parallel margins, delicate features, narrow pointed terminations, and pressure-flaked, single-faceted platforms. Trapezoidal cross-sections, caused by two or more long, parallel arrises on the dorsal surface, are also strong evidence that a flake is a true microblade. The most common violation of these diagnostic criteria is the classification of flakes that are long and narrow, but not straight-sided, as microblades, which is especially troublesome because these artifacts are not even blade-like in form. Flakes removed from genuine microblade cores will not display an ovoid outline. In addition, bipolar flakes are relatively thick, wide, and robust in comparison to microblades, commonly exhibiting triangular cross-sections and strong ripple marks on the dorsal flake scars as well as the ventral surface. Platforms may be crushed, while the lateral margins may be feathered. Bifacial trimming flakes may also appear blade-like if narrow in outline; however, the presence of lipped platforms, ovid outlines, and exclusively triangular cross-sections will help to distinguish these flakes from microblades.

DISCUSSION

The Cultural Significance of Microblades in the Subarctic
While the study of microblade artifacts from the northern Subarctic has great potential to expand our understanding of prehistoric life in northern Alberta, it

must also be emphasized that there are limitations to what these artifacts can tell us. At this time, only one microblade assemblage in the oil sands area has been dated, and even it has not been dated securely. Although microblades have been found in early occupation layers of sites in Alaska, it cannot be assumed that microblades represent an early occupation of the oil sands region, given that the wedge-shaped microblade core has persisted throughout much of the Holocene in the Yukon and Alaska. Likewise, the date obtained from the Bezya site cannot necessarily be considered representative of the age of the microblade assemblage at the site, much less of all microblade sites found in the oil sands region.

Our current perception of oil sands prehistory has been strongly influenced by a number of factors common to archaeological sites in the boreal forest. Well-drained, acidic soils promote the decomposition of organic artifacts, while also presenting thin A horizons that also quickly decompose once buried, preventing the identification of stratigraphic layers during excavation. Dry, sandy soils such as those seen in the oil sands are also especially susceptible to bioturbation, complicating any efforts at a detailed analysis of the spatial relationships between artifacts. This has led to a dependency on lithic materials, the analysis of which does prove valuable in areas near BRS outcrops, where extremely high rates of lithic reduction activities are apparent. However, the need to rely on lithic evidence has also produced a welter of hypothesized stone tool typologies, arrayed along a timeline that is impossible to evaluate without accurate dating methods. Given the collapsed stratigraphy common to the sandy boreal sites in the region, many attempts to date these sites rely on charcoal that cannot be definitively stratigraphically associated with the artifacts and therefore often yield results that are not necessarily representative of human occupation.

Relationships between microblade production and raw material selection strategies are significant not only to our understanding of why and how microblades were produced and used but also to our understanding of the prehistoric use of the landscape. Among the currently known archaeological localities in northern Alberta, the oil sands lease areas are unique for their large quantities of fine-grained, workable BRS, comprising 95% to 100% of the excavated materials in oil sands assemblages. The only known outcroppings of BRS in Alberta are located along the Athabasca River, in a small area near Fort McKay (Fenton and Ives 1990). BRS occurs in small percentages in assemblages from the Birch Mountains, 70 kilometres to the west, and along the Clearwater River, to the south (Ives 1993 and chapter 8 in this volume). The highest concentrations, however, occur within 30 kilometres of the source, in the Cree Burn Lake and Quarry of the Ancestors site localities to the north of Fort McKay (see chapters 9 and 10 in this volume).

Both localities seem to represent repeated, intensive land use and occupation, possibly over the entirety of the Holocene. In both locations, site assemblages consist primarily of BRS debitage, with artifacts now numbering in the hundreds of thousands. Tools are comparatively rare. Farther out from these occupation centres, sites become sparser, with lower proportions of BRS and higher proportions of tools (Ives 1993). For example, despite the presence of high, well-drained terrain, numerous sinkhole lakes, and proximity to a massive wetland complex ideal for moose and waterfowl hunting, sites such as Little Pond, situated farther north in the Fort Hills, are relatively rare and generally smaller, with a greater variability in raw material usage and a high percentage of isolated finds of fewer than ten flakes. This site distribution may reflect the seasonal rounds of historic Dene populations, who settled in relatively large communities in the summer and dispersed into small, mobile hunting groups in the winter. The two major site complexes, Cree Burn Lake and the Quarry of the Ancestors, may represent a traditional gathering area for prehistoric populations within the oil sands region. Located not far from both the Athabasca and Muskeg rivers, they are adjacent to a major transportation corridor, as well as to an abundant source of lithic raw material, and are surrounded by many small lakes and wetlands ideal for moose hunting.

The fact that microblade cores were most commonly produced from materials other than the omnipresent BRS could support the idea that microblade technology was a method of conserving desirable raw materials or, conversely, that flintknappers selectively used finer-grained, more workable materials to ensure more consistent and efficient microblade production. These are only the most obvious of a wide range of possibilities. While raw material conservation cannot be considered the sole purpose of microblade technology in the oil sands, there is clear evidence at Little Pond of attempts to conserve the fine-grained silicified mudstone used for microblades, most notably in the presence of exhausted cores exhibiting extensive use-wear and the occasional reworking of cores as scrapers or burins. It seems that lithic materials used at Little Pond were highly suited to microblade production, allowing for the systematic core-reduction patterns in evidence at the site. Because the material was finer-grained than the more common BRS, attempts were made to conserve and reuse the material.

This practice of material conservation would not, however, preclude the use of BRS at other sites, to replace worn microblades in composite tools, when no finer-grained materials were available. Fine-grained, highly workable BRS was used for the production of carefully shaped points and bifaces at many sites in the area, and given the ready availability of the material, it does not seem unlikely that it would have been extended to the purpose of shaping microblade cores. Indeed, a few microblades and a ridge flake of BRS have been found thus far, indicating that although finer-grained materials seem to have been preferred, at least some of the oil sands microblade makers were aware of the presence of locally available BRS.

Could it be that the many BRS blade-like cores reported in the oil sands represent an adaptation of the widely available, but less workable, local material to microblade production? It could be argued that these cores represent an adaptation of Denali microblade technology, involving less formalized reduction sequences and reduced platform preparation. Similar theories have been proposed to explain the method of microblade production seen on the Northwest Coast, where beach cobbles were readily available but often coarse-grained compared to the obsidian available in the far northern Alaskan coastal areas (Ackerman 1980). Cores of this tradition are not, however, comparable to the BRS cores from the oil sands. Although similar in showing little evidence of bifacial or unifacial shaping, Northwest Coast microblade cores often have tabular and conical shapes, including wide-fluted faces with high numbers of regular adjacent flute scars, resulting from unidirectional microblade removal from a single platform. In contrast, with the exception of the single handle core found at HhOv-304, oil sands BRS cores are irregularly shaped, have three or fewer adjacent scars that are rarely narrow or parallel, and often show evidence of percussive flake removal, with their flute-like scars originating from platforms on both ends of the core. In other words, these cores do not exhibit sufficient evidence of microblade technology, whether formal or informal, to be considered within the context of the known microblade traditions in North America.

Relationships to the Far Northwest of North America

The non-BRS core reduction patterns seen in the oil sands region have potential correlates in the microblade traditions of far northwestern North America. Most notable is the similarity between the cores from HhOv-449, HhOv-468, Little Pond, and Bezya to those of the Denali Complex. These cores share a wedge shape, bifacial and unifacial shaping techniques, and the removal of ridge flakes and platform tablets. Such features separate these cores from those of the Plateau, Northwest Coast, and Arctic Small Tool traditions, which generally lack ridge flakes, rely on less formal methods of platform production, and exhibit much wider-fluted faces. These characteristics are reflected in the high proportions of informal, tablet, boat-shaped, and conical forms. The associated presence of burin technology and the use of core-burins are also seen both in the Denali Complex and in the oil sands, but not in the Plateau and Northwest Coast traditions. However, cores from the oil sands do show some trends in production that differ slightly from the Denali types, including the tendency toward partial platform rejuvenation rather than the removal of a full core tablet, as well as the tendency toward thin-fluted faces exhibiting only two or three flute scars. The presence of a large, wide, fully formed core tablet at Little Pond is, thus far, the only exception to this pattern; however, the singularity of the artifact itself raises a number of questions. How the artifact reached the site, and why no other examples of this method of production were found at the site, cannot be easily explained. The occurrence of this tablet, which is much larger than the majority of microblade artifacts found in the oil sands and is unique in its production characteristics, seems to indicate that further variation in techniques of micro-blade production may be present in the oil sands area but as yet undiscovered. Finally, although they are technologically similar and contain burins and scrap-ers indicative of the classic Denali Complex of the early Holocene, the oil sands microblade assemblages studied thus far do not display the full array of Denali Complex artifacts, including lanceolate projectile points and foliate bifaces. These sites are thus not likely to belong within the classic Denali Complex as a continuous cultural group but instead may be said to exhibit technological simi-larities that suggest cultural communication and a shared adaptation to subarctic environments.

While the presence of microblade and blade-like technology has now been well established in the oil sands region and has been discussed by a number of researchers (Le Blanc and Ives 1986; Reeves and de Mille 2010; Wickham 2010; Younie, Le Blanc, and Woywitka 2010), further research may yet be possible to round out the current discussion, to clarify the role of BRS in microblade assem-blages, and to allow for the interpretation of stronger spatial and chronological relationships within both the oil sands region and the western Subarctic. As oil sands development, and therefore also heritage management survey and excava-tions, expand further from the centres of occupation near the highest density of locally available BRS, new microblade finds may be reported. This hypothesis is supported by the current distribution of known microblade assemblages rela-tively far from the Quarry of the Ancestors and by the tendency of these assem-blages to focus on lithic materials that are finer-grained than BRS. Furthermore, it may be that microblades and cores were in fact present in sites recorded during early surveys but went unrecognized or unreported because these surveys took place prior to the recent discoveries that have established the presence of micro-blade technology in the region. This idea could be tested through a physical review of curated collections, perhaps through random sampling. Such an approach may also be necessary to evaluate the distribution of microblade technology within areas too disrupted by oil sands industrial development to allow for new archaeological excavation. It is hoped that the framework presented here, and the questions of raw material usage and cultural associations briefly discussed within its context, may be used to guide such future study and will provoke further research into the role of microblade technology in oil sands prehistory.

REFERENCES

Ackerman, Robert E.

1980 Microblades and Prehistory: Technological and Cultural Considerations for the North Pacific Coast. In *Early Native Americans Prehistoric Demography, Economy, and Technology,* edited by David L. Browman, pp. 189–197. World Anthropology Series no. 13. Mouton Publishers, The Hague.

1996 Early Maritime Culture Complexes of the Northern Northwest Coast. In *Early Human Occupation of British Columbia,* edited by Roy L. Carlson and L. Dalla Bona, pp. 123–132. University of British Columbia Press, Vancouver.

Ackerman, Robert E., Kenneth C. Reid, James D. Gallison and Mark E. Roe

1985 *Archaeology of Heceta Island: A Survey of 16 Timber Harvest Units in the Tongass National Forest, Southeastern Alaska.* Center for Northwest Anthropology Report no. 3. Washington State University, Pullman.

Anderson, Douglas D.

1970 Microblade Traditions in Northwest Alaska. *Arctic Anthropology* 7(2): 2–16.

Bamforth, Douglas B., and Peter Bleed

1997 Technology, Flaked Stone Technology, and Risk. *Archeological Papers of the American Anthropological Association* 7(1): 109-139.

Bleed, Peter

2001 Cheap, Regular, and Reliable: Implications of Design Variation in Late Pleistocene Japanese Microblade Technology. In *Thinking Small: Global Perspectives on Microlithization,* edited by Robert G. Elston and Steven L. Kuhn, pp. 95–102. Archeological Papers of the American Anthropological Association no. 12. American Anthropological Association, Arlington, Virginia.

Clark, Donald W.

2001 Microblade-Culture Systematics in the Far Interior Northwest. *Arctic Anthropology* 38(2): 64–80.

Clark, Donald W., and Ruth M. Gotthardt

1999 *Microblade Complexes and Traditions in the Interior Northwest as Seen from the Kelly Creek Site, West-Central Yukon.* Occasional Papers in Archaeology no. 6. Heritage Branch, Government of the Yukon, Whitehorse.

Clarke, Grant M., and Brian M. Ronaghan

2004 *Historical Resources Impact Assessment and Mitigation Program, Muskeg River Mine Project: Final Report (ASA Permit 00-087).* 2 vols. Golder Associates Ltd. Copy on file, Archaeological Survey, Historic Resources Management Branch, Alberta Culture, Edmonton.

Ellis, Christopher J.

1997 Factors Influencing the Use of Stone Projectile Tips. In *Projectile Technology,* edited by Heidi Knecht, pp. 37–78. Plenum Press, New York.

Elston, Robert G., and P. Jeffrey Brantingham

2002　Microlithic Technology in Northern Asia: A Risk-Minimizing Strategy of the Late Paleolithic and Early Holocene. In *Thinking Small: Global Perspectives on Microlithization,* edited by Robert G. Elston and Steven L. Kuhn, pp. 103–116. Archeological Papers of the American Anthropological Association no. 12. American Anthropological Association, Arlington, Virginia.

Fenton, Mark M., and John W. Ives

1990　Geoarchaeological Studies of the Beaver River Sandstone, Northeastern Alberta. In *Archaeological Geology of North America,* edited by Norman P. Lasca and Jack Donahue, pp. 123–135. Centennial Special Volume 4. Geological Society of America, Boulder, Colorado.

Flenniken, J. Jeffrey

1987　The Paleolithic Dyuktai Pressure Blade Technique of Siberia. *Arctic Anthropology* 24(2): 117–132.

Goebel, Ted

1999　Pleistocene Human Colonization of Siberia and Peopling of the Americas: An Ecological Approach. *Evolutionary Anthropology* 8: 208–229.

2002　The "Microblade Adaptation" and Recolonizaton of Siberia During the Late Upper Pleistocene. In *Thinking Small: Global Perspectives on Microlithization,* edited by Robert G. Elston and Steven L. Kuhn, pp. 117–131. Archeological Papers of the American Anthropological Association no. 12. American Anthropological Association, Arlington, Virginia.

Goebel, Ted, Michael R. Waters, and Dennis H. O'Rourke

2008　The Late Pleistocene Dispersal of Modern Humans in the Americas. *Science* 319 (14 March): 1497–1502.

Gómez Coutouly, Yan Axel

2011　Identifying Pressure Flaking Modes at Diuktai Cave: A Case Study of the Upper Paleolithic Microblade Tradition. In *From the Yenisei to the Yukon: Interpreting Lithic Assemblage Variability in Late Pleistocene / Early Holocene Beringia,* edited by Ted Goebel and Ian C. Buvit, pp. 75–90. Texas A&M University Press, College Station.

Graf, Kelly E., and Nancy H. Bigelow

2011　Human Response to Climate During the Younger Dryas Chronozone in Central Alaska. *Quaternary International* 242(2): 434–451.

Green, D'Arcy, David Blower, Dana Dalmer, and Luc Bouchet-Bert

2006　*Historical Resources Impact Assessment and Mitigation, Albian Sands Energy's Muskeg River Mine and Shell Canada's Jackpine Mine: Final Report (ASA Permit 05-355).* 2 vols. Golder Associates Ltd. Copy on file, Archaeological Survey, Historic Resources Management Branch, Alberta Culture, Edmonton.

Hamilton, Thomas D., and Ted Goebel

1999　Late Pleistocene Peopling of Alaska. In *Ice Age People of North America: Environments, Origins, and Adaptations,* edited by Robson Bonnichsen and Karen L. Turnmire, pp. 156–199. Oregon State University Press, Corvallis.

Holmes, Charles E.

1984　The Prehistory of the Lake Minchumina Region, Alaska: An Archeological Analysis. PhD dissertation, Department of Anthropology, Washington State University, Pullman.

2001　Tanana River Valley Archaeology circa 14,000 to 9000 B.P. *Arctic Anthropology* 38(2): 154–170.

2011　The Beringian and Transitional Periods in Alaska: Technology of the East Beringian Tradition as Viewed from Swan Point. In *From the Yenisei to the Yukon: Interpreting Lithic Assemblage Variability in Late Pleistocene / Early Holocene Beringia,* edited by Ted Goebel and Ian Buvit, pp. 179–191. Texas A & M University Press, College Station.

Holmes, Charles E., Richard VanderHoek, and Thomas E. Dilley

    1996   Swan Point. In *American Beginnings: The Prehistory and Palaeoecology of Beringia,* edited by Frederick Hadleigh West, pp. 319–323. University of Chicago Press, Chicago.

Ives, John W.

    1993   The Ten Thousand Years Before the Fur Trade in Northeastern Alberta. In *The Uncovered Past: Roots of Northern Alberta Societies,* edited by Patricia A. McCormack and R. Geoffrey Ironside, pp. 5–31. Circumpolar Research Series no. 3. Canadian Circumpolar Institute, University of Alberta, Edmonton.

Keates, Susan G., Yaroslav V. Kuzmin, and Chen Shen (editors)

    2007   *Origin and Spread of Microblade Technology in Northern Asia and North America.* Department of Archaeology Publication no. 34. Archaeology Press, Simon Fraser University, Burnaby, British Columbia.

Knuth, Eigil

    1967   *Archaeology of the Musk-ox Way*. Contributions du Centre d'études arctiques et finno-scandinaves no. 5. École pratique des hautes études, Paris.

Kobayashi, Tatsuo

    1970   Microblade Industries in the Japanese Archipelago. *Arctic Anthropology* 7(2): 38–58.

Kuhn, Steven L., and Robert G. Elston

    2002   Introduction. In *Thinking Small: Global Perspectives on Microlithization,* edited by Robert G. Elston and Steven L. Kuhn, pp. 1–7. Archeological Papers of the American Anthropological Association no. 12. American Anthropological Association, Arlington, Virginia.

Larsen, Helge, and Jergen Meldgaard

    1958   *Paleo-Eskimo Cultures in Disko Bugt, West Greenland*. Meddelelser om Grønland 161, no. 2. C. A. Reitzels Forlag, Copenhagen.

Le Blanc, Raymond J., and John W. Ives

    1986   The Bezya Site: A Wedge-Shaped Core Assemblage from Northeastern Alberta. *Canadian Journal of Archaeology* 10: 59–98.

Maxwell, Moreau S.

    1985   *Prehistory of the Eastern Arctic*. Academic Press, Orlando.

McGhee, Robert

    1970   A Quantitative Comparison of Dorset Culture Microblade Samples. *Arctic Anthropology* 7(2): 89–96.

Morlan, Richard E.

    1970   Wedge-Shaped Core Technology in Northern North America. *Arctic Anthropology* 7(2): 17–37.

Odess, Daniel

    2005   Arctic Small Tool Tradition. In *Encyclopedia of the Arctic,* edited by Mark Nuttall, pp. 146–147. Routledge, New York.

Potter, Ben A.

    2008   Exploratory Models of Intersite Variability in Mid to Late Holocene Central Alaska. *Arctic* 61(4): 407–425.

Pyszczyk, Heinz W.

    1991   A Wedge-Shaped Microblade Core, Fort Vermilion, Alberta. In *Archaeology in Alberta, 1988 and 1989,* edited by Martin P. R. Magne, pp. 199–204. Archaeological Survey of Alberta Occasional Paper no. 33. Historic Resources Management Branch, Alberta Culture, Edmonton.

Reeves, Brian O. K., and Christy de Mille

    2010   Microblade Technologies from the Oilsands Regions of Northeastern Alberta. Paper presented at 43rd annual meeting of the Canadian Archaeological Association, Calgary, Alberta, 28 April–2 May.

Reimer, Paula J., Édouard Bard, Alex Bayliss, J. Warren Beck, Paul G. Blackwell, Christopher Bronk
        Ramsey, Caitlin E. Buck, Hai Cheng, R. Lawrence Edwards, Michael Friedrich, Pieter
        M. Grootes, Thomas P. Guilderson, Haflidi Haflidason, Irka Hajdas, Christine Hatté,
        Timothy J. Heaton, Dirk L. Hoffmann, Alan G. Hogg, Konrad A. Hughen, K. Felix Kaiser,
        Bernd Kromer, Sturt W. Manning, Mu Niu, Ron W. Reimer, David A. Richards, E. Marian
        Scott, John R. Southon, Richard A. Staff, Christian S. M. Turney, and Johannes van der
        Plicht
    2013   IntCal13 and Marine13 Radiocarbon Age Calibration Curves 0–50,000 years cal BP.
            Radiocarbon 55(4): 1869–1887.
Sanger, David
    1968a   The High River Microblade Industry, Alberta. *Plains Anthropologist* 13: 190–208.
    1968b   Prepared Core and Blade Traditions in the Pacific Northwest. *Arctic Anthropology* 5(1):
            92–120.
    1970   Mid-Latitude Core and Blade Traditions. *Arctic Anthropology* 7(2): 106–114.
Saxberg, Nancy
    2007   *Birch Mountain Resources Ltd. Muskeg Valley Quarry, Historical Resources Mitigation, 2004*
            Field Studies: Final Report (ASA Permit 05-118). 2 vols. Lifeways of Canada Ltd. Copy on
            file, Archaeological Survey, Historic Resources Management Branch, Alberta Culture,
            Edmonton.
Saxberg, Nancy, and Brian O. K. Reeves
    2004   *Birch Mountain Resources Ltd. Muskeg Valley Quarry, Historical Resources Impact*
            Assessment, 2003 Field Studies: Final Report (ASA Permit 03-249). Lifeways of Canada Ltd.
            Copy on file, Archaeological Survey, Historic Resources Management Branch, Alberta
            Culture, Edmonton.
Smith, Jason W.
    1974   The Northeast Asian–Northwest North American Microblade Tradition (Nanamt).
            Journal of Field Archaeology 1(3–4): 347–364.
Taylor, William E., Jr.
    1962   A Distinction Between Blades and Microblades in the American Arctic. *American*
            Antiquity 27(3): 425–426.
Unfreed, Wendy J., Gloria J. Fedirchuk, and Eugene M. Gryba
    2001   *Historical Resources Impact Assessment, True North Energy L.P. Fort Hills Oil Sands Project:*
            Final Report (ASA Permit 00-130). 2 vols. Fedirchuk McCullough and Associates Ltd. Copy
            on file, Archaeological Survey, Historic Resources Management Branch, Alberta Culture,
            Edmonton.
Vang Petersen, Peter
    1984   Chronological and Regional Variation in the Late Mesolithic of Eastern Denmark. *Journal*
            of Danish Archaeology 3: 7–18.
West, Frederick Hadleigh
    1967   The Donnelly Ridge Site and the Definition of an Early Core and Blade Complex in
            Central Alaska. *American Antiquity* 32(3): 360–382.
Wickham, Michelle D.
    2010   A Discussion of Wedge-Shaped Microblade Cores from Two Sites in Northern Alberta.
            Paper presented at the 35th annual meeting of the Archaeological Society of Alberta,
            Calgary, Alberta, 30 April–2 May.
Wickham, Michelle D., and Taylor Graham
    2009   *Historical Resources Impact Mitigation of the TransCanada Pipelines Ltd. Fort McKay*
            Mainline Expansion: Final Report and Post-construction Audit (ASA Permits 06-376 and
            07-266). Bison Historical Services Ltd. Copy on file, Archaeological Survey, Historic
            Resources Management Branch, Alberta Culture, Edmonton.

Wilson, Michael C., John Visser, and Martin P. R. Magne

    2011   Microblade Cores from the Northwestern Plains at High River, Alberta, Canada. *Plains Anthropologist* 56(217): 23–36.

Woywitka, Robin J., and Angela M. Younie

    2008   *Historical Resources Impact Mitigation, Fort Hills Energy Corporation, Fort Hills Oil Sands Project, HiOv-44, HiOv-47, HiOv-49, HiOv-50, HiOv-52, HiOv-87, HiOv-89, HiOv-104, HiOv-115, and HiOv-124: Final Report (ASA Permit 05-328)*. FMA Heritage Resources Consultants Inc. Copy on file, Archaeological Survey, Historic Resources Management Branch, Alberta Culture, Edmonton.

Wyatt, David

    1970   Microblade Attribute Patterning: A Statistical Examination. *Arctic Anthropology* 7(2): 97–105.

Wygal, Brian T.

    2011   The Microblade/Non-Microblade Dichotomy: Climactic Implications, Tookit Variability, and the Role of Tiny Tools in Eastern Beringia. In *From the Yenisei to the Yukon: Interpreting Lithic Assemblage Variability in Late Pleistocene / Early Holocene Beringia,* edited by Ted Goebel and Ian Buvit, pp. 234–254. Texas A&M University Press, College Station.

Yesner, David R., and Georges A. Pearson

    2002   Microblades and Migrations: Ethnic and Economic Models in the Peopling of the Americas. In *Thinking Small: Global Perspectives on Microlithization,* edited by Robert G. Elston and Steven L. Kuhn, pp. 133–161. Archeological Papers of the American Anthropological Association no. 12. American Anthropological Association, Arlington, Virginia.

Younie, Angela M., Raymond J. Le Blanc, and Robin J. Woywitka

    2010   Little Pond: A Microblade and Burin Site in Northeastern Alberta. *Arctic Anthropology* 47(1): 71–92.

4 Archaeological Methods

12 **Quarries** | Investigative Approaches in the Athabasca Oil Sands

GLORIA J. FEDIRCHUK, JENNIFER C. TISCHER, AND
LAURA ROSKOWSKI

Over the past several decades, the Athabasca oil sands region north of Fort McMurray has been shown to contain a remarkably dense concentration of archaeological sites. Starting in the 1990s, an upsurge in the number of proposed oil sands projects in the region produced a dramatic expansion of archaeological activity, most of it in the form of legally mandated historic resources impact assessments (HRIAs) and mitigative excavations, most often conducted by archaeological consultants hired by developers. This rapid escalation of activity has been associated with a certain degree of inconsistency in the investigative methodologies employed not only across time but also by the various consulting companies. It has also created enormous challenges for the Historic Resources Management Branch of the Alberta government, which must approve the methods to be adopted in a proposed field study. Current concerns regarding the cumulative effects of development have, in particular, raised questions regarding the methodologies presently in use. Evaluations of cumulative effects demand attention to contextual considerations that the procedures currently governing archaeological assessment and mitigation largely fail to capture.

Despite the number of assessment studies completed, the number of sites identified, and the number of mitigation programs conducted, it is clear that our understanding of the precontact use of the region—including the distribution of and relationship among sites, as well as the patterns of cultural development in the area—has not kept pace with commercial development. Consequently, we are ill prepared to evaluate the cumulative impact of development on the archaeological resources in the area. This chapter examines the challenges and shortcomings associated with the currently accepted methodologies and offers suggestions for a new direction in investigation techniques in the region. If we hope to improve our understanding of the precontact occupation of the region and of the cumulative effects of development on the archaeological record, our approach to assessment must enable us to collect data that are directly relevant to research questions, both general and specific.

Many of our current research questions pertain in some way to the complex of sites now known as the Quarry of the Ancestors, located on the east side of the Athabasca River about 50 kilometres north of Fort McMurray. The quarry was discovered in 2003 during an assessment conducted by Lifeways of Canada for Birch Mountain Resources, in connection with the company's Muskeg Valley Quarry project (Saxberg and Reeves 2004). The quarry features two exposures of Beaver River Sandstone (BRS), the dominant lithic material in archaeological assemblages from sites in the Lower Athabasca valley. One local source of this stone, the Beaver Creek Quarry (HgOv-29), was identified in the early 1970s (see Syncrude 1973, 1974; Sims 1974). However, the raw material available at that site proved to be a coarse-grained variety of BRS, whereas the BRS found in artifacts from other archaeological sites in the region is typically of higher quality. In particular, the Cree Burn Lake site (HhOv-16) produced a dense concentration of artifacts predominantly fashioned of much finer-grained BRS, which suggested to researchers that another source of the stone must be present somewhere in the immediate area (Ives and Fenton 1983; Fenton and Ives 1984). With the discovery of the Quarry of the Ancestors, such a source was found. Since then, the quarry—officially designated a Provincial Historic Resource in February 2012—has been the subject of several studies that have sought to determine the nature of the activities that took place at the quarry and its relationship to satellite sites (Tischer and Fedirchuk 2006; Saxberg 2007; see also Saxberg and Robertson, chapter 10 in this volume). Although it is possible that additional sources of fine-grained BRS will be identified, sites that lie in the immediate vicinity of the quarry, as well as sites further afield, may now be understood in the context of the Quarry of the Ancestors.

THE CURRENT APPROACH TO ASSESSMENT

We recognize three main problems with the assessment methodologies presently employed in the Athabasca oil sands region: the fragmented nature of archaeological studies, a preoccupation with numbers, and the nature of the predictive models used to guide field studies. The Quarry of the Ancestors complex, including potentially associated outlier sites, well illustrates the perceived problems with current approaches. A solution to these problems is essential, as they hamper our ability to understand the prehistory of the Athabasca oil sands region and to evaluate the cumulative effects of development on the archaeological record.

The Fragmented Nature of Assessment Studies

The first problem associated with the current approach is the fragmented nature of development-related assessment studies and of the body of data that results. As we have seen, as both the pace and the scale of oil sands development have escalated, so has the number of archaeological assessments conducted in the region. However, because the need for an HRIA arises only when a particular area is proposed for development, and because these areas are generally not contiguous, archaeological investigations are sporadic and scattered. In addition, these studies differ significantly in scale and investigative approach. Although the oil sands surface mines are often the subject of studies that cover a relatively broad expanse of terrain (given the large areas proposed for development), studies in the surrounding areas are often quite confined, undertaken only within the area of a proposed pipeline or road or transmission line. At the same time, these circumscribed areas—even though they may have relatively low archaeological potential—are often subjected to more intense scrutiny than are larger areas, such as those associated with a mine.

Fragmented assessment may also lead to the misidentification of site boundaries, generally resulting from the incomplete assessment of sites, which may in fact extend beyond the footprint of a particular project. Because testing is typically not conducted outside the project area, site boundaries are not always accurately delimited. Multiple assessments can thus result in an inflated number of recorded sites, often adjacent to or within close proximity of one another, with a new site defined in each study. Upon more intensive investigation, however, multiple sites may be found to represent a single large site or site complex. For example, the Cree Burn Lake site, HhOv-16, was originally recorded as a number of individual sites, which were eventually shown to coalesce into a single cultural entity. Inconsistencies in the cataloging of contiguous sites recorded under different study permits (including both the various catalogues themselves and the various methods used in cataloguing) also create real challenges in comparing and combining data from adjacent sites. These inconsistencies can obscure the true significance of particular sites, which may in turn pose problems for historic resources site management and cumulative effects assessment.

The Preoccupation with Numbers

The second issue is the current mitigation philosophy, which is directly relevant to investigations at or near the Quarry of the Ancestors. As it stands, the methodological emphasis falls on maximizing artifact recovery by excavating only those portions of a site that exhibit the highest density of artifacts. While this approach may work well enough at relatively small sites, which have a clear focal point of activity, the Quarry of the Ancestors is not a discrete site but a complex of sites. Although archaeological investigations of the quarry to date are relatively limited, it is assumed that the site complex includes not just stone extraction locales but also workshops, campsites, and hunting areas. Together, these sites may illustrate patterns of human activity that differ from those seen at other, more isolated archaeological sites in the area.

The assumption underlying the current approach appears to be that the greater the number of artifacts recovered, the greater the amount of information available, and hence the greater our knowledge. The inevitable result, however, has been the recovery of large volumes of manufacturing debris but with little contextual data. In the case of a complex site such as the Quatry of the Ancestors, excavating only areas that are predicted to contain a high density of artifacts typically results in the collection of only workshop-related data, simply because workshop activity produces a large volume of artifacts. The current focus in archaeological management on the areas of greatest artifact return forecloses opportunities to investigate peripheral areas associated, for example, with campsites—areas that, despite yielding fewer artifacts, may in fact provide information that has greater interpretive force,. If we hope to understand the relationships among quarries, workshops, and campsites, to say nothing of broader social and economic relationships, we need an approach that encourages the recovery of information on all identifiable aspects of precontact occupation, regardless of "artifact return."

A preoccupation with the recovery of large numbers of physical artifacts skews archaeological investigations in the oil sands area in other ways as well. Because boreal forest conditions are not conducive to the preservation of organic material, including datable faunal remains collected from hearths, and because natural processes such as tree throws and root action tend to obscure or disrupt organic remains, faunal materials are rarely recovered. However, although this is not a frequent occurrence, organic remains are known to be preserved at some boreal forest sites. Within the past decade, excavations in the Fort Hills (Woywitka et al. 2009), in areas north of the Ells River (Boland, Brenner, and Tischer 2009; Kjorlien, Mann, and Tischer 2009; Youell et al. 2009), and at sites to the north of the Quarry of the Ancestors (Roskowski, Landals, and Blower

440   Alberta's Lower Athabasca Basin

2008; Woywitka et al. 2008; Roskowski and Netzel 2011a, 2011b, 2012, 2015a; Bryant, Dalmer, and Balls 2011; Turney 2013) have produced datable material associated with hearths at sites interpreted to be campsites. A more careful and thorough approach to excavation would no doubt enhance our chances of discovering organic remains elsewhere. Further, because sites that offer clear stratigraphy are rarely encountered in boreal forest settings, little emphasis has been placed on identifying potential stratigraphic levels in oil sands sites, on the assumption that no such levels exist. The use of more refined excavation techniques might, however, allow for greater differentiation of stratigraphic levels, particularly in those portions of sites with less dense deposits of lithic artifacts.

A multitude of sites in the Athabasca oil sands region have already undergone mitigation using a methodology that privileges the excavation of areas perceived to contain the densest concentrations of artifacts. However, a different investigative approach could be applied in future mitigation efforts, while intact sites still exist in the region and are thus available for study.

The Nature of Predictive Models

Predictive models have been used to guide HRIAs and baseline studies for many large-scale oil sands projects. Such models—whether they are statistical in nature, based on a researcher's prior field experience and/or knowledge, or founded on a combination of scientific information and intuition—tend to reflect prevailing assumptions about where precontact archaeological sites are most likely to occur. Well-drained, elevated landforms are generally perceived to be of high potential, whereas saturated terrain is generally regarded as of low potential. Up to a point, predictive models founded on these assumptions work reasonably well. As Ives (1993, 20) noted, testing well-drained landforms near permanent watercourses for archaeological remains amounts to something of a "self-fulfilling prophecy": the best sources of food and wood for fires are typically located on well-drained landforms, and these landforms would have served as campsites throughout the year. However, while the current models are useful for identifying locales attractive for human habitation, they tend to overlook other considerations that might have influenced precontact use, such as the accessibility of lithic materials.

Although Light Detection and Ranging (LiDAR) imaging is increasingly used to map terrain, with a view to identifying areas of high archaeological potential, predictive models of the sort most commonly employed over the past decade or so generate maps through Geographic Information Systems (GIS) analysis by integrating data of various kinds about a particular expanse of terrain. The databases on which these models rely typically include information about vegetation, elevation, aspect, slope, and proximity to perennial water bodies. Of direct relevance to the area around the Quarry of the Ancestors complex was a model developed for Shell Canada's Muskeg River Mine Expansion project (Tischer 2004). Given the results of subsequent excavations at the site, such models offer a useful illustration of the limitations of current approaches to assessment.

The Quarry of the Ancestors

Prior to its designation as a Provincial Historic Resource, the Quarry of the Ancestors was protected under a Protective Notation (PNT) filed with Alberta Environment and Sustainable Resource Development. The Quarry of the Ancestors lies east of the Athabasca River and southeast of the Muskeg River, in an area characterized by three large north-south trending wetlands along the south and west boundaries. Black spruce, tamarack, and open areas dominated by willow are typical of the western wetland, while the remaining two wetlands are dominated by fens, crossed by beaver dams, that support grasses and sedges growing among large areas of open water (fig 12.1). Aspen and spruce occur on the uplands adjacent to the well-defined, aspen-covered margins of the fens. Generally featureless sandy uplands dominated by aspen and occasional spruce stretch to the north of the fens (fig 12.2). Along the western border of the quarry, the terrain deteriorates into a bog dominated by black spruce and tamarack; bog deposits generally consist of thick organics overlying sand, but wet clay is occasionally present below the organic layer. A small lake dominated by willow wetlands lies in the northwestern corner of the PNT. Complex terrain consisting of sandy ridges and knolls interspersed with poorly defined creeks, drainages, and areas of muskeg characterizes the eastern border of the quarry (fig 12.3).

Archaeological potential. A GIS predictive model of the archaeological potential of the area that includes the Quarry of the Ancestors was originally developed by archaeologists with the consulting firm of Golder Associates as part of an HRIA for Shell Canada's Jackpine Mine (Clarke 2002). This model was modified in 2004 for the HRIA carried out by FMA Heritage Resources Consultants Inc. in connection with Shell's Muskeg River Mine Expansion project (Tischer 2004). The model drew on available databases and incorporated information about vegetation, soil complexes, aspect, slope, and proximity to perennial water sources. To take into account the non-perennial water sources that occur in the Muskeg River Mine Expansion area, the weighting used by Golder for the Jackpine Mine, in which the

Figure 12.1. View northwest into the general region in which the Quarry of the Ancestors lies. In the foreground is the eastern wetland, which gives way to featureless aspen terrain to the north.

primary variable was proximity to water, was changed in the 2004 model to one in which vegetation (as a reflection of drainage) was the most heavily weighted variable. Overall, the archaeological potential illustrated in this 2004 model was subsequently considered to have been overestimated, in view of the results of the 2004 Muskeg River Mine Expansion study, and the model was subsequently altered slightly for use in 2005 archaeological studies. This adjusted model, overlain on the PNT boundaries as proposed in 2003 and in 2005, is shown in figure 12.4, with the archaeological site boundaries as established in 2005.

Modelling at this scale, which reflects the limitations of the available databases, does not capture small geographic features, and the archaeological potential is therefore visible only at a coarse level. Large expanses of generally well-drained areas in the Quarry of the Ancestors were thus classified as having moderate to high potential (see the pink areas in fig 12.4), while large blocks representing low-lying saturated areas were ranked as having low potential (the grey areas). However, a comparison of this model with aerial photographs indicated that the map did not correlate very well with the reality of the local terrain and drainage (which is one reason that LiDAR imaging has become an increasingly popular method of identifying landforms). For example, small terrain features that might have been attractive as sites for human habitation were often not illustrated in the map generated by the GIS model, which limited its utility as a predictor of archaeological value.

Figure 12.2. View to the northeast, along Canterra Road, showing the featureless, aspen-dominated terrain characteristic of the northern portion of the study area

In 2005, Shell Canada, the leaseholder for the area, engaged Lifeways of Canada to produce a map illustrating the predicted archaeological potential specifically for the Quarry of the Ancestors. Shell's request for an updated map was prompted by the results of the 2003 Muskeg Valley Quarry HRIA (Saxberg and Reeves 2004) that Lifeways had conducted for Birch Mountain Resources, during which the Quarry of the Ancestors was discovered. The Lifeways map (fig 12.5) was based on that HRIA and on additional shovel testing conducted during follow-up field studies in 2004, which aimed to further delineate site boundaries (Saxberg 2007). The map identified three areas of high value, outlined in red in figure 12.5. The area of high value located in the northwest of the quarry generally corresponds to the boundary of HhOv-305, although the eastern portion of the final site area was excluded from the high-value area. Similarly, the southeast area of high value roughly corresponds to the boundaries of HhOv-319, although the northern portion of the final site area was not included in the high-value area. These areas of high value were separated by an area of medium value (outlined in blue), where limited shovel testing had been completed (Saxberg 2007; Saxberg and Reeves 2004).

Figure 12.3. View southeast, along the eastern margin of the Quarry of the Ancestors site, showing complex terrain characterized by knolls surrounded by low terrain

As described in the report on the 2003 HRIA for Birch Mountain Resources (Saxberg and Reeves 2004), the area covered by site HhOv-305 is topographically diverse but includes a large, muskeg-covered landform, which the report suggested might represent an outcrop of BRS. The report described HhOv-305 as bounded on the west by the edge of this landform and a drop in elevation. The site area extends eastward into better-drained, flat terrain composed of sands and silts, to the north, and hummocky sands and boulder deposits, to the south. The HhOv-305 site area was defined on the basis of positive results from an intensive program of shovel testing, which skirted the edges of the landform on the west and the edges of the wetlands to the south, which intrude into the site area. The eastern boundary appears to have been more a matter of prediction, based on positive results from relatively fewer locations of intensive testing in the flatter terrain. Additional shovel testing in 2004 in the south, north, central, and eastern portions of the site was used to refine the boundaries and to identify additional areas of human activity (Saxberg 2007).

Site HhOv-319 lies generally southeast of HhOv-305. A small possible outcrop of BRS was identified adjacent to the northeast corner of a wetland area that

Figure 12.4. The Quarry of the Ancestors: FMA Heritage model of archaeological potential (2004), with areas of medium to high value shaded pink

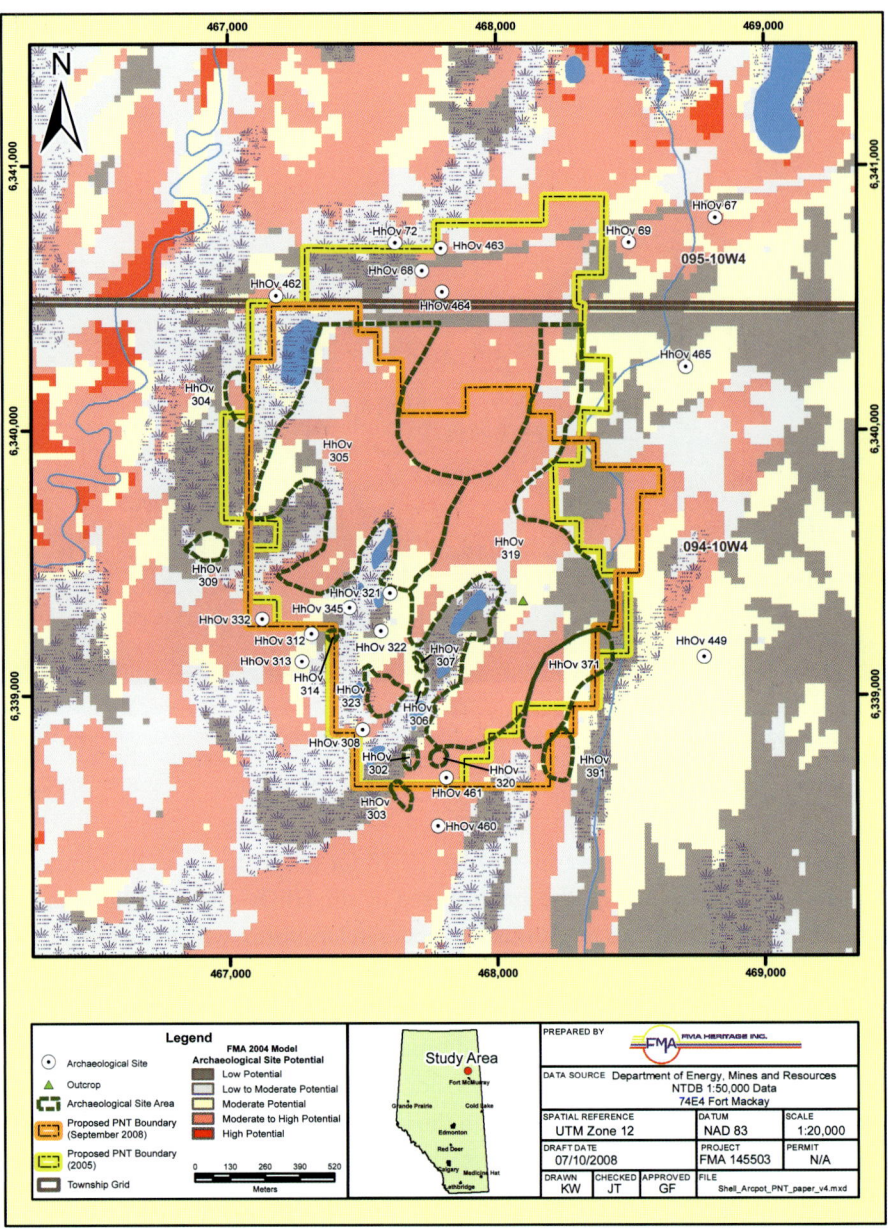

intrudes into the central portion of the site from the south. Elevation drops into wetter spruce forest, to the north, and into marsh and bog, to the south and east, generally defining the landform on which site HhOv-319 is located. Gently rolling terrain consisting of sandy ridges and knolls characterizes the area further

Alberta's Lower Athabasca Basin

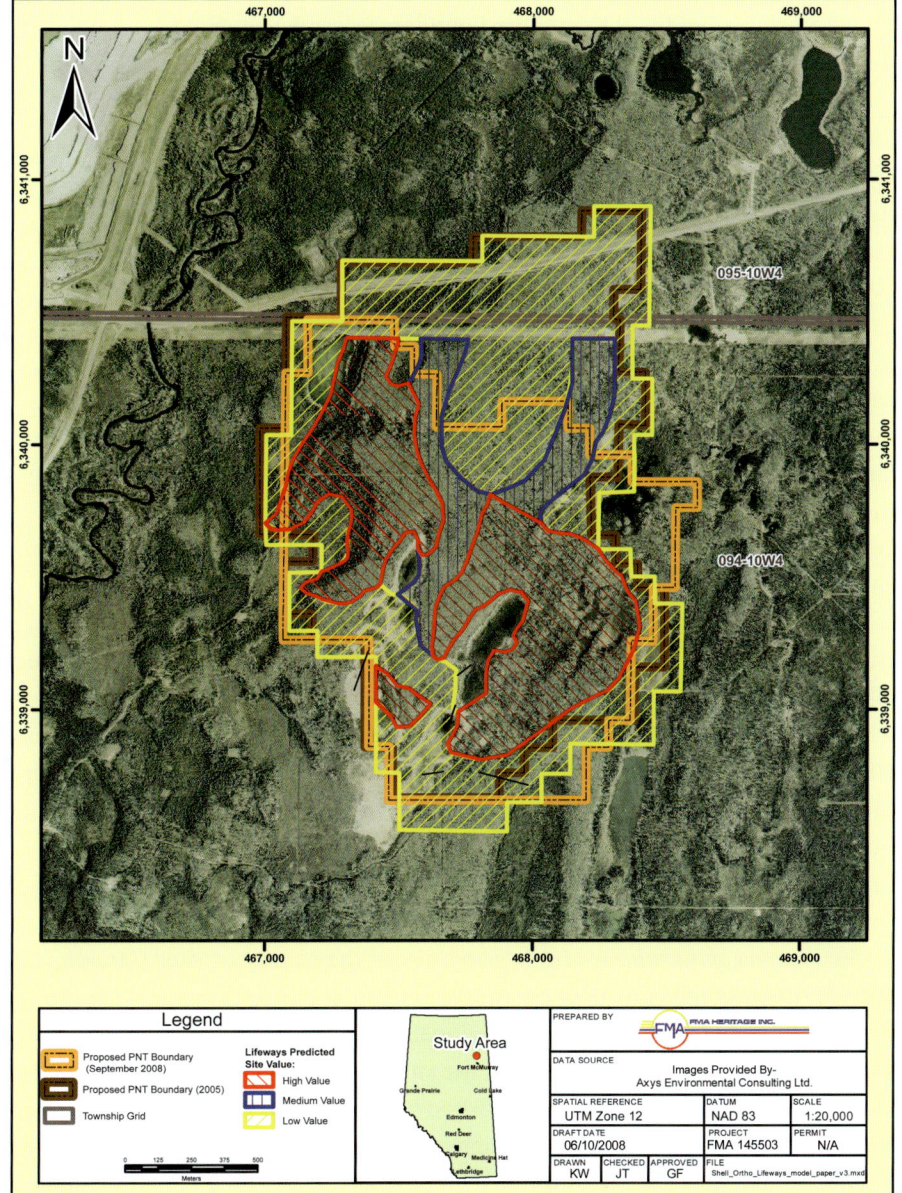

Figure 12.5. Revised Lifeways model of archaeological potential (2005), showing areas of high (red), medium (blue), and low (yellow) value

south, on the east side of the wetland. The final boundaries of the site HhOv-319 (see fig 12.4) are based more on prediction than on positive field results. Limited transects were excavated in 2003 along the edges of wetlands, which were assumed to mark the boundaries of the site in that area. Given the results of the shovel-testing program, the southern boundary is less well defined, although the frequency of artifacts does appear to diminish between HhOv-319 and sites to the south. The northern site boundary reflects a predicted presence of artifacts, rather than solid evidence of their existence. The 2005 map prepared by Lifeways excluded this part of HhOv-319 from the high-value area (see fig 12.5); the shovel-testing program in 2003 did not extend to the edges of the final site area as illustrated in figure 12.4, and additional testing to the north, conducted in 2004, yielded few positive results. In short, the final accepted boundaries of the site largely conform to the extent of the elevated landform.

The third area of high value illustrated on the 2005 map contains HhOv-323, a smaller site located south and west of the two large sites, HhOv-305 and HhOv-319. Site HhOv-323 is located on a sand and silt peninsula that extends into a wetland area. Although the assumed natural southern boundaries of this site (that is, the southern edges of the better-drained landform) were tested and found to contain cultural materials, the borders of the high-value area associated with this site were again mainly based on the extent of the elevated landform.

The area of medium value (outlined in blue in fig 12.5) is generally Y-shaped, with the base of the Y situated between the high value areas associated with sites HhOv-305 and HhOv-319. One arm of the Y flanks the east side of HhOv-305, whereas the other extends north and east and represents the northern portion of HhOv-319. Lifeways presumably determined this area to be of medium value on the basis of the number of positive shovel tests in 2003 in comparison to the number within the high-value areas. The remainder of the proposed PNT was ranked as low value (outlined in yellow in fig 12.5).

Verification of the value model. In 2005, also at the request of Shell Canada, FMA Heritage completed an intensive sampling program in the area of the proposed PNT, generally aimed at confirming the value boundaries suggested by the Lifeways of Canada map. Specifically, the study was charged with confirming the boundaries associated with the established high-value areas, verifying the boundaries of areas provisionally designated as of medium value, and providing recommendations based on additional low-density surveys of all the remaining areas within the PNT (Tischer and Fedirchuk 2006). The results of both the 2003 and 2004 Lifeways of Canada studies (Saxberg and Reeves 2004; Saxberg 2007) were available at the time of the 2005 study conducted by FMA Heritage.

To facilitate field investigations, the PNT was divided into forty-four discrete sampling locations that focused on areas in which gaps in the data were perceived to exist. These gaps were identified primarily on the basis of a review of the 2003 and 2004 Lifeways of Canada studies and consisted primarily of those areas in which limited testing had been conducted, particularly in the interior of landforms. To address the gaps, field sampling was completed along approximately 130 transects spaced 10 to 20 metres apart, within which a total of 1,370 shovel tests were carried out. Shovel tests were executed at 5- to 30-metre intervals (most commonly at 15-metre intervals) within each transect, with anywhere from 2 to 30 shovel tests per transect (fig 12.6). Although tests were conducted even when poorly drained terrain was encountered along a transect, the presence of standing water sometimes required a deviation from the intended path of the transect (see Tischer and Fedirchuk 2006, 39–41). This sampling program was consistent with the methodology previously employed by Lifeways of Canada in this area.

In the course of the 2005 study, it became apparent that the two models of archaeological potential developed for the study area (both the original Muskeg River Mine Expansion model prepared by FMA Heritage and the map developed by Lifeways of Canada) were inadequate for assessing the extent of the Quarry of the Ancestors cultural deposits. During the field reconnaissance, subsurface testing indicated that the sediments in certain areas were often saturated: many of the test sites filled with water during excavation (see Tischer and Fedirchuk 2006, 57–58, 67). Although 2005 was a particularly rainy year, the vegetation and soil conditions observed confirmed that many of these areas are continually saturated. Such water-logged areas had previously been assumed to be of low archaeological potential; however, numerous shovel tests conducted in areas initially deemed to be of low value yielded dense artifact concentrations, including tools. This resulted in the identification of new sites on the periphery of the 2005 PNT study area and also extended the boundaries of previously identified sites such as HhOv-305. Moreover, the additional shovel testing completed in these low-value areas revealed that several of the previously recorded sites were not, in fact, discrete localities. Testing of low-value areas between sites HhOv-304, HhOv-305, HhOv-319, and HhOv-323 indicated that cultural material is continuous among these sites and that the entire area is therefore best considered a single archaeological entity (see Tischer and Fedirchuk 2006, 89–91, 101–103). Without this testing of areas not predicted to be of high potential, these new sites, as well as the continuous distribution of artifacts between sites, might never have been recorded.

The results of the 2005 PNT study confirmed that areas of saturated terrain, formerly regarded as having low value, actually contained substantial archaeological materials. In addition, although shovel testing of high-value areas carried out in the 2005 FMA Heritage study did result in a large number of positive shovel tests, some portions of this high-value area, particularly in the northern

Figure 12.6. Shovel-testing program conducted in 2005 at the Quarry of the Ancestors

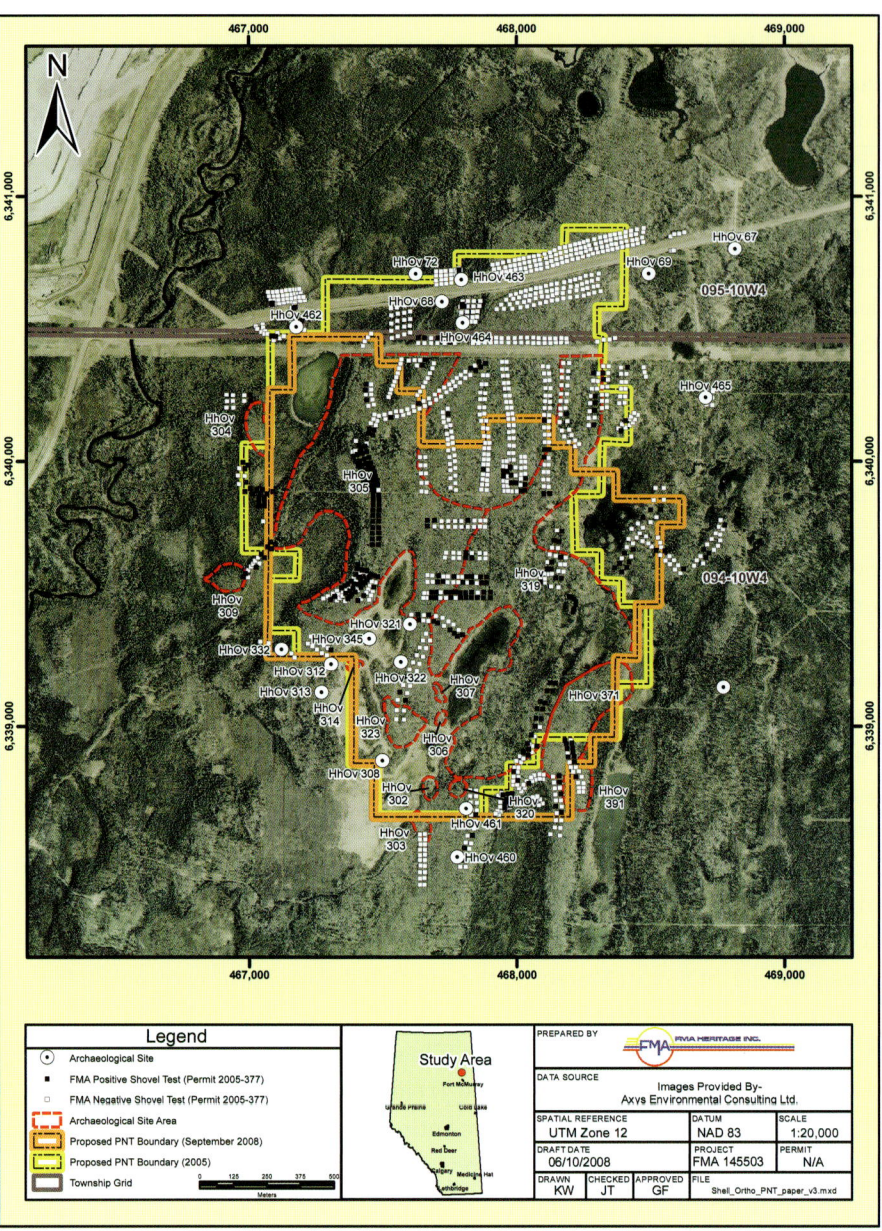

portion of sites HhOv-305 and HhOv-319, yielded a significant number of negative shovel tests (see fig 12.6). Such results suggest that the location of archaeological sites is influenced by other factors, in addition to terrain, which must be taken into consideration in determining archaeological potential.

Discussion. Both the 2003 Muskeg Valley Quarry study (Saxberg and Reeves 2004) and the 2005 PNT study (Tischer and Fedirchuk 2006) noted an unusually high percentage of positive shovel tests at the Quarry of the Ancestors in comparison to the results of other studies in northern Alberta. This observation is in all likelihood a direct reflection of the site type, that is, a site complex associated with the procuring and processing of a lithic resource. The method of extraction and the need to reduce and transport the extracted stone mean that quarry locales are characterized by the proliferation of lithic debris over a wide area. Given that the quality of BRS in and around the Quarry of the Ancestors varies, the selection of material for use was presumably preceded by significant percussive testing of the raw material, so as to identify promising locations for procurement. The extracted material would then have to be reduced to sizes suitable for transport. Activities associated with habitation would have resulted in the additional deposition of cultural materials both within and on the periphery of the quarry locales. We would thus expect the dimensions of the site complex to be extensive, and an intensive shovel-testing program would identify this debris, producing a high proportion of positive tests. The results of the two studies were thus consistent with these expectations.

Clearly, however, the location of the Quarry the Ancestors complex is not a function of features such as habitable terrain or proximity to water that are typically perceived as indications of high archaeological potential. Rather, it was the occurrence, by deposition and/or exposure, of a source of lithic raw material (namely, Beaver River Sandstone) that attracted precontact peoples to this area. However, current models of archaeological potential do not typically incorporate information about bedrock topography; among other things, the data that are readily available to the public tend to be insufficiently refined to be useful in such models. As a result, it is difficult to predict the presence of quarry sites on the basis of such models.

Relative to the predictive model, the field results in the Quarry of the Ancestors generally supported the ranking of areas such as ridges and the edges of creeks and lakes as high potential. What was unexpected was the discovery of cultural materials in areas considered to be of low archaeological potential, such as water-saturated black-spruce bogs (Tischer and Fedirchuk 2006, 69). Although these discoveries undoubtedly relate to the presence of lithic source materials in the area, the results also raise questions concerning local climatological and togographic conditions at the time of precontact use. Was the area drier in the past than it is today? If so, was the precontact use of the area a result of an aberrant situation involving short-term changes in temperature and/or rainfall, or was there a protracted period of drier conditions? Are the modern landscape and drainage patterns comparable to those at the time of site use, and how far can modern conditions reliably be used to predict precontact terrain use?

For example, site HhOv-449 is now situated in a bog just to the southeast of the Quarry of the Ancestors (see the lower right quadrant of fig 12.4). This site was identified during an HRIA completed for Shell Canada's Muskeg River Mine Expansion project (Tischer 2005), in the course of a post-impact assessment of an access road that was conducted as part of the study. The site is located in a bog characterized by black spruce in level terrain and lies 200 metres from the nearest perennial water source. Although in the 2004 model (Tischer 2004) the archaeological potential of the site was designated moderate on the basis of its elevation, overall the area was perceived, both in air photos and during field visits, to be of low potential. However, despite the fact that the area in which HhOv-449 is located boasts no features suggestive of high archaeological potential, the site proved to contain artifacts of a specialized nature, including a chert microblade core (Wickham and Graham 2009). One possible explanation for the existence of such a site in the middle of a bog is that the modern landscape is not a reliable guide to the ancient landscape. It can be assumed that drainage and terrain features would have differed significantly during various climatic periods. Changes in the landscape, such as an increase in the amount of drainage or the loss of trees as the result of a drier climate, would have significantly altered the attractiveness of the location. What is today a bog might once have been less saturated terrain, suitable for human habitation. An example like this confirms that the use of modern terrain features is not always useful in assessing archaeological potential.

To summarize, although the utility of current predictive models is limited in part by inadequately refined scientific databases, these models are hampered more significantly by a poor understanding of the full spectrum of past activities, including the cultural factors that influenced the choice of sites, and of past local environmental conditions and patterns of landscape use. Although predictive models may serve to guide initial investigations, the methodology must include testing the accuracy of these models by assessing selected areas considered to be of low potential, as well as by employing adaptive management studies to examine areas of differing archaeological potential (low, medium, and high) after surface vegetation has been cleared. The identification of a highly significant site such as the Quarry of the Ancestors complex has a major impact on the nature of archaeological investigations carried out within the Athabasca oil sands region, and it is crucial that we be able to detect the existence of such sites promptly.

The following section explores a different approach to understanding the nature of quarry sites and associated satellite sites. This approach suggests that historic resource investigations in the Athabasca oil sands area would be better served by assessing a quarry complex as a cultural whole rather than as a series of discrete sites. We need to view the Quarry of the Ancestors as a cultural nexus, to which all sites containing BRS are potentially related, and we need to employ more consistent and more rigorous mitigation techniques. Such changes in methodology would help us to answer broader questions regarding habitation, modes of subsistence, trade, and social relationships—answers that might well be relevant not only to the Athabasca oil sands area but all across the boreal region and perhaps even further afield.

CONTEXT AND INTERPRETATION: SHIFTING THE FOCUS

Most archaeological work in the Athabasca oil sands region is driven by a short-term need to decide whether a given archaeological site is important and to evaluate the consequences of its loss, rather than by research questions designed to enhance our knowledge base and, by extension, our ability to interpret the evidence and assign site value more accurately. Given the problems identified above with the current assessment approach, a more academically oriented approach might allow us better to understand the precontact patterns of social and economic use in the region. In an effort to identify the types of research questions, both general and specific, that need to be addressed, the following discussion presents some ideas regarding the nature of significant sites and the cultural factors that influence not only the archaeological record itself but also our interpretation of past lifeways that are relevant to the precontact history of the Athabasca oil sands region. The identification of the Quarry of the Ancestors site complex provides us with an opportunity to explore alternative assessment methodologies that could prove valuable in investigating a site type about which little information exists.

The archaeological record clearly demonstrates that, for well over 10,000 years, the indigenous inhabitants of North America were dependent on good workable stone material for the manufacture of tools and weapons. If raw lithic material was not available nearby, where it could be acquired in the course of the normal seasonal round, it had to be imported from areas beyond traditional local territories. On the basis of the somewhat limited information presently available, it appears that the Quarry of the Ancestors was primarily a site at which lithic material was quarried and reduced, with few discernible habitation areas (thus far) and with proportionally few finished tools left on site. Although habitation sites, whether temporary or long-term, must have existed within or very near the quarry, the site seems to have served principally as a source of BRS, representing a focal point of activity in a cultural network that linked it to many other sites in the region and even to areas beyond what is now northeastern Alberta.

With respect to the role of quarries in precontact cultures, three factors are of key importance: reduction technologies, economic interaction, and social organization. Reduction technologies include the extraction or selection of workable pieces of stone, the reduction of these nodules to bifacial blanks, and, to some degree, the subsequent manufacture of finished tools. Economic interaction is visible in the distribution of lithic materials across a broad area, as a function of transport and exchange mechanisms. Social organization is, however, a matter of inference. Our understanding of precontact cultures is grounded in interpretive explanations, based on the physical evidence, that help to account for observed patterns.

Reduction Technologies at Quarries

Despite the obvious importance of stone implements to human history, quarries have been the subject of relatively few detailed archaeological investigations. Sims (1974, iv) suggests that "the complexity of stone quarry deposits may be one reason for the dearth of archaeological studies concerning them." Similarly, Ericson (1984, 2) argues that the tendency to neglect quarries "most likely is the result of technical and methodological limitations imposed by a shattered, overlapping, sometimes shallow, nondiagnostic, undatable, unattractive, redundant, and at times voluminous material record." He goes on to comment: "When we consider the wealth of information on the varieties of human experience, our information on the activities at quarries and workshops ranks among the most abysmal" (1984, 8). This neglect is unfortunate. As Deal (2001, para. 7) observes, quarries and related workshop sites "can be viewed as the initial stages of a tightly integrated system that involves the selection, modification, distribution and consumption of lithic materials" and "can provide valuable information on quarrying procedures, tools and strategies for initial lithic reduction, and a wealth of analyzable debitage." In other words, beyond enhancing our understanding of reduction technologies, the study of quarries can yield insights into economic interaction and social organization.

The definition of quarries. When defining the term *quarry* in an archaeological context, one must first consider the nature of the source of raw material. In the literature on quarries, the specific type of locality under discussion varies widely, with the result that the term *quarry* can refer to a concentration of loose pieces of bedrock (known as "float") deposited in a particular area by glacial action, or to an outcrop of in situ bedrock, or to a vast territory of minable land in which numerous quarrying sites exist. In a discussion of the Hatch jasper quarry (located in central Pennsylvania), Andrews, Murtha, and Scheetz (2004, 63) distinguish between "prospects," or sources where tool stone exists "as surface 'float' material," and "formal quarries," which are characterized by quarry pits "indicating the prehistoric exposure of tool stone in primary bedrock contexts." As the term *prospect* implies, because stone was available on the surface, little effort would be required to identify and collect the raw material. In contrast, exploiting bedrock sources of stone would have involved significantly more time and organized effort. Overburden might need to be removed in order to access the bedrock source, specialized quarrying tools might need to be manufactured, and arrangements for food and shelter would need to be made to support the miners during extended quarrying activities. Thus, even though essentially the same activity—the procurement of lithic materials—is carried out at both prospects and quarry pits, the form in which a tool stone occurs influences the technologies employed during reduction, the length of stay required to extract raw material, and the number of knappers required to obtain a sufficient amount of raw material.

In addition, in the case of quarry sites that extend over a relatively large geographical area, the placement and size of individual quarrying locations, as well as the relationship of workshop sites to the central quarry area, are important aspects of site structure that need to be researched and documented. The Quarry of the Ancestors, for example, covers an area of approximately 200 hectares, with the boundaries of the original PNT defined by the extent of significant amounts of debitage. Within the main quarry complex, a number of individual sites have been delimited on the basis of artifact distributions observed during the HRIAs (Saxberg and Reeves 2004; Saxberg 2007). It is expected that further studies would result in the identification of additional areas of activity, which could be related to extraction or to subsequent lithic reduction or even to habitation. More intensive investigations, including detailed excavations in all areas of the complex, not merely in those exhibiting the densest concentrations of artifacts, could help us to isolate specific areas of activity and to identify the nature of the activities occurring at these different locales. The identification of such areas would also help to clarify the size of the area devoted to BRS extraction relative to the area occupied by the associated workshops and campsites.

Lithic resource procurement and reduction. In the Athabasca oil sands region, many lithic material types, including BRS, were readily accessible to precontact populations in the form of pebbles or cobbles found in lag gravel deposits observable on the ground or along stream beds. However, BRS also occurred either as float blocks—large pieces of stone that had become detached from the underlying bedrock formation—or as outcrops of in situ bedrock. Float blocks could have broken away from the bedrock and been transported to points further north during the massive flooding of Glacial Lake Agassiz that took place approximately 9,800 to 9,600 BP (Fisher and Smith 1994; see also Fisher and Lowell, chapter 2 in this volume). Such blocks of stone could occur as surficial boulders, or they could have been buried by fluvial deposits during the catastrophic flood event.

Within the Athabasca oil sands region, BRS was available in relatively small quantities from easily accessible locations such as gravel deposits along watercourses (see Gryba 2001), as well as from bedrock exposures at the Beaver River Quarry (HgOv-29), located on the west side of the Athabasca River to the south of Fort McKay. A significant amount of the BRS used in the Athabasca oil sands region, however, was probably extracted from the exposures of the stone at the Quarry of the Ancestors. Although extracting the stone from large blocks (whether attached to the bedrock or not) would have been considerably more labour intensive than obtaining the material from fluvial gravels, the quarry would obviously have provided a more reliable source of stone, of better quality and in greater quantities.

At the Quarry of the Ancestors, quarrying implements have not been explicitly identified as such. However, Saxberg (2007) notes the presence of hammerstones, anvils, and wedges from within the eastern half of the quarry (at HhOv-204 and HhOv-305), and it is possible that these tools were used not only for lithic reduction but for extraction as well. More extensive and more detailed excavations at sites within the quarry are likely to provide information regarding extraction techniques, which might have differed at specific locales within the quarry, depending on the quality and geological character of the stone. Extraction techniques might also be expected to vary relative to different occupations of the quarry by different cultural traditions.

Given that the Quarry of the Ancestors is now a provincially protected site, investigations will no longer be conducted by archaeological consulting firms. Thus, any additional research aimed at identifying the character of the sources of BRS available within the quarry, the types of technology employed, the time period during which the quarry was in active use and the cultural groups who used it, and the social and economic networks associated with quarry use is left to archaeologists at academic institutions or in government. In view of the costs associated with conducting research, however, it is unknown when, or if, further detailed studies will be undertaken at the quarry itself. In such circumstances, other avenues of investigation assume greater importance.

Ongoing development in the areas surrounding the Quarry of the Ancestors has led to the identification and excavation of a number of archaeological sites that could provide insight into the activities occurring at the quarry itself. In particular, a comparison of reduction techniques and associated waste material at workshops located within close proximity to the quarry (that is, at satellite sites) may be useful in answering questions regarding quarry access and use. Writing on lithic production systems, Ericson (1984) suggests that, out of the total production weight, typical quarry assemblages contain over 90% debitage, while, in a study of quarries in the Great Basin, Beck et al. (2002) cite figures as high as 98% debitage. These percentages compare well with ratios of debitage to tools in some of the satellite sites adjacent to the Quarry of the Ancestors, which have generally yielded between 96% and 100% debitage (Clarke and Ronaghan 2000; Green et al. 2006; Roskowski and Blower 2009; Roskowski, Landals, and Blower 2008; Roskowski and Netzel 2011b, 2015a; Tischer 2008; Woywitka et al. 2008). As would be expected, sites located further from the sources of BRS typically contain higher percentages of formed tools and fewer artifacts manufactured of BRS. The size of individual items of BRS debitage is also reduced (Kjorlien, Mann, and Tischer 2009; Roskowski and Netzel 2015b; Woywitka et al. 2009), suggesting that later stages of tool production occurred in these locations and implying that primary reduction took place at or very near the source of the raw material.

Given that most of the artifacts recovered from sites at the Quarry of the Ancestors are the by-products of lithic reduction (that is, debitage), questions naturally arise about the reduction process that was employed, how many tools were produced, and to what extent the tools were finished at or near the quarry. A careful analysis of this debitage is therefore critical. However, owing to the nature of archaeological consulting work, the analysis of debitage from sites that may be associated with the quarry has been somewhat limited. The required timelines for report submission and the cost of intensive debitage analysis dictate that bulk analysis must be conducted. Although informative at a general level, bulk analysis does not generally yield detailed information of the sort that can be obtained when each flake is analyzed individually—information about the stage of reduction, for example, and about the manufacturing processes employed for specific tool types. In contrast, when detailed analysis is conducted using the reduction sequences developed by Carr and McLearen (2005), the stages of reduction (early to late) can be identified even from seemingly uninformative piles of debitage. Sites within the Quarry of the Ancestors are generally unstratified and contain multiple, intermixed components, which makes it difficult to isolate the materials associated with a specific occupation. A detailed analysis of debitage could therefore prove particularly productive at sites located just outside of the quarry, which may represent only a single occupation.

Determining the degree of reduction that occurred at the quarry itself, in comparison to the reduction that took place both at nearby satellite sites and at sites increasingly distant from the main quarry, could potentially yield insights into possible restrictions on access to the quarry, as well as into the transportation routes and patterns of distribution of the raw material after extraction. Such an analysis could also allow us to make certain inferences about how indigenous groups used the site. For example, if a high degree of later-stage core reduction and tool production is identified at locales within the quarry itself, this would tend to indicate ongoing use of the site by long-term inhabitants. In contrast, if mainly early-stage core reduction and biface production is identified at satellite sites, this could indicate that reduction generally occurred further from the quarry, which would suggest that groups travelled to the area to acquire tool stone but then moved on relatively quickly.

Economic Interaction

Lithic material was most commonly transported in the form of bifacial blanks, an early stage in the production of formed bifaces. Evidence of the manufacture of bifacial blanks of BRS has been identified at many sites within the Athabasca oil sands region, suggesting transportation of the raw material—although the incidence of BRS decreases markedly beyond an approximately 30-kilometre radius of the Quarry of the Ancestors. At the same time, in addition to enormous quantities of BRS, artifacts made of non-local tool stones, including obsidian and various quartzites, cherts, and chalcedonies, have been recovered from sites within and associated with the Quarry of the Ancestors complex. The occurrence of such non-local materials, as well as of a variety of diagnostic projectile points suggestive of external influences, points to the presence of people from distant areas. While they were in the vicinity of the quarry, these travellers may have obtained BRS, in which case we would expect it to appear in relatively small quantities at other, more distant sites. Tracing non-local tool stones to their original sources would help us to identify the geographic areas in which precontact travellers originated. Conversely, assessing the extent to which BRS occurs in archaeological contexts outside the Lower Athabasca region might shed light on the significance of the Quarry of the Ancestors—on whether, for example, the availability of BRS was a major factor drawing people into the region or whether groups were in the area for other reasons and acquired BRS mainly because it was convenient.

Variability in material quality. Of importance in tracing economic interactions on the basis of the archaeological record is our ability to recognize specific lithic materials. For example, one of the best understood, and geographically extensive, archaeological trade networks is associated with Knife River flint, which originates in North Dakota but is recognized in archaeological assemblages from western Montana to western Pennsylvania and from Saskatchewan to northern New Mexico (Billick 1998). Knife River flint has distinctive characteristics and has been extensively described in the literature; thus, it is readily identifiable in archaeological site assemblages.

Other lithic materials are less easy to recognize and can therefore escape identification, particularly if the analyst is unfamiliar with the specific lithic type. In addition, tool stones that have been subject to similar geological processes may be similar in appearance despite originating in different areas. As Gryba observes in chapter 9 in this volume, the Dakota sandstone found in the Gunnison area of Colorado is visually almost indistinguishable from the coarser varieties of BRS recovered from the Athabasca oil sands region. Ives (1993) also noted that lithic materials similar to BRS have been identified in Montana and the Dakotas. Consequently, when one is conducting lithic analysis, care should be taken to consider all possible lithic sources. Microscopic examination of physical samples of particular stones, for purposes of comparison, together with a greater awareness of the character of specimens housed in collections throughout North America, substantially enhances our ability to pinpoint the occurrence of specific lithic types and thus to trace the routes of early trade networks.

The problem of identifying BRS is further complicated by the fact that the cortex observed on some examples of the stone, including artifacts found during the 2005 PNT study (Tischer and Fedirchuk 2006), consists of a very thick rind that in no way resembles the texture within the nodule. The fine-grained variety of BRS found at the Quarry of the Ancestors, which is also known as Muskeg Valley Microquartzite, is distinguished by its grey color and microcrystalline matrix, which supports somewhat larger, sand-sized quartz particles. The thick cortex observed on some nodules is tan in color, coarse in texture, and often lacking in quartz grains. Reduction of this thick cortex may result in flakes and shatter that, because they contain none of the grey, fine-grained material, are not immediately recognizable as BRS. Similar issues of identification have been noted with regard to the jasper found in Pennsylvania at King's Quarry:

Besides the extent of the prehistoric quarrying, another eye-opening aspect of the study is the variety of jasper present at King's Quarry. Although King's Quarry contains a notable amount of the typical opaque brown, yellowish brown and yellow jasper, some of the material includes other vivid colors as well as varying textures and lusters. Color variations include dark brown, black, gray, mahogany-like reddish brown, maroon, blue, white, buff, and-in rarer instances-deep green. In addition, many specimens are variegated or banded in two or more distinctive colors. The material also varies from opaque examples with a dull luster to translucent examples with shiny, glassy, or waxy luster.

An alarming aspect of this variation is that, if individual patches were flaked off of larger pieces and looked at as single specimens, many would be classified as non-jasper lithic types. For example, some of the translucent material is chalcedony, and some of the dark opaque materials would normally be classified as various cherts or flints. Indeed, there are relatively common examples where a single specimen of rock collected from the site contains a thick, solid mass of black chert/flint on one face and a mass of typical yellowish brown jasper on the opposing side. (Pennsylvania Historical and Museum Commission 2012)

Analogous problems can be anticipated in connection with BRS, which also varies in colour and texture, in addition to possessing a vastly different cortex.

The BRS collected from archaeological sites can vary greatly in texture and appearance, not only between one site and another but even within a single site. This variability in quality has been addressed by various investigators, usually in discussions of the sources of the material (see, for example, De Paoli 2005; Ives and Fenton 1983; Tsang 1998). The topic was first raised by Ives and Fenton (1983, 1985) in connection with studies at the Beaver Creek Quarry (HgOv-29). Since then, research and observations have confirmed that the BRS found in the Athabasca oil sands region varies widely in quality. Gryba (2001) reported, for example, that BRS cobbles found near the Fort Hills ranged from a very fine-grained material, with a texture approaching that of chert, to a very coarse quartzite. He has also noted a significant alteration in the quality of this stone as a result of heat treatment (see chapter 9 in this volume), which may require a rethinking of conclusions regarding the lithic materials observed at archaeological sites.

As is now well established, the BRS from the Beaver Creek Quarry (HgOv-29) is overall of low quality, with a relatively coarse grain, whereas artifacts recovered from workshops and other sites in the region are often made of high-quality, fine-grained material. The stone in these artifacts could have come from other sources, or it could be that the BRS acquired at Beaver Creek Quarry was subsequently heat-treated to improve its quality. In other words, while variations in the quality of the stone recovered within and between sites may reflect variations at the source from which the material was originally procured, they may also reflect human intervention, in the form of heat treatment. For this reason, it is important to become familiar with the effects of heat treatment on BRS, as well as to be alert to evidence of it.

Distribution. As studies undertaken to date in the oil sands region indicate, the area in which BRS represents the predominant lithic material used in tool manufacture is fairly small (see fig 12.7). This material is found in relative abundance at archaeological sites only as far south as Mildred Lake (Sims 1974), as far west as Joslyn Creek and the Ells River (Graham and Tischer 2009; Gryba and Tischer 2005), and as far north as the Fort Hills and McClelland Lake (Woywitka and Younie 2008a, 2008b). To the east, less information is available, as sites thus far identified are fewer and smaller, but BRS appears to constitute a significant portion of most assemblages near Kearl Lake (Unfreed and Blower 2005; Bouchet-Bert 2007). Figure 12.7 illustrates the core area of BRS use, an area in which a number of site assemblages contain a large percentage (anywhere from 50% to 100%) of BRS.

At sites in the Fort Hills, lithic assemblages are heavily dominated by BRS (Woywitka, chapter 7 in this volume; Woywitka et al. 2009). Farther to the northwest, along Asphalt Creek and Eymundson Creek, and to the north of these creeks, along the eastern slopes of the Birch Mountains, assemblages tend to contain very few artifacts made of BRS; quartzites, including Northern and Salt and Pepper varieties, are much more commonly recovered (Bryant 2004; Bouchet-Bert 2007; Gryba and Tischer 2008, 2009a, 2009b). In the Birch Mountains, Donahue (1976) noted a near absence of BRS, although relatively recent research just east of the Birch Mountains has identified artifacts manufactured of the stone (Roskowski, Netzel, and Tischer 2012; Foster 2013; see also Ives, chapter 8 in this volume). On the basis of this research, a suggested secondary area of BRS reduction can be described (see fig 12.7), in which a smaller proportion of each assemblage is BRS. Information for areas to the east of McClelland Lake and the south of Kearl Lake is minimal as these areas have not been subjected to intensive study and few archaeological sites have thus been identified.

Sites that contain relatively little BRS occur even in areas where BRS is by far the dominant lithic material. As Woywitka notes in chapter 7 of this volume,

Figure 12.7. Core and secondary areas of Beaver River Sandstone use

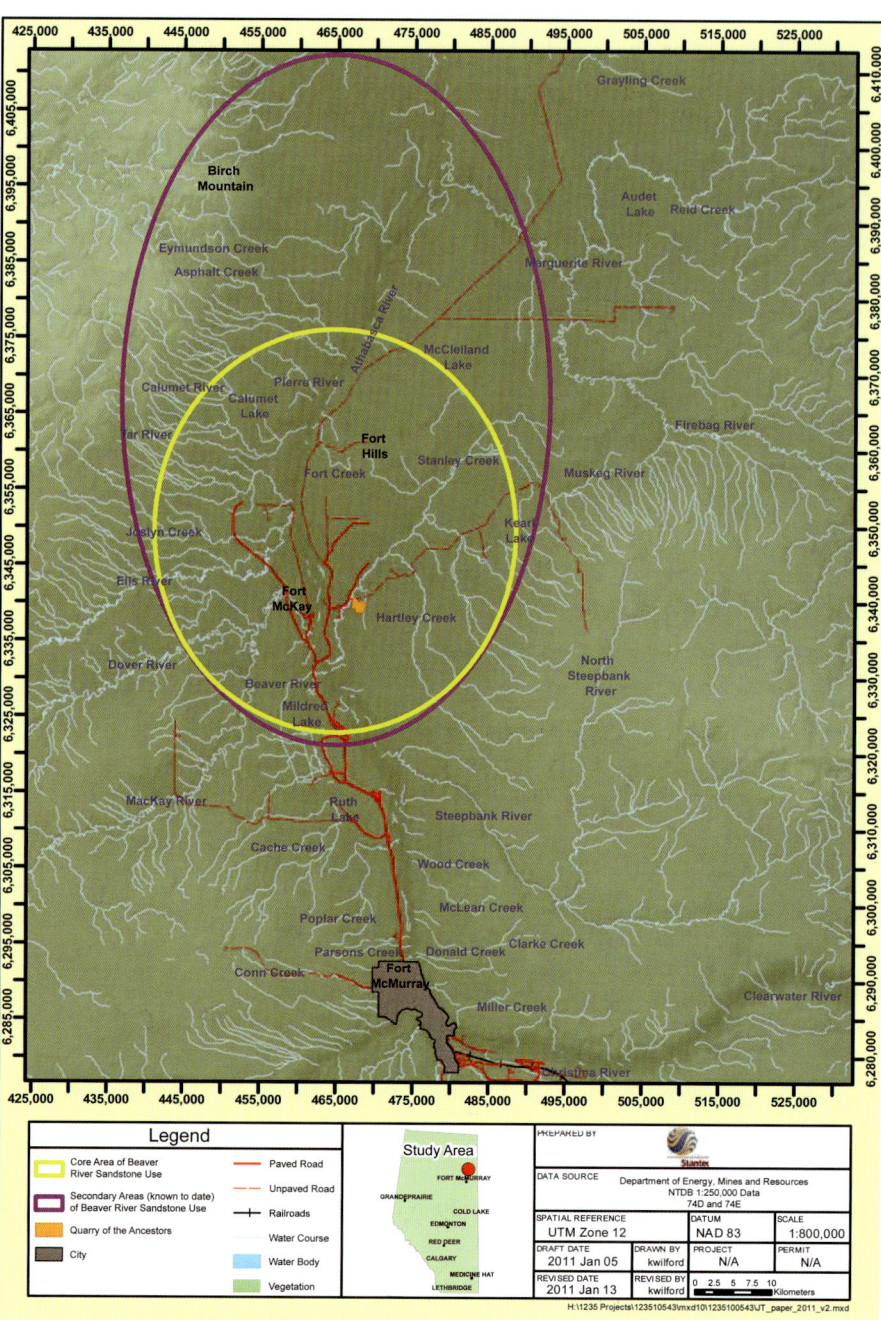

several sites in the Fort Hills contain significant quantities of non-BRS materials, notably two campsites, HiOv-104 and HiOv-126 (see table 7.1). In addition, BRS accounts for only about 3.5% of the assemblage at the Little Pond site (HiOv-89), although this appears to reflect the association of the site with microblade production. (As Woywitka points out, the same is true at the Bezya site, located a little further south, which is likewise associated with microblade technology.) A similar pattern, in which sites with very little BRS lie in close proximity to sites containing significant proportions of BRS, has been observed at Joslyn Creek, to the northwest of Fort McKay (Graham and Tischer 2009), as well as at sites south of the Fort Hills (Saxberg, Somer, and Reeves 2004). This variation could be attributed to the occupation of these sites by different cultural groups, or possibly by the same cultural group during different time periods or seasons, or it could reflect differing degrees of access to sources of BRS, including the Quarry of the Ancestors.

Outside the Athabasca oil sands region, evidence for the use of BRS drops off dramatically, as one would expect as the distance of sites from the source area increases. Examples of archaeological sites that contain BRS outside of the Athabasca oil sands region include specimens recovered from the Duckett site, on Ethel Lake (Fedirchuk and McCullough 1992), and from site GdOp-19, on Tucker Lake (Tischer 2002), both situated in the Cold Lake region. BRS has also been recovered from sites in the Barrhead area and at a several other sites in northern Alberta and northwestern Saskatchewan (see Fenton and Ives 1990; Somer 2007), including the Peter Pond Lake area of Saskatchewan, which lies southeast of the Alberta oil sands area (Young 2006). It may be that, in areas relatively distant from the stone's source, tools fashioned from BRS were rejuvenated and reused until they were completely worn out. The rejuvenation of tools leaves a very small archaeological footprint not likely to be discovered during a conventional shovel-testing program. Such rejuvenation might well have seemed preferable to the long-distance transport of newly quarried BRS.

Aside from considerations of distance, several other possible explanations exist for the scarcity of BRS in archaeological assemblages outside its core area of use. One is suggested by Saxberg (2007), who associates the almost exclusive use of a single tool stone with a low degree of residential mobility over an extended period of time. With regard to the Quarry of the Ancestors, she notes that "qualities of the stone tool assemblage regarding tool curation, tool maintenance and tool richness suggest long-term occupation of the area" (2007, 117; see also Saxberg and Robertson, chapter 10 in this volume). In this view, the very limited distribution of BRS at sites outside the Lower Athabasca region reflects the relatively circumscribed area in which precontact residents of the area moved about. However, the styles of certain projectile points recovered from sites in the oil sands region suggest that cultural relationships existed between the peoples who inhabited the area and those living in the Northern Plains (Gryba and Tischer 2009a; Saxberg 2007; Saxberg and Reeves 2003, 2004; Saxberg, Somer, and Reeves 2004; Syncrude 1973) and in the Barrenlands to the northeast (Clarke and Ronaghan 2000; Saxberg, Somer, and Reeves 2004; Tischer 2004). Such evidence of cultural exchange tends to imply at least some degree of mobility, unless we assume that these cultural influences were entirely the result of other, more mobile groups passing through and interacting with local residents.

Another possibility, already raised, is that, in view of the wide variation in the quality and appearance of BRS, artifacts discarded at distant locations may simply not be recognized by archaeologists, especially if they are unfamiliar with the lithic type (which is, after all, fairly unusual, as it occurs naturally in only a small area). As a result, occurrences of BRS may go unreported. The possible misidentification of BRS would obviously hinder attempts to reconstruct patterns of tool distribution beyond the area in which sources of the stone exist. As noted earlier, the amount of BRS debitage recovered from sites in the oil sands region far exceeds the number of finished tools made of BRS. One possible explanation is that groups visited the area seasonally, perhaps for social or cultural reasons as well as for the purpose of obtaining stone, reduced large amounts of BRS into blanks or preforms, and then transported them to other locations in western Canada or perhaps even further beyond. And yet, as we have seen, there are few recorded occurrences of BRS tools outside the Athabasca oil sands region. How far this lack of evidence reflects a failure to identify BRS correctly remains an open question. As Ives (1993, 2003) suggests, a greater familiarity with the geological character of lithic sources, together with an increased emphasis on recognizing and tracking BRS in archaeological assemblages, would undoubtedly provide information on trade, travel, and migration. Conversely, if we could confirm that BRS rarely occurs in archaeological assemblages beyond the oil sands region, this, too, would require an explanation, particularly in view of the wide distribution of other lithic types such as Knife River flint or Swan River chert.

Trade and mobility. As mentioned above, identifying distribution patterns of BRS in archaeological assemblages can shed light on early trade networks and group mobility. Beck et al. (2002) suggest that the central place foraging model can aid us in understanding the variability and observable economization in site assemblages, arguing that the distance between a group's home base (the central place) and the source of raw material will influence patterns of lithic behaviour. One important factor to be considered is "tool stone utility," that is, how useful a particular piece of stone is for the purpose of making tools. Tool stone utility is governed in part by the amount of usable stone in a nodule, as well as by how efficiently the stone can be worked—that is, by the proportion of usable stone that ends up in the finished tool rather than as waste flakes produced in the process of shaping the tool. Inherent impurities in the lithic material can also affect tool stone utility. Other considerations are the size of the nodule to be worked and the purpose of the tool: the lithic material used to fashion a tool must be appropriate to the tool's final function. Beck et al. (2002, 490) argue that "given this apparent functional requirement, an artisan will choose to travel a long distance to obtain suitable raw material and pass up unsuitable materials closer at hand," even though proximity to a quarry might otherwise seem an advantage. In addition, the distance from home to the "central place," as well as the time required for reduction, will influence subsequent choices relating to the degree of on-site processing and the long-distance transport of stone blanks.

With regard to tool stone utility, the overall quality of BRS is not exceptionally high in comparison to other locally available lithic materials, such as cherts or even fine-grained quartzites. In fact, projectile points recovered at sites in the region are often manufactured from materials other than BRS. This is true even at sites at which the assemblage consists almost entirely of artifacts made from BRS. For example, at site HhOv-319, one of the two large site areas at the Quarry of the Ancestors, all the tools were made of BRS and essentially all of the debitage was composed of BRS, but the artifacts collected also included a single side-notched projectile point, which was made of a grey quartzite (Saxberg and Reeves 2004). A similar pattern has been observed in assemblages elsewhere— for example, at HhOv-78 (Clarke and Ronaghan 2004), HhOw-10 (Bryant 2005), HiOu-49 (Somer 2005), and HhOv-212 (Green et al. 2006). This suggests that BRS was not always considered to be of sufficiently high quality for the manufacture of projectile points. In addition, many of the non-BRS projectile points exhibit extensive reworking and heavy wear, which suggests these non-BRS points were carefully conserved and discarded only when they were completely worn out.

Projectile points made from BRS have, however, been recovered from a number of sites in the Athabasca oil sands region, including those adjacent to the Quarry of the Ancestors. The overall quality of BRS used for these points is not especially high; many of the specimens do not exhibit wear, and those that do appear to have broken during manufacture or after limited use. It thus appears that, although projectile points were fashioned even from relatively low-quality BRS, these points may have been manufactured for short-term use and were discarded when higher-quality material was obtained. This pattern is particularly evident among projectile points dating to the Middle and Late Precontact periods. Some of the earlier specimens, which are fashioned from the highest-quality BRS, do exhibit extensive reworking and heavy wear. It may be, then, that supplies of the best stone were exhausted during the earliest occupations of the region.

The fact that exotic lithic materials, including Peace Point chert, Tertiary Hills clinker, and obsidian originating from several sources outside Alberta, have been recovered at sites in the Athabasca oil sands region suggests that precontact peoples travelled from great distances to the Lower Athabasca valley. Given that the Quarry of the Ancestors was used most intensively during early precontact times, it is possible that groups regularly travelled to the region during that period in order to acquire high-quality lithic raw material. Later visits to the quarry, however, made after most of the better-quality stone had been exhausted, must have been undertaken for other reasons—perhaps in connection with spiritual traditions or seasonal gatherings, or as part of the yearly round of hunting and gathering, or in connection with trade networks.

Direct comparisons of BRS assemblages in this region to assemblages of similar lower-quality lithic material from sites elsewhere would be useful. Over 35,000 artifacts made of Dakota sandstone, a microquartzite that exhibits characteristics similar to those of BRS, were collected from the Mountaineer site near Gunnison, Colorado, and assigned to the Folsom Tradition (Stiger 2006). A comparison of BRS assemblages with the Mountaineer site assemblage might help us understand how and why low-quality lithic sources were utilized. When linguistic connections exist between indigenous groups, such as the Dene of northern Alberta and the Navajo and Apache peoples of the southwestern United States, a comparative analysis of collections might also provide insight into questions of trade, mobility, and migration.

Social Organization

Quarries were essential to precontact peoples because their survival depended on stone tools and weapons. Assuming that the extraction of raw materials formed part of a group's seasonal round, gatherings at the locale of the Quarry of the Ancestors would have provided opportunities for trade, group hunting, and social and cultural activities, including ceremonial observances, the forging of marriage alliances, and games. To the extent that the area surrounding a quarry was perceived as the traditional territory of a specific group, and to the extent that lithic materials represented items of value in system of economic exchange, then the issue of access to the quarry becomes important. For example, writing about quarrying activities in the Scots Bay–Blomidon area of Nova Scotia, Deal (2001) states:

> If Scots Bay chalcedonies were being used in a lithic exchange system, it is likely that access was restricted at the local band level. While the ethnohistoric literature hints at a complex political hierarchy of chiefs and councils . . . the lowest level of this hierarchy is probably our most useful model for the precontact era. It consists of a local leader responsible for a group of related families who shared a specific summer camp. In historic times, each of these family bands controlled specific hunting territories around lakes and river courses (Speck 1922). Summer meetings were important for arranging marriages, settling disputes, and cooperative economic projects.
>
> In the precontact era, this would also be the ideal time for excursions to the quarry sites on the Fundy shore. Quarry blanks and some finished tools produced at Scots Bay were probably taken to summer camping areas. . . . If the quarry blanks were made intentionally for exchange with other local bands within the district, this exchange was most likely on a small scale. This may have involved infrequent exchanges between family bands along the borders of hunting territories or at contact points along major routes.

In an early description of the material culture of the Dene, an Athapaskan-speaking people whose forebears ranged throughout the northern boreal forest, Morice (1894, 65) recorded evidence of proprietary rights relating to quarry use:

> The material chosen in preference to fashion arrow or spear heads with was loose, broken pieces of the rock such as were found on the surface. Of course these were confined to a few locations only, wherein were situated sorts of quarries which were very jealously guarded against any person, even of the same tribe, whose right to share in the contents was not fully established. A violation of this traditional law was often considered a *casus belli* between the co-clansmen of the trespasser and those of the proprietors of the quarry.

Such traditional rights to specific quarries may have had their inception in very early times.

Given the importance of stone tools to precontact communities, it is not unexpected that quarries would have been accorded special status. At least in some instances, this status appears to have been expressed in ceremonies and rituals. For example, Scott and Thiessen (2005) describe the traditional rituals associated with extraction of catlinite, or pipestone, the material from which Plains groups fashioned ceremonial pipes. Not only were purification rites and propitiatory offerings integral ceremonial components of these undertakings, but there is some evidence for the archaeological presence of a sweat lodge at the catlinite quarry at Minnesota's Pipestone National Monument. Although, as far as we know, the BRS associated with the Quarry of the Ancestors is not a ceremonially significant material like catlinite, there is no reason to assume that the extraction of the stone was any less the occasion for ritual.

Rajnovich (1994) provides valuable insights into rituals and the symbolism associated with lithic material and source areas. She indicates that stone, particularly usable stone, contained "medicine." In eastern Canada, the significance of quarries was often proclaimed by the presence of nearby rock art. We can only speculate as to whether ceremonies were associated with precontact quarry use, but assuming that success in hunting depended on the "medicine" in the raw material of the projectile point, it is not unreasonable to expect that ceremonialism was associated with even the most mundane quarry. Topping and Lynott (2005) suggest that archaeologists may not be adequately alert to the placement and non-utilitarian significance of the artifacts recovered from quarry sites. That is, the apparent functionality of artifacts may serve to obscure their symbolic meaning. This theory could have considerable implications with respect to the observed lack of finished tools at BRS quarries.

Ives (1993) suggests that the large concentration of archaeological sites within the Fort McKay area in general is related to use of the Cree Burn Lake site complex as a seasonal gathering area, from which groups would disperse, creating numbers of satellite sites. The discovery of the Quarry of the Ancestors, so close to the Cree Burn Lake site, raises the possibility that such seasonal gatherings were also associated with activities at the quarry. As noted at the outset, however, excavations at the quarry have yet to identify habitation sites, much less evidence of social and cultural activities not directly related to subsistence.

Investigating archaeological sites in the oil sands region by conducting excavations that are spatially circumscribed as well as separated from one another could account for the current lack of evidence for activities other than lithic reduction at quarry sites. If areas related to social and cultural activities are present at these sites, as might be expected, these activity areas may be peripheral to the areas in which lithic reduction took place. Evidence of such activities could be manifest not only in campsite remains but also in archaeologically anomalous artifacts that might have had a ceremonial purpose or perhaps in unusual patterns of artifact distribution. If we hope to gain a more complete understanding of precontact cultures, increased efforts will need to be made to identify peripheral sites, away from the areas of densest concentrations of lithic debris, as well as to carefully interpret their contents.

CONCLUSIONS AND RECOMMENDATIONS

To date, archaeological investigations in the Athabasca oil sands region have generally focused on the assessment of those areas perceived to be of high archaeological potential, largely on the basis of the modern terrain, with less emphasis on how the ancient landscape and environment might have differed in ways that could alter these perceptions. Although palaeoenvironmental factors are written into predictive models and have been the subject of a number of studies, there is still a tendency to base field efforts on the current habitability of the landscape.

In addition, most HRIA studies define site boundaries in terms of the project footprint, with the result that adjacent areas remain unexplored. It is becoming increasingly obvious that the delimitation of sites should be archaeological in nature: it should not be based on an artificial boundary dictated by proposed plans for development. Failure to investigate the full extent of archaeological sites because of project specifications may have caused us to underestimate both the size and the significance of individual sites, which in turn makes it all but impossible to determine the true impact of development on historic resources.

In addition, because mitigation studies in the boreal forest region of Alberta typically focus more on excavating areas characterized by dense concentrations of artifacts (nearly all lithic, resulting from workshop activities) than on documenting and analyzing site structure, it is likely that significant portions of sites are being missed. Moreover, the artifact assemblages recovered through the excavation of lithic workshops tend to be dominated by non-diagnostic debitage, primarily flake fragments that, while they can shed light on lithic technologies, have little other interpretive value. Given the preoccupation with numbers, we often fail to excavate some or all of those portions of sites that are not directly related to lithic reduction. These may include campsite areas or areas that may have had spiritual significance. The excavation of larger areas, including those that contain fewer artifacts, would provide a more accurate picture of the activities that occurred across an entire site.

The identification of the Quarry of the Ancestors, a site of enormous significance to the precontact history of the Lower Athabasca valley, provides us with an opportunity to suggest that a revision of investigation methods is overdue. The complexity and size of the quarry, as well as the likelihood that this site is of pivotal importance to all sites containing BRS, requires a more theoretical and intensive approach to archaeological investigation, one that should be applied to all sites in the region. Such studies would be oriented toward explanation, seeking to answer questions such as: What was the full range of activities carried out at the Quarry of the Ancestors? What was the extent of the quarry proper, as distinct from associated sites? Were habitation sites located separately from the quarry, or did they lie within it? Is there evidence of specialized workshops that could shed light on the identity of those who used the quarry? Were rights of access to the quarry restricted to a particular group, who then traded bifacial blanks to other groups, and, if so, then how can we discern this arrangement archaeologically? Or was the quarry a destination shared by various groups who needed lithic material and thus a potential site for social interaction? How are sites in the oil sands region, including those that do not contain BRS, related to the Quarry of the Ancestors? Why is the distribution of BRS so strikingly limited? Did these patterns of resource procurement and use vary over time?

The current approach to assessment and mitigation studies in the Athabasca oil sands region has resulted in data being lost or fragmented, particularly data concerning site structure and activities not directly related to lithic technology. Given the number of sites that have been cleared for development, we already face serious challenges in assembling a complete archaeological record, except at those few sites that have been deemed sufficiently significant to preserve, that is, the Beaver Creek Quarry, the Cree Burn Lake site, and the Quarry of the Ancestors. Unless comprehensive and detailed investigations of other sites in the region are conducted, however, the relationship of these three sites to the broader cultural context will be all but impossible to determine and their relevance to the early history of the region therefore limited.

We thus recommend that, as far as possible, an altered approach to archaeological studies in this region be adopted. In addition to the required site inventory, it would be valuable if impact assessments were to address specific methodological or theoretical questions. At a most basic level, field methods should include some consideration of areas deemed to be of low archaeological potential. In addition, provided this is feasible, footprint-specific assessments should include a buffer zone, so that a larger, more cohesive area will be examined. Whenever possible, predictive models of archaeological potential should incorporate data regarding ancient climate and landscapes, rather than relying on modern features of the terrain. Spatially larger excavations would not only allow for better determination of the boundaries and structure of a site but would be more likely to identify areas of activity other than lithic workshops. Areas peripheral to the assumed centre of a site could yield crucial information about dwellings and day-to-day activities, ultimately offering insight into the material culture, the social organization, and the cultural practices of precontact groups. Although implementing even these relatively modest changes may not always be possible, such alterations to the present approach would result in a more complete and effective collection of data, enabling us to evaluate the significance of a site with greater accuracy and ultimately to arrive at a more complex understanding of precontact cultural traditions.

The Quarry of the Ancestors also provides us with an invaluable opportunity to investigate the relationship between a quarry and its associated sites, as well as the significance of a quarry to a broader region, helping us to identify larger patterns of precontact human occupation. As Heldal (2008) reminds us:

> Ancient quarries are the "forgotten" archaeological sites. . . .
> Within the totality of a quarry landscape, these landscapes can represent "lived" experiences or the embeddedness of significance relating to ancestry, kinship and practices within the landscape, such as quarrying, played out over centuries. Such places have important implications which have not previously been integrated into arguments about stone symbolism, source and use.

The Quarry of the Ancestors has the potential to transform our understanding not only of the nature of archaeological sites in the oil sands region but also of the links between the Lower Athabasca valley and the broader economic and cultural landscape in which precontact peoples lived and moved.

In the Athabasca oil sands region, a dense distribution of archaeological sites over a relatively small area overlaps almost directly with the distribution of minable oil resources. A large number of archaeological sites have been and will be destroyed by development. To date, the excavations completed at archaeological sites in the region have primarily preserved evidence of the lithic technology of the precontact peoples who inhabited the area. These investigations have, however, preserved very little information of the sort that would allow us to reconstruct the lives of these peoples and to understand the social, economic, and cultural relationships among them. Moreover, current approaches to assessment make it difficult, if not impossible, to address the cumulative effects of development on the archaeological record. If we hope to gain an adequate sense of the data that are being lost to development, the current assessment approach must be amended so that we will come away with more than just lithic debris.

REFERENCES

Andrews, Bradford W., Timothy M. Murtha, Jr., and Barry Scheetz

2004   Approaching the Hatch Jasper Quarry from a Technological Perspective: A Study of Prehistoric Stone Tool Production in Central Pennsylvania. *Midcontinental Journal of Archaeology* 29(1): 63–101.

Beck, Charlotte, Amanda K. Taylor, George T. Jones, Cynthia M. Fadem, Caitlyn R. Cook, and Sara A. Millward

2002   Rocks Are Heavy: Transport Costs and Paleoarchaic Quarry Behavior in the Great Basin. *Journal of Anthropological Archaeology* 21(4): 481–507.

Billick, W. T.

1998   Knife River Flint. In *Archaeology of Prehistoric Native America: An Encyclopedia,* edited by Guy Gibbon, pp. 427–428. Garland Publishing, New York and London.

Boland, Dale E., Bonnie Brenner, and Jennifer C. Tischer

2009   *Historical Resources Impact Mitigation, Total E&P Joslyn Limited Joslyn North Mine Project, 2008 Mitigation Studies (HhOw-18, HhOw-29, HhOw-32, HhOw-42, HhOw-43, HhOw-45, HhOw-46): Final Report (ASA Permit 08-208).* FMA Heritage Resources Consultants Inc. Copy on file, Archaeological Survey, Historic Resources Management Branch, Alberta Culture, Edmonton.

Bouchet-Bert, Luc

2007   *Historical Resources Impact Assessment for Shell Canada Energy's Ten-Year Footprints in the Pierre River, Jackpine Expansion Mining Areas, and Jackpine Mine, Phase 1: Final Report (ASA Permit 06-116).* Golder Associates Ltd. Copy on file, Archaeological Survey, Historic Resources Management Branch, Alberta Culture, Edmonton.

Bryant, Laureen

2004   *Historical Resources Impact Assessment and Mitigation, Fall 2003, Canadian Natural Resources Limited Horizon Oil Sands Project: Final Report (ASA Permit 03-269).* Golder Associates Ltd. Copy on file, Archaeological Survey, Historic Resources Management Branch, Alberta Culture, Edmonton.

2005   *Historical Resources Impact Assessment and Mitigation, Summer 2004, Canadian Natural Resources Limited Horizon Oil Sands Project: Final Report (ASA Permit 04-189).* Golder Associates Ltd. Copy on file, Archaeological Survey, Historic Resources Management Branch, Alberta Culture, Edmonton.

Bryant, Laureen, Dana Dalmer, Caroline Gray, and Vincent A. Balls

2011   *Historic Resources Impact Mitigation for Shell Canada Energy's Muskeg River Mine, Sharkbite Area, 2009 and 2010 (ASA Permit 10-093).* Golder Associates Ltd. Copy on file, Archaeological Survey, Historic Resources Management Branch, Alberta Culture, Edmonton.

Carr, Kurt W., and Douglas C. McLearen

2005   Recent Testing at the Kings Jasper Quarry, Lehigh County, Pennsylvania. Paper presented at the 70th annual meeting of the Society for American Archaeology, Salt Lake City, Utah, 30 March–3 April.

Clarke, Grant M.

2002   *Historical Resources Impact Assessment for Jackpine Mine, Phase 1: Final Report (ASA Permit 01-230).* Golder Associates Ltd. Copy on file, Archaeological Survey, Historic Resources Management Branch, Alberta Culture, Edmonton.

Clarke, Grant M., and Brian M. Ronaghan

    2000   *Historical Resources Impact Mitigation, Muskeg River Mine Project: Final Report (ASA Permit 99-073)*. Golder Associates Ltd. Copy on file, Archaeological Survey, Historic Resources Management Branch, Alberta Culture, Edmonton.

Clarke, Grant M., and Brian M. Ronaghan

    2004   *Historical Resources Impact Assessment and Mitigation Program, Muskeg River Mine Project: Final Report (ASA Permit 00-087)*. 2 vols. Golder Associates Ltd. Copy on file, Archaeological Survey, Historic Resources Management Branch, Alberta Culture, Edmonton.

Deal, Michael

    2001   Vignette: Distribution and Utilization of Scots Bay Chalcedony. Davidson Cove Site. http://www.ucs.mun.ca/~mdeal/Anth3291/DavidsonCove.htm.

De Paoli, Glen R.

    2005   Petrographic Examination of the Muskeg Valley Microquartzite (MVMq). Appendix A in Nancy Saxberg and Brian O. K. Reeves (2006), *Birch Mountain Resources Ltd., Hammerstone Project, Historical Resources Impact Assessment, 2004 Field Studies: Final Report (ASA Permit 04-235)*. Lifeways of Canada Ltd. Copy on file, Archaeological Survey, Historic Resources Management Branch, Alberta Culture, Edmonton.

Donahue, Paul F.

    1976   Alberta North, Project 75-8. In *Archaeology in Alberta, 1975,* edited by J. Michael Quigg and W. J. Byrne, pp. 42–50. Archaeological Survey of Alberta, Occasional Paper no. 1. Historic Resources Management Branch, Alberta Culture, Edmonton.

Ericson, Jonathon E.

    1984   Introduction: Toward the Analysis of Lithic Production Systems. In *Prehistoric Quarries and Lithic Production,* edited by Jonathon E. Ericson and Barbara A. Purdy, pp. 1–9. Cambridge University Press, Cambridge.

Fedirchuk, Gloria J., and Edward J. McCullough

    1992   *The Duckett Site: Ten Thousand Years of Prehistory on the Shores of Ethel Lake, Alberta*. Fedirchuk McCullough and Associates Ltd. Copy on file, Esso Resources Canada Limited, Calgary.

Fenton, Mark M., and John W. Ives

    1984   The Stratigraphic Position of Beaver River Sandstone. In *Archaeology in Alberta, 1983,* compiled by David Burley, pp. 128–136. Archaeological Survey of Alberta Occasional Paper no. 23. Historic Resources Management Branch, Alberta Culture, Edmonton.

    1990   Geoarchaeological Studies of the Beaver River Sandstone, Northeastern Alberta. In *Archaeological Geology of North America*, edited by Norman P. Lasca and Jack Donahue, pp. 123–135. Centennial Special Volume 4. Geological Society of America, Boulder, Colorado.

Fisher, Timothy G., and Derald G. Smith

    1994   Glacial Lake Agassiz: Its Northwest Maximum Extent and Outlet in Saskatchewan (Emerson Phase). *Quaternary Science Reviews* 13: 845–858.

Foster, Jean-Paul

    2013   *Historical Resources Impact Assessment, Teck Resources Limited Frontier Project, 2012 Studies: Final Report (ASA Permit 12-157)*. Stantec Consulting Ltd. Copy on file, Archaeological Survey, Historic Resources Management Branch, Alberta Culture, Edmonton.

Graham, James W., and Jennifer C. Tischer

    2009   *Historical Resources Impact Mitigation, Total E&P Joslyn Limited Joslyn North Mine Project, 2008 Mitigation Studies (HhOw-22, HhOw-30, HhOw-38, HhOx-9, HhOx-13): Final Report (ASA Permit 08-298)*. FMA Heritage Resources Consultants Inc. Copy on file, Archaeological Survey, Historic Resources Management Branch, Alberta Culture, Edmonton.

Green, D'Arcy, David Blower, Dana Dalmer, and Luc Bouchet-Bert

2006   *Historical Resources Impact Assessment and Mitigation, Albian Sands Energy's Muskeg River Mine and Shell Canada's Jackpine Mine: Final Report (ASA Permit 05-355)*. 2 vols. Golder Associates Ltd. Copy on file, Archaeological Survey, Historic Resources Management Branch, Alberta Culture, Edmonton.

Gryba, Eugene M.

2001   Lithic Resources Available to or Used by Precontact Knappers at the True North Energy L.P. Fort Hills Oil Sands Leases near Fort MacKay, in Northeastern Alberta. Appendix II in Wendy J. Unfreed, Gloria J. Fedirchuk, and Eugene M. Gryba, *Historical Resources Impact Assessment, True North Energy L.P. Fort Hills Oil Sands Project: Final Report (ASA Permit 00-130)*, vol. 2. Fedirchuk McCullough and Associates Ltd. Copy on file, Archaeological Survey, Historic Resources Management Branch, Alberta Culture, Edmonton.

Gryba, Eugene M., and Jennifer C. Tischer

2005   *Historical Resources Impact Assessment, Deer Creek Energy Limited Joslyn North Mine Project: Final Report (ASA Permit 05-094)*. FMA Heritage Resources Consultants Inc. Copy on file, Archaeological Survey, Historic Resources Management Branch, Alberta Culture, Edmonton.

2008   *Historical Resources Studies, Value Creation Inc. Terre de Grace Project: Final Report (ASA Permit 07-254)*. FMA Heritage Resources Consultants Inc. Copy on file, Archaeological Survey, Historic Resources Management Branch, Alberta Culture, Edmonton.

2009a   *Historical Resources Impact Assessment, UTS Energy Limited / Teck Cominco Limited, Equinox Project: Final Report (ASA Permit 08-265)*. FMA Heritage Resources Consultants Inc. Copy on file, Archaeological Survey, Historic Resources Management Branch, Alberta Culture, Edmonton.

2009b   *Historical Resources Baseline Studies, UTS Energy Limited / Teck Cominco Limited, Frontier Project: Final Report (ASA Permit 08-192)*. FMA Heritage Resources Consultants Inc. Copy on file, Archaeological Survey, Historic Resources Management Branch, Alberta Culture, Edmonton.

Heldal, Tom

2008   Significance of Ancient Quarries. QuarryScapes. http://www.quarryscapes.no/ancient_sign.php.

Ives, John W.

1993   The Ten Thousand Years Before the Fur Trade in Northeastern Alberta. In *The Uncovered Past: Roots of Northern Alberta Societies,* edited by Patricia A. McCormack and R. Geoffrey Ironside, pp. 5–31. Circumpolar Research Series no. 3. Canadian Circumpolar Institute, University of Alberta, Edmonton.

2003   Alberta, Athapaskans and Apachean Origins. In *Archaeology in Alberta: A View from the New Millennium*, edited by Jack W. Brink and John F. Dormaar, pp. 256–289. Archaeological Society of Alberta, Medicine Hat.

Ives, John W., and Mark M. Fenton

1983   Continued Research on Geological Sources of Beaver River Sandstone. In *Archaeology in Alberta, 1982,* compiled by David Burley, pp. 78–88. Archaeological Survey of Alberta Occasional Paper no. 21. Historic Resources Management Branch, Alberta Culture, Edmonton.

1985   *Progress Report for the Beaver River Sandstone Geological Source Study (ASA Permit 83-054)*. Copy on file, Archaeological Survey, Historic Resources Management Branch, Alberta Culture, Edmonton.

Kjorlien, Yvonne P., Leah Mann, and Jennifer C. Tischer

2009   *Historical Resources Impact Mitigation, Total E&P Joslyn Limited Joslyn North Mine Project, 2008 Mitigation Studies (HhOw-49, HhOw-54, HhOw-55, HhOx-10, HhOx-11,*

HhOx-15, HhOx-17, HhOx-18): Final Report (ASA Permit 08-166). FMA Heritage Resources Consultants Inc. Copy on file, Archaeological Survey, Historic Resources Management Branch, Alberta Culture, Edmonton.

Morice, A. G.

    1894 *Notes Archaeological, Industrial, and Sociological on the Western Dénés, with an Ethnographical Sketch of the Same*. Transactions of the Canadian Institute, vol. 4 (1892–93). Copp, Clark, Toronto.

Pennsylvania Historical and Museum Commission

    2012 The King's Quarry Site, 36LH2, Lehigh County. http://www.portal.state.pa.us/portal/server.pt/community/phmc_archaeology/2094/kings_quarry/410726.

Rajnovich, Grace

    1994 *Reading Rock Art: Interpreting the Indian Rock Paintings of the Canadian Shield*. Natural Heritage / Natural History Inc., Toronto.

Roskowski, Laura, and Morgan Blower

    2009 *Historical Resources Impact Assessment and Mitigation, HhOv-483 and HhOv-484, ATCO Electric Limited, Distribution Powerline Easement, Application No. 060077: Final Report (ASA Permit 06-515)*. FMA Heritage Resources Consultants Inc. Copy on file, Archaeological Survey, Historic Resources Management Branch, Alberta Culture, Edmonton.

Roskowski, Laura, and Morgan Netzel

    2011a *Historical Resources Impact Mitigation, Shell Canada Energy, Muskeg River Mine Expansion of RMS 10, Mitigation for Sites HhOv-87 and HhOv-200: Final Report (ASA Permit 09-168)*. FMA Heritage Resources Consultants Inc. Copy on file, Archaeological Survey, Historic Resources Management Branch, Alberta Culture, Edmonton.

    2011b *Historical Resources Impact Mitigation, Shell Canada Energy, Stage II Mitigative Excavation of HhOv-351: Final Report (ASA Permit 10-148)*. FMA Heritage Resources Consultants Inc. Copy on file, Archaeological Survey, Historic Resources Management Branch, Alberta Culture, Edmonton.

    2012 *Historical Resources Impact Mitigation, Shell Canada Energy Muskeg River Mine Expansion, Area 6, Stage I Mitigation of Sites HhOv-156 and HhOv-520: Final Report (ASA Permit 11-167)*. Stantec Consulting Ltd. Copy on file, Archaeological Survey, Historic Resources Management Branch, Alberta Culture, Edmonton.

    2015a *Historical Resources Impact Mitigation, Shell Canada Energy Jackpine Mine, New Compensation Lake, Sites HhOu-113 and HhOu-114: Final Report (ASA Permit 12-234)*. Stantec Consulting Ltd. Copy on file, Archaeological Survey, Historic Resources Management Branch, Alberta Culture, Edmonton.

    2015b *Historical Resources Impact Mitigation, Muskeg River Mine Expansion, Area 7 Gap, Staged Mitigation of Sites HhOv-374, HhOv-506, and HhOv-508, Shell Canada Energy: Final Report (ASA Permit 11-070)*. Stantec Consulting Ltd. Copy on file, Archaeological Survey, Historic Resources Management Branch, Alberta Culture, Edmonton.

Roskowski, Laura, Alison J. Landals, and Morgan Blower

    2008 *Historical Resources Impact Mitigation, Shell Canada Limited Albian Sands Muskeg River Mine Expansion Project, Mitigation for Sites HhOu-68, HhOu-69, HhOu-70, HhOu-94, HhOu-95, HhOv-378, HhOv-379, HhOv-380, HhOv-381, and HhOv-383: Final Report (ASA Permit 07-219)*. FMA Heritage Resources Consultants Inc. Copy on file, Archaeological Survey, Historic Resources Management Branch, Alberta Culture, Edmonton.

Roskowski, Laura, Morgan Netzel, and Jennifer C. Tischer

    2012 *Historical Resources Impact Assessment, Teck Resources Ltd. Frontier Project: Final Report (ASA Permit 10-123)*. FMA Heritage Resources Consultants Inc. Copy on file, Archaeological Survey, Historic Resources Management Branch, Alberta Culture, Edmonton.

Saxberg, Nancy

2007   *Birch Mountain Resources Ltd. Muskeg Valley Quarry, Historical Resources Mitigation, 2004 Field Studies: Final Report (ASA Permit 05-118).* 2 vols. Lifeways of Canada Ltd. Copy on file, Archaeological Survey, Historic Resources Management Branch, Alberta Culture, Edmonton.

Saxberg, Nancy, and Brian O. K. Reeves

2003   The First Two Thousand Years of Oil Sands History: Ancient Hunters at the Northwest Outlet of Glacial Lake Agassiz. In *Archaeology in Alberta: A View from the New Millennium,* edited by Jack W. Brink and John F. Dormaar, pp. 290–322. Archaeological Society of Alberta, Medicine Hat.

2004   *Birch Mountain Resources Ltd. Muskeg Valley Quarry, Historical Resources Impact Assessment, 2003 Field Studies: Final Report (ASA Permit 03-249).* Lifeways of Canada Ltd. Copy on file, Archaeological Survey, Historic Resources Management Branch, Alberta Culture, Edmonton.

Saxberg, Nancy, Brad Somer, and Brian O. K. Reeves

2004   *Syncrude Aurora Mine North, Historical Resources Impact Assessment and Mitigation Studies, 2003 Field Studies: Final Report (ASA Permit 03-279).* Lifeways of Canada Ltd. Copy on file, Archaeological Survey, Historic Resources Management Branch, Alberta Culture, Edmonton.

Scott, Douglas D., and Thomas D. Thiessen

2005   Catlinite Extraction at Pipestone National Monument, Minnesota: Social and Technological Implications. In *The Cultural Landscape of Prehistoric Mines,* edited by Peter Topping and Mark Lynott, pp. 140–154. Oxbow Books, Oxford.

Sims, Cort

1974   *Syncrude Lease 22—Beaver Creek Quarry Site (HgOv-29).* Report prepared for Syncrude Canada Ltd. Copy on file, Syncrude Canada Ltd., Edmonton.

Somer, Brad

2005   *Syncrude Aurora Mine North, Historical Resources Impact Assessment and Mitigation Studies, 2004 Field Studies: Final Report (ASA Permit 04-192).* 2 vols. Lifeways of Canada Ltd. Copy on file, Archaeological Survey, Historic Resources Management Branch, Alberta Culture, Edmonton.

2007   *Historical Resources Impact Assessment of a NuVista Energy Ltd. Pipeline in N1/2 8-40-28-W3M.* Lifeways of Canada Ltd. Copy on file, Saskatchewan Tourism, Parks, Culture and Sport, Regina.

Speck, Frank G.

1922   *Beothuk and Micmac.* Museum of the American Indian, Heye Foundation, New York.

Stiger, Mark

2006   A Folsom Structure in the Colorado Mountains. *American Antiquity* 71(2): 321–351.

Syncrude Canada Ltd.

1973   *Syncrude Lease No. 17: An Archaeological Survey.* Environmental Research Monograph 1973-4. Syncrude Canada Ltd., Edmonton.

1974   *The Beaver Creek Site: A Prehistoric Stone Quarry on Syncrude Lease No. 22.* Environmental Research Monograph 1974-2. Syncrude Canada Ltd., Edmonton.

Tischer, Jennifer C.

2002   *Historical Resources Impact Assessment, Final Report: Husky Energy, Tucker Lake Project (ASA Permit 01-201).* Fedirchuk McCullough and Associates Ltd. Copy on file, Archaeological Survey, Historic Resources Management Branch, Alberta Culture, Edmonton.

2004   *Historical Resources Studies, Final Report: Albian Sands Energy Inc. Muskeg River Mine Expansion (ASA Permit 04-249).* 2 vols. FMA Heritage Resources Consultants Inc. Copy on file, Archaeological Survey, Historic Resources Management Branch, Alberta Culture, Edmonton.

2005   *Historical Resources Studies, Final Report: Albian Sands Energy Inc.* Part I: *Historical Resources Impact Assessment and Historical Resources Mitigation, Muskeg River Mine Expansion (ASA Permit 05-297).* FMA Heritage Resources Consultants Inc. Copy on file, Archaeological Survey, Historic Resources Management Branch, Alberta Culture, Edmonton.

2008   *Historical Resources Studies, Final Report: Shell Canada Limited, Shell Sharkbite Project Area (ASA Permit 06-211).* FMA Heritage Resources Consultants Inc. Copy on file, Archaeological Survey, Historic Resources Management Branch, Alberta Culture, Edmonton.

Tischer, Jennifer C., and Gloria J. Fedirchuk

2006   *Historical Resource Studies Protective Notation 050083: Final Report (ASA Permit 05-377).* FMA Heritage Resources Consultants Inc. Copy on file, Archaeological Survey, Historic Resources Management Branch, Alberta Culture, Edmonton.

Topping, Peter, and Mark Lynott

2005   Miners and Mines. In *The Cultural Landscape of Prehistoric Mines,* edited by Peter Topping and Mark Lynott, pp. 181–191. Oxbow Books, Oxford.

Tsang, Brian W. B.

1998   The Origin of the Enigmatic Beaver River Sandstone. MSc thesis, Department of Geology and Geophysics, University of Calgary, Calgary.

Turney, Michael H. J.

2013   *Historic Resources Impact Assessment for the Cenovus Energy Inc. Pelican Lake Grand Rapids Project: Final Report (ASA Permit 11-241).* Golder Associates Ltd. Copy on file, Archaeological Survey, Historic Resources Management Branch, Alberta Culture, Edmonton.

Unfreed, Wendy J., and David Blower

2005   *Historical Resources Impact Assessment, Imperial Oil Resources Ventures Limited, Kearl Oil Sands Project Leases 6, 87, and 88A (Twps. 95 to 97, Rges. 7 to 8, W4M): Final Report (ASA Permit 04-375).* FMA Heritage Resources Consultants Inc. Copy on file, Archaeological Survey, Historic Resources Management Branch, Alberta Culture, Edmonton.

Wickham, Michelle D., and Taylor Graham

2009   *Historical Resources Impact Mitigation of the TransCanada Pipelines Ltd. Fort McKay Mainline Expansion: Final Report and Post-construction Audit (ASA Permits 06-376 and 07-266).* Bison Historical Services Ltd. Copy on file, Archaeological Survey, Historic Resources Management Branch, Alberta Culture, Edmonton.

Woywitka, Robin J., and Angela M. Younie

2008a   *Historical Resources Impact Mitigation, Fort Hills Energy Corporation, Fort Hills Oil Sands Project, HiOv-44, HiOv-47, HiOv-49, HiOv-50, HiOv-52, HiOv-87, HiOv-89, HiOv-104, HiOv-115, and HiOv-124: Final Report (ASA Permit 05-328).* FMA Heritage Resources Consultants Inc. Copy on file, Archaeological Survey, Historic Resources Management Branch, Alberta Culture, Edmonton.

2008b   *Historical Resources Impact Mitigation, Fort Hills Energy Corporation, Fort Hills Oil Sands Project, Stage I Mitigation, HiOv-97 and HiOv-98, Stage II Mitigation, HiOv-61 and HiOv-64: Final Report (ASA Permit 07-235).* FMA Heritage Resources Consultants Inc. Copy on file, Archaeological Survey, Historic Resources Management Branch, Alberta Culture, Edmonton.

Woywitka, Robin J., Angela M. Younie, Morgan Blower, and Alison J. Landals

2008   *Historical Resources Impact Mitigation, Shell Canada Limited Albian Sands Muskeg River Mine Expansion Project, Mitigation for Sites HhOv-384, HhOv-385, HhOv-387, HhOv-431, and HhOv-432: Final Report (ASA Permit 07-280).* FMA Heritage Resources Consultants

Inc. Copy on file, Archaeological Survey, Historic Resources Management Branch, Alberta Culture, Edmonton.

Woywitka, Robin J., Jennifer C. Tischer, Laura Roskowski, and Angela M. Younie

2009   *Historical Resources Impact Mitigation, Fort Hills Energy Corporation, Fort Hills Oil Sands Project, 2008 Mitigation Studies: Final Report (ASA Permit 08-163).* FMA Heritage Resources Consultants Inc. Copy on file, Archaeological Survey, Historic Resources Management Branch, Alberta Culture, Edmonton.

Youell, A. J., Jennifer C. Tischer, Morgan Blower, and Lauren Copithorne

2009   *Historical Resources Impact Mitigation, Total E&P Joslyn Limited, Joslyn North Mine Project, 2007 Mitigation Studies (HhOw-20, HhOw-21, HhOw-22, HhOw-24, HhOw-26, HhOw-27, HhOw-30, HhOw-38, HhOw-39, HhOw-40, HhOx-7): Final Report (ASA Permit 07-393).* FMA Heritage Resources Consultants Inc. Copy on file, Archaeological Survey, Historic Resources Management Branch, Alberta Culture, Edmonton.

Young, Patrick S.

2006   An Analysis of Late Woodland Ceramics from Peter Pond Lake, Saskatchewan. MA thesis, Department of Archaeology, University of Saskatchewan, Saskatoon.

13 Cumulative Effects Assessment | Evaluating the Long-Term Impact of Oil Sands Development on Archaeological Resources

BRIAN M. RONAGHAN

Alberta's oil sands represent one of the world's largest sources of petroleum. Existing development projects have already brought profound changes to the natural environment and to the social fabric of the region, changes that have generated widespread commentary. Despite fluctuations in the world economy and growing concerns about climate change, and despite the devastating wildfire that swept through Fort McMurray in the spring of 2016, development is likely to continue, further transforming the landscape. Numerous challenges are associated with understanding both the immediate implications and the long-term consequences of these changes, as well as with managing these changes effectively, and regional and national debate is ongoing with respect to how these challenges might successfully be met. Perhaps the chief question concerns the cumulative impact of these changes: what are the combined environmental and social effects of regional development over time, and how can they best be managed? As a formally defined method, cumulative effects assessment has relatively recent application and, given the complexity of the environmental systems to which it is applied, frequently falls short of ideals on several fronts. Adopting a broad regional perspective on the issue, this chapter explores the cumulative effects of oil sands development on a single dimension of the environment: archaeological resources.

Assessments of the combined incremental effects of oil sands development can be relatively formal or informal. Informally, people form impressions based on their personal experience and on information provided through a myriad of sources, from anecdotal accounts and stories in the media to government studies to published literature ranging from the sober to the polemical. The result is assessments that vary widely, are often imperfectly informed, and tend toward subjective interpretation. At the same time, such relatively informal assessments often reflect a regional scale of understanding that is either absent or poorly developed in more formal evaluations.

Formally, in Alberta and most other jurisdictions, cumulative effects analysis is undertaken as a key element of legal processes intended to aid regulatory decision makers in determining whether a specific project is in the public interest and complies with existing legislation. In Alberta, these analyses are carried out within the framework of an Environmental Impact Assessment (EIA), a procedure outlined in part 2, division 1, of the Environmental Protection and Enhancement Act (Alberta 2000c). The act stipulates that an EIA must include "a description of potential positive and negative environmental, social, economic and cultural impacts of the proposed activity, including cumulative, regional, temporal and spatial considerations" (Alberta 2000c, s. 49[d]). Included in the project-specific terms of reference issued for all EIAs are requirements that the effects of proposed activities on historic resources be described.

For a number of reasons, neither of the above approaches to assessment is especially well suited to the goal of evaluating the combined effects of oil sands development on archaeological resources. Although many valuable contributions have been made in informal assessments of cumulative impacts, such assessments frequently fail to appreciate the complexity of the issues, especially the interrelationships among various development activities, and often lack the scientific rigour necessary to support their conclusions. Moreover, because such assessments often have a rhetorical purpose, they are apt to rely on selective data—data originating with persons or agencies that highlight only the information and analysis that best support their own objectives. Even so, these relatively informal assessments avoid the often uncritical assumption of accuracy and objectivity that tends to characterize more formal approaches.

Formal cumulative effects assessments (CEAs) are one of the requirements for completion of an EIA. They generally take place individually, as part of the review processes that surround a specific project, and are based on information provided by the proponent of that project. Regional considerations focus on the project's additive impact—on how it will compound existing environmental change—as well as on the proponent's plans for mitigation, which are likewise project-specific. A consideration of future growth in non-project-related activities is typically not, however, integrated into the analysis. Furthermore, these assessments do not apply to entire geographic units—watersheds, for example, or landscapes officially defined as oil sands administrative areas or as natural subregions of Alberta (such as the Athabasca Plain: see Natural Regions Committee 2006, 149–52)—and they forecast only a short time into the future (Alberta Environment 2008). Other shortcomings involve an imperfect understanding of the baseline conditions against which environmental impacts are measured, as well as uncertainty surrounding the nature and extent of future development activities, how to ensure an accurate analysis of their effects, and how best to evaluate the overall impact of multiple activities that are insignificant when viewed in isolation (Kennett 1999). Frequently, analysis stumbles on the gap between the rigour required for accurate prediction and the imprecision of the data on which the analysis is based, especially with respect to relationships between activities and outcomes.

Aside from project-specific considerations, other drawbacks include the absence of a legislated or scientifically derived consensus concerning the thresholds of unacceptability for the impact of development on specific "valued ecological components" (VECs) and the lack of integrated policy instruments capable of addressing this impact (Kennett 1999). The latter issue continues to provoke much debate, and several concrete initiatives have been undertaken by the Alberta government in response to these shortcomings (Alberta Energy 2007; Alberta Environment 2008). It is now widely recognized that, given the environmental concerns that have already arisen, achieving sustainable development on a regional scale is a matter of considerable urgency.

In addition to these shortcomings, CEAs tend to focus on the impact of development on existing ecological and social systems that can respond positively to reclamation or remediation, rather than on historical systems, which cannot be recreated. Concepts such as "no net loss" of existing populations and "equivalent capability" vis-à-vis existing activities often feature in remediation plans, as does the notion of "adaptive management," which is intended to ensure a flexible remedial response to environmental effects through ongoing monitoring as development proceeds. These concepts, together with elaborate reclamation plans (in some cases phased in over the forty-year lifespan of a typical development activity), form the basis for cumulative effects assessments that enable a comparative approach, on the possibly unwarranted assumption that remediation efforts mandated by previous CEAs will be effective. Virtually all of these methods focus, moreover, on restoring pre-existing environmental conditions and on the management of social effects on a project-specific scale, while leaving long-term regional environmental effects and social consequences to government. These shortcomings are not limited to the Alberta context; rather, they are characteristic of the current application of CEAs generally.

Like an EIA, an assessment of cumulative effects seeks to weigh the negative and positive effects of development. Although the description of these effects must be based on a rigorous collection of information that is both accurate and complete, the evaluative component is often highly subjective, reflecting social values that are not always universally shared. Furthermore, activities, their effects, and linkages among them are typically complex, a fact that increases exponentially when multiple development projects, broad areas of terrain, and extended time periods are considered, as in the case of CEAs. Consequently, decision making in connection with regulatory approval necessarily involves efforts to balance competing values, a process that often entails compromise. This is not to argue, of course, that CEAs hold no value. On the contrary, and as is generally recognized, by providing information about a wide range of potential effects, they allow us to gain insight into the dynamic structures within which development occurs. Quite apart from the value of CEAs to regulatory decisions concerning specific projects, the analysis of complex systems, while inevitably somewhat tentative, enhances our ability to predict the impact of future growth on both social and ecological communities.

Outside of the regulatory context, less formal assessments, of which this chapter is an example, have more latitude, more freedom to adopt an explicit stance. However, to be credible, they must likewise be based on a complete and accurate description of the probable effects of development, to allow readers to reach informed conclusions, and should provide realistic evaluations that acknowledge the existence of contrasting perspectives and that recognize positive effects as well as negative ones. The assessment that follows seeks to balance the contrasting values associated with managing archaeological resources on a regional scale in the complex and continually evolving situation surrounding development of Alberta's oil sands. Definitive conclusions are difficult to draw in such shifting circumstances, but directions for improved resource management can at least be suggested.

Historical and archaeological resources are fragile and non-renewable. Damage to these resources is permanent and can be offset only through a limited number of measures, and then only partially. Furthermore, what we know, or can predict, about the location, extent, and overall character of these resources is significantly less reliable than is the case for other types of resources and is often gained only when their destruction is imminent.

On a national scale, the cumulative effects assessment process outlined by the Canadian Environmental Assessment Agency (1999, 2007) provides a framework for evaluating the combined effects of regional development on archaeological resources. Agencies of the Alberta government also provide general guidance on the scope and content of the CEA components of EIA reports (see, for example, Alberta Energy and Utilities Board, Alberta Energy, and the Natural Resources Conservation Board 2010), although these recommendations are not prescriptive. The approach described below combines the elements outlined in provincial documents but adheres more closely to the federal assessment framework. As applied in Alberta, this framework involves four basic steps. First, a scoping exercise identifies the character of regional archaeological resources, the VECs to be considered in the assessment, and the legislation enabling the management of these resources, as well as issues surrounding their identification and evaluation. Baseline conditions of the resource are reviewed in the scoping stage, and the temporal and spatial parameters of the assessment are also outlined. Second, past, present, and reasonably foreseeable development effects are identified, and procedures for mitigating predicted effects are reviewed. Third, the consequences of the planned development are evaluated both in isolation and in combination with existing and predicted future conditions, with particular attention to the effects deemed to be significant. At this stage, regional development effects are evaluated using criteria standard in the CEA process, focusing on their magnitude, the level of uncertainty involved in effective prediction and taking into consideration the positive effects of mitigation. Fourth, follow-up plans for the management of significant ongoing effects (including continued monitoring) are discussed, by way of a summary of the assessment.

The structure and conduct of CEAs are rooted in ecological studies, which bring together a wide range of disciplines that share an essentially biological base. A considerable literature has developed surrounding the processes applied in CEAs (see, for example, CEAA 1996), and an extensive and specialized terminology has arisen to facilitate discussion of various development effects on ecological systems. For example, effects can be direct, induced, linear, additive, discontinuous, exponential, or threshold; they can be experienced by VECs, populations, communities, or receptors; they can be measured through indicators, changes in key indicator species, loss of biodiversity, alienation, fragmentation or loss of connectivity, acidification, and many other parameters. While this terminology has value in certain contexts, I will not impose it extensively on this discussion. More broadly, however, effects can be positive, negative, or neutral and are typically considered in terms of their scope, duration, frequency, magnitude, and significance. These criteria structure the analysis of the CEA process later in this chapter.

ARCHAEOLOGICAL RESOURCES: THE LEGISLATIVE CONTEXT

Archaeological resources are protected and managed under the provisions of Alberta's Historical Resources Act. When the act was promulgated in 1973 (as the Alberta Heritage Act), it was the first provincial law of its kind in Canada, and it has since served as a model for the development of similar legislation across the country. The act defines an archaeological resource as "a work of humans that (i) is primarily of value for its prehistoric, historic, cultural or scientific significance, and (ii) is or was buried or partially buried in land in Alberta or submerged beneath the surface of any watercourse or permanent body of water in Alberta" (Alberta 2000a, s. 1[a]). Under the act, title to archaeological property is vested in the Crown (Alberta 2000a, s. 32). In the oil sands region, where, until relatively recently, Euro-Canadian presence was limited, the vast majority of known archaeological resources relate to occupation by hunter-gatherer groups over the millennia since the retreat of glacial ice around ten thousand years ago.

Among other things, the Historical Resources Act charges the minister of Alberta Culture with responsibilities for the preservation, orderly development and study, and interpretation of Alberta's historic resources. All archaeological studies in Alberta are accordingly conducted under the terms of an Archaeological Research Permit issued under Alberta Regulation 254/2002 (see Alberta 2000a, s. 30[1]; Alberta 2002). Permit conditions require the recording and reporting of all archaeological resources encountered or uncovered, the preparation and submission of reports, and the curation and submission of specimens and records associated with the work.

In addition, when the minister is of the opinion that an activity to be undertaken by any person will, or is likely to, result in the alteration, damage, or destruction of historic resources, he or she may order that person to carry out an assessment of the effects of the proposed activity and submit a report outlining those effects (Alberta 2000a, s. 37[2]). This assessment is referred to as an Historical Resources Impact Assessment (HRIA). Further, the minister may require the proponent of the planned activity to undertake salvage, preservation, or other necessary measures to offset or mitigate these effects. The discretionary aspects of the Historical Resources Act have an important influence on how assessment and mitigation programs and the required studies are structured. Should the existence of an historic resource of outstanding value, the preservation of which is considered to be in the public interest, come to the minister's attention, the Historical Resources Act further grants the minister the power to designate that resource as a Provincial Historic Resource (Alberta 2000a, s. 20). This status provides the highest level of protection afforded under the act and requires that the minister's written permission be obtained before any alteration can take place.

Together, these elements constitute the principal legislative framework for research on and management of archaeological resources throughout Alberta, including the oil sands region. The conservation objectives of the Historical Resources Act are generally supported in the provisions of the Environmental Protection and Enhancement Act, which, as we saw above, require that an EIA include an assessment of "cultural impacts" (Alberta 2000c, s. 49[d]). Additional support is provided in the EIA terms of reference issued by Alberta Environment and Parks for specific projects, which typically require that the EIA report contain a summary of the results of an HRIA, including proposed mitigation strategies.[1] Considerations pertaining to the conservation of historic resources also figure in the guidance documents and policies of other Government of Alberta regulatory approval agencies (see, for example, ERCB 2011; AUC 2009). In combination with historical factors, the agencies, processes, and programs established to fulfill the objectives of HRIAs, EIAs, and other approval processes have had a singular influence on the character of the baseline information available to us regarding the archaeological and other historic resources in the oil sands region and elsewhere.

Both HRIAs and EIAs include plans for mitigative activities intended to offset the predicted impact of proposed development, although they do so through different regulatory mechanisms. An HRIA provides recommendations to the minister of Alberta Culture by the proponent's consultant, in fulfillment of permit obligations, regarding the appropriate mitigative measures to be pursued. The delegated ministerial authority considers these recommendations from the standpoint of the Historical Resources Act and may accept, reject, or revise them. Clearance to proceed with development may simply be issued, or it may be withheld, or it may be made conditional on the fulfillment of further requirements. For example, a proponent may be obliged to revise development plans so as avoid specific areas, to complete scientific information recovery studies (usually excavations), or to make provisions for follow-up monitoring. The outcome of the HRIA process is communicated directly to the proponent and represents ministerial requirements the fulfillment of which is mandatory. If the conditions surrounding the development activity change, the minister may issue additional requirements at any time.

An EIA presents the proponent's plans for mitigating impacts in order to satisfy the requirements of the Environmental Protection and Enhancement Act and the Oil Sands Conservation Act, which are administered by Alberta Environment and Parks and by the Alberta Energy Regulator (formerly the Alberta Energy Resources Conservation Board), respectively. These plans represent commitments that are fundamental to a proponent's application for approval. They may prove satisfactory as they are, or one or both of the two regulatory bodies may require that the plans be modified in some way prior to granting approval under their respective pieces of legislation. Approval is typically contingent on regular reporting regarding the outcome of ongoing mitigation or monitoring procedures, and failure to comply may result in the rescinding or amendment of approval. In addition, EIAs invariably include summaries of historic resources mitigation recommendations, as well as commitments to abide by the requirements of the Historical Resources Act.

ARCHAEOLOGICAL RESOURCES IN NORTHERN ALBERTA: CHALLENGES TO ASSESSMENT

For reasons discussed in the introduction, the archaeological record in the boreal forest landscapes of northern Alberta consists almost exclusively of stone artifacts, given that the acidity of the soil accelerates the decay of organic materials. As a result, the possibility of radiocarbon dating is largely foreclosed. These conditions, coupled with the typically shallow burial and low frequency of diagnostic artifacts, make it difficult to establish the age of the evidence of prehistoric use of the region. The problem is compounded by the fact that diagnostic artifacts are not often recovered in the test excavations undertaken during the impact assessment stage of archaeological study. Impact assessment strategies generally employ rough measures of significance, such as site size and the relative density of artifacts, as means to determine management priorities, as well as to evaluate how serious the impact associated with a proposed project is likely to be and to identify mitigative strategies that might suffice to offset the anticipated impact. Test excavations accordingly tend to be spread out over a wide area and generally focus on the identification of site boundaries and on preliminary estimations of content. In the absence, however, of detailed excavations of the sort needed to recover time-sensitive artifacts, to pinpoint unusually dense concentrations of archaeological materials, or to gain insight into internal site structure, it is often not possible to establish the true significance of a site. Moreover, the significance of a particular find is often not evident until such excavations are conducted and/or regional comparative studies undertaken.

Given the ten-thousand-year time frame, coupled with the fact that the resources on which prehistoric groups subsisted were widely dispersed and considerable mobility was thus required to access them, evidence of the prehistoric human use of northern landscapes occurs throughout the oil sands region, in a myriad of locations, only some of which can currently be predicted. Written records that might provide sufficiently detailed insight into prehistoric lifeways to be broadly useful for determining the archaeological potential of specific areas are virtually non-existent. By the time Europeans actually arrived in northern Alberta, local First Nations groups had been involved in fur trade networks and had had access to European goods for several generations. While the arrival of newcomers probably had only a limited effect on basic subsistence practices, the records left by early explorers and fur traders suggest that the territorial distribution and patterns of movement of Aboriginal groups had already changed significantly from the situation that had prevailed throughout most of prehistory.[2] The changes brought about by epidemics of disease, competition for access to European commodities, and the ensuing depletion of natural resources have been treated in detail elsewhere (see, for example, Krech 1984; Ray 1974; Yerbury 1986). These factors, coupled with the numerous ecological changes experienced since the retreat of glacial ice in northern Alberta, point to a need for cultural adaptation and group mobility over time that renders historical records of little use for predicting the location of archaeological resources dating to the precontact period.

Furthermore, the legacy of Euro-Canadian colonization has had significant effects on our ability to reconstruct prehistoric lifeways through consultation with existing First Nations groups. Only in comparatively recent years have studies of traditional knowledge been undertaken in northern Alberta communities and elsewhere in the North. Most First Nations communities recognize the urgency of such studies, acknowledging that, with the passage of time, memories of traditional ways of life are increasingly forgotten or altered by the transformations brought about by Euro-Canadian colonization, by the creation of reserves and government efforts at assimilation, and subsequently by modern technology and social conditions. Traditional knowledge studies are valuable not only because they provide First Nations communities with a means to retain and pass on their distinctive culture but also because, if made widely available, they could foster an awareness of and respect for these cultures within the broader society. However, while the information gained through these studies may help us to identify locations habitually revisited or traditional foodstuffs and hunting practices, it does have temporal limitations. That said, archaeologists have been slow to recognize the potential of knowledge bases resident in local Aboriginal communities and, until recently, have not sought to incorporate this knowledge into assessment strategies. Relationships of trust between archaeologists and First Nations communities that would allow a mutually beneficial exchange of information have, unfortunately, been slow to develop.

Additional challenges are illustrated in the character of archaeological resources and the technological means available to identify their presence. Because the groups whose activities archaeologists seek to identify did not build permanent above-ground structures, virtually all of the evidence for their presence is now concealed by forest vegetation within near-surface sediment accumulations. Given the scarcity of existing soil exposures in the largely undisturbed terrain selected for development, for the most part this evidence can be discovered only through excavation. To cover the vast areas involved in proposed oil sands developments, archaeologists use whatever predictive information is at hand, supplemented by their professional judgment, to identify locations for exploratory test excavation. Testing most often occurs by hand, in the form of shovel tests, and only occasionally employs sediment screening.

Exploratory test excavations placed between or on the periphery of areas of prehistoric activity typically cover only a few square metres and thus may not detect evidence of archaeological remains that in fact lies nearby. Similarly, small, unscreened tests carried out in areas in which the evidence of prehistoric use consists of extremely small specimens may fail to discover the archaeological resources present. Variations in the skill and experience of individuals conducting the tests may also have a significant effect on the successful identification of resources. In essence, these samples consist only of the volume of sediment yielded by relatively circumscribed test excavations carried out in locations identified by educated guesswork and whose accuracy depends on the skills of the individual crew member collecting the sample. Many other disciplines employ exceedingly small sample sizes for use in prediction, but few of these samples entail the same degree of uncertainty imposed by variations in professional judgment and chance. During the late 1970s, efforts were made to implement a system of randomized testing, in an effort to provide samples suited to the techniques used in predictive statistical analysis (Conaty 1979). However, it quickly became apparent that once the bias associated with our understanding of human behavior patterns was removed, the minuscule sample sizes yielded by standard archaeological testing methods were inadequate on their own. They failed to identify archaeological resources in areas where they are known to occur, and they led to inaccurate predictions regarding project effects—results that did not meet the requirements associated with the impact assessment process.

In short, when one compares the accuracy of predictions relating to soils, vegetation, and even wildlife resources to the accuracy of predictions regarding the location and relative importance of archaeological materials, it quickly becomes apparent that serious challenges inhere in attempts to evaluate the impact of proposed development on archaeological resources, not only in northern Alberta but throughout Canada's boreal forest. Yet, despite these challenges, the sheer number of project-related assessment studies in the Athabasca oil sands region over the past three decades has enabled us to make considerable progress in understanding the character and distribution of the archaeological resources in the area.

THE SCOPE OF THE ASSESSMENT

In populated areas, the location of archaeological resources is often revealed simply by the presence of human beings: over time, local residents encounter archaeological evidence in the course of other activities, and their knowledge then serves as a guide for researchers. Even in the absence of human occupation, in areas where vegetation is sparse, such as deserts and open plains, traces of archaeological sites are often visible on the surface of the land. In remote, forested areas, however, such evidence is largely concealed. Our knowledge of the distribution and the significance of archaeological resources thus depends directly on the number of archaeological investigations that have previously taken place and on their degree of success. In the oil sands region, the archaeological database is very much a work in progress, one in which baseline archaeological data gradually accumulate as both academic research and project-specific impact assessments take place, and our understanding evolves accordingly. Whereas in the case of other resources cumulative effects can be measured against a reasonably well-established baseline, in the case of archaeological resources we are obliged to work with a continually changing information base.

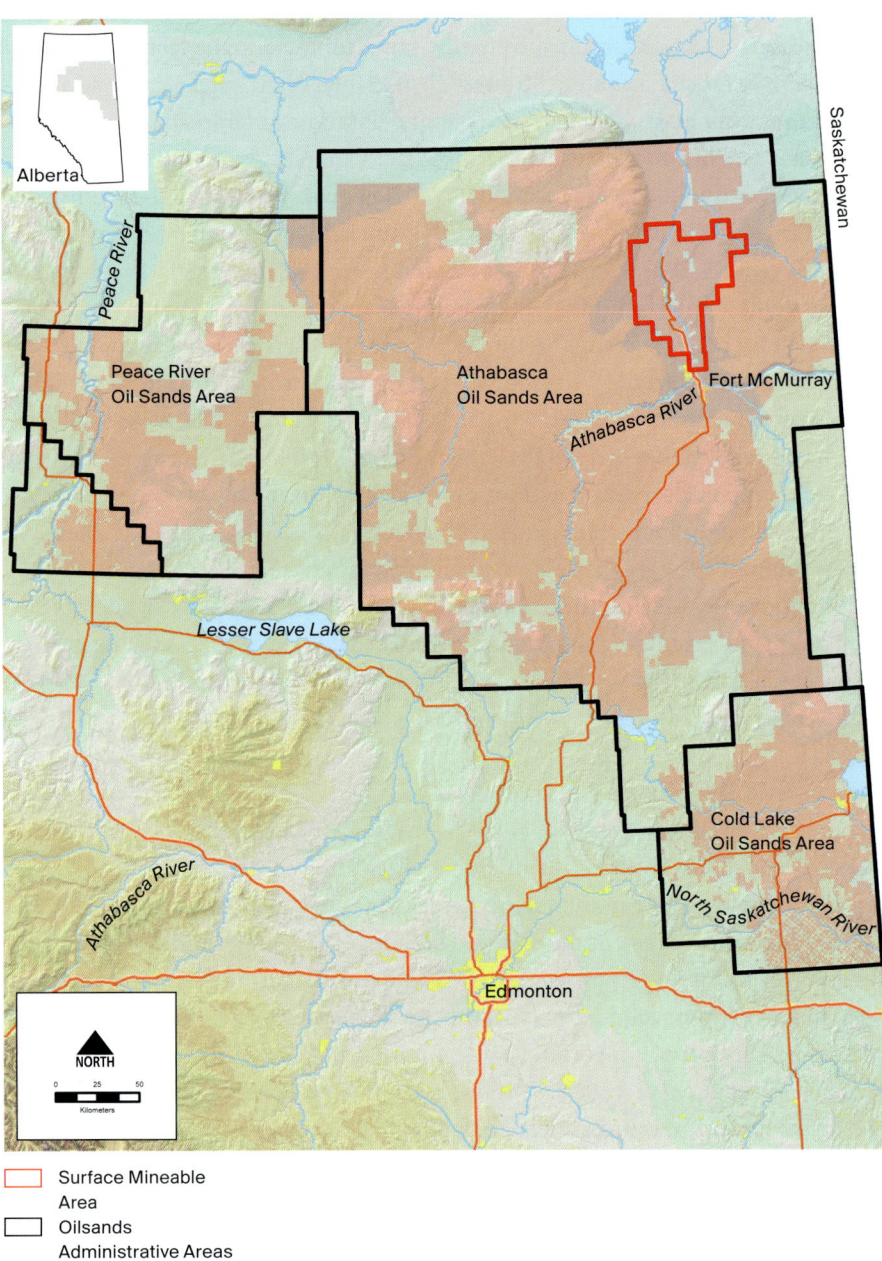

Figure 13.1. Alberta's oil sands region, showing the boundaries of the three administrative areas. The area outlined in red is the surface-minable zone as initially defined.

Despite the lack of comprehensive baseline archaeological information, in order to frame a discussion of the cumulative effects of oil sands development, the temporal and spatial scope of the assessment must first be established (as is true in any CEA). For the purposes of this chapter, the temporal scope is the full time span associated with the development of Alberta's oil sands. In this regard, most surface mining projects will span a period of approximately forty years; some are in mature stages of their operations, others are only in the planning stage, and still others remain to be planned. Existing in situ projects—those that extract bitumen from more deeply buried deposits—are often of shorter duration, but many of the bitumen resources available to such technology have yet to be specifically slated for development. Nevertheless, lease agreements with the province have been reached for most of the known high-value oil sands deposits. Although the future of Alberta's oil sands resources will depend on economic conditions and technological advances in alternative energy sources that cannot be predicted with any kind of accuracy, on broad estimate the full development process could last some eighty to ninety years.

The spatial parameters for this discussion are those established by the Alberta government for the purpose of framing planning initiatives and public consultation efforts in the oil sands region as a whole. On the basis of both geological and economic considerations, three areas have been defined: the Athabasca, Cold Lake, and Peace River Oil Sands areas (fig 13.1). Together, these cover an area of approximately 142,200 square kilometres and encompass all the subsurface heavy oil reserves identified to date. Within these boundaries, lease agreements have been entered into for the approximately 3,500 square kilometres of land originally designated as suitable for oil sands mining—the surface-minable area shown in figure 13.1. Located in the northeast portion of the Athabasca Oil Sands Area, the surface-minable zone is of particular interest for the present discussion, as it roughly corresponds with the Glacial Lake Agassiz flood zone described in several chapters in this volume. The broader area is included in the analysis, however, so as to account for future in situ development projects, as well as the indirectly related development that will likely take place.

THE HISTORY OF ARCHAEOLOGICAL INVESTIGATION IN THE OIL SANDS REGION

Academic Research Studies

Information on the archaeological resources of the oil sands region and northeastern Alberta more broadly is relatively recent and derives essentially from

Table 13.1 Radiocarbon dates from sites in oil sands administrative areas

Site	Oil Sands Area	Date (14C yr BP)	Materials	Diagnostic artifacts and potential association	Reference
GbOs-1 (Caribou Island Lake site)	Cold Lake	4,200 ± 140 GSC 660	Charcoal	Duncan point	Bryan 1987
GfPa-32 (Black Fox Island site)	Athabasca	1,220 ± 130 Beta 6549	Charcoal	Cree pottery	
GhPh-3	Athabasca	365 ± 55 S 518	Charcoal	No associated diagnostics	Gruhn 1981
GhPh-4	Athabasca	0 ± 80 GAK 1899	Wood	No associated diagnostics	Gruhn 1981
GhPh-7	Athabasca	410 ± 130 GSC 1140 1,190 ± 130 GSC 1034 1,150 ±160 GSC 1035	Charcoal	No associated diagnostics	Gruhn 1981
GhPh-11	Athabasca	2,770 ± 100 Beta 4267 1,760 ± 100 Beta 4268	Charcoal	Unknown	Fedirchuk 1982
HaPl-1	Athabasca	1,990 ± 70 GAK 5096 770 ± 70 DIC 1066	Bone	No associated diagnostics	Sims 1980
HcQh-6	Peace River	Eighteen dates ranging from 7,300 ± 110 to 3,160 ± 100	Sediment	No associated diagnostics	Bobrowsky, Damkjar, and Gibson 1988
HeOn-1	Athabasca	1,735 ± 100 S 1275	Charcoal	No associated diagnostics	Pollock 1978
HhOu-70	Athabasca	1,650 ± 40 Beta 248279	Composite calcined bone	Unknown	Roskowski, Landals, and Blower 2008
HhOv-16 (Cree Burn Lake site)	Athabasca	1,240 ± 60 TO-1439	Sediment	Unknown	Head and Van Dyke 1990
HhOv-73 (Bezya site)	Athabasca	3,990 ± 170 Beta 7839	Composite charcoal	Northwest Coast Microblade Tradition	Le Blanc and Ives 1986
HhOv-87	Athabasca	2,030 ± 40 Beta 277702	Bone	Possibly Late Taltheilei	Roskowski and Netzel 2011a
HhOv-156	Athabasca	3,970 ± 30 Beta 312092	Bone	No associated diagnostics	Roskowski and Netzel 2012
HhOv-184	Athabasca	1,640 ± 80 Beta 141288	Composite charcoal	Nezu (Cody) Complex), date considered unrelated to artifacts	Clarke and Ronaghan 2000s
HhOv-245	Athabasca	520 ± 40 Beta 229413	Composite calcined bone	No diagnostics	Wickham and Graham 2009
HhOv-256	Athabasca	4,740 ± 40 Beta 239181	Charcoal from hearth	No associated diagnostics	Wickham and Graham 2009
HhOv-351	Athabasca	1,910 ± 30 Beta 295837	Composite, calcined bone	No associated diagnostics	Roskowski and Netzel 2011b
HhOv-384	Athabasca	2,930 ± 40 Beta 248280	Composite calcined bone	Unknown	Woywitka et al. 2008
HhOv-387	Athabasca	1,900 ± 40 Beta 248281	Composite calcined bone	Unknown	Woywitka et al. 2008
HhOv-449	Athabasca	650 ± 40 Beta 229415	Composite charcoal	No associated diagnostics	Wickham and Graham 2009

Table 13.1 (*continued*)

Site	Oil Sands Area	Date (14C yr BP)	Materials	Diagnostic artifacts and potential association	Reference
HhOv-506	Athabasca	10 ± 30 Beta 298151 2,860 ± 30 Beta 312095 470 ± 30 Beta 312096	Bone	None: modern No associated diagnostics	Roskowski and Netzel 2015
HhOv-520	Athabasca	5,250 ± 40 Beta 312098	Bone	No associated diagnostics	Roskowski and Netzel 2012
HhOw-20	Athabasca	1,670 ± 40 Beta 244942	Calcined bone fragments	Side-notched projectile point, probably Late Prehistoric	Youell et al. 2009
HhOw-30	Athabasca	Modern Beta 188772	Calcined bone fragments	Possible Early Prehistoric point: no direct association	Bryant 2004
HhOw-37	Athabasca	1,300 ± 40 Beta 188773	Charcoal sample from hearth	No associated diagnostics	Bryant 2004
HhOw-45	Athabasca	2,320 ± 40 Beta 255737	Composite calcined bone	No associated diagnostics	Boland, Brenner, and Tischer 2009
HhOw-46	Athabasca	1,980 ± 40 Beta 255739	Composite calcined bone	No associated diagnostics	Boland, Brenner, and Tischer 2009
HhOw-55	Athabasca	280 ± 100 100 ± 40 Beta 255740	Bone	Mummy Cave Complex, Taltheilei, dates considered recent	Kjorlien, Mann, and Tischer 2009
HhOx-9	Athabasca	Modern (< 50 years)	Bone	Associated with historic component of site	Graham and Tischer 2009
HhOx-18	Athabasca	2,080 ± 40 Beta 255742	Calcined bone fragments	No associated diagnostics	Kjorlien, Mann, and Tischer 2009
HiOu-8	Athabasca	130 ± 40 Beta 258074	Calcined bone fragments	No associated diagnostics	Woywitka et al. 2009
HiOv-46	Athabasca	2,270 ± 40 Beta 258073	Calcined bone	No associated diagnostics	Woywitka et al. 2009
HiOv-70	Athabasca	1,710 ± 40 Beta 2580745	Calcined bone	No associated diagnostics	Woywitka et al. 2009
HiOv-126	Athabasca	Modern (< 50 years) Beta 258076	Calcined bone	Unknown	Woywitka et al. 2009
HjPc-4	Athabasca	3,640 ± 120 RL 533	Charcoal	No associated diagnostics	Archaeological Survey site records
HjPc-25	Athabasca	590 ± 40 BGS 2571	Composite charcoal	No associated diagnostics	Archaeological Survey site records
HkPa-4 (Eaglenest Portage site)	Athabasca	1,030 ± 100 DIC 720	Composite charcoal	Frank Channel Taltheilei	Ives 1977a
HkPa-12	Athabasca	865 ± 75 S 1962	Charcoal	No associated diagnostics	Donahue 1976
HkPa-13	Athabasca	2,030 ± 105 S 1973	Charcoal	No associated diagnostics	Archaeological Survey site records
HkPa-14	Athabasca	660 ± 70 S-2175 1,280 ± 95 S1974	Charcoal	No associated diagnostics	Archaeological Survey site records
HkPb-1	n/a	2,795 ± 85 S2174	Bone?	No associated diagnostics	Archaeological Survey site records

two types of study: research conducted by archaeologists based at universities, museums, and other such institutions and impact studies ordered under the management systems described above. Although far fewer in number, the former were some of the earliest to be carried out in the region.

Shortly after the 1973 proclamation of the Alberta Historical Resources Act, Paul Donahue, of the Archaeological Survey of Alberta, undertook surveys along the lower Peace and Athabasca rivers and also examined lakes within the Birch and Caribou Mountain uplands (Donahue 1976). Roughly simultaneously, as part of his master's research at the University of Alberta, Cort Sims (n.d. [1975], 1976) surveyed a portion of the Lower Athabasca River, the shores of Namur Lake in the Birch Mountains, the Peerless Lake–Graham Lake narrows, and the North Wabasca Lake outlet. John Pollock, who was also completing a master's degree at the University of Alberta, conducted surveys along the upper Slave River, in the far northern portion of the region (Pollock 1977), and along the Clearwater River and the shores of Gordon Lake, in the east-central portion of the region (Pollock 1976a, 1976b). Edward McCullough (1982) subsequently completed a comprehensive survey of the shorelines of Lac La Biche and surrounding areas for his master's research at the University of Calgary.

Each of these studies, while varying in productivity, resulted in the identification of a range of archaeological sites associated with major water bodies in the region. The materials recovered at these sites were typical of boreal forest assemblages, consisting largely of stone artifacts, concentrated in near-surface contexts, with few distinctive activity-related features, such as hearths or pits, and virtually no preservation of associated organic materials. Some of the regional sites that appeared to have relatively rich records of occupation or to hold the potential for a clear stratigraphic separation of their occupation sequences (a situation of great value for the reconstruction of regional chronology) were also the subject of research excavations. These include the Gardiner Lake Narrows site (Sims n.d. [1975]) and the Eaglenest Portage site (Ives 1977, 1985), both in the Birch Mountains; the Wetzel Lake site, in the Caribou Mountains (Conaty 1977); the Duckett site, located on Ethel Lake in the Cold Lake region (Fedirchuk and McCullough 1992; McCullough 1981b); and the Bezya site, an unusual inland site at which microblades were recovered (LeBlanc and Ives 1986).

While a relatively complete sequence of prehistoric regional occupation, beginning with fluted points, can be inferred as a result of these excavations, the details of that sequence are based exclusively on variations in the styles of projectile points, While a number of radiocarbon dates are available for the oil sands region (table 13.1), few are relevant to a clear understanding of prehistoric occupation in the region.

In the southern part of the region, in the Cold Lake area, a relatively accurate date of 4,200 ^{14}C yr BP for a Middle Prehistoric period Duncan occupation was obtained at the Caribou Island Lake site (GbOs-1), west of the town of Bonnyville (Bryan 1987). At Lac La Biche, in the Athabasca Oil Sands Area, Cree-related pottery styles at the Black Fox Island site were dated to roughly twelve hundred years ago (Learn 1986).

Further north, in the Athabasca Oil Sands Area, one date is from Bezya (HhOv-73), a site now lost to oil sands development (LeBlanc and Ives 1986). The date, of 3,990 ^{14}C yr BP, was obtained from a composite sample of small charcoal fragments, which may or may not relate to the Northwest Microblade component of this multi-component site. Nearby, a possible association with late Taltheilei material was dated to just over two thousand years ago (Roskowski and Netzel 2011a). A date of 1,670 ^{14}C yr BP, from HhOw-20, a small site on the lower flanks of the Birch Mountains (Youell et al. 2009), may reflect a Late Prehistoric occupation. In the Birch Mountains, at the Eaglenest Portage site (HkPa-4), a date of 1,030 ^{14}C yr BP, obtained from a composite charcoal sample, is thought to be an accurate indicator of a Frank Channel Taltheilei occupation roughly a thousand years ago.

The remainder of the dates available at the time of writing were either obtained at sites that lack relevant diagnostic artifacts or are too recent to pertain to the archaeological materials found at the site. The uncertainty surrounding possible associations between organic material and stone artifacts that occur in the same vicinity is a pervasive problem for the chronological assignment of the shallow, unstratified sites common in Alberta's oil sands region (as throughout Canada's boreal forest). Obtaining and dating organic material clearly associated with chronologically unmixed or stratigraphically separated archaeological occupations remains an important objective for ordering the prehistory of Canada's boreal forest.

Resource Management Studies

Athabasca Oil Sands Area. The presence of minable near-surface bitumen deposits in northeastern Alberta has been known since prehistoric times, but commercial exploitation of this resource did not begin until early in the twentieth century.[3] The evolution of the oil sands industry has been dealt with elsewhere (see, for example, Carrigy 1974; Hein 2000; McKenzie-Brown, Jaremko, and Finch 1993) and is a key topic at a major oil sands interpretive centre operated by the Alberta government in Fort McMurray. Broadly speaking, development can be divided into three major phases.

Initial efforts at bitumen extraction began in the 1920s and extended through the 1940s. Conducted on an experimental basis and on a relatively small scale, these efforts ultimately failed for reasons relating to extraction technology, coupled with transport difficulties and inadequacies in the processing technology. The second phase had its inception in the late 1960s, when commercially viable processes for recovering bitumen from near-surface minable deposits were developed and large-scale facilities became feasible. At this point, the provincial government became cognizant of the need to assess environmental effects in conjunction with the project approval processes. Environmental studies, including archaeological assessments, accompanied the approvals of the mine and plant sites proposed as part of the development of Syncrude Leases 17 and 22 (Losey and Sims 1973; Reeves 1977; Van Dyke and Reeves 1984). Preliminary studies (Conaty 1979; Ronaghan 1981a, 1981b) were also undertaken in connection with Shell's initial plans for development of their Lease 13, a joint venture called the Alsands project, which Shell has since reconfigured (with different partners) as the Muskeg River Mine. At roughly the same time, archaeological assessments took place for joint ventures by NOVA and PetroCanada, one called Canstar, west of the Athabasca River near the Birch Mountains (McCullough 1981a; McCullough, Wilson, and Fowler 1982), and the other, OSLO, located on the east side of the river in the Muskeg Mountain area (McCullough 1980c; McCullough and Fedirchuk 1989).

In addition, archaeological investigations were conducted for infrastructural developments that accompanied the proposed projects, including highway and bridge construction, pipelines, and power lines. These archaeological studies produced some very interesting results, including an unexpectedly high number and dense distribution of sites, and included evidence of stone quarry use on a very large scale. However, many of these project proposals were shelved during the recession of the 1980s, leaving only the Great Canadian Oil Sands (later Suncor) and Syncrude projects as operating mines.

The details of this rich pattern of prehistoric use did not begin to emerge until the third stage of oil sand development was initiated in the late 1990s. During this period, major project proposals were revived, principally because of the rising price of oil, which once again made profits and shareholder return feasible. As environmentally significant projects, including Syncrude's Aurora North and Shell's Muskeg River Mine, moved through the new environmental approval processes and into construction stages, detailed HRIAs were carried out and mitigation programs implemented. Owing to the extent and severity of the environmental effects associated with these projects, large-scale excavation programs were required, in the course of which large numbers of both newly identified and previously recorded archaeological sites produced large collections of artifacts, suggesting intense, ancient prehistoric use of this distinctive landscape. With the addition of proposals for several more surface mines both east and west of the Athabasca River in the surface-minable core of the oil sands region, archaeological studies intensified as impact assessment documents were prepared. These were augmented by studies associated with approvals for the development of infrastructural facilities such as roads, utilities, and sources of sand and gravel needed for construction activities. This process of development is, of course, ongoing and, given the pace and scale of activity in the region, requires considerable manpower and management expertise.

Since the passage of the Alberta Historical Resources Act in 1973, a total of 796 Archaeological Research Permits have been issued in the Athabasca Oil Sands Area (table 13.2). Of these, the vast majority (663) have been issued in connection with HRIAs, with only 44 issued purely for research purposes. Despite the difficulties associated with boreal forest archaeological investigation, within the roughly 93,000 square kilometres that constitute the Athabasca Oil Sands Area (Alberta Energy 2016b), 2,580 archaeological sites have been recorded (fig 13.2 and table 13.3). Six of these have been afforded ministerial protection as Provincial Historic Resources, while in 1,572 cases (roughly 61% of the total), the initial recording of the site and the recovery of sample artifacts were deemed sufficient to compensate for future impact.

The number and density of sites identified within the surface-minable area of the oil sands would not have been predicted on the basis of standard understandings of the prehistoric use of boreal forest environments in Canada. Perhaps as significant as the large number of sites is their content. Many consist of extremely dense deposits indicative of intense stone tool manufacturing and use, and, in several, the remains are believed to reflect the processing of in situ bedrock material for use elsewhere. Even in sites that reflect only tool manufacture and use, the recovery of one to two thousand items per square metre excavated is not uncommon.

Cold Lake Oil Sands Area. More deeply buried bitumen deposits in the Clearwater Formation in the Cold Lake area were explored geologically in the 1960s by Imperial Oil. However, initial archaeological work did not begin until the mid-1970s, and then only in conjunction with proposed larger development projects. These consisted largely of recovery operations employing cyclic steam stimulation, which, owing to their scale and complexity, require that EIAs and HRIAs be carried out in advance of approval. Wide-ranging archaeological studies associated with Imperial Oil's Cold Lake project lease areas (Reeves and

Table 13.2 Archaeological Research Permits issued in the oil sands region

Oil Sands Administrative Area	Impact assessment permits (HRIA)	Mitigation permits (HRIM)	Joint HRIA and HRIM permits	Research permits	Total permits
Athabasca	663	67	22	44	796
Cold Lake	256	16	5	25	302
Peace River	79	1	0	10	90

Table 13.3 Historic Resource Value of oil sands archaeological sites

Oil Sands Administrative Area	Provincial Historic Resources (HRV1)*	Significant sites (HRV3)	Sites warranting avoidance or further study (HRV4)	Sites of no further concern (HVR 0)	Site total
Athabasca	6	14	988	1,572	2,580
Cold Lake	4	3	159	517	683
Peace River	1	0	27	89	117

* HRV (Historic Resource Value) as assigned in Alberta Culture and Tourism, Listing of Historic Resources (http://culture.alberta.ca/heritage-and-museums/programs-and-services/land-use-planning/).

McCullough 1977) identified a series of both prehistoric sites and more recent cabin locations reflecting traditional use of the area by Aboriginal peoples. The archaeological sites clustered around the shores of existing fish-bearing lakes such as Marie Lake and Bourque Lake, to the northwest of Cold Lake, and Ethel Lake, to the west. Intervening areas, which were characterized by extensive low wetlands, produced limited evidence of prehistoric use: archaeological sites in these areas consisted of small campsites and isolated artifact finds. One significant site on the shores of Ethel Lake has been the subject of research excavations sponsored by Imperial Oil (Fedirchuk and McCullough 1992; McCullough 1981b). While not stratified, it revealed evidence of repeated occupation and use beginning with the Fluted Point Tradition as early as eleven thousand years ago.

At approximately 18,000 square kilometres (Alberta Energy 2016b), Cold Lake is the smallest of the three oil sands administrative areas. Because of the presence of conventional petroleum resources in this region, it did not experience as significant a recession in development activity as did the Athabasca region, which witnessed an associated decline in the number of archaeological investigations during the period between 1988 and 1994. However, some of the projects that proceeded in the Cold Lake area during this period (most of them

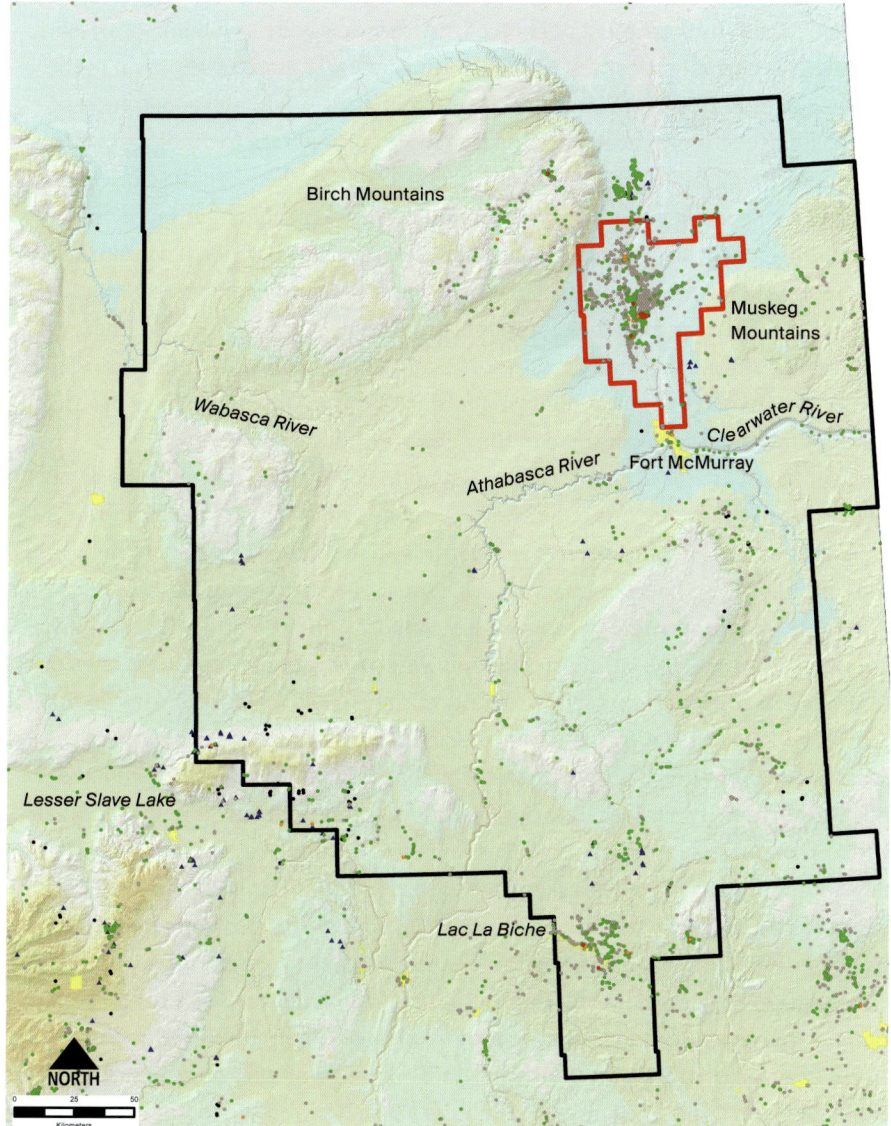

Figure 13.2. The distribution of archaeological sites in the Athabasca Oil Sands Area. "Historic Resources Value" (HRV) is a ranking assigned by Alberta Culture to all archaeological resources in the province. Value 1 refers to a site designated as a Provincial Historic Resource, and value 2 a site designated as a Municipal Historic Resource, while value 3 identifies a site that may warrant one of these designations in the future. Value 4 indicates that a site may require avoidance. Value 0 is assigned to sites deemed to be of relatively low significance.

HISTORICAL
RESOURCE VALUE
- HRV 1
- HRV 3
- HRV 4
- HRV 0
- Pending
- Unknown
- Athabasca Oil Sands Area
- Surface Mineable Area

situated in the southernmost portion of the territory) were smaller in scale, did not involve thermal processes, and were consequently granted approval in the absence of either EIAs or HRIAs. Since the mid-1990s, with demonstration of the feasibility of more recent advances in in situ production technology, such as steam-assisted gravity drainage (SAGD), coupled with the increasing price of oil, many EIA-approved projects have been initiated, and others have been proposed. Most of these projects proceed in stages, and almost all have required archaeological investigation prior to approval. Archaeological investigations undertaken in advance of developments proposed by Suncor, Norcen Energy, EnCana, CNRL, BlackRock Ventures, Husky Energy, Shell, and others, as well as numerous auxiliary developments including roads and pipelines, have produced results similar to those obtained in the early studies.

Significant archaeological resources tend to occur along lakeshores and in association with extant drainage systems. Areas subject to proposed development that are located between these features appear less likely to contain significant prehistoric archaeological sites, with sites relating to relatively recent use by Aboriginal residents more commonly encountered and recorded. (Particularly noteworthy among the latter are the remnants of cabins and other evidence of Aboriginal traditional use at the federally administered Cold Lake Air Weapons Range, which First Nations and Métis residents were forced to evacuate in 1953, during the Cold War.) These patterns of archaeological site distribution have been confirmed by assessments completed for infrastructural projects associated with oil sands development, as well as for projects not directly related to this development, that have been proposed within the boundaries of the Cold Lake Oil Sands Area. These include pipelines (McCullough 1980b; Ronaghan 1981c), transmission lines (Ronaghan 1982), roads and sources of granular material for construction projects (McCullough 1980a; McCullough and Fowler 1981), recreational parks situated along the shores of Cold Lake (Fedirchuk 1980b; Kowal 1990; Reeves 1976; Wood 1980), and subdivisions proposed within the towns of Cold Lake and Grande Centre (Fedirchuk 1980a; Newton 1980; Pollock 1981, 1982; Van Dyke 1980).

The North Saskatchewan River valley, which runs through the southern portion of the Cold Lake area, and the numerous fish-bearing lakes in the region are attractive for residential development and as tourist destinations. This commercial activity may augment the potential for impact to historic resources in the area and also complicates cumulative analysis, as it is not entirely clear how much of this potential for impact can be directly attributed to oil sands development. The same applies to the large tracts of privately owned agricultural lands, as well as major local development projects such as those within and surrounding the City of Cold Lake and the towns of Bonnyville, St. Paul, and Elk Point. The region also contains sites of historical significance to the fur trade, notably Fort George and Buckingham House (1792–1800, located not far southeast of Elk Point, and Fort de l'Isle (1799–1801), a site that consists of three trading posts clustered on an island in the North Saskatchewan. The Cree community of Frog Lake, the site of the famous uprising in 1885, during the North-West Resistance, is also located in the Cold Lake area.

A number of archaeological studies were carried out in the Cold Lake area prior to 1973, when Alberta Historical Resources Act became law. Since then, a total of 302 Archaeological Research Permits have since been issued for sites within the area, again with a majority (256) in connection with development management studies and only 25 for research purposes (see table 13.2). As a result, 683 archaeological resources have been identified within the Cold Lake Oil Sands Area to date (fig 13.3 and see table 13.3), of which four are protected as Provincial Historic Resources. In the case of 517 sites (roughly 76% of the total), initial recording and sample recovery have been deemed sufficient to offset any foreseeable future impact.

Peace River Oil Sands Area. Development of the relatively isolated oil sands deposits contained within the Peace River Formation and the Gething Formation, east of the Peace River valley and northwest of Utikima Lake, also began at a relatively early date. In 1979, development of a pilot project was initiated by Shell Canada, one that employed a cyclic steam stimulation process to extract bitumen from the company's long-held leases in the area. Archaeological studies preceded the original project in 1980, as well as subsequent stages of its expansion (Meyer 2002; Van Dyke 1984). Impact assessments have also been conducted for two stages of Shell's companion Carmon Creek project. In addition, several other projects have been undertaken in the area, but because they employ the primary cold extraction techniques standard for oil recovery throughout the province, they did not require approval through the EIA process and were not screened for their potential to disturb archaeological resources.

The archaeological studies conducted in connection with the thermal projects in the area have thus far been unproductive, although two relatively recent cabin sites have been identified in the proximity of areas slated for development. The fact that none of the thermal projects is located near major rivers or lakes reduces the likelihood that significant archaeological sites exist in their vicinity. Moreover, relatively modest impact zones are associated with the type and scale of development proposed, and none of the archaeological sites identified in the region to date is regarded as likely to be affected.

Figure 13.3. Cold Lake Oil Sands Area, showing the archaeological sites recorded to date

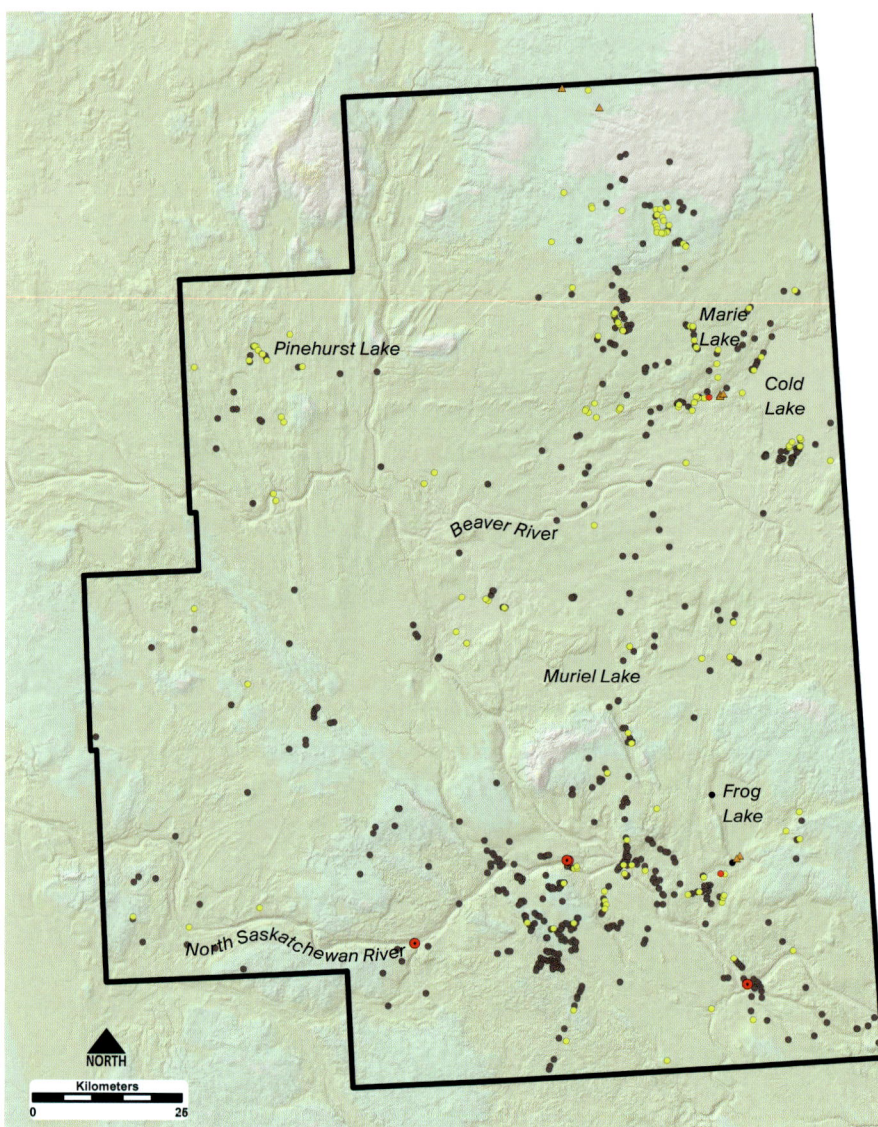

HISTORIC RESOURCES VALUE

- ● HRV 1
- ● HRV 3
- ○ HRV 4
- ● HRV 0
- △ Pending
- • N/A
- ☐ Cold Lake Oil Sands Area

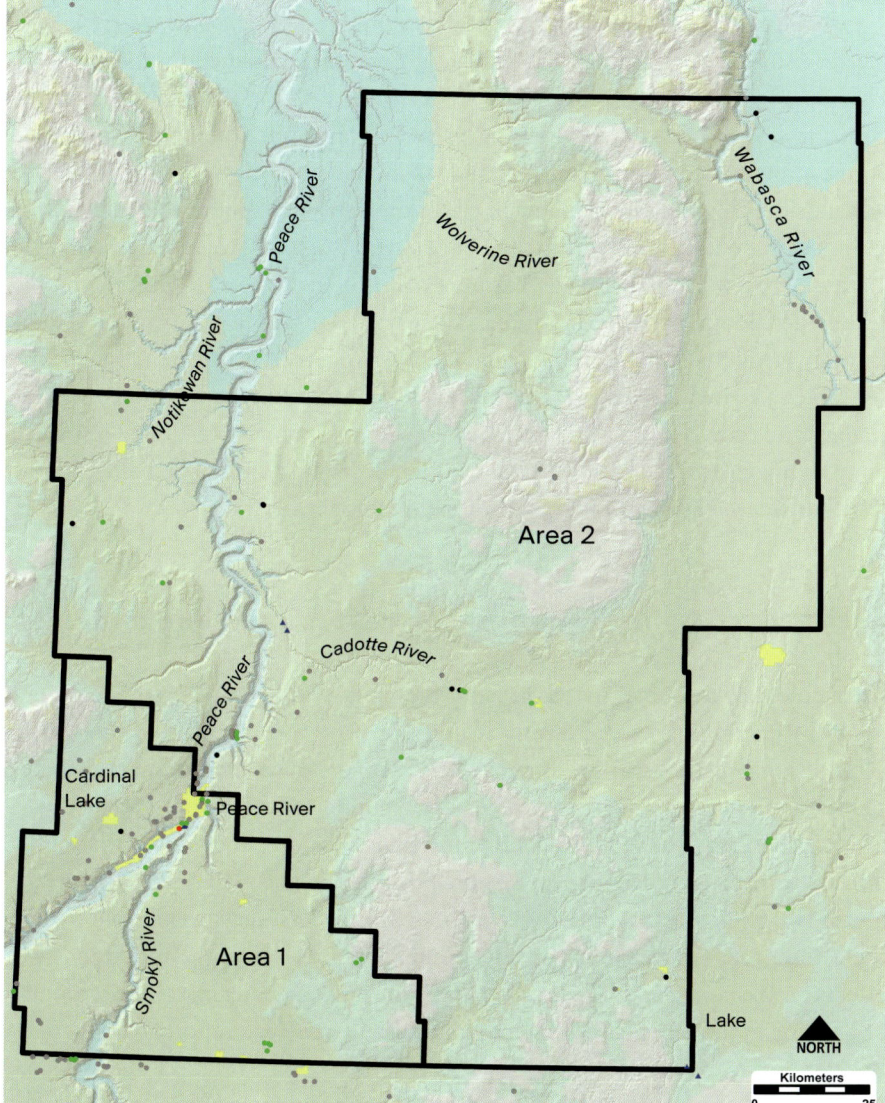

Figure 13.4. Peace River Oil Sands Area, showing the archaeological sites recorded to date

HISTORICAL
RESOURCE VALUE
- 🔴 HRV 1
- 🟠 HRV 3
- 🟢 HRV 4
- ⚫ HRV 0
- ▲ Pending
- ⚫ Unknown
- ▢ Peace River Oil Sands Area

The Peace River Oil Sands Area extends across roughly 29,000 square kilometres (Alberta Energy 2016b). Because the boundaries defined for the area encompass a significant section of the Peace River valley and the towns of Grimshaw, Peace River, and Manning, as well as large tracts of land subject to Forest Management Agreements, a range of archaeological studies not related to oil sands development have taken place. To date, 90 Archaeological Research Permits have been issued within the Peace River Oil Sands Area (see table 13.2). Of these, 79 represent impact assessments, mostly in connection with forestry-related development projects, while 10 were issued for research purposes. Thus far, only 117 archaeological resources have been recorded in the area (fig 13.4 and see table 13.3), including one Provincial Historic Resource, the 1819–1820 fur trading and provisioning post Fort St. Mary's II (also known as McLeod's Fort). For roughly 76% of the total (89 sites), initial recording and sample recovery were considered sufficient to offset any future impact.

THE EFFECTS OF OIL SANDS DEVELOPMENT: DIRECT AND INDIRECT

Archaeological resources in the oil sands region lie within near-surface sediments in locations scattered through the region. They are therefore vulnerable to activities that disturb these sediments. The recovery of bitumen from the oil sands entails some of the most extensive and complex development projects ever planned and executed in Alberta, and their physical effects constitute significant threats to archaeological resources located in and near development areas. The following discussion will evaluate the direct impact associated with the various types of oil sands development projects, as well as their indirect effects on the landscape. These effects must be measured against a background of ongoing natural processes that inevitably result in deterioration of archaeological resources over time.

Mention has already been made of the natural chemical processes that degrade the organic remnants of prehistoric activities. In addition, because vegetation established shortly after deglaciation has held sediments in place throughout the boreal region, and surfaces suitable for human use have thus remained the same for approximately ten thousand years, virtually all of the remnants of former occupation are held in the active root zone of current vegetation. Tree falls and root action that have taken place across the millennia, coupled with freeze-thaw cycles, have naturally altered the integrity of archaeological remains, as well as the relationships between these remnants of prehistoric human activity. The scale of this disruption is in some cases no less than the degree of disturbance that might be anticipated in the early stages of the oil sands development process that involve forest removal. This is mostly done when the ground is frozen and snow covered and typically has a limited effect on in-place mineral sediments.

Another natural disruptive process that warrants consideration is erosion, which, in the boreal forest, is largely limited to active drainage channels. Sites located in the immediate proximity of drainage channels may be altered or damaged by the erosion of stream banks during high-water episodes or by ice scouring during the winter. Unless a development project creates new drainages or alters the flow of existing drainages, which occurs rarely and then only in association with mine projects, these alterations are not usually related to development.

As I write, there are 131 active oil sands projects in Alberta (Alberta Energy 2016a). Of these, six are mining projects spread over several locations; the remaining projects use a variety of in situ recovery methods. The potential impact to historic resources differs significantly between these two types of projects. As of June 2007, an estimated 3,807 oil sands agreements were in place with the Alberta government, covering roughly 60,863 square kilometres of northeastern Alberta (Alberta Energy 2007, 8). To date, the areas containing the highest quality and largest quantity of bitumen deposits have been leased, totalling some 85,000 square kilometres, or roughly 61% of the three administrative areas combined (Alberta Energy 2016b).

Direct Effects

Mining projects. Once a lease has been signed, mining projects unfold in multiple phases. The earliest stages, beginning with exploratory activities designed to outline ore bodies and define the depths of overburden, do not require public approval processes but are screened for their possible impact on historic resources. In these early stages, flexibility exists with regard to the locations to be disturbed, and known archaeological sites must be avoided. However, subsequent to the approval of the project through the standard EIA and HRIA processes, more intensive and extensive disturbances take place. Forest clearance proceeds relatively quickly, as does the removal of overburden in plant and facilities locations, in early-stage mining areas and areas designated for the disposal of tailings (waste material from the separation process, suspended in water), and in areas that will contain "borrow pits," that is, sources of the sand and gravel needed in construction. In addition, in areas where overburden will be stored for use in later landscape reclamation, as well as along the numerous access roads required, top soil and peat (also reserved for reclamation purposes) are stripped away. All of these activities will consume any archaeological resources present.

Mine projects are required by law to maximize the recovery of the oil sands present within the project area (Alberta 2000b, s. 18[1]), but lands that do not contain deposits or that overlie low-grade deposits are also extremely valuable as facilities locations. As a result, virtually all the land within the area approved for mining will be disturbed, and all the archaeological resources that exist within the project area will be destroyed. Although this destruction is inevitable, opportunities for mitigation are available at certain stages in the operation. Mining itself tends to move relatively slowly over the landscape, but many other components of these projects proceed relatively swiftly, and the need for site security, along with procedural considerations such as seasonal impediments to activity, means that mitigative measures must be implemented fairly quickly after the project is approved.

In addition to the activities associated with the mining and processing of ore, these projects require auxiliary infrastructure that extends outside the area of the lease. This includes roads, additional sources of gravel and sand ("borrow"), supplies of electric power, gas, and water, and pipelines that transport partially processed bitumen to upgrading facilities or to refineries that are frequently located at a great distance from the development area. Planning and approval processes for these facilities often fall outside the publicly reviewed EIA process, but, because of their scale, most are screened for any potential impact on historic resources. Occasionally, some flexibility exists with regard to the location of ancillary facilities, and avoidance measures are therefore possible. Otherwise, unless an archaeological site is deemed highly significant, mitigative measures are generally regarded as sufficient.

As is evident from the closure plans associated with all EIA applications submitted for mining projects, once the forty or more years of mining operations conclude and reclamation of the landscape is complete, all but the most minor portions of proposed lease areas will have been disturbed. Although reclamation can help to reduce the negative impacts of development on many natural resources, the archaeological record cannot be restored. Moreover, reclamation activities may themselves have an impact on any archaeological resources that remain in project area.

In situ projects. Several types of bitumen recovery methods, known as "in situ" techniques, are applied to ore bodies that lie in buried at depths greater than 75 metres. These methods entail drilling in dispersed locations within a defined body of ore, the extraction of bitumen and its transport to central facilities that undertake varying degrees of initial processing, and then shipment of the partially processed product through pipelines to upgraders where it is converted to synthetic crude oil prior to further shipment to refineries for final processing. Currently only one SAGD project, Nexen's Long Lake Project, has an on-site upgrader.

Some of these more deeply seated deposits can be extracted without the use of heat by means of high-pressure pumps that draw primarily from vertically drilled wells. Such processes are suitable for the lighter, less viscous oil sands deposits in the southern Cold Lake and Peace River areas. Both economically and environmentally, these methods have advantages and disadvantages. On the one hand, they are less expensive than thermal methods, and they use less water. On the other, they have proportionally lower rates of bitumen recovery, and they produce proportionally greater amounts of sand, which requires disposal. Moreover, in terms of their impact on historic resources, the fact that such projects (which are relatively few in number) are exempt from the standard EIA process means that their effects on archaeological remains are assessed only if known sites occur in the proximity of the project area.

The most common extraction methods employed in the oil sands region involve the use of heat injected into the bitumen formation in the form of steam, sometimes with the addition of a solvent. The heat decreases the viscosity of the heavier bitumen, allowing it to flow into collection wells either vertically or horizontally drilled. Cyclic Steam Stimulation (CSS) uses vertical wells that cycle through periods of steam injection followed by pump-assisted production. This method, which is feasible for the thicker ore deposits near Cold Lake, achieves recovery rates of between 20% and 25% of the available bitumen. Drilling from pads that contain up to twenty wells has become the norm for this process. Steam-Assisted Gravity Drainage (SAGD) involves paired wells, both drilled horizontally into the deposit. Steam is injected into the upper part of the bitumen bearing formation along perforated horizontal pipes, again to decrease the viscosity of the bitumen, which then flows downward. A second horizontal, perforated pipe, located roughly 5 metres below the first, at the base of the formation, collects the mobilized bitumen, along with condensed water, which is then pumped to the surface. The injection and production wells are generally grouped in pairs on pads containing twelve or more such pairs. This process is suitable for the thinner deposits west of Cold Lake and in the portions of the Athabasca Oil Sands Area where overburden is too thick to make surface mining economically feasible. It typically achieves recovery rates approaching 60% of the available bitumen.

In both these processes, considerable water and energy in the form of natural gas are required to create the necessary steam. This entails the construction of facilities to produce the steam and above-ground pipelines to deliver it, along with pipelines needed to recover the heated bitumen-water mix for processing. Disturbances within the project area consist of large complexes of well pads and the necessary processing facilities, as well as sources of granular material or fill often required for building permanent facilities in watery terrain dominated by muskeg. Disturbances associated with above-ground pipelines and access roads, however, extend over considerably broader areas.

All in situ recovery scenarios proceed in a complex series of stages. Because the precise positions chosen for wells will determine production rates, detailed knowledge of the extent, depth, and quality of ore is essential. Very detailed exploration programs precede the planning of well-pad configurations. The numerous cut lines and test wells needed for these programs are generally developed in winter and often do not necessarily have severe effects on archaeological resources as the snow cover and frozen ground significantly inhibit disturbance to intact sediments from forest removal activity and vehicle traffic. These preliminary stages are not considered part of EIA-regulated activity and are screened in advance for their possible impact on historic resources only if known sites might be disturbed.

Subsequently, extraction and processing proceed in phases, which see the successive development and abandonment of specific "pay zones," throughout the area approved in the lease and over the life of the project. These stages of the project are subject to EIA review processes. However, because only the initial configuration of production facilities has been planned at this stage, only a small portion of the project's eventual zone of impact can be specified and examined for Historical Resources Act approval. It is rare for all the areas of archaeological potential within the territory covered by a lease to be examined at the public review stage. As a general rule, subsequent effects on archaeological resources must be predicted on the basis of a subsample of the lease, although developers may be obliged to carry out follow-up assessments as a condition of project approval.

Like mining projects, in situ projects also require the provision of water, natural gas, and electrical supply in the form of pipelines and transmission lines, as well as the construction of access roads and borrow pits to supply construction activities. With the exception of supply and disposal pipelines, this ancillary development generally takes place outside the area of the leases. For that reason, and because such construction activities do not directly involve the production of bitumen, they are exempt from the EIA process. Again, however, in view of their scale, most are screened for their potential impact on archaeological resources, and assessments are required when considered necessary.

The overall disruption of the landscape associated with in situ projects is considerably less than that associated with mining projects. A recent detailed analysis of the OPTI-Nexen Long Lake SAGD project, undertaken jointly by the Canadian Parks and Wilderness Society and the Pembina Institute, concluded that long-term clearing of 8.3% of the lease area would take place and that this level of disturbance would be considered representative of other projects of this nature (Schneider and Dyer 2006, viii).

In summary, the widely dispersed explorations that form the first phase of in situ projects can have a direct impact on archaeological resources, although this impact may be decreased by the typical winter staging of these activities. Direct impact may also occur at plant sites and in well-pad areas, as well as along roads and in pipeline and utility corridors. In contrast to mining projects, however, greater flexibility exists with regard to the placement of specific components of such projects, with the result that it is sometimes possible to avoid disturbing areas determined to be of archaeological significance.

Indirect Effects

As we have seen, oil sands development projects bring with them additional industrial and commercial activities that often result in significant new disruptions of the landscape. Examples include the construction of highways and bridges, along with their associated needs for borrow, materials lay-down areas, and so forth, as well as large product-supply operations that provide limestone and granular materials (sand and gravel). In addition, these projects tend to be accompanied by the construction of service facilities, such as the industrial parks east of Syncrude's Mildred Lake operation and east of the Athabasca River, along with hotels and industrial camps to house and feed temporary workers, new subdivisions and other forms of urban expansion, and the development of recreational facilities, including the expansion of existing parks. Because such ancillary development projects typically entail extensive disruption of the terrain, they pose a significant threat to any archaeological resources that are present. None of the above-mentioned projects is subject to EIA public review processes, but all require approval from relevant regulatory agencies, and, as a result of referral agreements with these agencies, most such projects are screened by Alberta Culture for their potential to damage or destroy archaeological remains. However, numerous small-scale developments—municipal roads, privately owned gravel pits, and the like—go forward without review.

At the same time, somewhat ironically, proposed oil sands and related development projects have another indirect effect on the archaeological record. When projects require HRIAs in advance of approval, numerous features of the landscape undergo close examination, archaeological resources, many previously unknown, are identified and recorded, assessment samples are collected, and reports are written. When avoidance measures can be implemented, they are frequently ordered, and sites are thus conserved for future study. When impact is inevitable, mitigative excavations are required to offset the damage, detailed notes are taken, representative samples of archaeological materials are recovered and analyzed, and the results are reported. All site records, reports, and artifact collections become part of the public record. The conservation of information and materials and an enhanced understanding of regional prehistory are benefits that accrue from legally mandated review and approval processes.

MITIGATION

The need for mitigative measures to offset the destruction of archaeological resources is determined by the assistant deputy minister of the Heritage Division of Alberta Culture, to whom ministerial authority is delegated on a project-by-project basis. Decisions pertaining to mitigation are based on a review of the results of HRIAs, which are generally conducted in advance of development approval. The archaeological component of the HRIA process, whether undertaken in conjunction with an EIA or independently, proceeds under a permit issued by the Archaeological Survey, a section of the Historic Resources Management Branch of Alberta Culture. This permit defines an area to be examined and ensures compliance with basic standards for field investigation, analysis, and reporting. The required studies are completed by professional archaeological consultants employed by the project proponent, and the resulting reports are subsequently submitted to and professionally evaluated by the government. If deficiencies exist, these must be addressed.

When proposed development plans threaten archaeological resources, both the significance of these materials and the severity of the proposed impact are evaluated. The results of these evaluations contribute to the management requirements, which are issued to the proponents as a condition of project approval. When an archaeological resource is deemed to have relatively limited significance as an historic resource, the information and samples collected during the assessment process are usually considered sufficient to allow development to proceed without further conservation measures. When an archaeological site appears to have moderate or relatively high significance, the requirements issued typically order that the site be avoided, if at all possible, or else that the proponent undertake mitigative measures, which consist in the recovery of a representative sample of the site. These measures involve standard archaeological excavation, the collection of artifacts and other materials, and subsequent analysis. Because the significance of an archaeological site is often only fully appreciated as information emerges through detailed sample recovery, mitigative procedures can and frequently do proceed in incremental stages.

When a resource is consumed either at the completion of the HRIA stage or after sufficient mitigation has been completed, the recovery and permanent preservation of the materials and information generated by initial archaeological studies is considered adequate compensation for the eventual destruction of the site during development. However, because archaeological excavation is a consumptive exercise that itself constitutes a form of impact, in situ conservation is often considered the preferable course of action for sites that have perceived value. Avoidance ensures that the site can be actively managed and remains intact for future study, as new interpretive techniques become available and an improved understanding of the archaeological record allows for better-directed research. Provided it is feasible to avoid the site entirely, this tends to be the option selected by proponents for a variety of reasons, not the least of which is cost.

If a resource of exceptional value is discovered, measures outlined in the Alberta Historical Resources Act can ensure its permanent preservation through its designation as a Provincial Historic Resource. The exceptional value of an historic resource can also be recognized through the National Historic Sites program administered by Parks Canada and the Historic Sites and Monuments Board of Canada, as well as by UNESCO's World Heritage Site program. The latter status has already been afforded to one archaeological resource in Alberta, Head-Smashed-In Buffalo Jump. Such designations are not, however, usually accompanied by requirements concerning the site's preservation, except insofar as damage or alterations to the site can result in the loss of its commemorative status.

THE CUMULATIVE EFFECTS ASSESSMENT

As a general rule, cumulative effects are discussed within an analytical framework based on certain basic attributes of these effects. Although the definitions of specific attributes can to some extent vary, the overall framework is accepted by most practitioners (see CEAA 1999). These attributes include the *direction* (positive or negative) of the effects and their *scope* (site-specific, local, regional, etc.), *duration* (long-, medium-, or short-term), *frequency* (once, sporadic, or continuous), *magnitude* (low, moderate, or high), and *significance* (negligible, significant, unknown), as well as the *confidence* (low, moderate, or high) vested in the assessment. The last two of these represent summary evaluations and will be considered in some depth below, although I have chosen to address the confidence issue first in order to underscore the largely impressionistic nature of the assessment of long-term oil sands development on archaeological resources. For the most part, CEA evaluations within each of the above categories reflect a prior consideration of the predicted outcome of mitigative action, with the remaining effects consequently considered "residual" (that is, effects for which no compensation can or will be undertaken). Owing to the high degree of uncertainty surrounding the evaluation of cumulative effects on archaeological resources, residual effects cannot usually be identified in any precise manner, although a discussion of them can conceptually inform future management direction.

Direction

The direction of the cumulative effects of oil sands development on archaeological resources is both positive and negative. Wherever surface disturbance occurs and archaeological resources are present, the impact is negative. As we have seen, in the case of the early-stage exploration activities carried out in connection with in situ projects, because these preliminary tests mostly occur on frozen, snow-covered ground and do not severely disturb mineral sediments, they may leave an archaeological site relatively unscathed. Subsequent stages of development, however, completely destroy any archaeological resources that occur within impact zones. These effects are most severely felt in areas where avoidance is not an option, that is, in the minable oil sands region within the Athabasca Oil Sands Area and in areas throughout the broader oil sands region where plant-related facilities are planned for in situ operations. In these areas, it can be predicted with relative certainty that virtually all archaeological resources will be consumed. Where avoidance is an option, such as in relation to the placement of production well pads, access roads, utility lines, and so forth, archaeological resources may be conserved. However, except when a highly significant resource is at risk (in which case the proponent can be ordered to avoid the site), the decision of whether to redesign the project so as to leave the site undisturbed or instead to undertake mitigative excavations is typically left to the proponent. In making such decisions, the proponent will consider a range of factors, among which the cost of conducting mitigative excavations is significant, if not necessarily decisive.

Nevertheless, there is a positive aspect to the direction of oil sands development effects as well. Had there been no oil sands development, archaeological activity in the area would almost certainly have been far more limited. While this is a matter of speculation, it seems very doubtful that any level of investigation equivalent to the 1,109 cultural resource management studies conducted to date within the oil sands administrative areas (752 in Athabasca, 277 in Cold Lake, and 80 in Peace River) would have taken place. In all likelihood, information relating to the prehistoric use of these areas would have been based almost exclusively on the relatively few early research studies conducted, some forestry-related impact assessment studies that might have taken place in each region, and a few other studies, possibly relating to recreational developments in the Cold Lake area, with its numerous lakes.

In particular, much has been learned about the prehistoric occupation of the Athabasca area as a result of industry-sponsored research studies. For example, without these studies, not a single chronologically sensitive artifact might have been recovered in the Lower Athabasca valley outside of the Birch Mountains, which attracted academic research interest in the early and mid-1970s and was the only area to see research excavations. Most of the archaeological chapters in this volume represent syntheses of information that has been gathered in the course of oil sands impact assessments and mitigation activities. This research has resulted in the identification of the densest concentration of archaeological sites thus far known to exist in the Canadian boreal forest. These sites, which lie within the minable oil sands area, reveal a pattern of intense human activity, which has been associated with a conjunction of geological events and changes in climate and vegetation. Such a pattern is unlikely to be repeated on such a scale anywhere else. In addition, immense quantities of cultural materials, chiefly stone artifacts related to quarrying activities, tool manufacture, and tool use, have been recovered and retained in provincial government facilities for future study. Because the scientific study of archaeological sites is continually evolving, and major collections are currently under analysis or have yet to be fully reported, it is difficult to predict what knowledge will eventually be gleaned from these materials. However, these collections owe their existence largely to impact assessments and mitigative efforts carried out in connection with oil sands development.

Scope

The impact of oil sands development on archaeological resources is felt at local, provincial, and national levels. The negative effects will be most acute in the minable oil sands region, where very few archaeological resources will remain if development is complete. To some degree, the samples recovered during mitigation programs will compensate for this loss, as will the preservation of three large site complexes—the Beaver River Quarry, the Cree Burn Lake site, and the Quarry of the Ancestors—as Provincial Historic Resources. Otherwise, however, the region's archaeological record would be erased.

At a provincial scale, the archaeological resources present in the surface-minable oil sands area represent some of the richest yet identified in northern Alberta. Although, as is the case with boreal forest sites throughout the north, organic remains and clear stratigraphic separations are rare at these sites, no other region is expected to contain the density of sites characteristic of the Lower Athabasca valley. The loss of these resources to development is therefore significant to our understanding of the prehistory of the province overall.

At a national scale, I am not aware of any other region in the Canadian boreal forest where a combination of geological processes related to deglaciation and shifts in climate and landscape could generate a pattern of dense prehistoric use equivalent to that identified in the Lower Athabasca valley. Along the northwestern margins of Lake Superior, a concentration of Lakehead Complex sites, dating to the Early Holocene period, has been correlated with outcrops of workable stone materials and glacial lake shorelines in the vicinity of Thunder Bay (Fox 1975; Hamilton 1996). Although these sites may provide some basis for comparison, their density, as well as the density of the materials they contain, appears to be considerably lower than that found in the oil sands. Our ability to reconstruct the early postglacial period in Canada's boreal forest regions will thus be significantly impaired if the archaeological resources in the minable oil sands area disappear.

Duration

This attribute is not considered especially relevant to archaeological resources. These resources are delicate and non-renewable: once an impact takes place, its effects are immediate and permanent. Archaeological remains are for the most part destroyed during the initial stages of development activity, such as the removal of overburden that precedes mining. Once the damage is done, continued disturbances over longer periods of time and at greater depths generally do not produce additional negative effects.

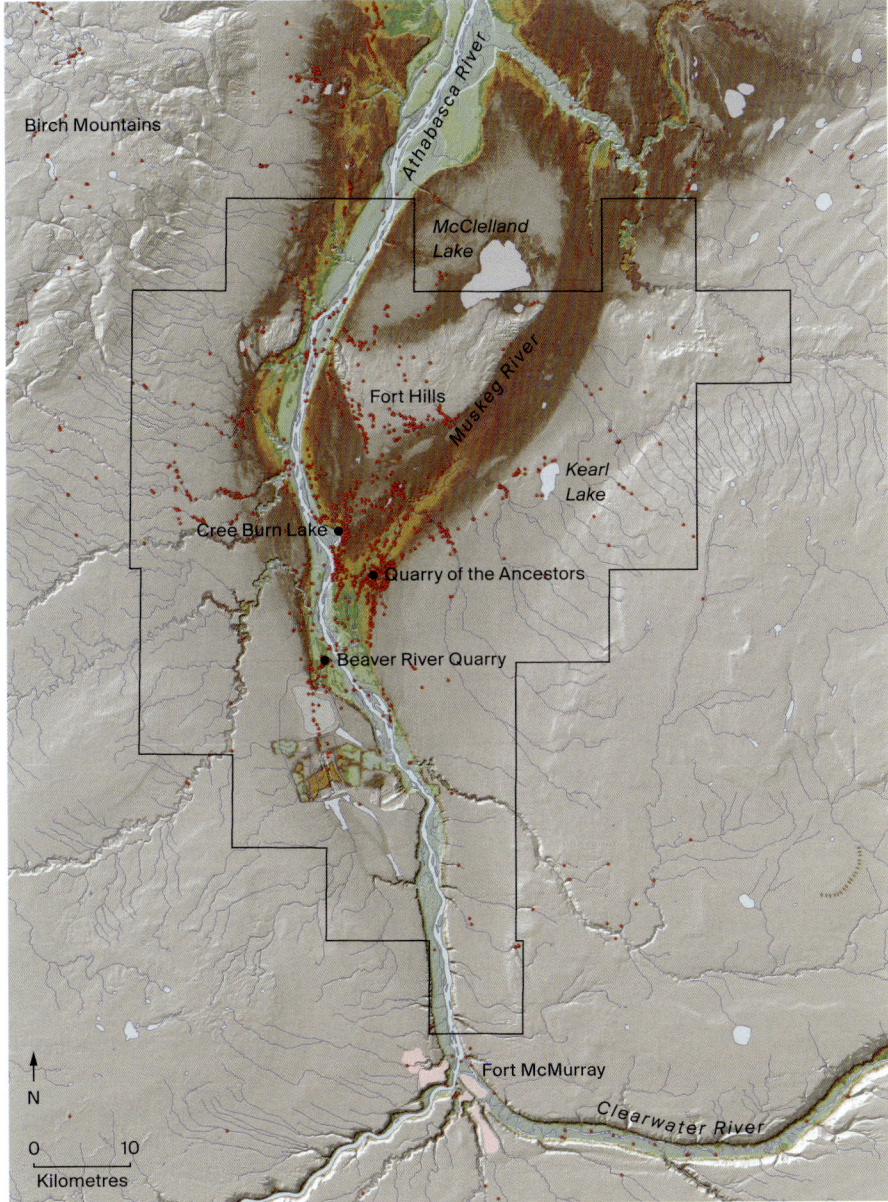

Figure 13.5. The surface-minable oil sands area, showing the archaeological sites recorded to date, including the three designated as Provincial Historic Resources

Frequency

Again, this attribute is relevant more to the impact of development on other environmental resources. Because archaeological materials are fragile, their destruction is typically final the moment it takes place.

Magnitude

As we have seen, oil sands development projects are complex, and they generate a multitude of physical effects. The sheer scale of these projects provides a rough indication of the magnitude of their combined impact on archaeological resources. However, especially in the longer term, considerable uncertainty surrounds specific plans for development, and we cannot know whether archaeological sites will prove to be present in the areas slated for development. Clearly, such a degree of uncertainty severely limits the accuracy with which quantitative predictions can be made.

As indicated earlier, in the surface-minable area, virtually all archaeological sites may be consumed in the course of development. Three major sites have, however, been designated Provincial Historic Resources and are currently preserved (fig 13.5). The first, the Beaver River Quarry, was identified in the 1970s, and for decades it was the only known in situ source of Beaver River Sandstone, the dominant lithic material in most regional archaeological sites. Because the stone found at that site is coarser in grade, however, than that from which most regional artifacts are made, the Beaver River Quarry is now considered to be a secondary source of the material.

The second, the Cree Burn Lake site, consists of a large number of focal points of activity spread over an area of approximately 1 by 2 kilometres that afford evidence of the processing and use of Beaver River Sandstone in a wide range of tasks. This site was designated a Provincial Historic Resource in 1999 but is now mostly contained within lands ceded to the Fort McKay First Nation in connection with treaty land entitlement claims. Although permanent preservation therefore cannot be guaranteed under provincial legislation, in view of the support provided by the Fort McKay First Nation for the initial designation of the site by the province, together with subsequent statements of band policy and the respect and value placed by the First Nation on its cultural and natural heritage, the site is likely to remain undisturbed.

The third site, the Quarry of the Ancestors, is apparently the first in situ bedrock source of fine-grained Beaver River Sandstone to be identified. The quarry complex encompasses twelve highly significant sites, as well as several less important ones, at which activities related to stone extraction and processing were carried out, with evidence of relatively continuous use for more than nine thousand years. The Quarry of the Ancestors is situated between two major oil sands mining projects and a large aggregate extraction development. The designation of the site and some of the surrounding natural landscape as a Provincial Historic Resource will preserve an important example of the pattern of prehistoric landscape and resource use associated with the minable oil sands area.

With these three exceptions, almost all of the 1,129 known archaeological sites within the surface-minable area of the Athabasca Oil Sands Area will likely be destroyed by development. To date, mitigative excavations have been undertaken at 312 (27.6%) of these sites. The size of the sample required at each site is based on information provided by initial assessment studies and on a consideration of professional recommendations, but it generally amounts to no more than 15% of the site. These samples nonetheless constitute more detailed and comprehensive effort at conservation than the samples initially collected as part of HRIAs. In addition, although many of the sites currently not slated for development may not need to be permanently avoided, they could remain intact indefinitely. As explained earlier, in the case of archaeological resources deemed to be of low value, only information about their location and the limited samples collected during HRIA procedures are retained, and the original sites are sacrificed when they lie in the path of development. There are currently 815 such sites in the surface-minable oils sands area.

In the remaining portions of the Athabasca Oil Sands Area, archaeological resources of significance appear to be less numerous. Those that have been identified are primarily located in the Birch Mountains uplands (see figure 8.2 in this volume), where the glacial deposits overlying the oil sands are exceedingly deep and considerations of cost thus tend to rule out oil sands development. The presence of both Wildland Provincial Park and large tracts of First Nations reserve lands is also likely to discourage development.

Other areas of concentration occur east of Muskeg Mountain along the Saskatchewan border, where impact assessments have been conducted in connection with recently proposed in situ projects in the upper Firebag River basin. Sites have also been recorded in the Gregoire Lake–Long Lake area, located in the east-central part of the region, and in the area north of the Cold Lake Air Weapons Range, where in situ projects are in operation or are proceeding through planning stages. In addition, largely as a result of research studies undertaken in the mid-1970s, sites have been identified in the southern portion of the Athabasca Oil Sands Area, around Lac La Biche, although no oil sands projects are proposed in the vicinity.

Outside of the surface-minable area, significant archaeological sites tend to be located in direct proximity to lakeshores and drainage margins. Oil sands development is unlikely to pose a significant direct threat to these sites, as the avoidance of harmful effects on surface water is not only desirable but generally required as a condition of project approval. Nevertheless, these sites could face indirect threats from population growth and the concomitant need for residential, commercial, and recreational development, of the sort currently underway in the Lac La Biche area. Adequate project referral and management mechanisms may be able to offset these effects, however, and would preserve the two designated Provincial Historic Resources in this area, Lac La Biche Mission and Portage La Biche.

Apart from prehistoric sites associated with bodies of water, archaeological sites of moderate value that are located outside the surface-minable area usually date to the historical period—remains of Aboriginal traditional use, such as abandoned and collapsed trappers' cabins and related features. The southern portion of the area, between Fort McMurray and Heart Lake, 70 kilometres northeast of Lac La Biche, has been designated as the wettest area in the Boreal Mixedwood ecoregion (Strong and Leggat 1981, 27–28). It is dominated by fen and muskeg that developed under moist or wet conditions, with mixedwood vegetation in limited upland areas. Such areas are not well suited to extended occupation, and efforts to recover significant evidence of prehistoric use have not been overly successful.

In the Cold Lake Oil Sands Area, we encounter a pattern of site distribution similar to that described above for the southern Athabasca region (see fig 13.3). Significant archaeological resources lie either along the shores of the numerous lakes in the area or along the banks of the North Saskatchewan River, a pattern that is especially clear in the central part of the region. As we have seen, the proximity of these sites to water affords them protection from oil sands development. The designated Provincial Historic Resources in the region, which date to the fur trade era and to the Frog Lake resistance in 1885, are also well protected, and development is further constrained by the freehold agricultural lands in the area. Moreover, in this southernmost of the three areas, oil sands projects chiefly employ standard (primary cold) extraction techniques, the disturbance footprint of which is much smaller than that of other in situ techniques. Indeed, none of the oil sands projects assessed to date has required mitigative excavations to offset direct impact. So far, in fact, the only development-related mitigative studies have been undertaken in connection with provincial park facilities and road construction, and, with one exception, these projects took place in the 1970s and 1980s, before the modern era of oil sands development.

Overall, given the dispersed nature of existing and planned in situ facilities in the region, the avoidance options available, and the application of mandatory assessment procedures through the standard project approval processes, the magnitude of the direct impact of oil sands development on archaeological resources is predicted to be modest. With its numerous lakes, well-developed infrastructure, and substantial existing population base, however, the region is likely to attract a growing number of people. Although the impact of population increase is difficult to forecast over a period of eighty to ninety years, residential and recreational development along lakeshores, only indirectly related to regional oil sands development, will probably pose a more significant threat to archaeological resources than oil sands projects themselves.

Archaeological investigations conducted as part of the assessment of proposed oil sands projects in the area have yielded relatively small numbers of archaeological sites of limited significance in proportion to effort expended but have served to clarify the distribution of archaeological sites. These studies suggest that, for the most part, standard avoidance and mitigative measures, if applied systematically, should be able to reduce the magnitude of direct oil sands development to acceptable levels. In addition, one of the regional developers, Imperial Oil, has sponsored a program of archaeological research and public education at a significant archaeological site, the Duckett site (GdOo-16; Fedirchuk and McCullough 1992; McCullough 1981b). The excavation of this site has made a substantive contribution to our knowledge, one that has provided a clearer understanding of regional the prehistory of the region.

In the Peace River Oil Sands Area, only limited development has thus far occurred, and the archaeological investigations that accompany development have likewise been limited. Moreover, the relatively few studies conducted to date have seen largely negative results. While this might suggest that resources in the area tend to be ephemeral, little of interpretive value has been recovered and few collections of any kind conserved. Most known sites are prehistoric in age and cluster along the Peace River; the one designated Provincial Historic Resource, Fort St. Mary's II, dates to the fur trade era. Almost without exception, sites warranting avoidance or further study are adjacent to river and creek margins or lakeshores. The two occupations thus far identified that have no association with water—a Métis settlement and a homestead—date to the historic period.

Given that proposed oil sands projects in the area are spread out over a large expanse of territory, that known prehistoric sites are located in the vicinity of water, and that direct effects of development on surface water are stringently controlled, the magnitude of direct negative effects on archaeological resources is expected to be relatively small. Future negative effects can probably be managed through the application of the standard assessment and mitigation procedures, which are applied not only to in situ oil sands projects and their related infrastructure (pipelines, power lines, and so forth) but also to forestry operations that take place within the area and to government infrastructure projects such as road construction and park development.

Considering the relatively modest potential for the development of the oil sands deposits in the Peace River area, population increases are also expected to be relatively modest, with the town of Peace River likely to absorb the majority of this growth. The recreational potential of this region is considerably less than that of the Cold Lake area, as the terrain is relatively level, the area contains no major lakes, and access throughout the region is constrained by extensive areas of muskeg. Although the long-term presence of agricultural development in the western portion of this area, serviced by the towns of Peace River and Grimshaw, has already had significant consequences for the integrity of archaeological resources in the area, these effects will probably not be significantly exacerbated by the indirect impact of oil sands development.

Confidence

The uncertainty associated with any cumulative effects analysis is perhaps the greatest constraint on the value of the conclusions that can be drawn. Even in fields of study where variables can be controlled with reasonable precision, uncertainty can surround the accuracy and depth of knowledge about baseline conditions, the relative applicability of the analytical processes employed, and the reliability of information concerning future effects. In the case of social science disciplines, particularly one that seeks to understand cultural systems that no longer exist and relies on evidence that is not immediately visible to the eye, a substantial degree of uncertainty is inherent in any analysis.

As we have seen, baseline archaeological information about specific areas of northern Alberta is largely absent until studies designed to assess the impact of development are undertaken. These studies are, however, carried out only in areas that are likely to be disturbed by individual projects. Information even from immediately adjacent areas is typically lacking, which not only limits our knowledge base but impairs our ability to engage in comparative analyses, except in the most general, speculative terms. Over time, regional baseline information will accumulate, such that, with enough study, we will be able to make decisions relating to conservation with greater confidence. At is stands, however, resource managers are often placed in a reactive position, called upon to determine the appropriate scope of a study and to evaluate the adequacy of the investigations, the significance of the archaeological resources in question, and the sufficiency of mitigation measures before the nature of the regional resource base is known. Given that the activities involved will consume the evidence, decisions cannot be revisited.

One source of uncertainty arises from the scope of the regulatory system associated with historic resources conservation. In comparison to the systems in place in other jurisdictions, Alberta's regulatory framework is generally regarded as one of the more comprehensive, as one might indeed infer from the description of that framework near the start of this chapter. All the same, it is not without certain limitations. While large-scale industrial, commercial, and recreational projects are subject to regulatory review, many smaller-scale projects are not, despite the fact that their collective impact could potentially compare with that of a major project. With regard to oil sands development in particular, exploratory tests conducted during the early phases of in situ projects, are not subject to assessment unless known archaeological resources are potentially affected, nor are projects that employ primary cold extraction methods. Especially in areas where few studies have taken place and few sites are already known to exist, this policy can have significant implications for the preservation of as yet undiscovered archaeological resources, even though the impact footprint of individual activities may be quite limited. Gaps in assessment may also occur in connection with small-scale projects that require approval only from municipal agencies. Developers must submit information about the proposed project to these agencies, but unless the municipal authorities have agreed to refer such information to Alberta Culture, so that a decision can be made about whether archaeological assessment should be undertaken, these projects will simply proceed without review. Even though the Alberta Historical Resources Act does apply on private land (in contrast to the situation in many other jurisdictions), similar gaps in assessment can arise in connection with relatively minor projects carried out on privately owned land, as these often do not require prior approval. Such limitations on the reach of regulatory processes can mean that cumulative effects assessments fail to consider the full impact of development on archaeological resources.

Even when archaeological assessments are required, significant uncertainty surrounds the design and execution of these studies. Archeological investigations proceed on the basis of predictions concerning the topographical locations and sedimentary contexts in which archaeological resources seem likely to occur. The assumptions on which these predictions are based may or may not be accurate and, while general principles often do apply, specific circumstances in the past may have influenced the distribution of sites in unsuspected ways, such as occurs in the surface-minable area of the oil sands. Granted, as local and regional baseline information accumulates, study designs achieve greater predictive accuracy. All the same, practical considerations, including cost, dictate the need for a selective approach, with the result that many areas do not receive examination. In addition, a considerable amount of uncertainty is inevitable in the application of archaeological techniques. After a location is chosen, the successful deployment of these techniques depends on a great number of variables, including professional judgments concerning the specific placement of exploratory excavations, the visibility of the materials sought (very small items are easily missed), and the skill and experience of the researchers responsible for identifying them. Together, the imperfect reliability of predictions concerning site locations and the imprecision inherent in excavation procedures may raise questions about whether the archaeological remains within a given area have been adequately identified.

The accuracy of site assessments is another area of concern. Archaeological sites vary enormously in size, complexity, and significance, and they require thoughtful assessment to ensure that their historic value is accurately recognized. The Historic Resource Values assigned to sites during assessment (for details, see the caption to fig 13.2) determine which archaeological resources will be protected and to what degree. If the presence of diagnostic specimens and/or datable materials, of dense concentrations of artifacts or other remains, or of other distinctive features goes unrecognized, significant resources may be consumed rather than conserved. Given that, inasmuch as archaeological evidence is buried, it can easily escape notice, such oversights may occur more often than might be hoped. When important sites are inaccurately evaluated, the cumulative effects of regional development will be underestimated.

Uncertainty also surrounds the mitigative processes undertaken to offset the impact of development. Avoidance is always the preferred option, but because we cannot accurately forecast when and where development will occur, the future integrity of such resources cannot be guaranteed. The designation of a site as a Provincial Historic Resource does ensure its preservation, but this status is afforded only to a few, highly valuable resources. For the most part, mitigative measures are applied. The resulting samples are intended to be representative, but how far this is true depends on the thoroughness of the initial evaluation of the site, on its size and complexity, and on the effectiveness of current archaeological methods. Existing techniques of analysis may not be sufficient to identify materials that would later be recognized as important sources of information, once newer techniques had been developed. Moreover, as we have seen, sample sizes typically amount to no more than 15% of the recognized site deposits, and decisions about where to conduct sample excavations must be made on the basis of informed judgment. Considerable uncertainty thus remains as to whether the results of mitigative measures adequately represent the archaeological resources that have since been consumed.

One of the greatest sources of uncertainty associated with any cumulative effects assessment lies with our ability to predict the location, timing, and scale of future development. As I write, economic conditions have entered a period of retrenchment, characterized by guarded optimism, lowered expectations, and a sharp decline in product prices. Additionally, if growing concerns about our carbon footprint result in the effective development and wide-scale adoption of alternative energy sources, forecasts of sustained growth in investment in oil sands development may assume a less aggressive trajectory. Because we cannot accurately predict how oil sands development will unfold, and on what schedule, uncertainty also surrounds the nature and timing of potential indirect effects of this development, such as those produced by population growth and the associated need for new infrastructure, housing, recreational facilities, and so on.

Finally, even the legal framework within which cumulative effects assessments are carried out is potentially unstable. Changes could be made to the regulatory systems that manage archaeological resources that would affect which development projects are reviewed and how management requirements are defined and implemented. The lack of established thresholds that might act as trigger mechanisms in the project review and approval stages further compounds the uncertainty that surrounds our efforts to evaluate the combined effects of regional development on archaeological resources.

Significance

The final component of a cumulative effects assessment consists of an evaluation of the significance of the combined effects of development within the area defined for study on particular elements of the ecological and socio-cultural environment, with specific emphasis on the additive effects of a proposed project. In many analyses, the accuracy of predicted outcomes is affected by the degree of confidence associated with the evaluation. While quantitative measures can be used to illustrate certain aspects of significance, qualitative evaluation is required to synthesize these complex issues and provide a direction for future decision making.

To evaluate the significance of the impact of oil sands development on archaeological resources, we must begin with an appreciation of the value of the resources that have been or will be affected. As we have seen, given that archaeological resources are concealed from view, their very existence, as well as their potential significance, generally becomes apparent only as a result of the assessment studies that accompany specific projects. The reliability of these studies is, however, constrained by the considerable degree of uncertainty surrounding the effectiveness of regulatory assessment systems, the archaeological methodologies currently in use, and the initiative and skills of those performing the work. In addition, we lack comprehensive baseline information that might help us to predict where sites may lie and that would lend context and depth to our understanding of their value.

On the regulatory side, despite existing tools for assessing impact, we have seen that many smaller-scale developments, especially those apt to be deemed either unrelated or indirectly related to oil sands development, are not screened, and their impact on archaeological resources thus goes unrecognized. The scale of this loss of information is impossible to quantify. In view of the relatively small footprint of the projects in question, however, we would expect it to be relatively limited, although not inconsequential.

In terms of procedures, the preference for study designs that have seen success and/or have been approved by regulators in the past tend to create a closed system, in which the results both reflect and reinforce the underlying assumptions, which may themselves go unquestioned. (For additional discussion, see chapter 12.) The tendency to cling to familiar principles and methods, rather than subject them to periodic scrutiny, can mean that significant resources go undiscovered and are subsequently lost. Similarly, the fact that variations often exist in where and how test excavations are carried out affects the comparability and/or the reliability of the results obtained. This can, in turn, lead to the assignment of landscape and resource values on the basis of information that is in fact misleading, whether because it is incomplete or simply inaccurate. In view of these factors, the question then becomes, how significant are the effects of these methodological issues with respect to combined effects of regional oil sands development?

Despite the limitations of current methods, chances are that large, complex resources—those that represent repeated, long-term use of the landscape and reflect broad, fundamental patterns of prehistoric land use—will be discovered in the course of standard archaeological assessments and mitigation procedures. It is the smaller, less multi-faceted, and possibly even task-specific sites that may remain undiscovered—or, if discovered, may go unsampled, except in a modest and unplanned fashion. By virtue of their very simplicity, however, and the relatively more focused set of activities that they reflect, these sites may be the very ones that could provide the most coherent and valuable information about specific prehistoric subsistence activities. The possibility that such sites will be missed and perhaps destroyed as a result of limitations built into current management procedures represents a significant unmitigated and ongoing negative effect of regional development. It may be offset to some degree by the potential for redundancy, in the form of repeated occurrences of sites devoted to similar uses, which together create a pattern that can be recognized by information gleaned from only a few such sites. However, patterns are defined by the number of occurrences, and redundancy needs to be demonstrated, not assumed.

Although it is not practical to design and execute studies that would identify and assess all resources, and a certain degree of loss must thus be anticipated, the discovery of a site that could fill in some of the current gaps in the boreal forest archaeological record remains an outstanding need, and one not confined to this region. Despite the quantity of sites identified to date, none possesses the qualities we seek: a stratigraphic separation of sequential occupations, which would enable us to trace cultural periods, and the preservation of organic remains, which would permit us to assign absolute dates to the occupations identified and to address questions regarding subsistence strategies more directly. The potential value of such a resource is widely recognized but not often explicitly considered in field strategies developed for impact assessments, largely because the circumstances that might produce such a site are rare and/or not yet fully recognized.

As we have seen, in the greater Athabasca Oil Sands Area, as well as in the Cold Lake and Peace River areas, the distribution of archaeological resources conforms to the pattern we would expect for the Canadian boreal forest, with concentrations of sites directly associated with ecologically rich and varied locations such as lakeshores and major drainage systems. Within the minable oil sands area, however, assessment and mitigation studies associated with ongoing and proposed development have identified a previously unanticipated concentration of rich and diverse evidence of prehistoric land use, one that is extremely ancient and, to our knowledge, unique in North America. The significance of these sites can be appreciated from several vantage points. From a scientific perspective, quite apart from the sheer quantity of the materials present, these sites have proven to contain a wide range of temporally sensitive artifacts, sufficient to enable the construction of a full postglacial chronological sequence of prehistoric human occupation (see Reeves, Blakey, and Lobb, chapter 6 in this volume). The record is also sufficiently rich and varied to allow functional interpretations of significant portions of the pattern of use observed. Furthermore, relatively unusual and unexpected technologies, most importantly in the form of microblades, are represented in both in situ resources and recovered collections. From an educational perspective, few areas in the Canadian boreal forest have seen such extensive studies or studies that have produced such valuable information on northern prehistoric lifeways. While some of this information is highly specialized, much of it can be presented in a way that allows it to contribute to public education—this volume being one such example. From a cultural perspective, the support for site preservation received from regional First Nations groups testifies to the significance that these communities attach to the archaeological evidence that exists in the area. It also serves to remind others of the depth of their past and the longevity and flexibility of their cultural traditions.

As several chapters in this volume have indicated, the process of deglaciation in the area created terrain attractive to human habitation, the remnants of which are reflected in the dense distribution of sites we see today. It was this same process that made surface mining possible, however, and for this reason these archaeological resources are especially vulnerable. As indicated earlier, questions surround the degree to which mitigative samples adequately represent the value of the resources that have been consumed by development. In addition, although archived reports, site records, and museum collections contain a wealth of information, this information is not widely circulated and, with the exception of publications like this, is largely unsynthesized. The potential value of these records and collections, while clearly significant, is difficult to quantify or predict without advance knowledge of future research interests and methods. While some might argue that a certain redundancy exists in collections of artifacts, which could limit their overall value, I doubt this could be conclusively demonstrated, and, in any case, the fact that these artifacts represent such a small proportion of the archaeological resources that have been consumed provides a defensible basis for their preservation. Few would argue, however, that these collections have a value equivalent to that of in situ resources.

The limitations of the materials conserved through mitigative procedures, coupled with the severity of the predicted impact in the surface-minable area, enhance the significance of the remaining in situ resources. While we are fairly certain that the core area of this dense concentration of sites has been examined, with the result that mitigation programs have completed and clearance for development has been issued, there remain some areas that have not been examined and can be expected to fill in missing pieces of the pattern. This raises a question germane to most cumulative effects analysis, namely, whether it is possible to establish a threshold beyond which impact is no longer acceptable.

In matters of safety and health, thresholds can be established on relatively firm grounds. Even in ecological matters, in cases where functioning systems and their interactions can be quantified, measures that predict population collapse, for example, can be developed. Archaeological resources do not, however, interact with existing ecological, social, economic, or cultural systems except in abstract, value-based terms. That is, the loss of irreplaceable historic resources may be considered utterly unacceptable to some, regrettable but inevitable to others, and inconsequential to many. When the preservation of these resources comes at the cost of making financially lucrative resources unavailable, negatively affecting income to the province and to corporations, their employees, and shareholders, one set of interests comes into play. Nevertheless, from a social and cultural perspective, a strong case can be made for the long-term value of the preservation of cultural heritage.

In view of the uncertainties outlined above, thresholds of acceptable losses for archaeological resources cannot be firmly established beyond general considerations of intrinsic value and representativeness. The intrinsic value of such resources is perhaps best recognized in terms of their relationship to the events and overall trends that serve to structure, characterize, and lend colour to human history. Like any process of understanding, reconstructing prehistory thorough archaeological study is itself a cumulative process: the more one knows, the better one can frame questions the answers to which will continue to advance that understanding. Resources that can address these questions thus take on special value. Given that archaeological resource management is reactive rather that proactive, the formulation of relevant research objectives or questions and the recognition of the resources that will best address them requires the continual review and synthesis of information, as well as the ability to respond flexibly. Although obstacles to this process exist, reasonable success in overcoming them could serve to keep the negative impact of oil sands development below a threshold that might ultimately be seen as unacceptable. Staying current about the evolution of research objectives is therefore a major imperative for those involved in the management of archaeological resources.

In the surface-minable oil sands area, key research questions cluster around a number of topics, including seasonal subsistence strategies, absolute dates for specific occupations, the cultural relationships between the prehistoric peoples who occupied the area and the peoples who inhabited the adjacent plains and subarctic regions, and the timing and functional integration of microblade technology into cultures with access to plentiful stone tool sources. Archaeological resources that have the potential to address these topics, as well as other evolving issues of importance, would be candidates for mitigative study or permanent preservation.

In the Cold Lake area, only one major excavation, that at the Duckett site, has taken place. As a result, more general regional research questions are appropriate. These would highlight the recovery of diagnostic specimens that would aid in confirming or extending current information bearing on local chronological sequences, on the distribution of archaeological sites, especially the apparent focus on water bodies of water, on the identification of patterns of lithic raw material use, and so forth. In the Peace River region, very little is known of the prehistoric archaeological record. Consequently, it is difficult to shape relevant research questions beyond those relating to the basic distribution and character of archaeological sites in the area. Today, the region offers only limited variability in environmental resources, which suggests that it may have had a comparatively low carrying capacity for the resources sought by prehistoric hunting groups. Assessment of this assumption may constitute a starting point for archaeological inquiry in this region.

In terms of thresholds pertaining to representativeness, the wholesale loss of archaeological sites deemed to be of low value has been identified as a shortcoming of existing mitigative strategies. Similar losses may be experienced in areas of apparent lower value located within or on the margins of recognized concentrations. Especially in view of the magnitude of the impact foreseen in the minable oil sands area, the possibility that these losses should be offset either through mitigative excavations or through the preservation of selected sites warrants consideration.

Finally, opportunities to preserve archaeological resources should be sought throughout the surface-minable oil sands area as a proactive long-term strategy for offsetting the anticipated cumulative effects of regional development. Such conservation could, for example, be considered for smaller sites that, in and of themselves, might seem of modest values but are considered representative of land use patterns that were once widely dispersed in the area.

CONCLUSION

As the population increases, and as expectations about the quality of life grow higher, human activities are having incrementally greater effects on both the natural environment and the social world. An increasing awareness of the fundamental connections within and between natural and social systems has resulted in a widespread recognition that, if we expect to achieve sustainability, we need to evaluate the effects of change holistically and over extended periods of time. Given the complexity of the conditions to which they are applied, cumulative effects assessments frequently fall short of their goals. Nevertheless, the systemic approach on which such assessments are based is universally acknowledged as vital to the long-term preservation of ecological and social integrity in concert with ongoing development activity.

As critics have pointed out (see Duinker and Greig 2006; Kennett 1999; Wenig 2002), cumulative effects assessments have generally been employed on a project-specific basis for purposes of compliance: such assessments are undertaken as part of the review and approval processes required by law. As presently applied, the CEA process is initiated by project proponents and has been one of the tools used to gain approval from regulatory agencies. This does not necessarily imply that study methods have been inappropriate, but the restricted scope and objectives of existing studies have limited their applicability on a regional and national scale, as well as their capacity to provide direction to regulators who consider the public interest in matters of development. In the oil sands region, efforts have been made to bring together regional stakeholders in a co-operative forum in order to discuss issues pertaining to the cumulative effects of development, but these efforts have not resulted in a consensus regarding management frameworks or impact thresholds.[4] However, subsequent initiatives on the part of the Alberta government have recognized the shortcomings of the CEA process and have sought to develop frameworks capable of addressing them (see, for example, Alberta Environment 2008).

In contrast to the project-specific approach, this chapter has adopted a broad regional perspective in its review of the cumulative effects of oil sands development on a single component of the environmental context: archaeological resources. Although its scope has been limited to the three administrative areas that encompass existing oil sands agreements, oil sands–related development in fact extends to areas farther south, along product delivery, utility, and transportation corridors, into what is now known as the industrial heartland east of Edmonton, where product-upgrading facilities are planned. Indeed, the effects of oil sands development extend farther afield, to wherever products are consumed and secondary employment is created. These wider-reaching effects have not been included in the current analysis, but they serve to highlight the challenges surrounding any credible consideration of the cumulative effects of even a single area of development activity.

As we have seen, considerable challenges are associated with the evaluation of cumulative effects in relation to archaeological resources, which we need to keep in mind when reviewing the results of an analysis. Incomplete baseline information against which impact can be measured is perhaps the most fundamental weakness of cumulative effects assessments. Other challenges specific to archaeological resources include the relative effectiveness of the methods adopted, the degree of uncertainty surrounding the results achieved, and the absence of clearly defined thresholds that could be used to determine when effects should be considered unacceptable.

In remote areas such as the oil sands, archaeological knowledge accumulates slowly and for the most part only as impact assessment studies take place. Cumulative effects must therefore be measured against a continually evolving information base. Similarly, detailed knowledge about the significance of many resources typically becomes available only when mitigative samples are recovered, which generally takes place immediately in advance of development approval. Consequently, the impact of oil sands projects, whether singly or in combination, becomes apparent only as development advances across the landscape, rather than beforehand. These concerns, coupled with the degree of uncertainty associated with the design of archaeological assessment studies, the methods they employ, and the conclusions they draw, limit the levels of confidence in the accuracy of cumulative effects assessments of archaeological resources.

As I have noted, the very conditions that created the dense distribution of sites we now observe in the surface-minable oil sands area also made it possible to strip-mine bitumen deposits. If development proceeds as planned, the evidence of this striking pattern of prehistoric human activity will disappear. Of the 1,129 archaeological sites recorded in this area to date, systematic samples of varying size have been recovered from 319, that is, roughly one-quarter of the known sites. These samples consist almost exclusively of stone tools and manufacturing debris, of which formed artifacts constitute only a small portion. Collections currently number approximately 1.5 million items. Although these collections constitute a highly significant resource from the perspective of future study and education, given that these samples represent somewhat less than 15% of the material available at those sites, and that the remaining three-quarters of the sites have not been considered sufficiently significant to warrant sample recovery, the scale of the loss of prehistoric cultural materials in this area can be readily appreciated.

The Alberta Historical Resources Act provides the legal means to ensure the adequate investigation of future effects of development and to maximize the scientific and educational benefits of mitigative sample recovery to the public. However, the scope of the act is discretionary. The effective application of its provisions thus requires formal policy and procedures that take into consideration a full range of development activities, including small-scale projects. It also requires the synthesis of information on archaeological resources as it accumulates. Such syntheses are necessary not only to further public education and an appreciation for the resources that have been or may be affected by development but also to provide analytical direction that will enhance the productivity and long-term value of future studies.

In the area of archaeological management, the ability to respond promptly and thoughtfully as our knowledge of site distribution and significance evolves is essential to the design of effective mitigative programs. The identification of regional research objectives firmly grounded in a regular review of new information would be a substantive contribution in this regard. At the impact assessment stage, the existence of established research goals would provide useful guidance, enabling us to develop more focused strategies for the discovery of sites that address gaps in our understanding of prehistoric land use patterns (such as seasonal variations or the use of wetlands), and/or that have potential for stratigraphic separation of occupations, and/or that contain preserved organic remains suitable for radiocarbon dating. At the impact mitigation stage, such objectives would aid us in determining what types of information and samples would prove most useful in addressing specific questions relating to culture historical developments, subsistence strategies, or exchange patterns, as well as to a range of other issues. Ongoing review and synthesis of accumulating information may thus be one of the most effective ways of offsetting the cumulative effects of oil sands development.

As this volume well illustrates, one positive effect of regional oil sands development has been knowledge. The synthesis and dissemination of this knowledge would be another productive result. With synthetic work and through the collections conserved, opportunities exist for heightening public awareness of the archaeological legacy of the oil sands area. This volume is an example of how this can be done for a relatively academic audience, but recent display and brochure work completed by one of the local industrial developers, Birch Mountain Resources, provides another example. The display, developed by professional archaeologists in consultation with the community, is located at the Fort McKay Elders Centre, where members of the Fort McKay First Nation now have an opportunity to enhance their understanding of the rich prehistoric past of their traditional lands. The presentation and celebration of our increasing understanding of prehistoric lifestyles and land use could be further pursued, not only in collaboration with First Nations communities but on a wider scale as well. For example, display material at the Oil Sands Interpretive Centre in Fort McMurray, which reaches a great many visitors to the area, could easily include a component featuring prehistory. Perhaps most important—although concrete plans do not exist in this regard—the inclusion of local prehistory in regional school curricula would help students, their family members, and their teachers to develop an appreciation not only for the past but for the ongoing human connection with the land and its resources.

Archaeological studies associated with oil sands projects are ongoing. New areas of prehistoric human activity will be discovered, further samples will be recovered, and additional sites may even be set aside for permanent preservation. With the thoughtful application of existing regulatory mechanisms, a regular synthetic review of the information collected, and co-operative efforts to bring the benefits of this work to the attention of the public, the projected loss of archaeological resources in the oil sands area may to some extent be redressed.

1   As part of these review processes, Alberta Culture and Tourism evaluates the required HRIA summary of EIA reports and comments to Alberta Environment and Parks (formerly Alberta Environment and Sustainable Resource Development) regarding the completeness of that component. Any changes required are transmitted to the project proponent as a "supplemental information request," a response to which is required before the EIA report can be accepted as complete.

2   Significant changes to group mobility on the Northern Plains had already been brought about by the acquisition of horses. In addition, smallpox reached the region at least by 1780–81 (see Houston and Houston 2000), devastating resident populations and often prompting survivors to relocate. For the Athabasca region, considerable information relating largely to trading activities with Aboriginal peoples and the day-to-day maintenance of trading posts can be found in the post journals available in the Hudson Bay Company Archives, particularly those for Fort Chipewyan. In addition, early explorers and traders such as Philip Turnor (Tyrrell 1934), Alexander Mackenzie (Lamb 1970; Mackenzie 1971 [1801]), Peter Fidler (MacGregor 1966), George Simpson (Rich 1938), Daniel Harmon (1904), and Cuthbert Grant (Duckworth 1989) have left us contemporary accounts of the Athabasca region and its resident First Nations.

3   In 1719, at York Factory on Hudson Bay, a Cree chief who had travelled to the Athabasca region brought Henry Kelsey a sample of "that Gum or pitch," explaining that it flowed from the banks of a river (Morton 1973, 134). In 1788, upon observing bitumen pools along the Athabasca River, Alexander Mackenzie noted that the indigenous inhabitants commonly used the substance in conjunction with spruce gum to seal canoes (Lamb 1970, 129).

4   These efforts were initiated in 1999 by CEMA (Cumulative Effects Management Association), a multi-stakeholder organization that comprised more than fifty members representing all levels of government, regulatory bodies, industry, environmental groups, Aboriginal groups, and the local health authority, all of which have an interest in protecting the environment in the oil sands region. Broadly speaking, CEMA's original objective was to achieve consensus regarding the identification of environmental limits and the legal frameworks needed to protect regional air and water quality, vegetation communities, and wildlife. Although numerous reports were prepared and some valuable contributions made, achieving consensus remained an elusive goal. In April 2016, CEMA ceased operation, owing to a lack of funding.

REFERENCES

Alberta

2000a   *Historical Resources Act.* Revised Statues of Alberta 2000, Chapter H-9. Alberta Queen's Printer, Edmonton. http://www.qp.alberta.ca/documents/Acts/h09.pdf.

2000b   *Oil Sands Conservation Act.* Revised Statutes of Alberta, Chapter O-7. Alberta Queen's Printer, Edmonton. http://www.qp.alberta.ca/documents/Acts/O07.pdf.

2000c   *Environmental Protection and Enhancement Act.* Revised Statutes of Alberta, Chapter E-12. Alberta Queen's Printer, Edmonton. http://www.qp.alberta.ca/documents/Acts/E12.pdf.

2002   Archaeological and Palaeontological Research Permit Regulation, Alberta Regulation 254/2002. Alberta Queen's Printer, Edmonton. http://www.qp.alberta.ca/documents/Regs/2002_254.pdf.

Alberta Energy

    2007   Oil Sands Consultations: Multistakeholder Committee Final Report. http://www.energy.
alberta.ca/OilSands/pdfs/FinalReport_2007_OS_MSC.pdf.

    2016a   Alberta's Oil Sands Projects and Upgraders. http://www.energy.gov.ab.ca/LandAccess/
pdfs/OilSands_Projects.pdf.

    2016b   Alberta's Leased Oil Sands Area. http://www.energy.gov.ab.ca/LandAccess/pdfs/
OSAagreeStats.pdf.

Alberta Energy and Utilities Board, Alberta Energy, and the Natural Resources Conservation Board

    2010   Cumulative Effects Assessment in Environmental Impact Assessment Reports Required
Under the Alberta Environmental Protection and Enhancement Act. http://aep.alberta.
ca/lands-forests/land-industrial/programs-and-services/environmental-assessment/
documents/CumulativeEffectsEIAReportsUnderEPEA-A.pdf.

Alberta Environment

    2008   Towards Environmental Sustainability: A Proposed Regulatory Framework for Managing
Environmental Cumulative Effects. http://www.assembly.ab.ca/lao/library/egov-
docs/2007/alen/162068.pdf.

AUC (Alberta Utilities Commission)

    2009   Rule 007: Applications for Power Plants, Substations, Transmission Lines, and Industrial
System Designations. http://www.auc.ab.ca/acts-regulations-and-auc-rules/rules/
Documents/Rule007.pdf.

Bobrowsky, Peter T., Eric R. Damkjar, and Terrance H. Gibson

    1988   *A Geological and Archaeological Study of HcQh-6, Peace River Alberta (ASA Permit 88-005)*.
Copy on file, Archaeological Survey, Historic Resources Management Branch, Alberta
Culture, Edmonton.

Boland, Dale E., Bonnie Brenner, and Jennifer C. Tischer

    2009   *Historical Resources Impact Mitigation, Total E&P Joslyn Limited Joslyn North Mine Project,
2008 Mitigation Studies (HhOw-18, HhOw-29, HhOw-32, HhOv-42, HhOw-43, HhOw-
45, HhOw-46): Final Report (ASA Permit 08-208)*. Copy on file, Archaeological Survey,
Historic Resources Management Branch, Alberta Culture, Edmonton.

Bryan, Alan L.

    1987   *Final Report of a Test Excavation at the Caribou Island Site, East-Central Alberta (ASA
Permit 86-047)*. Copy on file, Archaeological Survey, Historic Resources Management
Branch, Alberta Culture, Edmonton.

Bryant, Laureen

    2004   *Historical Resources Impact Assessment and Mitigation, Fall 2003, Canadian Natural
Resources Limited Horizon Oil Sands Project: Final Report (ASA Permit 03-269)*. Copy on
file, Archaeological Survey, Heritage Resources Management Branch, Alberta Culture,
Edmonton.

Carrigy, M. A.

    1974   Historical Highlights. In *Guide to the Athabasca Oil Sands Area,* edited by M. A. Carrigy,
pp. 173–186. Information Series no. 65. Alberta Research, Edmonton.

CEAA (Canadian Environmental Assessment Agency)

    1996   Cumulative Environmental Effects Cross-referenced Annotated Bibliography. Canadian
Environmental Assessment Agency, EA Enhancement and Intentional Affairs Team,
Ottawa.

    1999   *Cumulative Effects Assessment Practitioners' Guide*. Prepared by the Cumulative Effects
Working Group (George Hegmann, Chris Cocklin, Roger Creasey, Sylvie Dupuis, Alan
Kennedy, Louise Kingsley, William Ross, Harry Spaling and Don Stalker) and AXYS
Environmental Consulting Ltd. Canadian Environmental Assessment Agency, Ottawa.
http://www.ceaa-acee.gc.ca/default.asp?lang=En&n=43952694-1.

2007 Framework for Addressing Cumulative Environmental Effects in Federal Environmental Assessments. In *Reference Guide: Addressing Cumulative Environmental Effects.* Canadian Environmental Assessment Agency, Ottawa. http://www.ceaa-acee.gc.ca/default. asp?lang=En&n=9742C481-1&offset=4&toc=show.

Clarke, Grant M., and Brian M. Ronaghan

2000 *Historical Resources Impact Mitigation, Muskeg River Mine Project: Final Report (ASA Permit 99-073).* Copy on file, Archaeological Survey, Historic Resources Management Branch, Alberta Culture, Edmonton.

Conaty, Gerald T.

1977 Excavation of the Wentzel Lake Site, Project 76-11. In *Archaeology in Alberta, 1976,* compiled by J. Michael Quigg, pp. 31–36. Archaeological Survey of Alberta Occasional Paper no. 4. Historic Resources Management Branch, Alberta Culture, Edmonton.

1979 *Alsands Lease Archaeological Survey: Final Report (ASA Permit 79-056).* Copy on file, Archaeological Survey, Historic Resources Management Branch, Alberta Culture, Edmonton.

Donahue, Paul F.

1976 *Archaeological Research in Northern Alberta.* Archaeological Survey of Alberta, Occasional Paper no. 2. Historic Resources Management Branch, Alberta Culture, Edmonton.

Duckworth, Harry W. (editor)

1990 *The English River Book: A North West Company Journal and Account Book of 1786.* McGill-Queens University Press, Montreal and Kingston.

Duinker, Peter N., and Lorne A. Greig

2006 The Impotence of Cumulative Effects Assessment in Canada: Ailments and Ideas for Redeployment. *Environmental Management* 37(2): 153–161.

ERCB (Energy Resources Conservation Board)

1984 Order No. OSA 1: An Order Declaring the Athabasca Oil Sands Area. http://www.ercb.ca/ orders/oilsands/osa1_Athabasca.pdf.

2011 Directive 056: Energy Development Applications and Schedules. 1 September. https:// www.aer.ca/documents/directives/Directive056_April2014.pdf.

Fedirchuk, Gloria J.

1980a *Historical Resources Impact Assessment, Parts of N 1/2, Sec. 7, Sec. 18, SW 1/4, Sec. 17, Twp. 63, Rge. 1, W4th and SE 1/4, Sec. 13, and Part of NE 1/4, Sec 13, Twp. 63, Rge. 2, W 4th, Cold Lake Alberta (ASA Permit 80-100).* Copy on file, Archaeological Survey, Historic Resources Management Branch, Alberta Culture, Edmonton.

1980b *Historical Resource Mitigation, GcOm-18, Cold Lake, Alberta (ASA Permit 80-123).* Copy on file, Archaeological Survey, Historic Resources Management Branch, Alberta Culture, Edmonton.

1982 *Evaluative Investigations, GhPh-11, Ka Kittoo Wak Site (ASA Permit 81-126).* Copy on file, Archaeological Survey, Historic Resources Management Branch, Alberta Culture, Edmonton.

Fedirchuk, Gloria J., and Edward J. McCullough

1992 *The Duckett Site: Ten Thousand Years of Prehistory on the Shores of Ethel Lake, Alberta.* Fedirchuk McCullough and Associates Ltd. Copy on file, Esso Resources Canada Limited, Calgary.

Fox, W. A.

1975 The Palaeo-Indian Lakehead Complex. In *Canadian Archaeological Association, Collected Papers, March 1975,* edited by Peggie Nunn, pp. 28–53. Research Report no. 6. Historical Sites Branch, Division of Parks, Government of Ontario, Toronto.

Graham, James W., and Jennifer C. Tischer

2009 *Historical Resources Impact Mitigation, Total E&P Joslyn Limited Joslyn North Mine Project, 2008 Mitigation Studies (HhOw-22, HhOw-30, HhOw-38, HhOx-9, HhOx-13): Final*

Report (ASA Permit 08-298). Copy on file, Archaeological Survey, Historic Resources
Management Branch, Alberta Culture, Edmonton.

Gruhn, Ruth

1981   *Archaeological Research at Calling Lake, Northern Alberta.* Mercury Series no. 99,
Archaeological Survey of Canada, National Museum of Man, Ottawa.

Hamilton, J. Scott

1996   *Pleistocene Landscape Features and Plano Archaeological Sites upon the Kaministiquia River
Delta, Thunder Bay District.* Lakehead University Monographs in Anthropology no. 1.
Department of Anthropology, Lakehead University, Thunder Bay

Harmon, Daniel W.

1904   *A Journal of Voyages and Travels in the Interior of North America.* George N. Morang & Co.,
Toronto.

Head, Thomas H., and Stanley Van Dyke

1990   *Historical Resources Impact Assessment and Mitigation, Cree Burn Lake Site (HhOv-16)
Jct. S.R. 963 to Gravel Pit Source "A" in NW 29-95-10-4 (ASA Permit 88-032).* Copy on
file, Archaeological Survey, Historic Resources Management Branch, Alberta Culture,
Edmonton.

Hein, Frances J.

2000   *Historical Overview of the Fort McMurray Area and Oil Sands Industry in Northeast
Alberta.* Earth Sciences Report 2000-05. Alberta Energy and Utilities Board and Alberta
Geological Survey, Edmonton.

Houston, C. Stuart, and Stan Houston

2000   The First Smallpox Epidemic on the Northern Plains: In the Fur-Traders' Words.
Canadian Journal of Infectious Diseases 11 (2): 112–15.

Ives, John W.

1977   The Excavation of HkPa4, Birch Mountains, Alberta. In *Archaeology in Alberta, 1976,* compiled by J. Michael Quigg, pp. 37–44. Archaeological Survey of Alberta Occasional Paper
no. 4. Historic Resources Management Branch, Alberta Culture, Edmonton.

1985   *A Spatial Analysis of Artifact Distribution on a Boreal Forest Archaeological Site.*
Archaeological Survey of Alberta Manuscript Series no. 5. Historic Resources
Management Branch, Alberta Culture, Edmonton.

Kennett, Steven A.

1999   *Towards a New Paradigm for Cumulative Effects Management.* CIRL Occasional Paper no. 8.
Canadian Institute of Resources Law, University of Calgary, Calgary.

Kjorlien, Yvonne P., Leah Mann, and Jennifer C. Tischer

2009   *Historical Resources Impact Mitigation, Total E&P Joslyn Limited Joslyn North Mine Project,
2008 Mitigation Studies (HhOw-49, HhOw-54, HhOw-55, HhOx-10, HhOx-11, HhOx-15,
HhOx-17, HhOx-18): Final Report (ASA Permit 08-166).* Copy on file, Archaeological
Survey, Historic Resources Management Branch, Alberta Culture, Edmonton.

Kowal, Walter A.

1990   *Historical Resources Impact Assessment, English Bay Golf Course and Family Resort at Cold
Lake, Alberta: Final Report (ASA Permit 90-088).* Copy on file, Archaeological Survey,
Historic Resources Management Branch, Alberta Culture, Edmonton.

Krech, Shepherd

1984   *The Subarctic Fur Trade: Native Social and Economic Adaptations.* University of British
Columbia Press, Vancouver.

Lamb, W. Kaye (editor)

1970   *The Journals and Letters of Sir Alexander Mackenzie.* Cambridge University Press for the
Hakluyt Society, Cambridge.

Learn, Kathleen C.

1986   Pottery and Prehistory of Black Fox Island: Technical Patterns in a Cultural Perspective.
MA thesis, Department of Anthropology, University of Alberta.

Le Blanc, Raymond J., and John W. Ives

    1986   The Bezya Site: A Wedge-Shaped Core Assemblage from Northeastern Alberta. *Canadian Journal of Archaeology* 10: 59–98.

Losey, Timothy C., and Cort Sims

    1973   *Syncrude Lease No. 17: An Archaeological Survey (ASA Permit 73-004).* Copy on file, Archaeological Survey, Historic Resources Management Branch, Alberta Culture, Edmonton.

MacGregor, James G.

    1966   *Peter Fidler, Canada's Forgotten Surveyor, 1769–1822.* McClelland and Stewart, Toronto.

Mackenzie, Sir Alexander

1971 [1801]   *Voyages from Montreal on the River St. Laurence Through the Continent of North America to the Frozen and Pacific Oceans in the Years 1789 and 1793.* Reprint, M. G. Hurtig, Edmonton.

McCullough, Edward J.

    1980a   *Historical Resources Inventory and Assessment, Esso Resources Canada Ltd., Cold Lake Project Commercial Development Area and Medley River Gravel: Final Report (ASA Permit 80-068).* Copy on file, Archaeological Survey, Historic Resources Management Branch, Alberta Culture, Edmonton.

    1980b   *Historical Resources Inventory and Assessment, Husky Oil Ltd., Cold Lake to Lloydminster Pipeline: Final Report (ASA Permit 80-044).* Copy on file, Archaeological Survey, Historic Resources Management Branch, Alberta Culture, Edmonton.

    1980c   *Historical Resources Inventory and Assessment, NOVA-PetroCanada Oil Sands Joint Venture, Core-Hole Drilling Program, B.S.L. Nos. 52, 20, 78, 88, 89, and 5 (ASA Permit 80-133).* 5 vols. Copy on file, Archaeological Survey, Historic Resources Management Branch, Alberta Culture, Edmonton.

    1981a   *Historical Resources Impact Assessment, Canstar Oils Sands Ltd., Calumet Construction Camp and Athabasca River Access Road (ASA Permit 81-094).* Copy on file, Archaeological Survey, Historic Resources Management Branch, Alberta Culture, Edmonton.

    1981b   *The Duckett Site (GdOo-16): An Evaluative Study (ASA Permit 80-155).* Copy on file, Archaeological Survey, Historic Resources Management Branch, Alberta Culture, Edmonton.

    1982   *Prehistoric Cultural Dynamics of the Lac La Biche Region.* Archaeological Survey of Alberta Occasional Paper no. 18. Historic Resources Management Branch, Alberta Culture, Edmonton.

McCullough, Edward J., and Gloria J. Fedirchuk

    1989   *Historical Resources Phase I: Baseline Study, OSLO Project (ASA Permit 89-052).* Copy on file, Archaeological Survey, Historic Resources Management Branch, Alberta Culture, Edmonton.

McCullough, Edward J., and C. M. Fowler

    1981   *Historical Resources Inventory and Assessment, Cold Lake Project Off-Site Facilities (ASA Permit 81-098).* Copy on file, Archaeological Survey, Historic Resources Management Branch, Alberta Culture, Edmonton.

McCullough, Edward J., Michael C. Wilson, and C. M. Fowler

    1982   *Historical Resources Studies, Canstar Oil Sands Ltd., Bitumous Sands Leases 33, 92, and 95 (ASA Permit 81-129).* Copy on file, Archaeological Survey, Historic Resources Management Branch, Alberta Culture, Edmonton.

McKenzie-Brown, Peter, Gordon Jaremko, and David Finch

    1993   *The Great Oil Age: The Petroleum Industry in Canada.* Detselig Enterprises, Calgary.

Meyer, Daniel A.

    2002   *Historical Resources Impact Assessment, Shell Peace River Complex Proposed Expansion Project: Final Report (ASA Permit 02-183).* Copy on file, Archaeological Survey, Historic Resources Management Branch, Alberta Culture, Edmonton.

Morton, Arthur S.

    1973   *A History of the Canadian West to 1870-71.* 2nd ed. Toronto: University of Toronto Press, Toronto.

Natural Regions Committee

    2006   *Natural Regions and Subregions of Alberta.* Compiled by David J. Downing and Wayne W. Pettapiece. Government of Alberta Publication no. T/852. https://www.albertaparks.ca/media/2942026/nrsrcomplete_may_06.pdf.

Newton, Barry

    1980   *Historical Resources Impact Assessment, Part of the SW 1/4, Sec. 24, Twp. 63, Rge. 2, W4M, Townsite of Cold Lake, Alberta: Final Report (ASA Permit 80-065).* Copy on file, Archaeological Survey, Historic Resources Management Branch, Alberta Culture, Edmonton.

Pollock, John

    1976a   *Early Cultures of the Clearwater River Area, Northeastern Alberta (ASA Permit 76-040).* Copy on file, Archaeological Survey, Historic Resources Management Branch, Alberta Culture, Edmonton.

    1976b   *Preliminary Archaeological Inspection of the Graham-Peerless Lake Areas, 1976: Final Report (ASA Permit 76-059).* Copy on file, Archaeological Survey, Historic Resources Management Branch, Alberta Culture, Edmonton.

    1977   *Prehistoric Settlement, Material Culture, and Resource Utilization of the Slave River Area, Northeastern Alberta (ASA Permit 77-030).* Copy on file, Archaeological Survey, Historic Resources Management Branch, Alberta Culture, Edmonton.

    1978   *Early Cultures of the Clearwater River Area, Northeastern Alberta.* Archaeological Survey of Alberta Occasional Paper no. 6. Historic Resources Management Branch, Alberta Culture, Edmonton.

    1981   *Historical Resources Impact Assessment, Part of SW 1/4, Sec. 2, and SE 1/4, Sec 3, Twp. 65-R.2-W4MER, Cold Lake, Alberta: Final Report (ASA Permit 82-131).* Copy on file, Archaeological Survey, Historic Resources Management Branch, Alberta Culture, Edmonton.

    1982   *Historical Resources Impact Assessment, Nelson Heights Subdivision, Part of the NW 1/4, Sec. 23, Twp. 63, Rge. 2, W4M, Town of Cold Lake, Alberta: Final Report (ASA Permit 81-089).* Copy on file, Archaeological Survey, Historic Resources Management Branch, Alberta Culture, Edmonton.

Ray, Arthur J.

    1974   *Indians in the Fur Trade.* University of Toronto Press, Toronto.

Reeves, Brian O. K.

    1976   Cold Lake Area Survey. Unpublished manuscript in the possession of the author.

    1977   *Historical Resources Impact Assessment, Syncrude Canada Ltd., Western Portion of Lease No. 17 (ASA Permit 77-087).* Copy on file, Archaeological Survey, Historic Resources Management Branch, Alberta Culture, Edmonton.

Reeves, Brian O. K., and Edward J. McCullough

    1977   *Historical Resources Overview and Preliminary Assessment, Cold Lake Lease (ASA Permit 77-093).* Copy on file, Archaeological Survey, Historic Resources Management Branch, Alberta Culture, Edmonton.

Rich, E. E. (editor)

    1938   *Journal of the Occurrences in the Athabasca Department, by George Simpson, 1820 and 1821, and Report.* Publications of the Hudson's Bay Record Society no. 1. Champlain Society, Toronto

Ronaghan, Brian M.

    1981a   *Final Report: Historical Resources Impact Assessment, Fort McMurray Energy Corridor, Fort Hills Townsite and Airstrip (ASA Permit 80-091).* Copy on file, Archaeological Survey, Historic Resources Management Branch, Alberta Culture, Edmonton.

1981b   *Final Report: Historical Resources Impact Assessment of Selected Portions of the Alsands Lease 13 (ASA Permit 80-091).* Copy on file, Archaeological Survey, Historic Resources Management Branch, Alberta Culture, Edmonton.

1981c   *Cold Lake Bitumen Pipeline System, Esso Resources Leming Pilot Project, Cold Lake to Strathcona Refinery: Final Report (ASA Permit 81-034).* Copy on file, Archaeological Survey, Historic Resources Management Branch, Alberta Culture, Edmonton.

1982   *Historical Resources Impact Assessment, Bonnyville–Ethel Lake Transmission System (ASA Permit 82-100).* Copy on file, Archaeological Survey, Historic Resources Management Branch, Alberta Culture, Edmonton.

Roskowski, Laura, and Morgan Netzel

2011a   *Historical Resources Impact Mitigation, Shell Canada Energy, Muskeg River Mine Expansion of RMS 10, Mitigation for Sites HhOv-87 and HhOv-200: Final Report (ASA Permit 09-168).* Copy on file, Archaeological Survey, Historic Resources Management Branch, Alberta Culture, Edmonton.

2011b   *Historical Resources Impact Mitigation, Shell Canada Energy, Stage II Mitigative Excavation of HhOv-351: Final Report (ASA Permit 10-148).* Copy on file, Archaeological Survey, Historic Resources Management Branch, Alberta Culture, Edmonton.

2012   *Historical Resources Impact Mitigation, Shell Canada Energy Muskeg River Mine Expansion, Area 6, Stage I Mitigation of Sites HhOv-156 and HhOv-520: Final Report (ASA Permit 11-167).* Copy on file, Archaeological Survey, Historic Resources Management Branch, Alberta Culture, Edmonton.

2015   *Historical Resources Impact Mitigation, Muskeg River Mine Expansion, Area 7 Gap, Staged Mitigation of Sites HhOv-374, HhOv-506, and HhOv-508, Shell Canada Energy: Final Report (ASA Permit 11-070).* Copy on file, Archaeological Survey, Historic Resources Management Branch, Alberta Culture, Edmonton.

Roskowski, Laura, Alison Landals, and Morgan Blower

2008   *Historical Resources Impact Mitigation, Shell Canada Limited Albian Sands Muskeg River Mine Expansion Project, Mitigation for Sites HhOu-68, HhOu-69, HhOu-70, HhOu-94, HhOu-95, HhOv-378, HhOv-379, HhOv-380, HhOv-381 and HhOv-383: Final Report (ASA Permit 07-219).* Report on file, Archaeological Survey, Historic Resources Management Branch, Alberta Culture, Edmonton.

Schneider, Richard, and Simon Dyer

2007   *Death by a Thousand Cuts: Impacts of In Situ Oil Sands Development on Alberta's Boreal Forest.* Canadian Parks and Wilderness Society and the Pembina Institute, Edmonton. http://pubs.pembina.org/reports/1000-cuts.pdf.

Sims, Cort

n.d. [1975]   An Archaeological Survey of the Namur Lake Area in Northeastern Alberta. Unpublished manuscript in possession of the author.

1976   *Report of an Archaeological Survey of the Athabasca River, 1976 (ASA Permit 76-005).* Copy on file, Archaeological Survey, Historic Resources Management Branch, Alberta Culture, Edmonton.

1980   Models for Explaining Material Culture Form and Distribution at Two Archaeological Sites in Northern Alberta. Draft report on file (CRM 194), Archaeological Survey, Historic Resources Management Branch, Alberta Culture, Edmonton.

Strong, W. L., and K. R. Leggat

1981   *Ecoregions of Alberta.* ENR Technical Report T/4. Alberta Energy and Natural Reserves, Resource Evaluation and Planning Division, Edmonton.

Tyrrell, J. B. (editor)

1934   *Journals of Samuel Hearne and Philip Turnor, Between the Years 1774 and 1792.* Champlain Society, Toronto.

Van Dyke, Stanley

1980   *Historical Resources Impact Assessment, Cold Lake Residential Reserve, NW 1/4 and SW 1/4 14-63-2-W4M: Final Report (ASA Permit 80-023)*. Copy on file, Archaeological Survey, Historic Resources Management Branch, Alberta Culture, Edmonton.

1984   *Historical Resources Impact Assessment, Peace River In-situ Pilot Project, Peace River Expansion Project Water Supply Pipeline (ASA Permit 84-071)*. Copy on file, Archaeological Survey, Historic Resources Management Branch, Alberta Culture, Edmonton.

Van Dyke, Stanley, and Brian O. K. Reeves

1984   *Historical Resources Impact Assessment, Syncrude Canada Ltd. Lease No. 22 (ASA Permit 84-053)*. Copy on file, Archaeological Survey, Historic Resources Management Branch, Alberta Culture, Edmonton. Published in 1985 as Environmental Research Monograph 1985-4, Syncrude Canada Ltd., Edmonton.

Wenig, Michael M.

2002   Cumulative Effects: Oil, Gas, and Biodiversity. *LawNow* 27(2) (October–November): 24–28. http://www.cirl.ca/system/files/LawNow2002OctNovMW.pdf.

Wickham, Michelle D., and Taylor Graham

2009   *Historical Resources Impact Mitigation of the TransCanada Pipelines Ltd. Fort McKay Mainline Expansion: Final Report and Post-construction Audit (ASA Permits 06-376 and 07-266)*. Copy on file, Archaeological Survey, Historic Resources Management Branch, Alberta Culture, Edmonton.

Wood, William J.

1980   *Historical Resources Inventory, Cold Lake Provincial Park (ASA Permit 79-189)*. Copy on file, Archaeological Survey, Historic Resources Management Branch, Alberta Culture, Edmonton.

Woywitka, Robin J., Angela M. Younie, Morgan Blower, and Alison Landals

2008   *Historical Resources Impact Mitigation, Shell Canada Limited Albian Sands Muskeg River Mine Expansion Project, Mitigation for Sites HhOv-384, HhOv-385, HhOv-387, HhOv-431, and HhOv-432: Final Report (ASA Permit 07-280)*. Copy on file, Archaeological Survey, Historic Resources Management Branch, Alberta Culture, Edmonton.

Woywitka, Robin J., Jennifer C. Tischer, Laura Roskowski, and Angela M. Younie

2009   *Historical Resources Impact Mitigation, Fort Hills Energy Corporation, Fort Hills Oil Sands Project, 2008 Mitigation Studies: Final Report (ASA Permit 08-163)*. Copy on file, Archaeological Survey, Historic Resources Management Branch, Alberta Culture, Edmonton.

Yerbury, Colin

1986   *The Subarctic Indians and the Fur Trade, 1680–1860*. University of British Columbia Press, Vancouver.

Youell, A. J., Jennifer C. Tischer, Morgan Blower, and Lauren Copithorne

2009   *Historical Resources Impact Mitigation, Total E&P Joslyn Limited, Joslyn North Mine Project, 2007 Mitigation Studies (HhOw-20, HhOw-21, HhOw-22, HhOw-24, HhOw-26, HhOw-27, HhOw-30, HhOw-38, HhOw-39, HhOw-40, HhOx-7): Final Report (ASA Permit 07-393)*. Copy on file, Archaeological Survey, Historic Resources Management Branch, Alberta Culture, Edmonton.

Contributors

Alwynne B. Beaudoin is head curator of Earth Sciences and curator of Quaternary Environments at the Royal Alberta Museum, in Edmonton. She holds a BSc from Leeds University and MSc and PhD degrees from the University of Western Ontario, all in physical geography. Following an interval of university teaching, she moved to Alberta in 1986, to work for the Archaeological Survey, transferring to Royal Alberta Museum in 1991. Her work concentrates on the investigation of the landscapes and environments of Alberta during the past twelve thousand years, especially as these relate to the province's human history. This research involves the examination of plant remains, especially seeds and pollen, and the analysis of soils and sediments. Over the past thirty years, she has worked in many parts of Alberta, including the Canadian Rockies, although her most recent work has focused on the region around Edmonton, the prairies, and the Cypress Hills area.

Janet Blakey is an assistant project archaeologist with Lifeways of Canada, based in Calgary. Since joining Lifeways in 2001, she has worked on numerous projects throughout western Canada. Blakey specializes in geoarchaeological methods, in faunal analysis, and in public outreach. She is actively involved in a volunteer role with the Archaeological Society of Alberta and has served for the past six years on the executive of the ASA's Calgary Centre, most recently as president.

Luc Bouchet holds an MA in archaeology from the University of Calgary, where he adopted a palaeoenvironmental approach to the role of climate in human adaptation to the landscape. Having noted the central place of stone in archaeology, he decided to learn the world's second-oldest profession and undertook an apprenticeship as a stonemason, which ultimately led him to the Burgundy forests of France, where he worked with stonecutters and masons on a project that involved building a castle from scratch using only the tools and methods available in the thirteenth century. Obliged to recognize that the demand for castles, cathedrals, and even Roman aqueducts was at low ebb, he decided to return to Alberta to work in cultural resources management. After a decade spent criss-crossing our beautiful province as a consulting archaeologist, he now works as a firefighter. Although he no longer studies the past, he can often be spotted in the woods hunting with a bow and arrow.

James A. Burns is curator emeritus of Quaternary Palaeontology at the Royal Alberta Museum, in Edmonton. During his tenure at the museum, from 1983 to 2006, he oversaw the development of the largest collection of ice age fossils in Canada. Now based in Winnipeg, he is an active caver and director of the Speleological Society of Manitoba and is currently studying an early postglacial cave faunule from the northern Interlake region of Manitoba. The author or co-author of more than fifty academic papers and popular articles, Burns is presently at work on a history of the Manitoba Museum. By way of diversion, he writes murder mysteries with his wife, Sheilla Jones, and sings in a Celtic men's choir.

Grant M. Clarke is an Aboriginal consultation advisor with the Historic Resources Management Branch of Alberta Culture and Tourism. An archaeologist with more than twenty-five years' experience in the field, Grant holds a BA and MA in anthropology and archaeology from the University of Saskatchewan. His master's thesis focused on faunal analyses of material from a twelve-hundred-year-old archaeological site in central Saskatchewan. Prior to joining the Alberta government, he was an archaeological consultant, working chiefly with Golder Associates Ltd. out of the firm's Calgary, Yellowknife, and Edmonton offices. He has held archaeological permits for numerous historical resources impact assessments and mitigation studies, as well as environmental impact assessments, at sites throughout Alberta, the Northwest Territories, Nunavut, and Saskatchewan, and has been involved in many projects in the oil sands region. His technical experience includes faunal analysis, lithic tool analysis, traditional knowledge and use studies, and the design of archaeological predictive models.

Gloria J. Fedirchuk holds a BA and MA in archaeology from the University of Calgary and a PhD in anthropology from the University of New Mexico. After a brief period of university teaching, she entered the field of contract archaeology, founding Archaeological Heritage Consultants Ltd. and then, with Edward McCullough, Fedirchuk McCullough and Associates Ltd., which subsequently became FMA Heritage Inc. As principal of FMA, Fedirchuk participated in archaeological consultation programs throughout the Canadian Prairies and the Northwest Territories. Her interests lie especially with Dene history and with precontact cultural development more generally, and she has worked with First Nation elders on traditional land use and the perceived effects of development.

Timothy G. Fisher is professor in and chair of the Department of Environmental Sciences at the University of Toledo. He is interested in the evolution of glacial lakes and landscapes associated with the most recent glacial cycle. Some of his current research examines the timing of the development of sand dunes in the southern Great Lakes basins, in which connection he co-edited, with Edward Hansen, Geological Society of America Special Paper no. 508, *Coastline and Dune Evolution Along the Great Lakes* (2014).

Duane G. Froese is a professor in the Department of Earth and Atmospheric Sciences at the University of Alberta, where he is also Canada Research Chair in Northern Environmental Change. His current research focuses on the development and interpretation of records of past environmental change in northwestern Canada and Alaska and, in particular, on the impact of past and present climate change on permafrost. His interests also include the chronology of volcanic ash beds (tephra), which can serve to constrain the age of relict permafrost, ice age mammals, ancient DNA records, and other evidence of past environmental change.

Eugene M. Gryba is a Calgary-based archaeological consultant. Born in a log cabin at Cormorant Lake, in northern Manitoba, he became interested in archaeology and Pleistocene geology while growing up on the family farm in the Swan River valley, not far from the Upper Campbell beach of Glacial Lake Agassiz. He worked at the Saskatchewan Museum of Natural History, in the Department of Anthropology at the University of Manitoba, and at what is now the Royal Alberta Museum, in Edmonton, before earning his BA at the University of Alberta in 1972 and his MA from the University of Manitoba in 1975. Gryba has carried out archaeological surveys and excavations in southwestern Manitoba, southern Saskatchewan, and much of Alberta, including the oil sands area, as well as in northeastern British Columbia, the southwestern Yukon, and Texas. The author or co-author of numerous consultants' reports and papers published in Canadian and international journals, he is also a self-taught flint knapper. His research interests include the Fluted Point Tradition and microblade technology.

John W. (Jack) Ives is currently the Faculty of Arts Landrex Distinguished Professor in the Department of Anthropology at the University of Alberta and the executive director of the Institute of Prairie Archaeology. His most recent research focuses on the migration of Navajo and Apache ancestors from subarctic Canada, on terminal Pleistocene and early Holocene archaeological sites in western Canada, and on the Besant-Sonota phenomenon on the northern Plains. During the 1970s and 1980s, Ives worked extensively in the boreal forest and Plains regions of Alberta and Saskatchewan, which included seven field seasons in the Birch Mountains and Lower Athabasca region, and he continues to supervise graduate students working in both areas. From 1979 to 2007, he served with the Archaeological Survey of Alberta, the Royal Alberta Museum, and the Historic Resources Management Branch, with senior management responsibilities as Alberta's Provincial Archaeologist for twenty-one years and extensive cross-ministry experience in Aboriginal policy initiatives. During this time, he dealt with a wide variety of regulatory issues pertaining to historical resources in Alberta's Lower Athabasca region, particularly in connection with oil sands development. Ives is the recipient of three Alberta Premier's Awards.

Raymond J. Le Blanc received his PhD from the University of Toronto in 1983. After brief stints with the Alberta and federal governments, working with the Archaeological Survey of Alberta and the Archaeological Survey of Canada, he joined the staff of the Department of Anthropology at the University of Alberta in 1987. His research has focused on the subarctic and arctic regions of western North America, particularly the northern Yukon and the Mackenzie Delta, in the Northwest Territories. Over the course of his teaching career (from which he retired in 2013), he supervised twenty graduate students at the MA or PhD level on topics ranging in geographic scope from the western Prairies to northern Québec and Newfoundland-Labrador.

Murray Lobb is a senior archaeologist with Amec Foster Wheeler Environment and Infrastructure, in Calgary. He earned his BSc and MA from the University of Calgary and, during his seventeen years in historical resources management, has conducted fieldwork in the Northwest Territories, Saskatchewan, Manitoba, and Alberta. His research interests include historical archaeology and landscape archaeology, in regions that extend from the Canadian Subarctic and the boreal forest to the Rocky Mountains and the northern Plains. He has worked on many projects that incorporate GIS (Geographic Information Systems) technology, including mapping, predictive modelling, cluster (point pattern) analysis, and 3D visualization.

Thomas V. Lowell is a professor in the Department of Geology at the University of Cincinnati. He is interested in the nature and causes of past climate change and the interaction of climate and glaciers, with a particular focus on the evolution of the southern Laurentide Ice Sheet from the last glacial maximum to the Holocene. His current research includes investigations of the timing and pattern of warming at the end of the most recent ice age via work on glacier records from the southern hemisphere.

Brian O. K. Reeves is the founder and former president of Lifeways of Canada Ltd., Alberta's first archaeological consulting firm, established in 1973, and professor emeritus of archaeology at the University of Calgary. He has been practicing archaeology in Alberta since 1962 and consulting for industrial and transportation project developers in the Athabasca oil sands region since 1973. Many of the large-scale historic resources overviews, impact assessments, and mitigation studies undertaken by Lifeways' project archaeologists over the past forty years have provided the baseline archaeological information needed to interpret precontact First Nations cultural history and its palaeoenvironmental correlates in the Athabasca oil sands region. These studies have also helped to establish the region's key significance to the archaeological study of Canada's northwestern boreal forest.

Elizabeth C. Robertson is currently a private archaeological and ethnographic conservator and an adjunct professor in the Department of Archaeology and Anthropology at the University of Saskatchewan. She holds a BSc and a PhD in archaeology from the University of Calgary, as well as a Master of Art Conservation (MAC) from Queen's University, in Kingston, Ontario. Her professional and research interests focus on the archaeology of western Canada's boreal forest region and on the use of geoarchaeological methods to improve archaeological interpretation within and beyond this region. Her geoarchaeological work has included geomorphic and palaeoenvironmental examinations of archaeological site formation processes in Alberta's Cypress Hills, as well as in central Saskatchewan. She has also undertaken archaeometric studies of the stone used in precontact tool production to better understand how heat treatment was applied to improve the workability of various lithic raw materials.

Brian M. Ronaghan recently retired from his position as director of the Archaeological Survey of Alberta, the government regulatory agency responsible for the study and conservation of archaeological resources in the province. He holds a BA in history and a BA and MA in archaeology, all from the University of

Calgary. In addition to his more than twenty years as a regulatory compliance officer, he led archaeological research and compliance studies as a professional consultant in western Canada and the northwestern United States, undertaking field surveys, analysis, excavations, and traditional use studies, authoring numerous reports and publications, and taking part in several public hearings. During his tenure in government, he was responsible for the conduct of all archaeological studies, initially in one region and later for all of Alberta, and for the issuance of all archaeological permits and the management of archaeological sites and reports pertaining to them. He also participated in regulatory approval processes for major developments, undertook consultation with Indigenous groups and industry organizations, and developed management policy. His interests principally reside in archaeological field and analytical methods and in conservation management policy for heritage resources.

Laura Roskowski is a senior archaeologist at Stantec Consulting Ltd., in Calgary. After receiving a BSc in geology, anthropology, and history from Eastern Michigan University, she completed her master's research at the University of Calgary, focusing on geoarchaeology in the parklands of Saskatchewan. Over the past twelve years, she has worked as a consultant, conducting archaeological and traditional land use assessments, as well as directing large mitigative excavation programs. She specializes in boreal forest archaeology but enjoys working in a variety of regions.

Nancy Saxberg is a senior archaeologist with Amec Foster Wheeler Environment and Infrastructure, based in Calgary. Prior to joining Amec, she was associated with Lifeways of Canada, during which time she directed the initial excavations at the Quarry of the Ancestors. For more than fifteen years, she has also been working at Edmonton's Rossdale site (FjPi-63), one of the locations of Fort Edmonton.

Jennifer C. Tischer is currently the managing leader for Stantec Consulting Ltd.'s Calgary-based Heritage Resources group. She holds a BSc in anthropology and a MA in archaeology, both from the University of Calgary, where her master's thesis research focused on a faunal analysis of a bison kill site in southern Alberta. In her consulting role, Tischer has experience in historical resources impact assessments, baseline studies, and mitigation programs, as well as traditional land use and environmental impact assessment reporting for projects in western and northern Canada. She has extensive field experience in all parts of Alberta and the North, with most of her fieldwork in recent years taking place in Nunavut.

Stephen A. Wolfe is a research scientist with the Geological Survey of Canada, at Natural Resources Canada, and an adjunct research professor at Carleton

University, in Ottawa. His research currently focuses on cold-climate processes and landforms within the discontinuous to continuous permafrost zone of the Canadian Subarctic and on climate change in northern and western Canada. His interests include drought and wind erosion, as well as the origin and evolution of sand dune fields in western Canada and their associations with past climate, geomorphic processes, and future environmental change.

Robin J. Woywitka is a cultural land use analyst at the Alberta Archaeological Survey, in Edmonton, where his main focus is the use of geospatial technology in cultural resource management. In addition, he assumes regulatory duties in the northeastern part of the province. He has worked in the Lower Athabasca region since the late 1990s and is currently completing a PhD in earth and atmospheric sciences, examining site formation processes and the broader geoarchaeology of the minable oil sands area.

Robert R. Young is an associate professor in the Department of Earth and Environmental Sciences at the University of British Columbia's Okanagan Campus. He holds an MSc from the University of Alberta and a PhD from the University of Calgary, both in physical geography. His research interests lie with Quaternary glacial processes and landscapes and with the reconstruction of the physical and biological systems of past environments through sedimentary, stratigraphic, and geomorphological evidence. In his teaching, he seeks to inspire students by sharing his own fascination with the influence of the Pleistocene Epoch on the present day.

Angela M. Younie has worked as a professional archaeologist in Alberta, Yukon Territory, Alaska, and California for more than a decade and is currently a senior archaeologist at Far Western Anthropological Research Group, Inc., based in Davis, California. She received her MA in anthropology from the University of Alberta in 2008, where her research focused on microblade technology in the oil sands region, and her PhD in anthropology from Texas A&M University in 2015. Her dissertation research involved field study and excavation of an early precontact site in the interior of Alaska and incorporated lithic analysis, museum collections research, and consultation with community and local tribal governments. Her work has included the development of field school programs for underprivileged youth in collaboration with the Tanana Chiefs Conference and the Rural Alaska Honors Institute, the management of archived collections, together with database design, and field recordings of traditional use sites with the help of tribal elders.

Index of sites

General Index

Aboriginal communities: animal food sources, 129–36; knowledge resources, 487–88; land use, 151, 286, 500, 501. *See also* Fort McKay First Nation

Ackerman, Robert E., 184, 409

adzes, 181, 193, 206, 210, 211

aeolian landforms, 61–62, 72, 78, 79, 248, 368

Agate Basin point styles, 143, 144–45, 146, 174, 181, 187, 256, 297, 310

aggregate deposits, 56, 58, 516

Alaska archaeology, 294, 384, 386, 389, 401, 405

Alberta government, 5, 18, 480–81, 483, 484–86, 491, 520–21, 529, 530

alder (*Alnus*), 37, 99, 101–2, 103, 106, 107, 291

Alook site (HaPl-1), 199, 212, 213

Andrefsky, William, Jr., 378

Andriashek, Laurence D., 48

animal processing, 210, 264–66, 273, 301

animal species: evidence of, 128–36, 179, 201, 209–10, 266, 307, 440; landscapes for, 108, 294–95. *See also specific animals*

antler tools, 183, 212, 384, 402

archaeological methods: application, 14, 18–20, 426, 441–42; challenges, 6–7, 321–22, 429–30, 488–89, 521–22, 524

Archaeological Research Permits, 484, 497, 501, 504, 510

archaeological resources: definition and nature of, 6–7, 489; disturbances to, 504–6; early excavations, 491, 494–95; protection and mitigation, 20–21, 216, 437–38, 484–86, 496–97, 506, 508–23, 528–31; valuation of, 523–28

archaeological sites: Borden numbering, 10; dating of, 292–93, 302, 321–22, 495; investigative approaches, 440–53; and landforms, 69, 72; location patterns, 108, 115–17, 133, 137, 271, 319–20, 365, 500, 513–14, 516–20. *See also* historical period sites; *specific sites and complexes*

Archaic period, 360, 373, 382, 385, 386, 387–89. *See also* Middle Prehistoric (Precontact) period

Arctic Small Tool Tradition, 166, 199, 301, 409–10, 414

arthropods in deposits, 35–37

artistic activities, 383–84

assessment studies: archeological approaches, 18–20, 438–42, 453, 469–72, 486–89, 491, 526–31; in development projects, 5–6, 249–50, 495–98, 500–501, 504, 505–11, 512–13, 516–24; legislation and regulations, 480–81, 483–86. *See also* cumulative effects assessment; Cumulative Effects Assessment (CEA) framework

Athabasca oil sands region: development, 20, 216, 479, 491, 495–96, 517; geology and landscape, 118–19, 121–23, 426

Athabasca River, 3, 56, 245, 268

Athabasca River valley: flood events in, 25–28, 39, 49, 55, 77, 121; human occupation in, 198; landscape, 61, 249, 268

Atkinson, Nigel, 48

Avonlea technologies, 207–8, 209, 210

Bamforth, Douglas B., 319, 380

Bannatyne, Barry B., 352

Barrenlands: artifacts, 209, 259, 261–62; groups, 141, 142, 198, 201, 205, 277, 464

basal thinning technique, 171

Bayrock, L. A., 48

Clearwater-Lower Athabasca spillway: features, 70, 77–79, 245; formation of, 29–30, 39, 49–50, 53, 55–60, 63, 87, 121

Clearwater River, 45, 48–49

climate patterns: adaptations to, 185, 190, 205, 214, 389, 487; Birch Mountains, 37–38; in flood zone, 125, 135; Fort McMurray region, 45; Holocene, 269; in Kearl Lake core, 100–108; Quarry of the Ancestors, 368–69, 451–52. *See also* Hypsithermal interval; Younger Dryas interval

Cody Complex, 131, 144–45, 184, 185–87, 298, 320, 344, 349, 365. *See also* Nezu Complex

Cody knives, 178, 181, 184

Cold Lake, 320, 495, 497–98, 500–501, 517–19, 527

Collateral Point Complex, 145

Cree Burn Lake Complex, 140, 144, 146, 170, 185, 187–89, 189–90, 343, 426–27

Cree Burn Lake–Kearl Lake lowland, 12–13, 69, 73–80

Cree Burn Lake site (HhOv-16), 137, 147, 274, 339–40, 341, 371, 438, 439, 516

Cree groups, 215, 285–86, 308, 495, 501

Cree Lake Moraine, 51, 86–87, 291

Cretaceous age formations, 41n2, 45, 85, 118, 338

cultural indicators, 381–84, 387–88

Cummins Quarry, 382, 387, 388, 389

cumulative effects assessment, 437–38, 479–82, 516, 520–27

Cumulative Effects Assessment (CEA) framework, 19–20, 483–84, 511–12, 528–29

Cyclic Steam Simulation (CSS), 507–8

D2 artifact cluster, at Eaglenest Portage site (HkPa-4), 15, 312–15, 320

Dakota sandstone, 353, 459, 466

dart points, 141, 170, 183, 192–93, 196, 198, 201, 209–11, 306

De Paoli, Glen R., 167, 337–38, 340, 341, 342

Deal, Michael, 454, 467

debitage: analysis of, 457–58; Birch Mountains, 301, 310, 313; Fort Creek Fen Complex, 177; Fort Hills area, 251, 254, 264, 265–66, 270, 274; Quarry of the Ancestors, 371, 374–75

deciduous woodlands, 100, 103, 104, 106–7, 108, 125, 127

deer, 129, 136, 179, 188, 203, 384

deglaciation: Birch Mountains, 287, 289, 291–92, 293; environments, 46–50, 51–52, 61–63; Fort Hills area, 267–68; Kearl Lake area, 86–87;

process, 12, 26–27; Quarry of the Ancestors, 366

Denali Complex, 183–84, 262–63, 277, 295, 408–9, 414, 423–24, 428–29

Denbigh Flint Complex, 410

Dene groups, 141, 142, 205, 212–13, 214, 308, 427, 466, 467

dog species, 133, 179, 209

Donahue, Paul, 161, 204, 206, 210, 286, 301, 342, 461, 494

Don's Moraine, 87

drainage divide shifts, 62–63

drills, 181

Dry Creek site, 295–96

Duckett site (GdOo-16), 318, 344, 365, 463, 494, 519, 527

Duke, Daron G., 381

Duncan points, 199, 495

dunes: forested, 61. *See also* sand dunes

Dyke, Arthur S., 50

Eaglenest Lake: environment, 107, 302, 307, 366; pollen record, 37–38, 83, 96, 102, 104, 105, 289, 291–92

Eaglenest Portage site (HkPa-4): artifacts, 174, 184, 189, 204, 210–11, 296, 304–6, 495; D2 cluster at, 15, 312–15, 320; excavations, 165, 286, 295, 307, 309, 494

Early Cayley Series points, 304, 306

Early Prehistoric (Precontact) period: complexes, 140; technologies, 141–46, 170, 259, 293–98, 310, 318–20. *See also* Palaeoindian period

Eden points, 181, 184, 185, 186, 344

Edmonton area: glacier margin and, 46; mammoth fossils, 32

education and public awareness, 5, 20–21, 519, 525–26, 530, 531

EIA. *See* Environmental Impact Assessment (EIA)

elephant species, 128–29, 184

elk rack scores, 35

elk (wapiti), 37, 38, 129–30, 135, 294. *See also* wapiti skull discovery

Elson, John A., 49

Emerson Phase, 28, 31

engravers, 183, 188, 193, 203, 383–84

Environmental Impact Assessment (EIA), 480, 485–86

epidemics, 214, 487

Ericson, Jonathon E., 454, 457

erosion impacts, 25, 119, 505

Farrell Creek site (British Columbia), 304

Fedirchuk, Gloria J., 371

Fenton, Mark M., 136-37, 321, 335, 336, 338-40, 344, 351, 370-71

fire activity, 100-101, 104-5, 107, 128, 266

Firebag Hills Complex, 140-41, 166, 168, 199-205, 206

Firebag Moraine, 51, 55, 86-87, 291

Firebag River flow channel, 70, 77, 79

First Nations. *See* Aboriginal communities; Fort McKay First Nation

fish, 134, 209, 210, 213, 294, 302, 307-8

Fisher site (GbPo-1), 185-86

Fisher, Timothy G.: on flood event, 30-31, 78, 119; on Kearl Lake core, 89; on landforms, 70, 87, 248, 320

Flach, Peter D., 340

flood deposits, 30, 49, 54, 70, 77-79, 366, 370-72

flood events: chronologies, 364; Clearwater-Lower Athabasca spillway, 49-50, 55-60, 70, 77-79; landscape scouring, 25, 39-40, 58; subglacial, 11, 26-28, 119. *See also* Lake Agassiz flood

flood zone: artifacts, 171-72, 174; ecosystems, 125-28, 128-36, 139; human occupation of, 150, 178; quarries of, 137, 365

Fluted Point Tradition, 171, 228n6, 498

fluted points, 293-94, 319, 320

FMA Heritage Resources Consultants, 442, 448-50

Folsom technologies, 319

food sources, 3, 129-36, 186, 187, 213, 270-71, 273, 302, 307-8, 382

forested dunes, 61

Fort Creek Fen, 245, 268, 271

Fort Creek Fen Complex, 140, 143, 166, 170, 174-79

Fort Hills: archaeological surveys and sites, 14-15, 249-51; area features, 70, 243-49, 248; artifacts, 171-72, 174, 199, 203-4, 209, 254-57, 259-67, 380; Beaver River Sandstone in, 371, 461; human occupation, 198, 211-12, 268-76, 277-78

Fort Hills Energy, 250

Fort Hills Moraine, 51

Fort McKay First Nation: land use patterns, 286; support for heritage, 516, 531; traditional food survey, 129-30, 131, 133, 134-35

Fort St. Mary's II, 504, 519

fossils: dearth of vertebrate, 25, 39-40; Kearl Lake core, 88. *See also* wapiti skull discovery

Frederick points, 187

Frog Lake, 501, 518

fur harvests, 133-34

fur trade encounters, 3, 285-86, 308, 487

Gardiner Lake Narrows site (HjPd-1), 164, 174, 189, 286, 309, 494: artifacts, 143, 184, 198, 204, 210, 296-98, 301, 305; human occupation, 308

Glacial Lake Agassiz, 28, 29, 49-50, 53, 55, 178, 185, 320. *See also* Lake Agassiz flood

Glacial Lake Churchill, 12, 50, 54-55, 56, 61, 63

Glacial Lake McConnell, 28-29, 30, 48-49, 121, 178, 245

Glacial Lake McMurray, 48, 52, 53, 58, 61

Glacial Lake Peace, 48, 53

Glacial Meadow Lake, 50, 53

glaciation episodes, 45-46, 48. *See also* deglaciation

glacioisostatic adjustments, 48, 56, 60, 62-63

Gordon, Bryan H. C. A., 142, 192

Gowen site artifacts, 192, 386

grasses and grasslands (*Gramineae*), 96, 99, 100, 104, 105, 126, 289, 291

gravers, 183, 188, 193, 203, 383-84

greasewood (*Sarcobatus*), 99, 100, 104, 107, 125, 126

Great Lakes Palaeoindian period, 178, 382

Gregoire Lake site (HcOs-1), 196, 210

grey quartzite artifacts, 168, 189, 196, 204, 209, 210, 212, 295, 301-2, 310

Gros Roche site (HeOn-1), 209

grouse, 133

Gryba, Eugene M., 167, 209, 249, 460

Guthrie, R. Dale, 294, 366

Gypsy Lake site (HcOn-3), 210, 215

hare, 133, 266

Hass Lake core, 62

hearths, 147, 209, 265, 301, 307, 378, 440-41

heat treatment of stone, 302, 319, 333-34, 344-51, 352-53, 378, 460-61

Heldal, Tom, 471

Hell Gap points, 172, 259, 310

Heron Eden site (EeOi-11), 185, 187

Hickman, Michael, 105, 106

Hidden Creek site (GjPx-6), 207, 211, 212

high technology forager theory, 318-19

Historic Resource Values, 522

historical period sites, 500, 501, 504, 518, 519

Historical Resources Impact Assessment (HRIA), 485–86, 508, 510–11

Holocene period, 269–70, 275–76

HRIA. *See* Historical Resources Impact Assessment (HRIA)

human occupation: Birch Mountains, 15, 285–86, 307–8; Chartier Complex, 211–13, 214; Early Precontact, 170, 171, 174, 178–79, 181, 185–87, 189–90; evidence of, 6–7, 8, 115–16, 150–51, 487–88, 494–95; Fort Hills Project area, 266–67, 268, 270–78; landscapes for, 13–14, 69, 79–80, 107–8, 118, 135–36, 139, 293–95, 321; Middle Precontact, 198, 205; Quarry of the Ancestors, 365–66, 369, 374–75, 385–89; Taltheilei peoples, 142–43. *See also* migrations (seasonal); mobility

hunting patterns: Birch Mountain, 286, 308; Chartier Complex, 213; Fort Hills area, 267, 270–71; landscapes for, 131–36, 427; Nezu Complex, 179, 181, 184, 186–87; Plains groups, 257

Hutton, Michael J., 100, 105

hydrocarbons, 32, 41, 41n2, 347, 348

Hypsithermal interval: defined, 368; effects of, 37, 88, 102–3, 104–6, 108, 125–26, 382–83, 387

Ice Free Corridor environment, 294–95

Ice Mountain Microblade Tradition, 177, 179, 183

ice wedges, 367–68

Imperial Oil, as sponsor, 519

in situ projects, 491, 505, 506–9, 512, 517

Ives, John W.: on archaeological methods, 441; on artifacts, 189, 199, 210–11, 414; on Beaver River Sandstone, 335, 336, 338–40, 370, 371, 377, 464; excavations and studies, 161, 163, 468

jack pine, 101, 104–5, 107, 126, 136

Jackpine Creek excavations, 203

Jimmy Allen points, 188

Jones, George T., 381

Karpinsky site, 297

Kathol, C. P., 48

Kearl Lake area, 13, 85–86, 87, 96–97, 99–108, 126–27, 368

Kearl Lake core sample, 87–91, 91–106, 125

Keatley Creek microblades, 385

Kehew, Alan E., 58

Keyhole Lake site, 259, 273

King's Quarry jasper, 459–60

Kisis Complex, 214–15

Knife River flint, 184, 185, 186, 459

knives, 178, 181, 184, 201

Knudson, Ruthann, 183

Kuhn, Steven L., 379

La Loche House (HdOj-1), 198, 199

Labelle, Jason, 353

Lac La Biche, 3, 495, 517–18

Lake Agassiz flood, 7–8, 30–31, 33–34, 39–40, 99, 119–21, 268, 292; dating of, 50, 53–55, 70, 78–79

Lake Athabasca, 3, 30, 49

Lake Nezu, 365

Lake One Dune site (IgPc-9), 168, 196, 210, 211

Lake Tyrell, 48

Lake Wagtufro, 55, 56

lanceolate points: Birch Mountains, 295–97; Chartier Complex, 206, 209; classification of, 140–42, 143–46, 294; Early Precontact, 170, 171, 172, 174, 176, 177, 187–89, 190; Fort Hills area, 256–57, 259; Middle Precontact, 201, 203, 204

Late Pleistocene Athabasca braid delta, 49, 52, 61, 121, 245, 249, 251, 268, 271

Late Prehistoric (Precontact) period: human occupations in, 150, 264–66; technologies, 141–42, 303–6, 466

Laurentide Ice Sheet: airflows off of, 61; depths in Northern Alberta, 26; movements of, 12, 28, 46, 51, 53, 63, 119, 287, 291

Le Blanc, Raymond J., 201, 262, 414

legislation and regulations, 480–81, 483–86

Lichti-Federovich, Sigrid, 105

LiDAR (Light Detection and Ranging) imaging, 73, 77, 441

Lifeways of Canada, 167, 349, 438, 444, 448

limaces, 298

lithic technologies. *See also* stone tool production: defined, 359–60; with organic components, 402, 403–4; Quarry of the Ancestors, 373–77, 378–86, 389–90

Little Pond site (HiOv-89): activity at, 273, 276–77, 427; artifacts, 17, 204, 254, 262, 414–15, 423–24, 428, 463

Lobb, Murray, 140

Lofty Lake, 84, 102–3, 105

Lord, Mark L., 58

Lower Athabasca basin geography, 3

Lusk points, 187, 310

Lynott, Mark, 468

Oxbow points, 192–93, 196, 199, 260, 277, 301

palaeoecological records, 83–85
Palaeo-Eskimo groups, 384, 410
Palaeoindian period: complexes, 170, 360; food sources, 132, 135–36; peoples, 386–87, 388; technologies, 318–19, 365, 373, 382. *See also* Early Prehistoric (Precontact) period
palaeowinds, 61–62, 63, 97
Papua New Guinea microliths, 385
paraglacial environments, 61–63
Paulen, Roger C., 39
Pawlowicz, John G., 39
Peace Point site (IgPc-2), 168, 306
Peace River chert, 185, 196, 466
Peace River oil sands area, 501, 504, 519–20, 527–28
peat accumulations, 78, 127, 249, 273
peat deposits, 33–34, 39, 301–2
peat moss (*Sphagnum*), 103, 104, 105, 106, 126
Pelican Beach site (HkPa-14), 302
Pelican Lake artifacts, 260, 262, 277
permafrost zones, 61–62
Peter Pond Lake, 195–96, 198, 463
pine (*Pinus*): pollen, 37, 90, 292. *See also* jack pine
Plains groups, 468. *See also* Northern Plains groups
Plateau Tradition, 409
Pleistocene age deposits, 25, 32, 119
Pointed Mountain Complex, 259, 298, 300
pollen analysis: Eaglenest Lake, 37, 289, 291–92; Fort Hills area, 272–73; interpretation of, 102–3, 104–6, 107, 124; Kearl Lake, 79, 89–91, 96–100
Pollock, John, 161, 210, 215, 494
Pond, Peter, 3–4, 285
Poplar Lake, 58
Porcupine Hills, 26
Porter, James, 286, 287, 308
Prairie Side-Notched points, 206, 304, 306
Prather, Colleen, 105
Preboreal oscillation, 29
predictive methods, 441–52, 488–89, 521–22
Pre-Dorset Tradition, 199, 201, 203, 301
Prentiss, Anna Marie, 385
Prest, Victor K., 50
proglacial lakes, 7–8, 12, 48–49, 50, 53–55
projectile points: Birch Mountains, 295–301, 304–6; Chartier Complex, 206, 209, 210, 211; and cultural exchange, 464; diagnostic forms, 141, 293–94, 303–4, 318; Early Precon-

tact, 171–74, 176, 178, 181, 188–89; Fort Hills Project area, 254–62, 266–67, 268, 269, 277; Middle Precontact, 193, 196, 201, 203, 204–5; patterns of use, 465–66; Quarry of the Ancestors, 373–74; study methods, 163–66. *See also* lanceolate points
protein residue analysis, 266–67, 271, 365–66
ptarmigan, 267
public awareness and education, 5, 20–21, 519, 525–26, 530, 531

quarries, 454–55, 466–69. *See also specific quarries*
Quarry of the Ancestors: access to, 123, 178, 185, 198, 274–75; archaeological methods, 440–41, 442–53, 470–71; artifacts, 184, 189–90, 196, 201, 203, 212, 270, 343, 373–77, 379, 416–18, 420–21; environment, 362–64, 365, 366–69; excavations, 8–9, 167, 341–42, 359, 438, 455, 516; exploitation of, 16–17, 138–39, 360, 378–79, 380, 381–91, 426–27, 453–54, 456–58
quartzites: artifacts, 171–72, 181, 203, 210, 211, 261–62, 264–66, 373; varieties and distribution, 167–68, 351–53, 461. *See also specific varieties*

rabbit, 133, 179
radiocarbon dating: of mammoth bones, 31–32; of sediments and deposits, 48, 51, 88–89, 146–50, 263–66, 269–70, 289, 301–3; technique and measurement, 9–10; of wapiti skull, 34
Rajnovich, Grace, 468
Reardon, Gerard V., 370
reduction technologies, 454. *See also* stone tool production
Reeves, Brian O. K., 69, 83, 135, 137, 138, 341
regulations and legislation, 480–81, 483–86
Reimchen, T. H. F., 48
relief images, 73
Rhine, Janet L., 49
Richardson, Dr. (of Franklin expedition), 132
Richardson Moraine, 51
Ritchie, James C., 97, 102
Robertson, Louis, 50
Ronaghan, Brian M., 187, 201, 210, 249, 303, 309
Ronaghan's Ridge, 74, 365, 371

SAGD. *See* Steam-Assisted Gravity Drainage (SAGD)
St. Germain, Paul ("Buffalo Head"), 285–86